Quaternary Plant Ecology

6286254 [5]

QUEEN'S UNIVERSITY BELFAST
Geography Department Library
This book must be returned by
the latest date stamped below,
unless previously renewed,
otherwise a fine will be imposed.

~~5 JAN 1988~~
~~11 APR 1988~~
~~23 NOV 1988~~
~~14 DEC 1988~~
10 JAN 1989

Quaternary Plant Ecology

The 14th Symposium of The British Ecological Society
University of Cambridge, 28–30 March 1972

edited by

H.J.B. Birks
Botany School
University of Cambridge

and

R.G. West
Botany School
University of Cambridge

Blackwell Scientific Publications
OXFORD LONDON EDINBURGH MELBOURNE

© 1973 Blackwell Scientific Publications
Osney Mead, Oxford,
3 Nottingham Street, London W1,
9 Forrest Road, Edinburgh,
P.O. Box 9, North Balwyn, Victoria, Australia.

All rights reserved. No part of this publication
may be reproduced, stored in a retrieval system,
or transmitted, in any form or by any means,
electronic, mechanical, photocopying, recording
or otherwise without the prior permission of
the copyright owner.

ISBN 0 632 09120 7

First published 1973

Published in the U.S.A. by
Halsted Press, a Division of
John Wiley & Sons, Inc.
New York.

Set by Oliver Burridge Filmsetting Ltd,
Crawley, Sussex
printed by Compton Printing Ltd,
Aylesbury, Bucks
and bound by
Webb, Son & Co.,
London and Ferndale, Glamorgan

Johs. Iversen

The programme of this Symposium was to have contained a lecture by Dr Johs. Iversen on 'The dynamics of the forest ecosystems of Draved in ancient and recent times'. His absence was felt with a deep sadness. Johs. Iversen was a young and keen ecologist when he entered the field of vegetational history and he treasured the world of living plants throughout his life-long study of past vegetations. He showed us that former changes of vegetation can only be understood through an intimate knowledge of the present, and that the vegetation of today must be viewed in terms of its past. The great vegetational cycles of the Quaternary, the heliophytes of the late-glacial, the post-glacial forest immigration, the land-clearance by agricultural man, and the evolution of present forest ecosystems were some of the subjects of his penetrating studies, which made him a leader in Quaternary plant ecology, and for which he was honoured amongst others by the university where this symposium was held. He will be remembered also because of his modesty and the generosity with which he passed his knowledge on to others. The loss of Johs. Iversen at a time when he was still in the middle of his work is mourned by all his scientific colleagues who had ever met him and discussed common problems with him. The volume containing the papers of this symposium, which he had looked forward to attending, is dedicated to his memory.

Contents

ix Preface

x Acknowledgements

1 Introduction
R.G. West, University of Cambridge

Section 1 Methodological Problems

7 Multivariate techniques applied to palynological problems: a review
Edward J. Cushing, University of Minnesota, U.S.A.

9 Calibration of absolute pollen influx
Margaret Bryan Davis, Linda B. Brubaker and T. Webb, III, University of Michigan, U.S.A.

27 Discussion on methodological problems

Section 2 Pollen Dispersal and Sedimentation

31 Local and regional pollen deposition
C.R. Janssen, State University, Utrecht, The Netherlands

43 Pollen budget studies in a small Yorkshire catchment
Rona M. Peck, University of Hong Kong

61 Pollen in a small river basin: Wilton Creek, Ontario
Adèle A. Crowder & D.G. Cuddy, Queen's University, Kingston, Ontario, Canada

79 Absolute pollen frequencies in the sediments of lakes of different morphometry
Winifred Pennington, Freshwater Biological Association and University of Leicester

105 Discussion on pollen dispersal and sedimentation

Section 3 Pollen Representation

109 The differential pollen productivity of trees and its significance for the interpretation of a pollen diagram from a forested region
Svend Th. Andersen, Geological Survey of Denmark, Copenhagen, Denmark

117 Pollen dispersal and deposition in an area of Southeastern Sweden – some preliminary results
Björn E. Berglund, University of Lund, Sweden

131 The use of modern pollen rain samples in the study of the vegetational history of tropical regions
J.R. Flenley, University of Hull

143 Modern pollen rain studies in some Arctic and Alpine environments
H.J.B. Birks, University of Cambridge

169 Discussion on pollen representation

Section 4 Plant Macrofossil Assemblages

173 Modern macrofossil assemblages in lake sediments in Minnesota
Hilary H. Birks, University of Cambridge

191 Discussion on plant macrofossil assemblages

Section 5 Vegetational History and Community Development

195 Rates of change and stability in vegetation in the perspective of long periods of time
W.A. Watts, Trinity College, Dublin

207 Recent forest history and land use in Weardale, Northern England
B.K. Roberts, Judith Turner, and Pamela F. Ward, University of Durham

223 The anthropogenic factor in East Anglian vegetational history: an approach using A.P.F. techniques
R.E. Sims, University of Cambridge

237 A comparison of the present and past mire communities of Central Europe
Kamil Rybniček, Czechoslovak Academy of Sciences, Brno, Czechoslovakia

263 Discussion on vegetational history and community development

Section 6 Limnological History

267 Preliminary studies of Lough Neagh sediments I. Stratigraphy, chronology and pollen analysis
P.E. O'Sullivan, F. Oldfield, and R.W. Battarbee, New University of Ulster

279 Preliminary studies of Lough Neagh sediments II. Diatom analysis from the uppermost sediment
R.W. Battarbee, New University of Ulster

289 The impact of European settlement on Shagawa Lake, Northeastern Minnesota, U.S.A.
John P. Bradbury and Jean C.B. Waddington, University of Minnesota, U.S.A.

309 Discussion on limnological history

Section 7 Summation

313 Summing up: an ecologist's viewpoint
M.C.F. Proctor, University of Exeter

315 List of participants

319 Author index

323 Subject index

Preface

In the last ten years there has been an increasing interest and concern about many of the basic methodological problems involved in the interpretation of Quaternary palaeoecological data resulting from studies of fossil pollen grains and spores and of macrofossils such as seeds, fruits, and leaves. The aim of the symposium was to assess our current knowledge and understanding of some of these problems and to discuss the applications of this knowledge to the interpretation of Quaternary ecological questions in an attempt to present an integrated picture of the more outstanding recent advances in the field of Quaternary plant ecology.

Quaternary palaeoecology must, by necessity, use the present to model the past. There is, however, an acute lack of rigorous, quantitative studies of present-day processes and factors acting on organisms or parts of those organisms that are finally preserved as fossils in sediments. Although these processes are recognisable at the present day and their importance in the past is barely in doubt, they have, until recently, received little attention from palaeoecologists, with the result that much of the current research of palaeoecologists who seek to reconstruct past communities, biotas, and environments concerns itself with the description and quantification of contemporary processes such as fossil transportation, preservation, diagenesis, and representation. Pollen analysts are thus increasingly investigating the present-day geographical variation in pollen sedimentation in a variety of depositional environments in an attempt to find suitable quantitative models for interpreting fossil pollen assemblages in terms of past vegetation and environment. It is this general concern with the quantification of the relationships between pollen and vegetation in both relative and absolute terms that provided the principal theme for many of the contributions.

The papers were grouped into seven sections. The first considered methodological problems, in particular the handling of quantitative data and the interpretation of such data in terms of absolute plant numbers. The second section dealt with modern processes of pollen dispersal and sedimentation, and ranged in scale and process from local aerial pollen deposition, through processes of pollen dispersal by streams into lakes, to the complex patterns of deposition within lakes of different morphometries. The third section was concerned with modern pollen representation studies from which pollen analysts attempt to find some functional relationship between pollen frequencies and plant numbers. The papers presented considered pollen representation in temperate forests, in tropical regions, and in arctic and alpine areas. The fourth section considered problems of modern dispersal and representation of plant macrofossils in lakes and the application of such data to interpreting fossil seed assemblages.

The fifth section reviewed the application of many of the techniques considered in the previous sections to Quaternary plant ecological problems, such as stability and rates of change of ecosystems with time, interpretation of the rates and magnitudes of forest clearance, and the floristic composition of plant communities in space and time. The apparent juxtaposition and mixture of species for which there are no modern analogues and the extreme flexibility of response of individual species to environmental change are of considerable ecological interest, and they clearly merit detailed study as ecological phenomena that would not be readily predicted from observations of present-day ecosystems. The solutions to such problems represent one of the fields where palaeoecology can contribute most to modern ecological theories and concepts.

The sixth section considered palaeolimnological problems, with particular reference to documenting in quantitative terms the changes in populations of lake biota over the last few centuries. In the seventh and final section an ecologist and a palaeoecologist presented what they considered to be the outstanding problems raised by the speakers, and summarised some of the principal conclusions arising from the symposium.

H.J.B. Birks
December 1972

Acknowledgements

The symposium was held in the Botany School and the Zoology Department of the University of Cambridge by kind permission of Professor P.W. Brian and Professor T. Weiss-Fogh and accommodation for participants was provided by Sidney Sussex College and Clare College. Dr H.J.B. Birks, Dr R.G. West, and Dr Hilary H. Birks assisted with the planning of the meeting and with the editing of this volume. Dr Margaret B. Davis, Professor B.E. Berglund, Dr H.J.B. Birks, Dr E.J. Cushing, Professor Sir Harry Godwin, Professor G.F. Mitchell, and Dr R.G. West acted as chairmen of sessions. Mr L. Rymer organised the exhibits at the soiree; Dr Hilary H. Birks, Mr J.H.C. Davis, and Mr J. Dodd recorded the discussions: Mrs Joy Deacon, Mr Roger Andrew, and Dr W.W. Black were responsible for the domestic arrangements; and Mrs Mary E. Pettit and Miss Ruth Braverman carried much of the organisational and secretarial burden.

Introduction

R.G. WEST *Subdepartment of Quaternary Research, University of Cambridge*

The objectives of carrying out research in Quaternary palaeobotany are diverse. The geologist may be interested in stratigraphical palaeobotany and correlation, the geographer in relative land/sea level changes, the archaeologist in man's environment in the past, the biologist in adding a fourth dimension to his study of biota. This diversity of approach to Quaternary palaeobotany is a measure of the success of the subject in solving or clarifying problems in all these fields. Palynology has been especially successful, and it is this success which has whetted the appetite for further refinements of approach of the kind that are discussed in this volume.

The botanical objectives of Quaternary palaeobotany are an improved knowledge both of the history of floras in space and time and of the history of vegetation in space and time, each considered in relation to the present flora and vegetation. The floristic objectives attained by identification and correlation include the history of species, their biogeography in the past, and their evolution and extinction, both regional and world-wide. The vegetational or ecological objectives range from the general to the particular. We may be interested in the history of zonal vegetation, the history of plant successions, such as hydroseres, or of particular plant communities. We may be interested in these matters as events covering a few years or millennia. The questions asked of palaeobotany naturally tend to get more particular as knowledge increases, so we now seek to trace the history of species aggregates in plant communities known at the present time.

The diversity of objectives of palaeobotanists means that methods of investigation will vary according to the objective. Nevertheless, we can divide the process of investigation into three stages: the getting of results, their expression, and their interpretation. Contributors in this volume discuss all these stages and the processes of technique and thought which they involve. The subject is clearly a very wide one and the contributions are able to focus only on particular problems of importance to the development of the subject. It will, then, be useful to discuss briefly the subject in a wider context against which we may view the contributions to this volume.

The major ecological problem is that of relating fossils to the vegetation which generated them. We are interested in the processes which govern the production of fossil assemblages from living vegetation, a subject which has been called taphonomy by palaeontologists (Rolfe & Brett 1969). An outline scheme of the processes involved is shown in Fig. 1. The analysis of these processes can be carried out by experiment,

Figure 1. Processes involved in the production of fossil assemblages from living vegetation.

either treating the processes as a whole by relating pollen- or macro-trap contents to the vegetation which generated them or by the analysis of particular processes. The hope is that generalisations or assertions of principle can be made which will assist in the interpretation of fossil assemblages in terms of the vegetation which produced them, according to the classic Huttonian approach through 'actualism'. The question that follows is how far such generalisations or principles can be applied to situations in the past when the conditions of the modern experiment did not necessarily obtain, for example differences between present and past in species frequency, vegetation physiognomy, or atmospheric circulation. We return to this point later on, after making some general observations on processes of dispersal, deposition, and interpretation.

Two extremes of environment for dispersal are tundra and forest. Conditions in the atmosphere for dispersal may be very different between the two. The structure and pattern of the vegetation is different, as seen in the mosaic of plant communities likely to be found near sites of fossil deposition and preservation. A relatively uniform tree-dominated vegetation will contrast with a tundra vegetation with a variety of microhabitats and concomitant local variation in pollen rain. The separation of local and regional components in pollen assemblages may be far more difficult in the tundra than the forest environment. A further difference is in pollen and macro productivity of the two environments. Low pollen production of tundra vegetation often results in the increased importance of pollen which has suffered long distance transport. Sites of deposition in the tundra environment often favour the preservation of a wide variety of macroscopic remains. Such a variety has not been often noticed in sites normally studied in forested stages, though they may be approached in the fluviatile environment. Points of difference of dispersal processes in the two environments are brought out by several authors in this volume.

Environments of deposition are similarly varied. Fossil pollen and macros are sediment, and their deposition will follow the principles of sedimentology. The four chief depositional environments are terrestrial, limnic, fluviatile, and marine (including estuarine). Each of these environments offers very different conditions for the deposition of pollen and macro assemblages. Processes common to each environment may operate until the pollen or macro assemblage lands on a receptive surface, such as soil, water, or a bog surface. Table 1 is a simple analysis of the nature of fossil assemblages according to their derivation in each of these four environments.

As an example of a terrestrial receptive surface we may take the surface of a raised bog. In this environment there may be preservation of the pollen rain (pr) with little transformation. The macro assemblages are related strongly to the plant communities forming the peat. They are residual fossil communities (Fagerstrom 1964). If it were easy to extract all pollen from these sediments, they would be the ideal sediments for the determination of annual pollen rain. In fact, pollen is more easy to extract from limnic deposits, but the origin of the pollen assemblage in limnic sediment is unfortunately not so clear-cut.

In the limnic environment the pollen rain (pr) is transformed ($t[pr]$) by the various processes which take place between reception at the water surface, and the final deposition of the assemblage as sediment. The relative importance of these processes (such as selective transport, recirculation) is related partly to the morphology of the lake and partly to the morphology of the surrounding catchment. If the lake is closed in an area of low relief, the pollen rain on the soil of the catchment area (cpr) may not bias [pr]. In an area of high relief, a closed lake basin will receive a component of cpr transformed by dispersal and weathering processes ($t[cpr]$). If

Table 1. Derivation of assemblages

Environment	Pollen			Macrofossils		
	pr	$t[pr]$	$t[cpr]$	RFC	MFA	TFA
Terrestrial	+			+		
Limnic, closed low relief		+			+	
closed high relief		+	+		+	+
open low relief		+	+		+	+
open high relief		+	++		+	++
Fluviatile		+	++		+	++
Marine			+		+	++

- ++   component likely to be larger than others in environment
- pr pollen rain at receptive surface
- cpr pollen rain on catchment area
- t transformation in dispersal after reception
- RFC   residual fossil community
- MFA   mixed fossil assemblage
- TFA   transported fossil assemblage
  (classes of Fagerstrom 1964)

the ratio of the lake surface area to catchment area is small, then $t[cpr] > t[pr]$. In a lake with a throughput of water, then the significance of $t[cpr]$ may increase with rate of flow, until in a fluviatile environment it will predominate or even become the sole origin of pollen in sediment. In a fluviatile regime pollen sedimentation will be determined more closely than in limnic environments by the factors which determine the lithofacies.

The macrofossils in a closed lake basin will be redistributed from local communities. With a steep relief and/or throughput of water there will be an increased representation of terrestrial and more distant communities.

In the marine environment $t[cpr]$ will be the most important component, the more so the more distant from the coast. The predominant macro assemblage will be the transported fossil assemblage of Fagerstrom (1964).

In considering the magnitude of $t[cpr]$ the processes of hillside erosion must be significant. The pattern of erosion (Kirkby 1969) is related to vegetation cover, soil type, relief, and rainfall distribution. Much pollen will be deposited with rain, and be moved by overland flow to lakes and rivers and by channel erosion of soils containing pollen. Thus $t[cpr]$ may be a complex of pollen of different age. In a humid climate with much vegetation, overland flow is likely to be minimal, with infiltration and throughflow being important processes in soil water movement. In a semi-arid climate, overland flow may be important as a process for moving pollen, as it may also be in tree-less regions in a more humid climate. Thus the vegetation type and the climatic conditions will cause variation in the size and make-up of $t[cpr]$.

As you will have noticed, I have been discussing these matters as if facts were known about them. In reality, little is known, and what is required are experiments measuring the size of these components of pollen and macro assemblages in different lake and river types and varying catchment types. Experiments of this kind are reported in this volume.

The fossil assemblage is, then, a sample of pollen and macros, sampled from the pollen and macro rain by diverse natural processes of varying importance in each environment. This fossil sample is further sampled by the investigator and it is this further sample which is the basis for interpretation of vegetational history. Statistical treatments of such data for the improvement and increased objectivity of interpretation are also considered in this volume.

Improvements in interpretation can be gained by knowing the time interval in which a sample accumulated, allowing the measurement of numbers of grains deposited on a unit surface area of sediment per year; this matter is also discussed in this volume. At the other end of the chain of events leading to the deposition of the fossil assemblage, improvements can be gained from a knowledge of pollen production by plant communities and species. This again is discussed in the volume. Developments in both these aspects allow improved interpretation of the constitution of past vegetation. The need here is for more knowledge of the phytosociology of both natural and anthropogenic plant communities of the present time, as well as of the pollen rain and macro assemblages they generate.

An important further problem concerns the pattern and constitution of past plant communities within the vegetation. The discernment of pattern and constitution in plant ecology results from phytosociological analyses of various sorts, with a sampling grid used to determine pattern characteristics. A similar approach is necessary in the analysis of past vegetation. The analysis of a grid of sites has led to knowledge of regional differentiation of vegetation. The closer the grid the more detail will be apparent. A grid within a single catchment may be expected to yield the most detailed knowledge of pattern of communities, and this approach is now being applied. The three-dimensional pollen diagrams prepared by Turner (1970) from a raised bog in Ayrshire show that it is possible, using such methods, to distinguish local and regional patterns of vegetation clearance. In limnic and fluviatile environments the analysis of pollen, macros, and sediment in a grid of sites will also lead to better knowledge of pattern and constitution of plant communities.

In the last decade there have been an increasing number of experimental studies of processes involved in the production of fossil assemblages. Even though each site studied for its fossils may be unique in terms of its pollen and macro catchment, sedimentary processes, and vegetation history, generalisations may be expected from the experimental evidence which are applicable to other sites, both Quaternary and pre-Quaternary. Comparisons of results from sites of diverse characteristics with different sources of pollen and macros will allow a better understanding of the processes at work in each site in the past, with a consequent improvement of interpretation of the history of vegetation patterns.

The contributors to the volume all bring forward recent research into the sort of problems I have discussed, and it is hoped that the resulting statements of the present state of affairs will bring together the wide variety of approaches at present being applied, and stimulate us all in our own research into Quaternary plant ecology.

References

FAGERSTROM J.A. (1964) Fossil communities in paleoecology: their recognition and significance. *Bull. geol. soc. Am.* 75, 1197-216.

KIRKBY M.J. (1969) Infiltration, throughflow, and overland flow. Erosion by water on hillslopes. In *Introduction to fluvial processes* (ed. by R.J. Chorley), pp. 85-107. Methuen, London.

ROLFE W.D.L. & BRETT D.W. (1969) Fossilization processes. In *Organic Geochemistry* (ed. by F. Eglinton & M.T.J. Murphy), pp. 213-44. Longman, London.

TURNER J. (1970) Post-Neolithic disturbance of British vegetation. In *Studies in the vegetational history of the British Isles* (ed. by D. Walker & R.G. West), pp. 97-116. Cambridge University Press, London.

Section 1
Methodological problems

Multivariate techniques applied to palynological problems: a review

EDWARD J. CUSHING *Department of Botany, University of Minnesota, Minneapolis, U.S.A.*

Abstract

The problems of stratigraphical correlation, long-distance pollen transport, reconstruction of plant communities, palaeoclimatic reconstruction, and classification of surface samples are considered, and the application of multivariate statistical techniques to these problems is reviewed.

Calibration of absolute pollen influx[*]

MARGARET BRYAN DAVIS,[1] LINDA B. BRUBAKER[2]
AND THOMPSON WEBB III[3] *Great Lakes Research Division,
University of Michigan, Ann Arbor, Michigan, 48104 U.S.A.*

Introduction

Vegetation has left a fossil record, unrivalled in stratigraphic completeness and detail, in the form of pollen grains preserved in sediments. Interpretation of this record is made difficult, however, by the many complex factors affecting the quantities of pollen that are deposited and preserved at a given site. Although no one denies that pollen is ultimately a reflection of vegetation, disagreements are frequent on the correct reading of vegetation from the pollen record.

Palynologists have traditionally presented their results as pollen diagrams that show the percentages of the pollen types at different stratigraphic levels in the sediment. The pollen percentages represent the percentages of species in the vegetation, once differences in productivity, propensity for dispersal, resistance to decay, etc. have all been taken into account. These corrections can be done under special circumstances where dispersal, for example, is not a problem (Andersen 1970, 1973). More frequently pollen assemblages are compared to modern assemblages recovered from surface sediments (Wright 1967, Davis 1967a, Ritchie & Lichti-Federovich 1967). Interpretation rests on the assumption that where pollen assemblages are the same, the vegetation that produced them is the same.

Recently a number of investigators have measured pollen influx, hoping for a more straightforward interpretation of the fossil pollen record. Pollen influx is the yearly rate of pollen accumulation on the sediment surface. Influx values are calculated from the concentration of pollen in the sediment, corrected at each level for the time represented by a unit thickness of sediment. The time per unit thickness, or deposition time, is usually calculated from the differences in age between radiocarbon-dated levels in the profile.

Despite the hope for a more precise interpretation, influx has been interpreted only in general terms. Most authors have simply used the increase or decrease in influx from level to level to indicate a vegetation increase or decrease of the same magnitude. This was the kind of interpretation given to the early diagram from Rogers Lake, Connecticut, the first site for influx calculation (Davis & Deevey 1964).

The authors stated in their final paragraph that the assumptions basic to even this simple interpretation had not yet been investigated:

'The amount of pollen deposited at one point on one lake bottom is not necessarily proportional to that deposited over the whole surface of the lake, nor is the latter necessarily related in any systematic way to the abundance and distribution of vegetation.' They went on, however, to state optimistically that the method was worth pursuing because, if influx *was* an index of the absolute abundance of plants on the landscape, 'it would allow a completely objective interpretation of results such as those from Rogers Lake'.

In this paper we have tested the basic assumptions concerning sedimentation of pollen and the relation of influx to the vegetation. Sediments in Michigan contain a sharp pollen horizon dating from the time of settlement. This horizon provides a concrete reference for measurements of modern influx. The amounts of pollen deposited since that time can be measured and expressed as annual pollen influx. This measurement permits comparison of influx at different points within the same lake. Influx can also be measured at different lakes surrounded by similar vegetation. In this way we have arrived at a measure of the variations in influx caused by factors other than vegetation. Having thus delimited aspects of change in influx which do not reflect vegetation, we have gone on to calibrate the remaining variance in pollen influx by comparing modern influx to quantitative measures of the modern vegetation. The calibration can be used for interpretation of fossil influx values, in a manner analogous to the use of modern pollen spectra for the interpretation of fossil assemblages. Finally we have used our calibration of influx to reinterpret the Rogers Lake data, in somewhat the manner envisaged by Davis and Deevey in 1964. The results, while preliminary and incomplete, suggest the potential of future research in increasing the information to be derived from measurements of pollen influx.

[*]Contribution number 163 from the Great Lakes Research Division, University of Michigan.
[1]Present address: Department of Biology, Yale University, New Haven, Conn.
[2]Present address: Trumbull Regional Campus, Kent State University, Warren, Ohio.
[3]Present address: Department of Geology, Brown University, Providence, R.I.

Methods

MEASUREMENT OF MODERN POLLEN INFLUX

Introduction

Surficial lake sediments in Michigan show a striking change in pollen composition resulting from recent settlement. Logging and large-scale farming have dramatically changed the vegetation throughout this area, both by changing the species composition of the forest and by removing large tracts of forest. Increased percentages of pollen from ragweed (*Ambrosia*) and other agricultural weeds appear in sediment deposited after settlement; percentages of tree pollen decrease (Fig. 1). In areas that

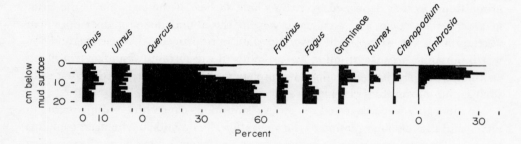

Figure 1. Pollen diagram from surface sediments from Blind Lake, Washtenaw County, Michigan. Above 12 cm depth percentages of *Rumex* (sorrel) and *Ambrosia* (ragweed) pollen increase and tree pollen decreases, indicating the inception of forest clearance and agriculture in southern Michigan in the 1830's. Increased percentages of *Quercus* (oak) pollen near the surface reflect local farm abandonment and partial reforestation in the last 50 or 60 years.

are still forested, tree pollen percentages reflect successional stages following logging.

The number of pollen grains above the settlement horizon provides an estimate of pollen accumulation over a known time interval, because the date of logging and settlement can be determined from the historical record. The pollen numbers can then be expressed as average pollen influx—the numbers of grains deposited per unit area per year. 'Modern pollen influx', as the term is used in this paper, refers to average pollen influx since the time of settlement.

Pollen influx measured in this way represents an integration of pollen produced by the vegetation since settlement; in some regions the landscape has remained farmland throughout this interval, but in others the vegetation has been changing, passing through successional stages as the forest has been allowed to regenerate. The pollen diagram in Fig. 1 shows the effect of partial regrowth of forest following farming.

Laboratory and field methods

Cores 0.5 to 1 m long were collected from lakes by a piston sampler (Davis & Doyle 1969) or by a modified Phleger free-fall corer with a check valve at the upper end (Davis *et al.* 1971). In the laboratory the cores were extruded in 2 cm- or 5 cm-thick segments by pushing a piston up from the lower end. Small subsamples were used for pollen analysis to locate the settlement horizon. Subsequently all samples above this level were combined for a quantitative assay of pollen in the core above the settlement horizon. These data were expressed as numbers of grains accumulated per cm^2 of core area and divided by the number of years since settlement. The result is pollen influx—the numbers of grains deposited per cm^2 per year.

We recognise that the first rise in ragweed pollen marking the settlement horizon may reflect regional changes in vegetation rather than local events. Therefore the historical record of events in the region as a whole rather than events in the immediate vicinity of the lakes has been used for setting a date for the pollen horizon. Generally we have used a date 10 to 20 years after the first logging in the region, but before the peak logging period. The estimates range from the 1830's in southeastern Michigan to 1870 for the northern Lower Peninsula. 1900 is the approximate date for logging in the Huron Mountains region in the Upper Peninsula (John Case, Marquette, Michigan, personal communication). It should be noted that an error of 10–15 years in estimating the age of the settlement horizon would only introduce a 10% error to our final estimate of pollen influx. Because of the inherent variability of influx, an error of this magnitude is unimportant. Nor will problems in dating the horizon affect results where we are comparing influx in different parts of the same lake.

The 29 lakes used in this study are described in Table 1; locations are shown in Fig. 2.

Measurement of Modern Vegetation

Major forest regions of Michigan are shown in Fig. 3. There is an easily discernible difference between the vegetation of southern Michigan, where the forests are exclusively deciduous, and the vegetation farther north, where the forests include both coniferous and deciduous species. On well-drained soils in the south, oak (*Quercus*) and hickory (*Carya*) are common, while clay-rich soils support beech (*Fagus*) and maple (*Acer*). Soils developed from the ancient lake beds of the glacial

Table 1. Characteristics of the lakes used as sampling sites in this study. For locations see Fig. 1.

Lake name	Location Latitude	Longitude	Size (in hectares)	Maximum depth (in metres)	Depth of water where core was collected (in metres)
1. Canyon	48°52′ N	87°54′ W	1.4	23	22
2. Mountain	48°52′ N	87°54′ W	314	21	9
3. Rush	48°52′ N	87°54′ W	132	83	80
4. Trout	48°52′ N	87°54′ W	14	20	19
5. Lily	48°52′ N	87°54′ W	1.3	2	2
6. Lancaster	45°37′ N	84°42′ W	21	15	9
7. Healy	44°26′ N	86°00′ W	18	15	6
8. One	45°03′ N	84°42′ W	1	2	1
9. Carpenter	45°11′ N	85°17′ W	15	9	6
10. Crystal	43°59′ N	86°19′ W	51	15	14
11. Shupac	44°49′ N	84°29′ W	44	30	18
12. Budd	44°01′ N	84°48′ W	71	10	7
13. Rifle	44°25′ N	83°58′ W	74	22	13
14. Esau	45°19′ N	83°28′ W	121	9	8
15. Clear	43°41′ N	85°24′ W	53	12	8
16. Duck	43°21′ N	86°05′ W	73	no data	6
17. Otter	43°13′ N	83°28′ W	28	18	17
18. Holland	43°14′ N	85°03′ W	32	15	7
19. Lee	42°11′ N	85°07′ W	47	14	14
20. Bryam	42°48′ N	83°48′ W	53	15	8
21. Blind	42°26′ N	84°01′ W	25	23	23
22. Pine	42°35′ N	85°24′ W	27	10	6
23. Whitmore	42°27′ N	83°45′ W	270	19	17
24. Crooked	42°05′ N	86°11′ W	45	15	8
25. Frains	42°20′ N	83°38′ W	6	10	8
26. Murray	42°20′ N	83°38′ W	7	12	12
27. Rush	42°14′ N	86°11′ W	48	18	17
28. Taylor	42°35′ N	85°42′ W	5	6	6
29. Sodon	42°35′ N	83°18′ W	2	18	13

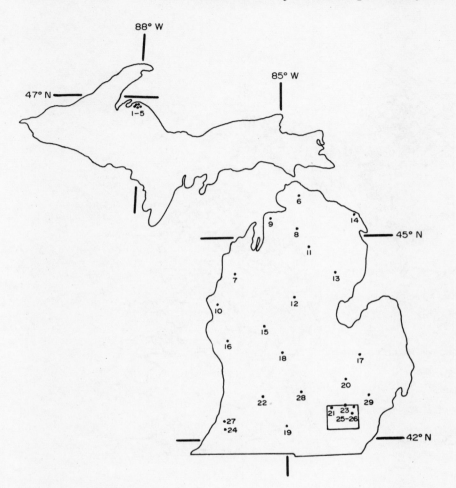

Figure 2. Map of the locations of the lakes sampled. The location of Washtenaw County is outlined. Table 1 provides a key to the numbers used on this map.

Great Lakes support forests dominated by elm (*Ulmus*) and ash (*Fraxinus*). Most of the deciduous forest region is presently used as farmland, with ploughed fields, meadows, and occasional woodlots. Only 15–20% of the land area is forested in southern Michigan (Ostrom 1967).

Farther north in the Lower Peninsula, forests on sandy soils are rich in pine (*Pinus*), and where there has been extensive logging and burning, poplar (*Populus*) and birch (*Betula*). Oak is abundant in some areas, apparently as the result of fire in areas where pine was abundant before logging. On the finer-grained till soils beech, birch, maple, and hemlock (*Tsuga*) form a community known as 'northern hardwoods'. In the Huron Mountain region of the Upper Peninsula, where our sampling sites are located, rich soils support forests that are also mapped as northern hardwoods. These include birch, maple, and hemlock, with occasional spruce (*Picea*) and fir (*Abies*) trees. The area is about 30 miles (50 km) west of the limit for beech.

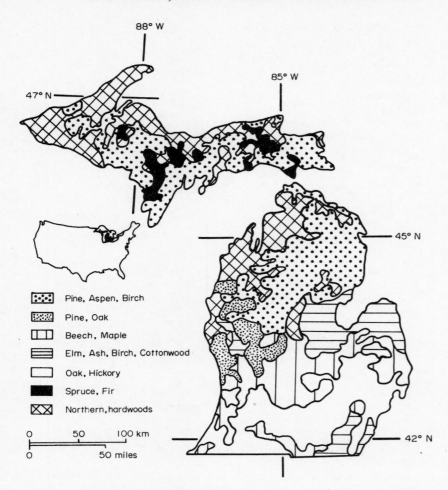

Figure 3. Map of the forest regions of Michigan. Forest is continuous in some areas, and represented by scattered woodlots in others. (Adapted from Chase *et al.* 1970, Küchler 1964, Marschner 1946, Veatch 1959).

Areas of thin soils over bedrock and sandy soils support concentrations of pine trees. Oak is rare, growing as occasional trees on rocky or xeric sites or in areas that have been burned (Bourdo 1955).

Estimates of abundance of tree species in Michigan are based on the Forest Inventory, completed in 1966 by the U.S. Forest Service. A tape made available by the North Central Forest Experiment Station in St. Paul, Minnesota contained the raw data for this phase of the study. The tape provided a listing of 2,600 Bitterlich plots (Grosenbaugh 1952) taken in proportion to forest density across all of Michigan.

Sixteen hundred of these plots were located in the Lower Peninsula, but only a third of these are found in the southern half of this region, and many of these 500 plots contain 10 or fewer trees. The plots from more densely forested land, on the other hand, contain as many as 50 trees.

We calculated the basal area per acre for selected species for forest land within a 16-township area surrounding each site. [Basal area refers to cross-sectional area of all stems more than 1 inch (2.5 cm) in diameter 5 feet (1.6 m) above the ground.] The basal area estimate was corrected for the percentage of non-forest land (which was assumed completely treeless for purposes of calculation) within the counties in which the townships fell. The result is an estimate of the absolute abundances of trees as basal area per unit area of landscape, for a region of 576 square miles (c 1,500 km^2) which surrounds the lake in which modern pollen influx was estimated. The area approximates a circle with a 14 mile (22 km) radius. Where a part of the area fell in Lake Michigan, Lake Huron, or Lake Superior, this part was not included in the calculations, giving a smaller total area in the sample.

The size of the area was chosen to approximate the size of a Michigan county, the unit used by the Forest Service in their inventory. Intuitively it seemed an appropriately-sized source area for the pollen. Size was also influenced by the necessity of including a large enough sample of trees. This aim was not accomplished in some of the southern sites, where the area in some cases contained as few as 9 plots with fewer than 100 trees total. Accordingly basal area estimates from southern lower Michigan are not as accurate as those further north. The area was not large enough to include all the trees contributing pollen to the sites; our results showed that pollen for some species does reach the sites in measurable quantities from forests at greater distances. Determination of the size of an appropriate area for each genus awaits more detailed study. We felt that the level of precision of our forest samples was probably sufficient for our purposes. Precision is lacking in any case because the estimate of trees is for the contemporary landscape, whereas the pollen influx with which it is to be compared represents average pollen influx over the last 100 to 150 years.

Basic assumptions: variations in pollen influx caused by sedimentary processes and errors in measurement

An assumption basic to the use of pollen influx is that influx at the coring site is proportional to pollen input to the lake. This relationship must remain constant through time if changes in influx are to provide a faithful record of changes in input, which in turn reflect vegetation. Sedimentary processes affecting influx are therefore of great importance. If they cause uneven sedimentation of pollen, they will cause different influx in different parts of the lake. Worse still, if they change in their effect through time, the relationship will change between pollen influx at the coring site and pollen input to the lake. It will then be difficult to distinguish changes in influx due to sedimentation from changes that reflect the vegetation.

Measurements of pollen influx at a number of sampling stations within lake basins were made to see whether influx was uniform from place to place. For the sake of simplicity, total influx is the unit of study. Our observations include modern influx and fossil influx. Within wide limits of variation that we are able to define, the results show that pollen influx is uniform within lake basins, implying that the rates of influx are controlled primarily by input and are proportional to input to the lake. Exceptional cases are considered where changing patterns of sedimentation disturb the relationship between pollen influx and pollen input to the lake.

Recent investigations on pollen input to lakes emphasise the complex relationship between pollen input and vegetation (Tauber 1967, Berglund 1973). While we recognise the importance of this relationship, we have not considered it in detail in this paper. We have used lakes within a standard size-range (except in the Upper Peninsula) in an attempt to standardise the relationship between input and vegetation. We recognise, however, that there may be important differences in the control of input from site to site.

VARIATION IN POLLEN INFLUX WITHIN LAKES

Modern influx

Series of short cores were taken from two lakes in Washtenaw County (nos. 21 & 25, Fig. 2) to measure variation from one sampling point to another within the same lake basin. One of these lakes, Blind Lake, is located in the oak-hickory forest region (now mostly farmland and second-growth oak forest), and the other, Frains Lake, is located in the beech-maple region (now 85% farmland), 15 km from the edge of the oak-hickory region. Both lakes are described in detail by Davis *et al.* (1971).

Blind Lake has a steep-sided basin, extending to 23 m depth. A series of 7 cores were taken down the slope of this basin along a line from a shoal area near the lake's centre to the centre of the basin (a map is shown in Davis *et al.* 1971). Fig. 1 shows one of the pollen diagrams from this lake.

The resulting estimates of pollen influx are shown in Table 2. Influx ranges from 7,500 to 22,800 grains $cm^{-2} year^{-1}$ for water more than 2 m deep. Pollen influx in the deepest part of the basin (23 m) is approximately 3 times as high as in shallower (5 m) water. Cores from very shallow water (< 1.2 m) range from 1,300 to 5,600 grains $cm^{-2} year^{-1}$.

Table 2. Pollen influx at a number of sampling stations in two lakes from Washtenaw County, Michigan.

	Water depth where sample collected (metres)	Influx pollen grains $cm^{-2} year^{-1}$
Blind Lake	0.45	3,500
	1.22	5,600
	0.92	1,300
	5.3	7,500
	16.8	8,400
	19.6	14,500
	23	22,800
Frains Lake	0.7	>1,200
	2.0	11,500
	4.0	10,800
	4.0	9,200
	4.7	8,100
	6.5	11,000
	7.1	13,500
	7.1	17,500
	8.0	9,100
	8.3	20,600
	9.0	63,600
	9.5	120,000

At Frains Lake a series of cores were taken from all parts of the lake except a small deeper area (> 9 m) in the very centre of the basin. A map of the lake, showing sampling locations, is given in an earlier publication (Davis *et al.* 1971). The basin is symmetrical and oval in shape, sloping gradually to about 9.0 m depth. A very small central area slopes more steeply to maximum depth about 9.8 m; this area will be discussed separately under 'Changes in the pattern of sediment accumulation.' Ten cores from water more than 2 m and less than 9 m deep yield influx values ranging from 8,000 to 21,000 grains $cm^{-2} year^{-1}$ (Table 2). The highest values are approximately 3 times the lowest.

In general, deeper parts of these two lake basins accumulate pollen faster than shallower areas. The total range of variation is manageable, however, and, as will be discussed below, similar in magnitude to variation between lakes. Of course cores of sediment used for fossil study sample deposition through time at one point in a lake basin. We believe that the variations due to sedimentary factors at one point within a basin may be considerably less than the total range of variation within the basin that we have measured here.

Sediment in shallow water (<2 m) accumulates many fewer pollen grains. We have not included these sediments in our generalisations because sediment cores would always be taken from deeper water in lakes like these. Variation greater than 3× may be found in lakes of very different size and morphometry. The maximum we have observed (see 'Changes in pattern of sediment accumulation') is 10×. A greater understanding of the sedimentary processes affecting influx should lead to generalisations about the kinds of lakes that are most suitable (i.e. least variable) for pollen influx studies (Pennington 1973).

Fossil Influx

Another way of looking for variation within lake basins is through comparison of pollen influx diagrams from different parts of a lake. Two cores were collected from the two basins of Rogers Lake, Connecticut. The southern basin core was analysed completely (Davis & Deevey 1964, Davis 1967b, 1969), while the north basin core was used for the analysis of only a single metre's thickness. Three radiocarbon dates from the latter were interpreted to indicate more or less uniform deposition time (years per cm of sediment thickness) for the sediment, which accumulated between 7,500 and 8,500 years ago (Davis 1969). Pollen influx over this interval was compared with pollen influx measured in the south core. The results were identical (Fig. 4). Deposition time for the south core had been based on 24 radiocarbon dates (Davis 1967b). Later, when additional radiocarbon dates became available (Stuiver 1967), the deposition time for the south core was revised slightly. The revision changed the apparent deposition time in this part of the core, either because the additional radiocarbon dates detected a real change in deposition time (Stuiver 1967) or because an atmospheric anomaly influenced radiocarbon dates in this time range (Tauber 1970, Kendall 1969). When the pollen influx measurements in the south core were revised using the new deposition time estimate, the values between the two cores no longer corresponded as closely (Fig. 4).

The difference in influx between the two cores from Rogers Lake is similar in magnitude (about 3×) to the within-basin variation measured at Frains Lake and Blind Lake. At Rogers Lake, however, the difference can be accounted for by interpretations given to radiocarbon dates.

The problems in measuring deposition time are difficult to assess, as they are different at each site. The error resulting from radiocarbon dates that are too few to pick up short-term changes in deposition time can only be determined by obtaining an ever-increasing number of dates, until errors in the radiocarbon method itself limit the accuracy of deposition time measurements. Perhaps laminated sediments will prove useful for comparative studies to determine the errors in pollen

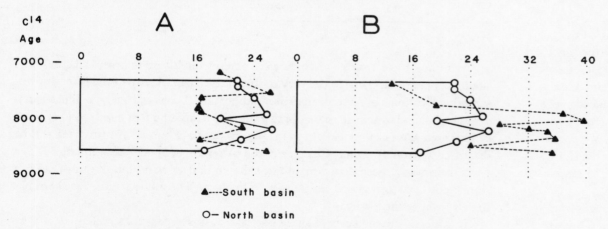

Figure 4. Comparison of total pollen influx between 7,500 and 8,500 years before present at two sampling sites in Rogers Lake, Connecticut. A: first interpretation, B: second interpretation. For explanation see text.

influx that result from the use of radiocarbon dates to determine deposition time. We believe, however, that the variability inherent in influx data due to sedimentary processes makes it pointless to attempt total precision in determining the deposition time of the sediment matrix.

CHANGES IN THE PATTERN OF SEDIMENT ACCUMULATION WITHIN A LAKE BASIN

Under certain circumstances, for reasons that are still unknown, the pattern of distribution of sediment within a lake can change, and remain in a new configuration over a long time interval. Apparently resuspension of sediment in shallow water, and its transport and redeposition in deeper water, can become more effective, decreasing the sediment accumulation rate in shallow water and increasing the accumulation rate in deeper water. Resuspension and redeposition of sediment moves sediment as a whole, including its contained pollen grains (Davis 1972). Consequently the pattern of pollen influx changes in the same manner as total sediment accumulation. At Frains Lake, for example, the sedimentation pattern appears to have been nearly uniform throughout the basin in early- and mid-post-glacial time. Just prior to settlement, however, deposition in the lake centre was 2–3 times greater than near the shore. After settlement there was an even stronger differential between shallow and deep water, with accumulation of clay-rich sediment in the lake centre 10 times as great as over most of the rest of the basin (Kerfoot 1972, Davis unpublished data). Pollen influx was also 10 times greater (Table 2). The results of this change in pattern show up in a sediment core from the deep area in the centre of the basin in the following way. Pollen influx was 25,000 grains cm^{-2} $year^{-1}$ during most of postglacial time (Kerfoot 1972). After 2,000 years before present however, both sediment and pollen accumulation increased 2–3 fold. After settlement in 1830 pollen influx increased to 3–5 times the mid-post-glacial rate.

A similar change is reported for sediment from Rutz Lake in Minnesota. Pollen influx increased in a core from the centre of the lake by a factor of 2 or 3 beginning about 500 years ago and increased markedly (to about 5×) after settlement in the 19th century. This change is also attributed to a change in sediment distribution within the basin, since the pollen concentration in the sediment remained constant, while the total amount of sediment accumulating at the coring site increased (Waddington 1969).

Changes in the distribution patterns of sediment are not always recognisable in the fossil record by a change in sediment lithology. A change in pollen influx, without an accompanying change in pollen percentages, is presumptive evidence for a change in sediment distribution. However, movement of shallow-water sediment to deep water could in some cases cause a change in pollen percentages (Davis et al. 1971) as well as in influx. Sedimentary processes in lakes are still poorly understood, so we know of no way to avoid sites where this phenomenon has occurred. Rapid changes in sedimentation rate detected from radiocarbon dates indicate that there have been changes in sedimentary processes. Sites that show sudden changes should be carefully evaluated for those changes in sediment accumulation and pollen influx that are unrelated to events outside the lake.

Basic assumptions: relationship between pollen influx and vegetation. Influx rates characteristic for major vegetation formations

If pollen influx reflects vegetation, then influx should be the same at sites surrounded by similar vegetation, and different at sites surrounded by different vegetation. We have looked at total pollen influx in order to test this idea. The next section of the paper considers influx for individual genera.

VARIATION WITHIN REGIONS OF SIMILAR VEGETATION

To measure the similarity of pollen influx at sites surrounded by similar vegetation, we compared modern pollen influx at a number of sites in Michigan. The sites were arranged in groups according to the forest region in which they are located (Figs. 2 & 3). The variance within each group is a measure of differences in total pollen influx unrelated to vegetation differences. The variance results from random processes, or from conditions of sedimentation or pollen input that are unique to each lake. We believe that minor vegetation differences account for only a small part of the variance, although the forest regions are not completely uniform, and several of the sites are located near vegetation boundaries.

The results are summarised in Table 3. Within most groups of lakes there is a modal value for total influx, with 3 or 4 lakes closely similar. A few lakes have influx values that differ by a factor of 3, or 6 in extreme cases. The main point to be drawn from these observations is that the variations between sites surrounded by similar vegetation are similar in magnitude to those we have observed within a single lake. The variations are similar in magnitude to variations caused by sedimentological factors and cannot be distinguished from them. It appears that the inherent variability of influx is high (3× at least); within these limits, however, lakes surrounded by similar vegetation record similar total influx values.

Table 3. Total pollen influx and tree pollen influx at 29 sites in Michigan. Sites are listed in numerical order (Table 1).

Vegetation	Total pollen influx grains cm^{-2}year^{-1}			Tree pollen influx grains cm^{-2}year^{-1}		
	Mean	Standard deviation	Values	Mean	Standard deviation	Range
Northern Hardwoods, Upper Peninsula $n^* = 5$	34,200	27,400	49,200 14,200 75,300 19,700 12,400	29,400	27,400	11,000–70,200
Northern Hardwoods Lower Peninsula $n = 5$	40,500	23,000	28,400 25,100 24,200 46,200 78,300	31,400	16,900	17,700–58,300
Pine Region $n = 6$	46,800	28,700	44,300 34,800 49,200 17,500 101,000 33,800	38,000	24,700	13,900–84,100
Oak-Hickory Region $n = 8$	33,700	17,100	46,900 22,500 27,100 35,200 22,800 15,000 31,100 68,800	25,100	15,000	7,700–54,600
Beech-Maple & Ash-Elm Regions $n = 5$	18,300	2,200	20,600 15,600 20,600 17,300 17,400	11,100	2,200	8,100–13,600

*n = number of lakes

VARIATION BETWEEN REGIONS OF DIFFERENT VEGETATION

Characteristic values for total pollen influx are associated with each of the major vegetation formations. These values, which are summarised in Table 4 (see also Craig 1972 and Pennington 1973), are sufficiently different from one another (without overlap) that they serve as indicators of the presence of a particular kind of vegetation. For example, modern measurements with traps in tundra (Ritchie & Lichti-Federovich 1967), and a number of fossil influx measurements from pollen zones where the assemblage indicates tundra, show that tundra produces few pollen grains. Fossil influx in tundra ranges between 500 and 2,000 grains cm^{-2}year^{-1} (Table 4). These values can be considered typical of tundra, and an indication of treeless conditions.

Fossil and modern studies also imply that boreal woodland and spruce forest are unproductive of pollen, although influx rates are typically higher than tundra. Modern measurements range from 300 to 1,200; fossil from 5,000 to 15,000.

In the fossil record, influx rates for mixed conifer-deciduous forest or deciduous forest range from 16,000 to 40,000. There is a tendency for mixed forest with white pine (*Pinus strobus*) to have a higher influx than deciduous forest.

Modern influx measurements from the deciduous (oak-hickory, beech-maple) and mixed coniferous-deciduous (pine, pine-oak, and northern hardwoods) forests of Michigan range from 14,000 to 101,000 (Table 3). Most of the values are similar to those recorded from these vegetation types in the fossil record—16,000 to 40,000. Differences that may exist among the subdivisions of the mixed and deciduous forest formations in Michigan are masked by the high variance of the observations within each region. There is some tendency for higher values within the pine region, although the difference is not statistically significant. Lower values were observed in southern Michigan (oak-hickory and beech-maple) where the landscape is only partially forested. The differences in total tree pollen are not as large as would be expected, however, considering the actual difference in numbers of trees between northern Michigan and southern Michigan.

Influx from prairie vegetation is still inadequately quantified. Very low values were observed at a fossil site in the northern prairie in Manitoba (Ritchie 1969). A prairie-border situation in Minnesota, however, gave high fossil values (Waddington 1969, Table 4). We suspect, from our modern measurements from unforested land in Michigan, that the difference in total pollen production is small or undetectable between prairie-border vegetation and forested areas.

Total pollen influx is useful for identifying the major vegetation formations from fossil evidence. However, our modern measurements, with their high variance, indicate that total influx lacks sufficient sensitivity to distinguish vegetation differ-

Table 4. Pollen influx, fossil and modern, in North America and Europe.

Vegetation (inferred)	Hirst spore trap	Fossil	Locality	Author	Influx grains cm^{-2}year^{-1}
tundra	×		Northern Canada	Ritchie & Lichti-Federovich 1967	5–800
tundra		×	Rogers Lake, Conn.	Davis 1969	500–1,000
tundra		×	Lake of the Clouds, Minn.	Craig 1972	1,600–5,000 (est.)
tundra		×	Blelham Bog, England	Pennington & Bonny 1970	100–150, 400
tundra, with spruce at lower elevations		×	Buckle's Bog, Md.	Maxwell & Davis 1973	1,000–2,000
forest-tundra	×		Northern Canada	Ritchie & Lichti-Federovich 1967	275–2,000
spruce woodland (open spruce-lichen forest)	×		Northern Canada	Ritchie & Lichti-Federovich 1967	3,000–12,000
spruce woodland		×	Rogers Lake, Conn.	Davis 1969	10,000–15,000
spruce woodland		×	Frains Lake, Mich.	Kerfoot 1972	5,000
spruce woodland		×	Buckle's Bog, Md.	Maxwell & Davis 1973	5,000–10,000
pine savanna		×	Singletary Lake, N.C.	Whitehead 1967	1,000–1,800
birch woodland		×	Blelham Bog, England	Pennington & Bonny 1970	2,000
spruce forest		×	Manitoba	Ritchie 1969	2,000–5,000
boreal forest		×	Manitoba	Ritchie 1969	5,000
boreal forest		×	Rutz Lake, Minn.	Waddington 1969	5,000
boreal forest		×	Lake of the Clouds, Minn.	Craig 1972	14,000
birch-alder forest		×	Rutz Lake, Minn.	Waddington 1969	35,000
mixed forest		×	Rogers Lake, Conn.	Davis 1969	40,000
mixed forest		×	Canyon Lake, Mich.	Davis & Beiswenger, unpublished data	40,000
mixed forest		×	Lake of the Clouds, Minn.	Craig 1972	20,000–40,000
spruce-pine forest		×	Frains Lake, Mich.	Kerfoot 1972	40,000

Table 4. (continued)

Vegetation (inferred)	Hirst spore trap	Fossil	Locality	Author	Influx grains cm^{-2}year^{-1}
deciduous forest with conifers		×	Mirror Lake, N.H.	Davis unpublished data	40,000
deciduous forest		×	Rogers Lake, Conn.	Davis 1969	20,000–30,000
deciduous forest		×	Frains Lake, Mich.	Kerfoot 1972	25,000
oak-pine forest		×	Singletary Lake, N.C.	Whitehead 1967	10,000
deciduous forest with prairie openings		×	Rutz Lake, Minn.	Waddington 1969	15,000–20,000
prairie with some forest		×	Rutz Lake, Minn.	Waddington 1969	15,000–20,000
prairie		×	Manitoba	Ritchie 1969	1,000–2,000

ences at the community level (Table 3). The fossil material (Table 4) is less variable, indicating it might be possible, for example, to distinguish mixed conifer-deciduous forest and purely deciduous forest by total influx alone. The lower variance for the fossil material is probably a result of the method of presentation; the fossil rates given in Table 4 are modal rates throughout a pollen zone or an entire profile, not the extremes that might be observed over a short (100-year) time interval.

Calibration of pollen influx for individual species and genera

Influx for individual tree species varies from one region to another in a characteristic manner. In order to relate these differences to vegetation, we have compared modern influx for the 29 sites in Michigan with a quantitative measure of the modern vegetation. The comparison is not rigorous in that influx is averaged over the last century or so, while the trees are inventoried from the present landscape. Nevertheless the comparison offers an opportunity to see whether influx shows a correlation with population size.

The 29 sites were grouped according to forest region. Figures 5–8 show for each site the influx for each of the major pollen types and the absolute abundance of trees of the corresponding genus or species on the landscape surrounding the site (see Methods).

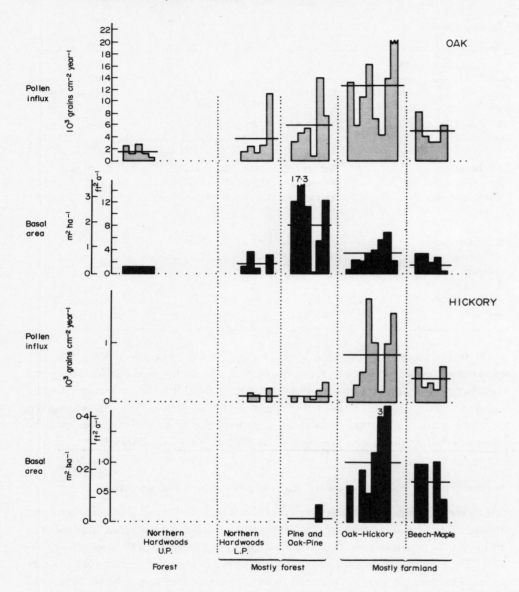

Figure 5. Modern pollen influx for oak and hickory at 29 lakes in Michigan (in numerical order left to right from Table 1), compared to basal area per unit area of landscape in the vicinity of each lake. Mean values for each group of lakes indicated by horizontal line.

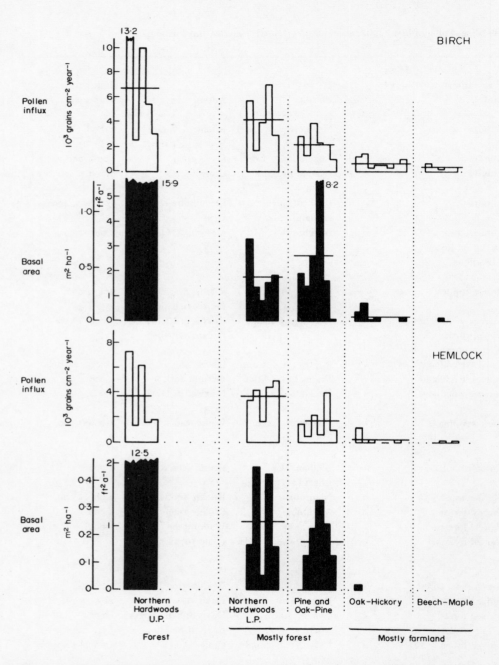

Figure 6. Modern pollen influx for birch and hemlock, compared to basal area on the modern landscape.

Calibration of absolute pollen influx 19

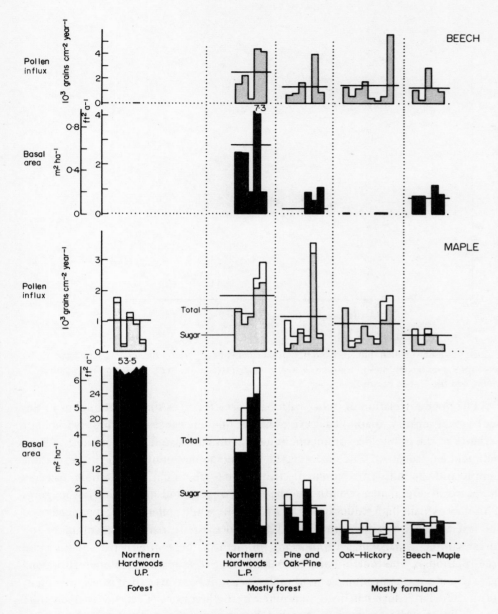

Figure 7. Modern pollen influx for beech and maple, compared to basal area on the modern landscape.

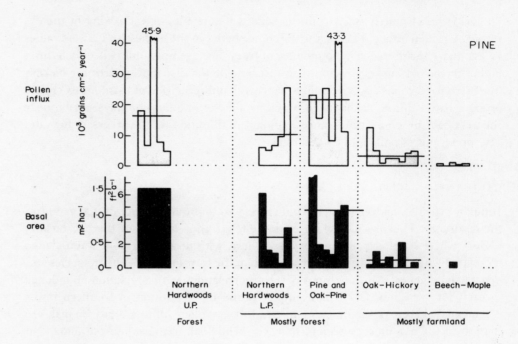

Figure 8. Modern pollen influx for pine, compared to basal area on the modern landscape.

GENERAL TRENDS

Forest regions in Michigan can easily be distinguished by the characteristic differences in influx for individual genera. Oak and hickory influx (Fig. 5) is high in the south; birch, hemlock, beech and maple influx is high in the north (except that beech is absent at the Upper Peninsula sites; Figs. 6 & 7); pine shows maximum influx values in the pine region (Fig. 8). These differences from region to region are clearly significant: they show that vegetation differences at the community level will be easily discernible in the fossil record.

A second generalisation concerns pollen dispersal. Hemlock, beech, hickory, and pine all have range limits within the state of Michigan. Dispersal of pollen occurs well beyond the range limits, causing measurable influx where the trees are absent in the vegetation. A gradient of influx can be seen to the north beyond the range of hickory (Fig. 5); with data like these it might be possible to calibrate pollen influx sufficiently to estimate the distance of a fossil site from the edge of a past species boundary.

Dispersal of pollen away from the source area results in a lessening of the extremes of pollen influx. That is, the difference between the highest and lowest values is less than the difference in abundance of trees. For example, influx is only 8 times higher in the oak-hickory region than just outside the distribution area for hickory (northern hardwoods region). Consequently the interpretation of fossil influx should proceed with caution. A change in influx by a factor of 8 does not necessarily mean the trees have become 8 times more abundant. Calibration will be necessary for valid interpretations of pollen influx diagrams.

CALIBRATION WITH VEGETATION

Influx for each forest tree in each region is related to the abundance of that genus on the landscape. Hickory influx, for example (Fig. 5), is highest in the oak-hickory region, where basal area for hickory is highest, with decreasing pollen abundance correlated with decreasing basal area in each of the forest regions. Other species and genera show general, although not perfect, correlations with vegetation abundance in each forest region. Oak, for example, shows highest influx in southern lower Michigan, where oak is abundant in the forest. Pollen influx values for oak are highest in the oak-hickory region, however, while oak trees are most abundant in the pine and pine-oak forest regions (Fig. 5). Beech also shows poor correlation with absolute abundance in southern lower Michigan, but a clear maximum of influx in the northern hardwoods region where basal area reaches a maximum. The same is true for maple (Fig. 7). Birch shows a northward increase, which is clearly related to increasing abundance, although not perfectly correlated with it (Fig. 6).

Pollen influx in each forest region shows general patterns that are related to population size; therefore we would expect a correlation at individual lakes that could be demonstrated statistically, with high influx values when a tree is abundant on the surrounding landscape, and low influx when a tree is rare. Our data provide an opportunity to look for this relationship and to attempt calibration of influx for individual genera in terms of numbers of trees. If a statistical correlation exists, the regression equation could be used for direct, and objective, interpretation of fossil pollen influx values.

Least-squares linear regressions were run for all the genera shown in Figs. 5–8, comparing influx to the basal area per unit area on the landscape surrounding each sampling site. Significant correlations were absent in most cases due to the wide scatter of points. The influx was then regressed on the log of the basal area; this gave a slightly better result for hemlock, but not for other genera. In fact, the wide scatter of points makes it impossible to judge whether a linear or curvilinear relationship is more appropriate for the data. As an example a scatter diagram is shown for pollen influx for birch plotted against basal area for birch trees (Fig. 9).

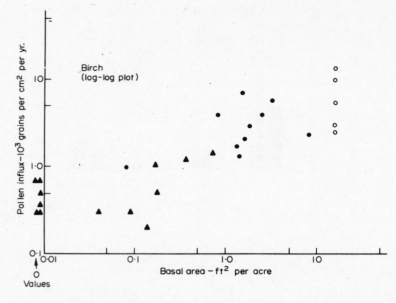

Figure 9. Modern pollen influx for birch, plotted against basal area on the modern landscape surrounding each site. Sites in southern lower Michigan (triangles), northern lower Michigan (dots), and the Upper Peninsula (circles).

The poor correlation of influx with basal area for trees that are abundant in the south (oak, hickory, maple) has several rather obvious causes. The first is the high variance of the influx measurements due to factors unrelated to vegetation, such as sedimentary processes. The second is the lack of contemporaneity for the vegetation sample and the influx measurement. In the pine-oak region, for example, oak has increased in abundance recently, as the result of repeated fires following logging. This may explain high values in the forest sample, while pollen influx, averaged over the last century, remains quite low. The third is the statistical inaccuracy of the forest sample, a particularly important problem in the southern part of the state (see Methods). The fourth is the probability that trees growing in open situations have larger crowns, producing more pollen per unit basal area than trees growing in forest. The latter idea has been mentioned in the literature (Faegri & Iversen 1964 p. 39). These authors point out, too, that dispersal may be more effective from trees in open situations, resulting in a larger proportion of grains reaching sites of deposition.

Our data suggest that differences in pollen contribution per tree may also be important when semi-forested regions, such as northern lower Michigan, are compared with densely forested regions, such as upper Michigan. The areas in northern lower Michigan are 40–80% forested; in upper Michigan 95% forested. Several genera which are abundant in both areas (maple, birch, hemlock) show either a drop in influx at the more northerly sites, or an increase that is much less than the increase in basal area per acre. Lowered pollen contribution per unit basal area because of lowered production and dispersal is a reasonable explanation. Post-glacial fossil material, which was presumably produced at a time when the landscape was densely forested, shows influx values similar to those modern values we have measured in most regions of Michigan. This similarity suggests that the relationship between influx and population density is not linear, but asymptotic, as the density of trees increases, to values similar to those we have measured in modern material from Michigan. It is unfortunate that we do not have more measurements from upper Michigan, to see what the relationship between pollen influx and basal area would be for a large group of samples from different kinds of forest, all on landscape that is 80–100% forested.

Several different measures of abundance were used in an attempt to find a population measure better correlated with pollen influx than basal area. Density of forest trees, as number of stems greater than 5 inches (12.5 cm) in diameter per acre, was used as a measure of population size. The correlation with pollen influx was no better than basal area, however. The basal area of forest trees on forested land only, ignoring now the unforested land, showed some correlation with pollen influx, but no better than absolute abundance on the entire landscape. We concluded from these attempts that our data are insufficient for a rigorous test of the relationship between population size and pollen influx. They demonstrate, however, that a relationship exists, and provide some suggestions about the nature of that relationship.

Further work should be done to calibrate pollen influx. In order to avoid the problems of disturbance, successional changes over the last century, and open *vs* forested landscape, we would suggest working out influx rates from sediment immediately underlying the settlement horizon by means of one or more radiocarbon dates at a deeper level. These influx rates, averaged over the last few centuries, could be compared with estimates of population density based on presettlement witness-tree data collected by early land surveyors. Although each observation would require a great deal of work, results might emerge with fewer observations than in our sample; we would expect lower variance in influx measurements averaged over several centuries prior to settlement than in the short-term, postsettlement sediment. The calibration would be for forested areas, comparable to the kind of forest that has existed in most of eastern North America during post-glacial time.

Application to the fossil record at Rogers Lake

Comparisons of modern pollen influx and vegetation in Michigan can be used to interpret fossil pollen influx diagrams. Previous interpretations of the diagram from Rogers Lake in Connecticut, for example, used influx only as a measure of changes in abundance: an increased influx value was assumed to represent a similar increase in abundance, etc. Now we can go a step further and attempt to estimate what those abundances were, by comparison with Michigan.

The influx diagram from Rogers Lake for the major tree pollen types (Fig. 10) shows the sequence from tundra (zone T), spruce woodland (A), mixed forest (B), and deciduous forest (C).

POPULATION SIZE

Spruce is the first tree to show a major increase in pollen influx. It rose 11,500 years ago, increasing to maximum values about 9,500 years ago. Influx was maintained at low, but significant values of 300 grains $cm^{-2} year^{-1}$ until about 8,000 years ago. This last value is a slightly lower influx than occurs in upper Michigan (600 grains) where spruce occurs at a frequency of 5 $feet^2$/acre. Spruce influx averages about 200 grains in 10 lakes in lower Michigan where spruce makes up 0.21 $feet^2$/acre.

Pine influx was about 200 grains $cm^{-2} year^{-1}$ during the tundra period. This is only half the influx measured in the beech-maple region of southern Michigan, where pine is essentially absent. Undoubtedly pine was also absent from Connecticut during this period. In zone A influx rises to 4,000–5,000, rates similar to the oak-hickory region. This estimate of abundance fits well with a previous interpretation based on percentages. Modern assemblages that resemble the fossil assemblage of zone A are found in northern Quebec, a region of open spruce woodland where pine is 65 km from the sampling localities (Davis 1967a). The sharp rise in pine influx 9,000 years ago to values of about 25,000 grains represents an increase in population size to values similar to the pine and pine-oak regions of Michigan—about 5 $feet^2$/acre. According to pollen influx, pine was more abundant in Connecticut 9,000 years ago than it is currently in the Huron Mountains region in upper Michigan. This interpretation is also compatible with the previous interpretation based on assemblages, which compared the forests of zone B to modern forests of the northern Great Lakes region. Generally these forests contain more pine than the Huron Mountains region, where pine has been artificially reduced in frequency as a result of logging (L.B. Brubaker, unpublished data). After about 8,000 years ago, the pine population in Connecticut decreased again to values similar to the oak-hickory region of Michigan. Pine is now uncommon in southern Connecticut.

A weakness in the interpretation of the pine data is the lack of species identification. At Rogers Lake, pollen morphology indicates that most of the pine pollen in zone A is jack pine or red pine (*P. banksiana* or *P. resinosa*), while most of the pine pollen in zone B is from white pine (*P. strobus*; Davis 1969). Preliminary data collected by one of us (L.B. Brubaker) suggest that white pine is a much more prolific

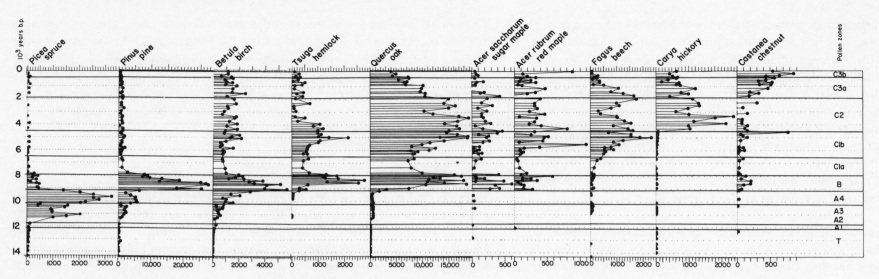

Figure 10. Fossil pollen influx at Rogers Lake, Connecticut. Major pollen types are shown, plotted against radiocarbon age of the sediments (After Davis 1969).

pollen producer than the other species; an attempt should be made to separate the species and to calibrate their pollen influx separately. Most of the pine now present in Connecticut is an eastern species, pitch pine (*P. rigida*), which is not represented in the Michigan calibration.

Birch influx suffers from the same problem, as a number of species are involved, including sweet birch (*Betula lenta*) which is presently abundant in Connecticut but which does not occur in Michigan at all. The species that occur in the earlier part of the sequence are likely to be paper birch (*B. papyrifera*) and yellow birch (*B. lutea*), the major species on which the calibration is based. Accordingly a comparison with Michigan for the older pollen zones is not unreasonable. Birch trees were absent in Connecticut before about 12,000 years ago. The numbers then increased to a density similar to the oak-hickory region (i.e. rare), increasing 9,000 years ago to abundances similar to those now found in the northern hardwoods region.

Hemlock was absent until 9,000 years ago when it suddenly immigrated to Connecticut and increased to densities similar to the pine region. About 4,000 years ago hemlock decreased to values similar to those found just south of the pine region: average basal area in oak-hickory samples is 0.1 feet^2/acre; present abundance of hemlock in southern Connecticut 0.05 feet^2/acre (Davis & Goodlett, unpublished data). It seems unlikely that hemlock was ever as abundant in Connecticut as it is now in the northern hardwoods region of Michigan.

Oak was apparently absent prior to 12,000 years ago. Between 12,000 and 9,000 years ago oak pollen influx was only about one-half the value observed in our samples from upper Michigan, where oak is rare. Oak was either absent, or present in extremely low numbers. Nine thousand years ago oak trees increased to an abundance greater than occurs now in the oak-hickory region of Michigan. The abundance may have been similar to the oak-hickory region of Michigan prior to settlement. This population size (which is higher, incidentally, than in the forests of the northern Great Lakes region with which zone B has been compared on the basis of assemblage) was maintained until about 2,000 years ago, when it began to decline. Influx was then similar to the pine and pine-oak regions of Michigan. We are uncertain what this means in terms of population size, as pollen influx and basal area were poorly correlated in our modern samples from this part of Michigan.

Maple was absent until 9,000 years ago, when it increased suddenly to attain abundance similar to forests in Michigan. It was probably never as abundant in Connecticut as it is now in the northern hardwoods region of Michigan. This interpretation is imprecise, but is probably the best that can be made given the poor correlation between influx and population size in our data for maple.

MIGRATION OF SPECIES

Beech, hickory, and chestnut (*Castanea*) represent an interesting sequence of post-glacial immigrations to Connecticut. Pollen is usually interpreted as representing the presence of a species, once a certain minimum percentage (the so-called rational limit) has been exceeded. The rational limit is different for each species, depending upon the productivity of that species and also upon the prevalence of heavy pollen producers in the vegetation (Faegri & Iversen 1964 p. 118). The rational limit can be defined much more easily using absolute pollen influx, since the rest of the vegetation has no effect on the value. For example, hickory pollen occurs as occasional pollen grains throughout the core from Rogers Lake. The influx rates remain very low through early post-glacial time (C-1a). These rates are probably not significantly different from the zero rates we observed in upper Michigan, 250 km beyond the range limit for hickory. 5,500 years ago, hickory pollen influx rose to 200–300 grains cm^{-2}year^{-1}, rates slightly higher than those in the northern hardwoods region of the Lower Peninsula, which is just north of the range limit for the species. These rates are still below the rational limit, but indicate that hickory was migrating toward Connecticut. Between 5,000 and 4,000 years ago, influx rose steeply to 1,000–2,000 grains cm^{-2}year^{-1}. This rise exceeds the rational limit, indicating that the trees actually grew near Rogers Lake, because rates in the oak-hickory region of southern Michigan today average 800 grains cm^{-2}year^{-1}.

Immigration of beech to Connecticut is documented by a more gradual increase in influx. Pollen is essentially absent from the tundra zone. Influx is 50–100 in the spruce zone (A) and in zone B, similar to modern values in the northern hardwoods of upper Michigan, 50 km west of the range limit for beech. Influx rose gradually to 200 grains cm^{-2}year^{-1} 8,000 years ago, a rate lower than any recorded in Michigan within the range of beech. Apparently this is still below the rational limit for this species. Beech influx began to rise again 6,000 years ago, reaching 1,500 grains cm^{-2}year^{-1} 5,000 years ago. Beech was then as abundant as it is today in the northern hardwoods region of lower Michigan. Additional data are needed from sites close to the range boundary of beech to define the rational limit more precisely.

Chestnut would be interesting to study in the same manner, but its virtual extinction as the result of disease 40 years ago makes this impossible.

RECONSTRUCTION OF COMMUNITIES

Pollen percentages give precise estimates of past vegetation where modern analogues exist to which fossil material can be compared. Where there are no modern analogues, the percentages can only be interpreted through the difficult procedure of correction for differences in production. Pollen influx, however, provides a means for reconstructing these communities. Population size can be estimated separately for each genus, and the vegetation reconstructed by adding them together. For example pollen zone C-1a at Rogers Lake has no known modern analogue. The pollen assemblage was first interpreted to represent a cool, 'mesic' type of deciduous forest with abundant hemlock, similar to the northern hardwoods (Deevey 1939). Interpretations of influx, however, suggest that oak and pine abundances were similar to the oak-hickory region of Michigan, and hemlock similar to the pine region. Hickory and probably beech were absent; maple was apparently less common than it is today in southern Michigan. The forests so described are different from any that now grow in Michigan. Certainly they are considerably less 'mesic' than assumed in the early interpretations.

Summary and conclusions

Modern influx measurements have been made in Michigan lakes using an identifiable pollen-horizon in the sediments 70–140 years old. Pollen deposited since that time can be measured and expressed as average yearly influx. These measurements of modern influx have been used to interpret fossil influx in a manner entirely analogous to the use of surface samples for comparison with fossil pollen assemblages.

Pollen influx in the deep basin of a small lake varies by a factor of about 3 from one part of the basin to another. Exceptional cases show differences by a factor of 10. The high variance of influx measurements is apparently inherent, largely the result of sedimentary processes. High variance has handicapped us in making generalisations. Nevertheless it does not prevent the use of pollen influx. Pollen influx becomes uninterpretable only under unusual circumstances, where patterns of sedimentation change through time, altering the relationship between pollen input to the lake and influx to the sediments.

Pollen influx, within its wide variance, is functionally related to the vegetation surrounding the lake. The total influx reflects the total pollen productivity of the vegetation in that influx assumes characteristic values in each of the major plant formations, permitting identification of tundra, boreal forest, and mixed and deciduous forest. Pollen influx for individual species and genera permits identifications of vegetation at the community level. Influx for a given species is generally highest where that species is abundant and lowest where that species is rare.

Calibration of pollen influx was attempted by comparing influx for individual genera to quantitative measures of modern forests in the vicinity of the sampling sites. Although some sort of quantitative relationship exists, we were unable to define it precisely through statistical correlation. The high variance of our data was a handicap. We believe, however, that the amount of forest cover was a critical factor affecting the relationship between basal area per acre and pollen influx for several, if not for all, genera. This finding is important because many studies of modern pollen and its relationship to vegetation are based on areas of disturbed and partially open vegetation. Our data suggest that the pollen contribution per tree is higher in open and in semi-open vegetation than in completely forested landscape. The near-similarity we found in the influx of tree pollen in regions of farmland and in regions of forest corroborates this generalisation.

Data on pollen influx in Michigan were used to reinterpret a fossil influx diagram, giving approximate population estimates for the influx recorded in each pollen zone. The results are somewhat imprecise, but they represent an advance over previous subjective interpretations, and they illustrate the value that influx calibration will have for interpretation of the fossil record.

The results of this study are largely preliminary, pointing to problems that should be investigated. Nevertheless the prospect that influx values can eventually be interpreted as a census of trees is exciting, reaffirming the potentiality of pollen analysis as a precise instrument for studying the history of vegetation.

Acknowledgements

We gratefully acknowledge Jane M. Beiswenger and James Ogg, who assisted with field work and pollen counting and provided discussion of the results throughout the course of this study. We are grateful to the North Central Forest Experiment Station for making the forest inventory data tape available to us. This work was supported by the National Science Foundation, Grants GB 5320 and GB 7727, and by the Institute of Science and Technology, University of Michigan.

References

ANDERSEN S.TH. (1970) The relative pollen productivity and pollen representation of north European trees, and correction factors for tree pollen spectra. *Dan. geol. Unders.* Ser. II, **96**, 99 p.

ANDERSEN S.TH. (1973) The differential pollen productivity of trees and its significance for the interpretation of a pollen diagram from a forested region. (this volume).

BERGLUND B.E. (1973) Pollen dispersal and deposition in an area of southeastern Sweden—some preliminary results. (this volume).

BOURDO E.A. JR. (1955) *A validation of methods used in analyzing original forest cover*. Unpublished Ph.D. thesis. University of Michigan.

CHASE C.E., PFEIFER R.E. & SPENCER J.S. JR. (1970) The growing timber resource of Michigan, 1966 North Central Forest Exp. Station, St. Paul, Minn. *U.S.D.A. Forest Service Resource Bulletin NC-9*.

CRAIG A.J. (1972) Pollen influx to laminated sediments: a pollen diagram from northeastern Minnesota. *Ecology* **53**, 46–57.

DAVIS, M.B. (1967a) Late-glacial climate in northern United States: a comparison of New England and the Great Lakes region. In *Quaternary Paleoecology* (ed. by E.J. Cushing & H.E. Wright), pp. 11–43. Yale University Press, New Haven & London.

DAVIS M.B. (1967b) Pollen accumulation rates at Rogers Lake, Conn., during late and post-glacial time. *Rev. Palaeobotan. Palynol.* **2**, 219–30.

DAVIS M.B. (1969) Climatic changes in southern Connecticut recorded by pollen deposition at Rogers Lake. *Ecology* **50**, 409–22.

DAVIS M.B. (1972) Redeposition of pollen grains in lake sediment. *Limnol. Oceanogr.* (in press).

DAVIS M.B. & DEEVEY E.S. (1964) Pollen accumulation rates: estimates from late-glacial sediment of Rogers Lake. *Science N.Y.* **145**, 1293–5.

DAVIS M.B., BRUBAKER L.B. & BEISWENGER J.M. (1971) Pollen grains in lake sediments: pollen percentages in surface sediments from southern Michigan. *Quaternary Research* **1**, 450–67.

DAVIS R.B. & DOYLE R.W. (1969) A piston corer for upper sediments in lakes. *Limnol. Oceanogr.* **14**, 643–8.

DEEVEY E.S. (1939) Studies on Connecticut lake sediments. I. A postglacial climatic chronology for southern New England. *Am. J. Sci.* **241**, 717–52.

FAEGRI K. & IVERSEN J. (1964) *Textbook of pollen analysis*. Munksgaard, Copenhagen.

GROSENBAUGH L.R. (1952) Plotless timber estimates ... new, fast, easy. *J. For.* **50**, 32–7.

KENDALL R.L. (1969) An ecological history of the Lake Victoria basin. *Ecol. Monogr.* **39**, 121–76.

KERFOOT W.C. (1972) *Cyclomorphism of the genus* BOSMINA *in Frains Lake, Michigan*. Unpublished Ph.D. thesis. University of Michigan.

KÜCHLER W.A. (1964) Potential natural vegetation of the conterminous United States. *Am. Geogr. Soc. Spec. Publ.* **36**, 116 p.

MARSCHNER F.J. (1946) *Original forests of Michigan* (Map redrawn by A.D. Perejda). Wayne State University Press, Detroit.

MAXWELL J.A. & DAVIS M.B. (1973) Pollen evidence of Pleistocene and Holocene vegetation on the Allegheny Plateau, Maryland. *Quaternary Research* **2**, 506–30.

OSTROM A.J. (1967) Forest area in Michigan counties, 1966. *North Central Forest Exp. Station, St. Paul, Minn. Research Note NC-38*, 4 p.

PENNINGTON W. (1973) Absolute pollen frequencies in the sediments of lakes of different morphometry. (this volume).

PENNINGTON W. & BONNY A.P. (1970) Absolute pollen diagram from the British late-glacial. *Nature, Lond.* **226**, 871–3.

RITCHIE J.C. (1969) Absolute pollen frequencies and carbon-14 age of a section of Holocene Lake sediment from the Riding Mountain area of Manitoba. *Can. J. Bot.* **47**, 1345–9.

RITCHIE J.C. & LICHTI-FEDEROVICH S. (1967) Pollen dispersal phenomena in Arctic-Subarctic Canada. *Rev. Palaeobotan./Palynol.* **3**, 255–66.

STUIVER M. (1967) Origin and extent of atmospheric C^{14} variations during the past 10,000 years. In *Radio-active dating and methods of low-level counting*, pp. 27–40. Int. Atom. Energy Agency, Vienna.

TAUBER H. (1967) Differential pollen dispersion and filtration. In *Quaternary Paleoecology* (ed. by E.J. Cushing & H.E. Wright), pp. 131–41. Yale University Press, New Haven and London.

TAUBER H. (1970) The Scandinavian varve chronology and C^{14} dating. In *Radiocarbon Variations and Absolute Chronology* (ed. by I.U. Olsson), pp. 173–96. Almqvist & Wiksell, Stockholm.

VEATCH J.O. (1959) *Presettlement forest in Michigan* (Map). Dept. of Resource Development, Michigan State University, East Lansing.

WADDINGTON J.C.B. (1969) A stratigraphic record of the pollen influx to a lake in the Big Woods of Minnesota. *Geol. Soc. Am. Special Paper* **123**, 263-82.

WHITEHEAD D.R. (1967) Studies of full-glacial vegetation and climate in southeastern United States. In *Quaternary Paleoecology* (ed. by E.J. Cushing & H.E. Wright), pp. 237-48. Yale University Press, New Haven and London.

WRIGHT H.E. (1967) The use of surface samples in Quaternary pollen analysis. *Rev. Palaeobotan./Palynology* **2**, 321-30.

Discussion on methodological problems

Recorded by Dr Hilary H. Birks,
Mr J.H.C. Davis and Mr J. Dodd

Dr Andersen pointed out the difficulties of comparing surface samples from within a stand of vegetation with samples from lake sediments, considering the transport of the pollen to the site of deposition, and the subsequent amalgamation of spectra from a mosaic of vegetation communities. Dr Cushing agreed that this was a problem, and said that Adam had considered these points, and had put them forward as one explanation of the discrepancies between the matching of the samples by canonical analysis. In his case, he considered that the problem was minimised, as the surface samples were of soil taken from sparse desert communities in which the internal variance was not very high. Dr Janssen then questioned the value of using multivariate techniques in palaeoecology, considering that the factors of transport, deposition, and representation were as yet relatively unknown. Dr Cushing stressed that there were few cases where statistical techniques could be used to predict and test against a model. As yet there are no models for the pollen variation within a core, or within surface samples which take into account the processes of deposition, differential preservation, etc. Most multivariate techniques have been used for data handling and arrangement. However, Mosimann has considered models to which statistical techniques can be applied, for example, the model of stability within samples, or of a steady trend. New techniques have to be developed by statisticians to test the models against reality, and these techniques may turn out to be more important and satisfying than the current methods of data handling.

Dr Andersen recalled that Dr Davis had shown an aerial photograph of a lake surrounded by agricultural land with patches of forest, and wondered whether there would be a larger pollen influx if the landscape was entirely forested. Dr Davis replied that, although 95% of the landscape was agricultural, the pollen influx was within the same order of magnitude as pre-clearance influx values, although there could be a threefold error. Perhaps trees produced more pollen in open situations, such as at woodland margins, and that in a lake about 1,000 m diameter the pollen was well dispersed. The pollen percentages were similar in nearby lakes, and also in fossil spectra from before 1800. One could only conclude that variations in the forest cannot be detected in such a lake over a distance of 15–20 km. Although the pollen influx into a very small lake would be affected by the local vegetation, these larger lakes give a more integrated, regional picture. More detailed ecological information would undoubtedly be gained from pollen influx studies in very small lakes. Dr Andersen said that there were no such lakes surrounded by natural forest in Denmark. He added that isolated trees do indeed have a much larger pollen production than those inside the forest, and that Dr Davis's lake was probably receiving pollen from a much wider source area than in the past. Dr Davis again mentioned the similarity of modern percentages and influx with those of pre-1800. However, lakes in the north of Michigan have more variable influx and percentages within natural forest. Perhaps this is due to different patterns of sedimentation or local differences in forest composition between species with good or poor pollen representation, such as *Quercus* or *Picea*. In Frains Lake, the sediment distribution changed after forest clearance, with a greater pollen influx to the centre. This may have been due to increased windspeed after clearance. Professor Faegri doubted the importance of windspeed, as pollen dispersal trajectories decreased so near to the source, but Dr Andersen pointed out that pollen transport was proportional to windspeed. Dr Davis thought that the size and morphometry of the lake was probably more important. Dr Osmaston wondered whether there was much variation in influx over a lake basin, to which Dr Davis replied that it depended upon the morphometry of the lake bottom. Sedimentation was uniform in flat-bottomed lakes, but could be highly variable in steep-sided basins. This was illustrated in Frains Lake, which used to have a depression in the centre. This began filling rapidly about 2,000 years ago, by movement of sediment from the edges. Today the lake was flat-bottomed, and there were no great differences in the modern sedimentation rate. These changes could be due to windspeed or morphometry, or to a change in the balance of limnological factors.

Dr Moore wondered how justifiable it was to apply modern influx values of species such as *Fagus* and *Carya* to late-glacial situations. The present north-south vegetational transect in Michigan is mature and stable, whereas in late-glacial times, the forests were unstable as species immigrated, and the structure of the woodland may have been very different. Dr Davis replied that modern influx had been measured in open and closed forest, and at different distances from the limits of the species distribution. In five different types of forest, the total influx was very similar, although it was probably less in Boreal Forest. For a single species, flowering may be affected by open or closed conditions. However, the level of accuracy of influx measurements was such that such differences probably could not be detected.

Dr Cushing commented that the effect of lake morphometry on sedimentation rate could be detected in the sediments of Lake of the Clouds in the coniferous forests of northern Minnesota, from which Mr Alan Craig had prepared a pollen influx diagram, and Mr Albert Swain had studied the sediments laid down over the last 1,000 years in detail. The lake is 500 m across and is flat bottomed except for a 30 metre deep hole in the centre, in which the sediments are laminated. Radiocarbon dating has proved the laminations to be annual. The laminations increase in thickness from 0.5 mm to 1.0 mm for periods of 20–50 years, so although the pollen concentration remained the same, the influx doubled. This is explained by the movement of sediment from the shallow margin to the centre. The disturbances correlate both historically and by charcoal counts in the sediment with forest fires, the lamination thickness increasing immediately after a forest fire. One could make the hypothesis that, when the forest cover was reduced, there was an increased windspeed leading to increased turbulence in the water, and hence to sediment movement. Perhaps if sufficient cores were analysed, the sediment accumulation over the basin could be integrated, and perhaps an influx budget in the past be reconstructed. Dr Davis agreed that this would be theoretically possible, but would be a great deal of work, and probably not worth the effort.

Section 2
Pollen dispersal and sedimentation

Local and regional pollen deposition

C.R. JANSSEN *Laboratory for Palaeobotany and Palynology,
State University, Utrecht, The Netherlands*

Introduction

Perhaps as long as pollen analysis exists the local over-representation of pollen grains will be a frequently observed and much debated phenomenon. Often changes in the frequencies of pollen in a sediment core occur simultaneously with a change in the character of the parent material. This is generally true for N.A.P. but the curves for the arboreal species may also be affected when these species are constituents of the local vegetation. Pollen analysis of sediment cores from lakes that, except for submerged aquatics, do not have a pollen-producing vegetation is therefore often preferred for the reconstruction of vegetational history from pollen diagrams. Faegri and Iversen (1964, p. 102) are well aware of the problem and they say 'the local pollen production of peat bogs does not influence the N.A.P. only, but also may influence the A.P. spectra in many unforeseen ways. In ordinary work peat should not be analysed if sediment is available'.

In the literature the concepts of the local and regional pollen deposition and the corresponding vegetation are defined in different ways. The local vegetation sometimes includes only the vegetation at the site itself, but sometimes this concept may be enlarged to include the entire vegetation of a stand or a peatland, even to the point where local vegetation can mean everything in a wide area around the coring site. Clearly the meaning varies with the scale of the region that is considered for an investigation of the development of the vegetation.

It soon was recognised that over- and under-representation depends, among other things, on the dispersal of the various pollen types. This resulted in Germany between 1930-1940 in the following scheme (reviewed by Firbas 1949):

1 örtliche Niederlschlag (local deposition)
   a Moorgehölzniederschlag (deposition of elements from the bog)
   b Randgehölzniederschlag (deposition of elements from the marginal area around the bog)
2 Umgebungsniederschlag   deposition from distances less than 570 m.
3 Nahflugniederschlag   deposition from 500 m—10 km.
4 Weitflugniederschlag   deposition from 10 km—100 km.
5 Fernflugniederschlag   deposition from distances over 100 km.

In this scheme the 5 categories that are distinguished are based on fixed distances from the source and problems will clearly arise when basins of varying size are compared. For this reason we may need a more flexible system.

DISPERSAL OF POLLEN GRAINS

Pollen grains that are not dispersed by insects or shot away by some mechanism of the parent plant may be dispersed like all other small particles in the atmosphere. For this process Sutton's (1953) equation is applicable. According to this equation the dispersion rate depends on a number of factors, including the turbulence of the atmosphere, the wind speed and wind direction, the weight of the particles, and the height and strength of the pollen source. The Sutton equation shows that the pollen concentration is inversely proportional to the distance from the source.

Tauber (1965) has calculated from his model of pollen dispersal that in the case of light pollen 50% of the grains present above the crown of a forest comes from distances within 250 m, that 75% comes from within 5 km, and that only 5% of the pollen comes from distances between 5 and 10 km. In front of a forest edge the influence of long-distance transport is greater the larger the distance from the forest. In the case of light pollen at 500 m from the forest edge 50% of the pollen comes from distances over 7 km, at 1 km from the edge 50% of the pollen comes from distances over 10 km, and at 5 km from the edge 50% of the pollen comes from distances over 30 km. This demonstrates that there is a strong gradient of decreasing deposition with distance from the source. Tauber has also made these calculations for heavy pollen grains, and his results show that in this case the gradient is even steeper.

Experiments have also shown that there is a steep gradient with increased distance from the source. Raynor and Ogden (1965) have measured, at varying heights above the ground, the concentration of *Ambrosia* pollen at various distances from an artificial source. Like Tauber they show that the dispersal curve is highly exponential. Wright (1952) measured the dispersal distances from isolated trees and came to much the same result.

BACKGROUND POLLEN

Unlike the experiments by Raynor and Ogden we seldom deal in nature with just one source, as the pollen that is deposited at a point generally comes from many additional sources, mostly at much larger distances. The dispersion of pollen grains from these sources results in a background pollen concentration. It is, for our purpose, interesting to know at what distance from a local source the influence of this source is so slight that it does not stand out against the background concentration. Raynor and Ogden (1965), and Raynor, Ogden and Hayes (1968) have measured this in their experiments with an artificial pollen source of *Ambrosia*. They have shown that the pollen concentration falls to background values within 100–200 m from the source, depending on the strength of the source, but at any rate already within a rather short distance.

Surface sample studies have confirmed that the dispersal distance from a local source is short. For example, Turner (1964) found that the influence of a pine stand was noticeable only within a distance of c 300 m from the stand. In summary it may be stated that the pollen curve that shows the dispersal from a local source is highly exponential in such a way that within rather short distances the background concentration exceeds that from a local source.

The regional pollen deposition

In the discussion above only one pollen type has been considered. In the reconstruction of vegetation types by means of palynology, however, very little can be concluded from just one type. The basis of any palaeoecological approach must be the entire assemblage of pollen types.

When analysing surface samples in a not too large area from sites away from the direct influence of local sources (e.g., in medium-sized lakes) it soon becomes apparent that the pollen types in the assemblage are present in the samples in a roughly constant proportion. This may be explained by assuming that the rate of deposition is more or less constant for all these types or if there is a change it is not differential for the pollen types concerned. The constant rate of deposition for all the types regardless of the distance from the source may be most expected when we are dealing with distances with a background concentration, that is in the trajectory of the dispersal curve where there is only a slight decrease in the pollen concentration with large distance. We may term the deposition here the *regional pollen deposition* and the vegetation that acts as the main source as the *regional vegetation*.

It is possible to characterise the vegetation at the integration level of the formation by means of regional pollen assemblages. This was recognised in 1940 by Aario in Finnish Lappland and by several others. The most detailed piece of work of this kind has been done by Lichti-Federovich and Ritchie (1968) in Manitoba. Ritchie and Lichti-Federovich took surface samples in lakes of uniform size in a range of what they call landform-vegetation units in which the vegetation can be characterised at the integration level of the formation and whose areal extent is determined by geomorphology and associated physiogeographic features. From north to south in Manitoba they distinguish a tundra, a forest-tundra on uplands, closed coniferous forests, a mixed coniferous-deciduous forest, and a forest-tundra on lowlands.

The vegetation within these formations is certainly not homogeneous, as in all these areas peat bogs, upland forests, floodplains, lakes, etc. are present, each with their own type of vegetation. Lichti-Federovich and Ritchie's work and that of others show, however, that in spite of these vegetational differences the regional pollen assemblages in each formation have a characteristic pollen composition that differ clearly from the assemblages in adjacent formations and that these assemblages may be interpreted in terms of vegetation at the level of the formation.

It is difficult to explain why the proportions between the regional pollen types is constant within a formation. Perhaps it is a result of the repetition of similar stands (Janssen 1970), but the reason may also be that the regional pollen is simply the component that is carried upwards by strong air currents, for instance those associated with rain storms, where it is mixed thoroughly before it settles down by rain or other factors.

The concept of a regional pollen deposition in connection with the landform vegetation unit may be useful for the determination of the horizontal extent of the regional pollen assemblage. Even when the vegetation is not known we may expect a change in the composition of the pollen assemblage when there is a change in the landform.

Pollen is, of course, also transported across formation boundaries where it constitutes an extraregional component (McAndrews 1968). We may expect a mixture of the pollen assemblages of two adjacent formations in the transitional area. We do not know the extent of the belt where we find such a mixture, but because of the exponential decrease of the dispersion curve I would expect the transitional belt to be quite narrow. There are, in the pollen analytical literature, some slight indications

Figure 1. Vegetation map of the Myrtle Lake peatland, simplified from Heinselman (1970). The main transect is shown by a solid black line.

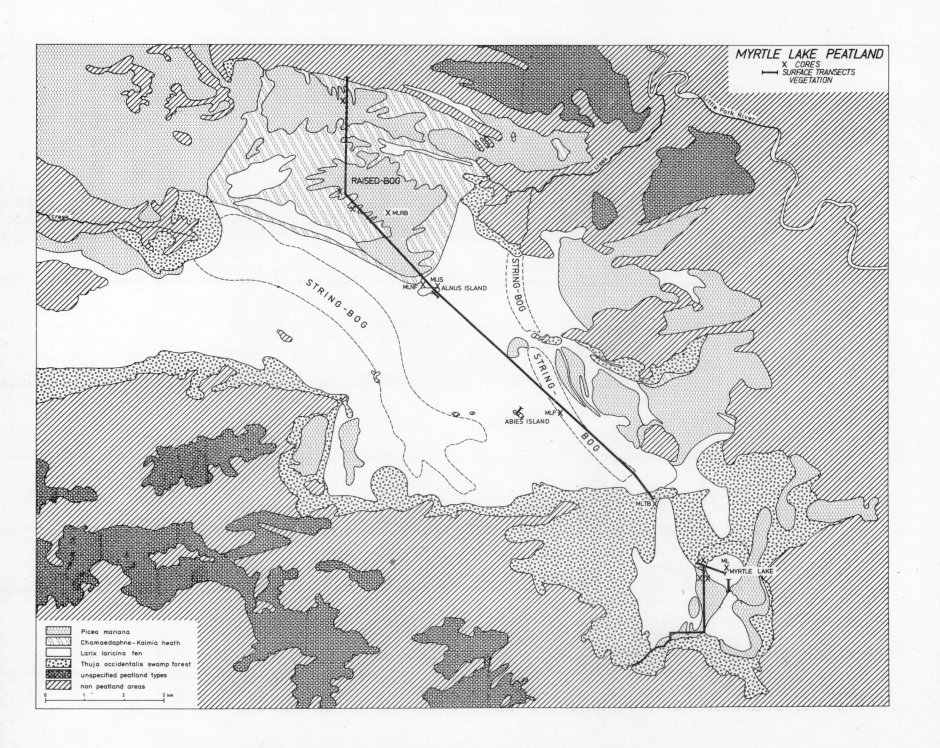

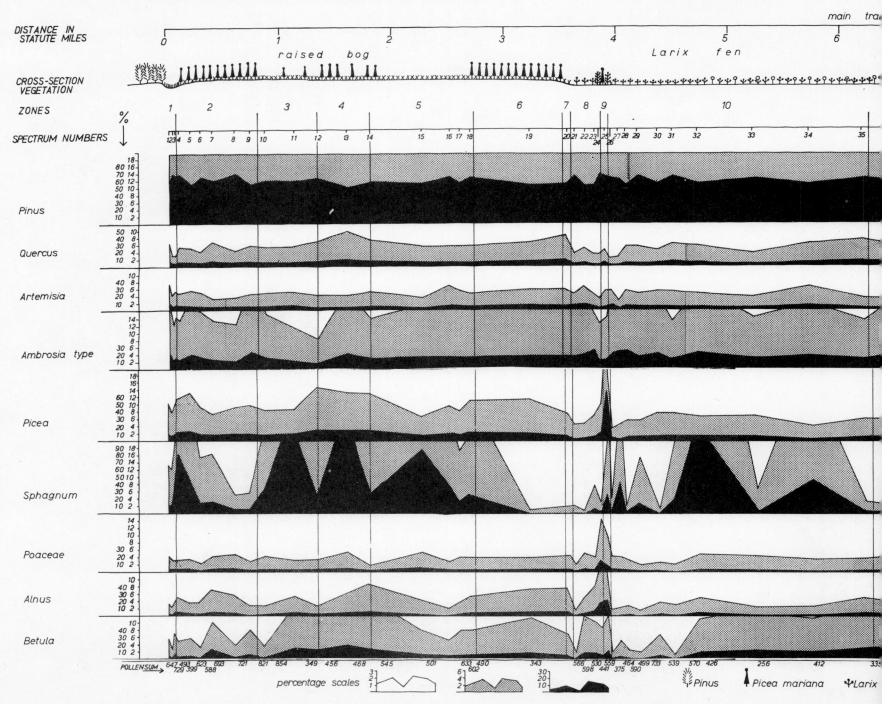

Local and regional pollen deposition

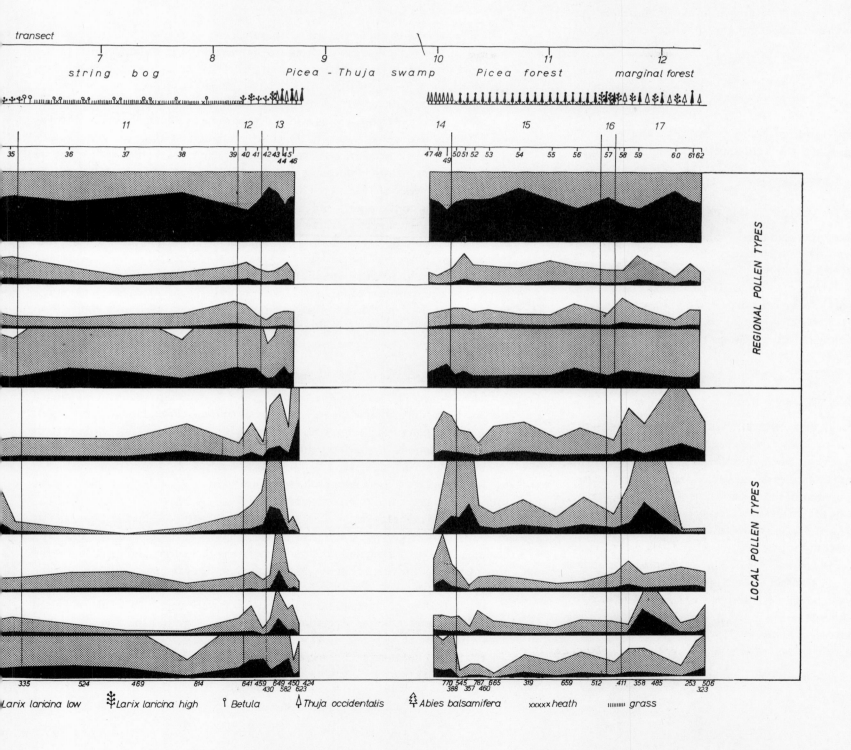

that this notion is true. Smith (1964) reported differences in pollen diagrams from two sites *c* 5 km apart (Cannons Lough and Fallahogy), and Trautman (1957) reports similar features for sites *c* 10 km apart.

The local pollen deposition

So far we have dealt with the composition of pollen assemblages away from the immediate influence of local pollen sources. What happens if we approach such a source?

SURFACE SAMPLE STUDIES IN THE MYRTLE LAKE PEATLAND

The Myrtle Lake peatland is situated in a characteristic landform-vegetation unit (*sensu* Ritchie) in the Lake Agassiz peatlands in northern Minnesota, that is characterised by extensive peatlands between a number of rivers interspersed by sandy beaches from the various stages of retreat of the late-glacial Lake Agassiz. Fig. 1 shows a simplified vegetation map of the Myrtle Lake peatland redrawn from Heinselman (1970). Surface samples were collected along a 18 km long transect (solid black line in Fig. 1) across the peatland, and also along short transects across vegetational gradients.

Short description of the vegetation along the long transect

In the north the transect runs through an extensive raised-bog covered by stunted *Picea mariana* and *Chamaedaphne calyculata—Kalmia polifolia* heath. At the southeastern tip of this raised-bog there is an area where the mineral soil comes close to the surface (*Alnus* island in Fig. 1). Here *Picea mariana* and *Larix laricina* show a vigorous growth, whereas *Alnus rugosa* is abundant in the understorey of the forest together with many herb species characteristic of rich swamp forests. Towards the south the transect runs through an extensive *Larix laricina* poor fen. Near MLTB (see Fig. 1) a ridge of mineral soil close to the surface of the peatland occurs and here we find much taller trees of *Picea mariana* and *Larix laricina*. Also *Thuja occidentalis* is abundant here. From MLTB to Myrtle Lake the transect is interrupted. The transect begins again in dense *Thuja occidentalis* forest, north-west of Myrtle Lake and is followed by *Picea mariana* forest. Near the southern margin of the peatland we find increasing amounts of *Fraxinus nigra* and *Thuja occidentalis*.

Figure 2. Selected pollen curves from the main transect across the Myrtle Lake peatland.

Pollen assemblages along the long transect

As before we will consider the entire assemblage of pollen types. Fig. 2 shows a number of selected pollen curves from the long transect. In the upper part of this diagram the curves of the pollen types from plants that do not occur along the transect, that is the types that will contribute to the regional pollen deposition, are shown. The significance of the upper part of the diagram is that the regional pollen curves do not show any special trend. This is essentially the same as what is shown by Ritchie's work. In this case the vegetation of the entire peatland and its surroundings can be characterised by the proportions of the regional pollen types regardless of where we take the samples, provided of course that the samples were collected away from the immediate influence of a local pollen source.

This conclusion supports the hypothesis that the rate of deposition of the regional pollen types is everywhere the same within the peatland. The alternative explanation would be to assume that the deposition varies from place to place, but that it is not differential for the various types. In this case there is no repetition of stands along the transect to explain the constant proportion of the regional pollen types. An important part of these types may, however, have an extraregional rather then a regional origin. Most of the pollen of *Ambrosia* and *Artemisia* comes from areas outside the Lake Agassiz peatlands. At such a large distance from the source this may result in constant proportions of the types along the relatively short distance of the transect.

The frequencies of the local pollen types (lower part of Fig. 2) have been calculated on the basis of the sum of the regional types. As mentioned before it is assumed that the rate of deposition of the regional types is constant throughout the peatland. If this is true then in surface sample studies these types may be profitably used as a basis for the calculations of the frequencies of the local pollen types. In this way an 'absolute' pollen diagram may be simulated. The values of the local types calculated in this way are, of course, not absolute in a strict sense, but at least the basis for the calculation is presumed to be constant.

In contrast to the regional pollen types the curves for the local pollen types such as *Picea*, *Sphagnum*, grasses, *Alnus*, and *Betula* in Fig. 2 fluctuate wherever there is a change in the vegetation along the transect. The pollen curves show maxima at the *Alnus* island, at the ridge of mineral soil in the middle of the peatland, and near the marginal swamp forest at the southern end of the transect.

To study the local fluctuations in more detail surface samples were also collected along short transects across vegetational gradients. Figs. 3 and 4 show selected pollen curves from one such short (150 m) transect in the centre of the Myrtle Lake peatland where a small outcrop of rock is located a little off-side the main long transect (*Abies* island in Fig. 1).

Short description of the vegetation along the short transect

The upland vegetation consists mainly of *Thuja occidentalis*, *Abies balsamifera*, *Betula papyrifera*, and *Picea glauca* but part of the outcrop is not covered by vegetation. The transect starts underneath *Thuja*, then runs north-east over rock and through a small stand of *Abies*. At the edge of the island there is a single tree of *Pinus resinosa*. The island is surrounded by a marginal swamp forest of mainly *Larix laricina* and *Thuja occidentalis* with *Alnus rugosa* in the understorey which gradually passes into *Larix* poor fen. On top of the diagram of Fig 4 a summary of the vegetation along the transect is given in terms of presence and absence of the species. The species have been arranged according to their occurrence along the transect. In this way 8 synecological groups can be established.

Pollen assemblages along the short transect

The basis for calculation of the frequencies is again the sum of the regional pollen types, that is all the pollen types from species that do not occur along the transect but in this case *Pinus* is excluded because of its presence along the transect. Fig. 3 shows the curves for the regional dominant pollen types. As in the main long transect they do not show any striking trends. In Fig. 4 the curves for the local pollen types are shown. It is not my purpose at present to discuss fully the trends of all the curves. This will be done together with other surface sample and modern vegetation studies in a later paper.

From Fig. 4 the following general characteristics may, however, be extracted:

1 The maximum values of the local pollen types are found at the source itself.
2 The fluctuations of the pollen values right at the source itself may be large because part of the pollen deposition is deposition by rain drops. At the source itself lumps of immature pollen are sometimes found in the samples indicating that parts of inflorescences were incorporated into the substrate (macroscopic deposition, cf. Andersen 1970).
3 For most of the pollen types there is no transition between the deposition at the source itself and the regional deposition. Apparently most pollen types do not spread beyond the site where they are released. This applies particularly to herb pollen types. Most of these types are found only at the source, i.e. their regional value is zero. Because of this phenomenon the local pollen assemblages often show many more pollen types than the regional assemblages. In this case where we are dealing with a forested region only a few pollen types from peatland herbs are found in the regional assemblage in spite of the fact that the peatlands have a large areal extent in this region. Among the peatland herbs that are found in the regional assemblage are *Typha latifolia*, *Spiraea* type and *Thalictrum*, not surprisingly since these species probably produce a relatively large amount of pollen grains, and are all rather tall.

The large variety of pollen types in the local assemblage is advantageous for the reconstruction of the local plant community because it allows us to use combinations of pollen types as indicators of the local plant communities. This is done in Fig. 4 where the pollen types have been arranged according to their maximum occurrence along the transect. In this way 8 topographical groups can be established. The composition of pollen types of these topographic groups closely resembles the species composition of the 8 synecological groups shown on the top of the diagram. The plant communities present along the transect can thus be retraced by combinations of topographic groups of pollen types.

4 In a few cases there is a transition between the local pollen values and the regional pollen values i.e., these types show what I have called earlier the extralocal effect (Janssen 1966). In Fig. 4 this effect is clearly shown for *Pinus*, *Picea*, and *Abies* where between spectra 16-25 the pollen values are transitional between the local and regional values. The curve for *Alnus* shows a huge local peak between spectra 15-19 but does not show a clear extralocal trajectory towards the right side of the transect. *Alnus* pollen is released in the understorey of the rich swamp forest where it may be easily filtered in the shrub-layer of the vegetation. On the left side of the local *Alnus* maximum an extralocal trajectory seems to show up clearly. The *Abies* stand at the margin of the peatland is devoid of shrubs and the presence of the extralocal effect here could be explained by the absence of filtering. However, the increased values of *Alnus* in the open rock area on the island may also be the result of convective air currents during sunny days that will prevent the deposition of regional pollen, combined with a compensatory air current from the margin of the island, the very place where *Alnus rugosa* is abundant.

If this is true, then the assumption that the deposition of regional pollen types is constant along this transect is false. In this case the regional deposition must be less above the island, resulting in increased relative values of the local pollen types. This might also be the explanation for the fact that the values for *Abies*, *Picea*, and *Pinus* are quite irregular in the open area on the island and sometimes do not show any decrease with distance from the source. To resolve these problems pollen traps that will provide us with absolute numbers of pollen grains per unit time will be necessary.

Local and regional pollen deposition along other transects in Minnesota

The characteristics of the regional and local pollen deposition discussed so far are

Figure 3. Selected regional pollen curves from the short transect across the *Abies* island.
Figure 4. Selected local pollen curves from the short transect across the *Abies* island. Explanation of topographical pollen groups and synecological species groups in the text.

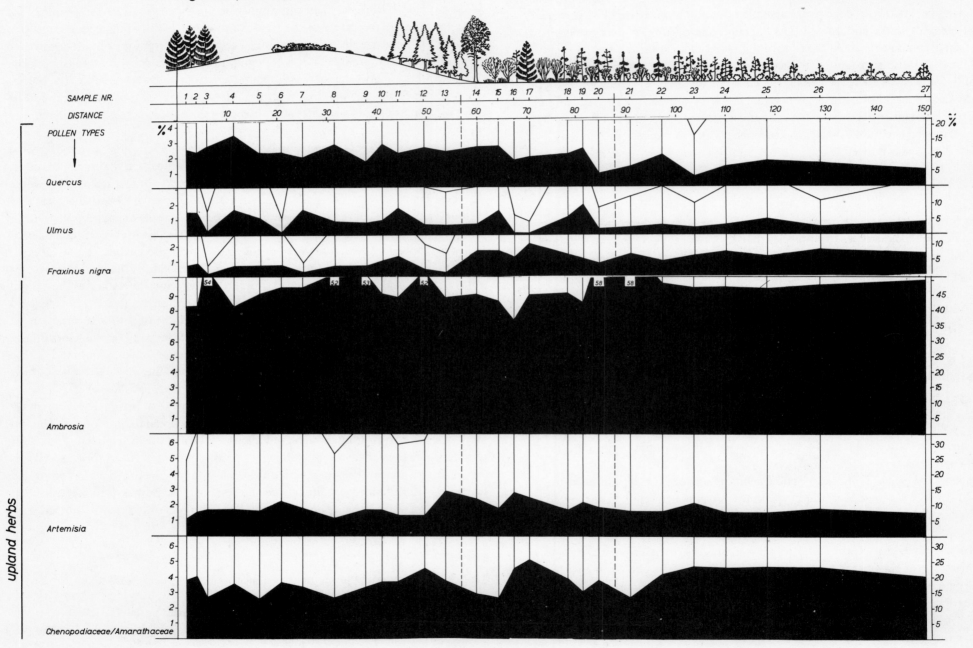

Local and regional pollen deposition   39

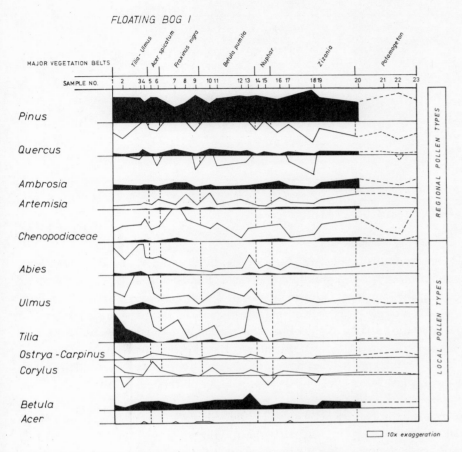

Figure 5. Regional pollen curves and local tree pollen curves from a short transect approaching a *Tilia-Ulmus* forest in Itasca State Park, Minnesota (after Janssen 1966).

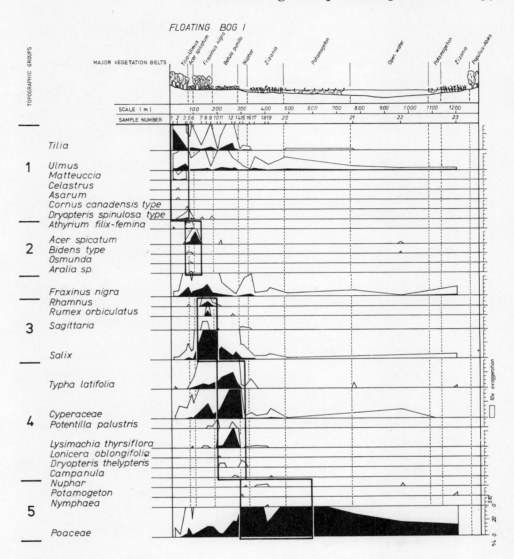

Figure 6. Local pollen curves from a short transect approaching a *Tilia-Ulmus* forest in Itasca State Park; pollen curves rearranged from Janssen (1966).

also found in other short surface transects in Minnesota. In Fig. 5 some pollen curves are shown (values calculated upon a pollen sum including the regional pollen types) from a short transect approaching a *Tilia americana—Ulmus americana* forest in the Itasca area of north-west Minnesota. The diagram shows again the constant proportions between the regional pollen types and extralocal effects.

Fig. 6 is a pollen diagram showing herb pollen types, re-arranged on the basis of the pollen topography. Again it indicates the possibility of establishing synecological groups from such a topographical arrangement. For details of the vegetation see Janssen (1966).

CONCLUSIONS

The examples shown above are all from forested regions outside the mountains. We need many more data from other locations such as in mountainous regions and from non-forested areas. From the studies of short transects that we have so far (a total of twelve of which two are in the mountains), the following picture begins to emerge:

1 A local effect is often observed when the species is present at the site of pollen collection. For trees it almost never fails to occur. In all the transects the absence of such an effect has only occurred once.

2 The presence of an extralocal trajectory is much rarer, however. It is found mostly when pollen is emitted from trees and in the majority of cases when entire stands are involved. Only twice has an extralocal effect for an element occurring in the understorey of the forest been observed. This agrees with the Sutton equation of dispersion of small particles in the atmosphere, for according to this the dispersal distance depends on the height and strength of the source. Also when pollen is released underneath the canopy of a forest it belongs to Tauber's (1965) trunk-space component where it is more easily filtered by herb-and shrub-vegetation. This applies even more for herbs, where filtering in the 'stem' space of the herb layer must be an important explanation for the very short dispersal distances for many herb pollen types. The wind speed is also low near the forest floor (Andersen 1970).

3 The distance over which an extralocal effect is present does not exceed a few hundred metres from the pollen source.

In conclusion, it seems possible to distinguish four kinds of pollen deposition tied to the phenomena of pollen dispersion.

a *Local pollen deposition*. The values of the pollen types are high and often quite irregular and the pollen assemblage shows many types, which enables the recognition of minor vegetation types.

b *Extralocal pollen deposition*. Pollen values are slightly higher than the regional values over a distance of only a few 100 m from the local source. This is only present for a few types in specific locations.

c *Regional pollen deposition*. The deposition of pollen in specific proportions in an area of a formation, and only large vegetation types are recognisable.

d *Extraregional pollen deposition*. The pollen deposition from outside the area of the formation where the samples are collected.

Applications in palaeopalynology

We have seen that the concepts of local and regional pollen deposition can be tied to the pollen dispersion curve of a species. The consequence of this is that there is no knowledge of the regional pollen deposition when the local- and extra-local deposition is not known. This has been recognised for a long time. The opposite, however, is also true: we cannot judge the local vegetation from a pollen assemblage when the regional deposition is not known, not even when macrofossils prove that a species was locally present, because then we still do not know the local contribution to the total pollen

	Bog D. center lake Mc. Andrews, 1966	Bog D. margin lake Erdtman, 1943	Nicolet Creek, narrow valley Shay 1971	Stevens Pond, small pond Janssen, 1967
Post-settlement time	83.6	91.5	over 90	84.7
Pre-settlement time	82.7	92.0	92-95	92.0
Present vegetation around site	primeval pine forest. sites in protected area			fields. cuttings at settlement time

Figure 7. Average pine pollen percentages based upon an upland tree pollen sum in a restricted area in northern Minnesota in 4 different locations.

deposition of that species.

In recent work this problem can be solved, as has been shown by several workers, by a comparison of samples in different locations followed by a comparison of these samples with the known actual distribution of the vegetation. In palaeopalynology the actual distribution of the vegetation is the very thing we are trying to reconstruct from pollen analytical data. Insight in this matter can be gained, as we all know, by a comparison of a sufficient number of properly dated pollen diagrams.

This is a common procedure for regional diagrams and there is no need to discuss this. It also applies, however, to the interpretation of local pollen assemblages. Two examples may clarify this:

1 Fig. 7 shows the average percentage of pine calculated on the basis of an upland tree pollen sum in a small area in northern Minnesota. In this area large scale lumbering of pine started when white man settled the land around 1900. Bog-D and Nicolet Creek are sites in a protected area where lumbering of pine was prevented. As a consequence there is no change in the pine percentages before and after settlement. There is, however, a difference between samples from the centre of the Bog-D pond, those from the margin of Bog-D, and the values from Nicolet Creek, a site in a narrow valley. The samples from the latter two sites come from locations close to the pine-covered upland today and they thus show extralocal pollen values. The samples from the centre of the Bog-D pond at a much larger distance from the upland show regional pollen values. The samples from the small Stevens Pond, presently surrounded by fields, show a marked difference between pre- and postsettlement time. The values in presettlement time equal the extralocal values at Bog-D and Nicolet Creek suggesting that pine was present around Stevens Pond at that time. The pollen values in postsettlement time at Stevens Pond equal those of the regional values at Bog-D Pond. The drop in the values for pine at the transition between pre- and

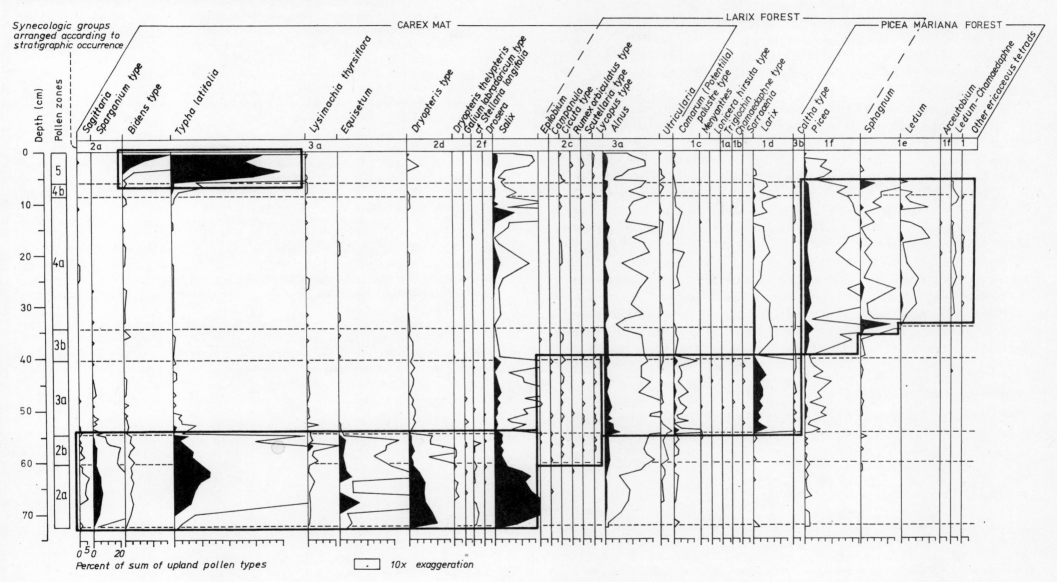

Figure 8. Stratigraphical arrangement of local pollen types at Stevens Pond, showing groups of pollen types.

postsettlement time at Stevens Pond therefore may be interpreted as the result of an actual removal of pine in the immediate surroundings of Stevens Pond.

2   Fig. 8 shows a stratigraphical arrangement of local pollen types from a core of a small peat bog in a region where we have the regional control of pollen analysis of cores from several larger lakes (McAndrews 1966). As may be expected a large variety of local pollen types is present, many of which are absent in the regional pollen assemblages.

It is possible to arrange the local pollen types stratigraphically into groups, much in the same way as the topographic arrangement in the recent transect diagrams shown above. In this way the vegetational successions are easily visualised. In this particular case the combinations of local pollen groups can be compared with those in recent surface samples from vegetations in formations that are still present today.

In Europe this fortunate condition does not exist any more. But suppose we did not know the regional deposition and a comparison with surface samples from modern vegetational analogues is not possible, a stratigraphical arrangement of the local pollen types would still be useful.

If we compare in Fig. 8 the stratigraphic pollen groups with the composition of the species groups in recent plant communities, we arrive at much the same conclusion on the development of the local mire vegetation. Clearly we owe this to the fact that the dispersal distance of many pollen types is so short, with the result that when these types are present in the samples we may, by deduction and ecological reasoning, conclude the local occurrence of the plants in the vegetation.

Acknowledgements

Thanks are extended to Dr H.J.B. Birks, Cambridge and Dr M.L. Heinselman, St Paul for their assistance in the field and to the Minnesota Department of Conservation for technical assistance.

References

AARIO L. (1940) Waldgrenzen und subrezenten Pollenspektren in Petsamo, Lappland. *Ann. Acad. Sci. Fenn. A* **54**(8), 1-120.

ANDERSEN S.TH. (1970) The relative pollen productivity and pollen representation of north European trees, and correction factors for tree pollen spectra. *Danm. geol. Unders.* Ser. II, **96**, 99p.

ERDTMAN G. (1943) *An introduction to pollen analysis*. Waltham, Mass. U.S.A.

FAEGRI K. & IVERSEN J. (1964) *Textbook of pollen analysis*. Munksgaard, Copenhagen.

FIRBAS F. (1949) *Waldgeschichte Mitteleuropas, I: Allgemeine Waldgeschichte*. Gustav Fischer, Jena.

HEINSELMAN M.L. (1970) Landscape evolution, peatland types, and the environment in the Lake Agassiz Peatlands Natural Area, Minnesota. *Ecol. Monogr.* **40**, 235-61.

JANSSEN C.R. (1966) Recent pollen spectra from the deciduous and coniferous deciduous forests of northwestern Minnesota: a study in pollen dispersal. *Ecology* **47**, 804-25.

JANSSEN C.R. (1967) A postglacial pollen diagram from a small *Typha* swamp in northwestern Minnesota, interpreted from pollen indicators and surface samples. *Ecol. Monogr.* **37**, 145-72.

JANSSEN C.R. (1970) Problems in the recognition of plant communities in pollen diagrams. *Vegetatio* **20**, 187-98.

LICHTI-FEDEROVICH S. & RITCHIE J.C. (1968) Recent pollen assemblages from the western interior of Canada. *Rev. Palaeobotan. Palynol.* **7**, 297-344.

MCANDREWS J.H. (1966) Postglacial history of Prairie, Savanna, and Forest in Northwestern Minnesota. *Mem. Torrey bot. Club* **22**(2), 72p.

MCANDREWS J.H. (1968) Pollen evidence for the protohistoric development of the 'Big Woods' in Minnesota (U.S.A.) *Rev. Palaeobotan. Palynol.*, **7**, 201-11.

RAYNOR G.S. & OGDEN E.C. (1965) Twenty-four-hour dispersion of ragweed pollen from known sources. *Brookhaven National Laboratory, Publ.* **957**, (T-398).

RAYNOR G.S., OGDEN E.C. & HAYES J.V. (1968) Effect of a local source on ragweed pollen concentrations from background sources. *J. Allergy* **41**, 217-25.

SHAY C.T. (1971) *The Itasca Bison Kill site, an ecological analysis*. Minnesota Historical Society, St. Paul.

SMITH A.G. (1964) Problems in the study of the earliest agriculture in northern Ireland. *Rep. VI Intern. Congr. Quater. Warsaw*, 1961, Sect. 2, 461-71.

SUTTON O.G. (1953) *Micrometeorology*. McGraw-Hill, London.

TAUBER H. (1965) Differential pollen dispersion and the interpretation of pollen diagrams. *Danm. geol. Unders.* Ser. 11, **89**, 69p.

TRAUTMAN W. (1957) Natürliche Waldgesellschaften und nacheiszeitliche Waldgeschichte des Eggegebirges. *Mitt. flor.-soz. ArbGemein.* N.F. **6/7**, 276-96.

TURNER J. (1964) Surface sample analysis from Ayrshire, Scotland. *Pollen Spores* **6**, 583-92.

WRIGHT J.W. (1952) Pollen dispersion of some forest trees. *U.S. Forest Service, North-Eastern Forest Exp. Station*, Paper 46.

Pollen budget studies in a small Yorkshire catchment

RONA M. PECK *The Botany School,
University of Cambridge, and Department of
Geography and Geology, University of Hong Kong*

Introduction

The available literature on water transport of pollen is rather sparse, although investigations into certain aspects of this phenomenon were made on the River Volga by Federova as early as 1952. Since then, data have increased mainly through marine palynological studies, as exemplified by Muller (1959) on the Orinoco Delta, Traverse and Ginsburg (1966) on the Bahama Bank, and other work by Groot (1966), Stanley (1969), Rossignol (1961), Koreneva (1966), and Cross, Thompson and Zaiteff (1966). Further work on freshwater environments has been fragmentary, but includes a recent survey by Grichuk (1967) on alluvial terraces in Russia, work by Horowitz (1970) on spring water content, and two short-term stream-water surveys in England by Evans (1970) and Ingram et al. (1959).

Two major conclusions result from these studies. The first is that there is evidence for extreme long-distance transport of pollen by water: (a) by marine currents over distances of 1,500–2,500 km, as proposed by Muller (1959) and Rossignol (1961), and (b), in freshwater where Federova (1952) found pollen transport to exceed 1,400 km. Secondly, there is evidence of an association between sedimentation of pollen and fine mineral particles in marine depositional environments (Stanley 1969, Muller 1959). A further fundamental factor emerged from the studies of both Ingram et al. (1959) and Federova (1952), which indicate that freshwater flood-flows contain higher amounts of pollen than do low discharges. So far, however, only Muller (1959) has attempted to estimate the proportion of pollen derived through water transport to the site of deposition.

In view of the increasing numbers of palynological investigations made on limnic sediments, and the previous emphasis on large bodies of flowing water, a site and sampling scheme were chosen to allow both an assessment of the importance of freshwater transported pollen, and a full investigation of the mechanics of pollen transport by water. Use of artificial lakes (reservoirs), with known dates of construction, rather than natural lakes, enables the calculation of pollen influx values to the sediment without recourse to C^{14} dating.

The Oakdale catchment

TOPOGRAPHY AND VEGETATION

The Oakdale catchment in north-east England is an east-west trending valley, 3 km in length, in which there are two small compensating reservoirs, labelled L_1 and L_3 in Fig. 1. The valley is incised into a plateau of Tertiary age (Versey 1937), at approximately 250 m height. The area is formed of sandstone and calcareous beds of the Mid-Jurassic Estuarine Series, capped by Moor Grit at the eastern end, and underlain with Alum Shales. Slope gradients seldom exceed 8° on the Moor Grit plateau, but gradients down to both reservoirs on the Estuarine Beds are of the order of 12–15°. The upper reservoir lies at a height of 200 m and the lower at approximately 130 m. Recent unconsolidated deposits are limited to a small tongue of boulder clay which protrudes up-valley from the Vale of York as far as the lower reservoir. The minerogenic soils on the Estuarine beds are thin and well-draining on the hillslopes, but some development of a peaty humic deposit 15–20 cm thick occurs on the plateau. This layer, due to repeated heather burning, is subject to some localised rill erosion, but other than this, as streams run over bedrock, secondarily derived pollen through stream bank erosion is at a minimum.

The distribution of the major vegetation associations within the catchment are shown in Fig. 1. Table 1 lists typical species in each defined nodum, with their Domin cover-abundance values (after McVean & Ratcliffe 1962). These are taken from quadrat data compiled in a survey completed during the sampling year.

The Callunetum is a species-poor nodum, developed on the plateau above 220 m altitude. It interdigitates downslope with a *Pteridium*-dominated association, whose upward edge is determined by factors of exposure and competition (Elgee 1914). On the moor, in seepage zones either side of Jenny Brewsters Gill (Fig. 1), patches of *Juncus effusus*-dominated mire exist, which is largely replaced beside the water courses at lower altitudes by calcareous rich flushes with *Carex* spp. and *Equisetum palustre* as co-dominants. Several areas of *Agrostis-Festuca* grassland remain, parti-

Table 1. Vegetational data—species and cover values (Domin scale) from quadrats

DECIDUOUS/CONIFEROUS WOOD		CONIFEROUS WOOD		MIXED OAKWOOD		AGROSTIS–FESTUCA GRASSLAND	
Acer pseudoplatanus	6	*Acer pseudoplatanus*	2	*Acer pseudoplatanus*	1	*Agrostis tenuis*	6
Fagus sylvatica	4	*Betula pubescens*	1	*Alnus glutinosa*	7	*Aira praecox*	3
Crataegus monogyna	1	*Fagus sylvatica*	1	*Corylus avellana*	6	*Cirsium vulgare*	1
Fraxinus excelsior	8	*Larix laricina*	10	*Quercus petraea*	4	*C. arvense*	1
Larix laricina	7	*Pinus sylvestris*	2	*Sorbus aucuparia*	2	*Dactylis glomerata*	8
Sambucus nigra	1	*Sorbus aucuparia*	1	*Athyrium filix-femina*	6	*Galium saxatile*	2
Circaea lutetiana	3	*Dryopteris dilatata*	4	*Blechnum spicant*	1	*Juncus effusus*	4
Epilobium montanum	1	*D. filix-mas*	1	*Dryopteris dilatata*	4	*J. squarrosus*	3
Fagus sylvatica	3	*Pteridium aquilinum*	9	*D. borreri*	4	*Poa annua*	2
Fraxinus excelsior	3	*Endymion non-scriptus*	4	*Thelypteris limbosperma*	6	*Sagina procumbens*	1
Geum urbanum	1	*Holcus lanatus*	2	*Ajuga reptans*	3	*Trifolium repens*	3
Rhododendron ponticum	1	*Oxalis acetosella*	3	*Anthoxanthum odoratum*	3	*Veronica serpylifolia*	1
Rubus fruticosus	5	*Rubus fruticosus*	1	*Carex rostrata*	3		
		Trientalis europea	1	*C. sylvatica*	4	JUNCUS EFFUSUS MIRE	
				Cardamine pratense	2	*Carex curta*	+
				Cirsium dissectum	1	*C. echinata*	2
				C. palustre	2	*Galium palustre*	1
				Crepis paludosa	1	*G. saxatile*	2
				Dactylorchis fuchsii	1	*Hydrocotyle vulgaris*	+
				Endymion non-scriptus	2	*Juncus effusus*	10
				Filipendula ulmaria	2	*Potentilla erecta*	1
				Galium palustre	3	*Potamogeton polygonifolius*	+
				G. saxatile	3	*Viola palustris*	2
				Holcus lanatus	5	*Sphagnum recurvum*	8
				Lysimachia nemorum	2		
				Mentha aquatica	2		
				Oxalis acetosella	4		
				Potentilla erecta	2		
				Primula vulgaris	2		
				Ranunculus repens	4		
				Rubus fruticosus	1		
				Teucrium scorodonia	2		
				Trientalis europea	2		
				Urtica dioica	2		
				Valeriana dioica	2		
				V. officinalis	2		
				Viola palustris	2		
				Sphagnum palustre	2		
				S. squarrosum	2		

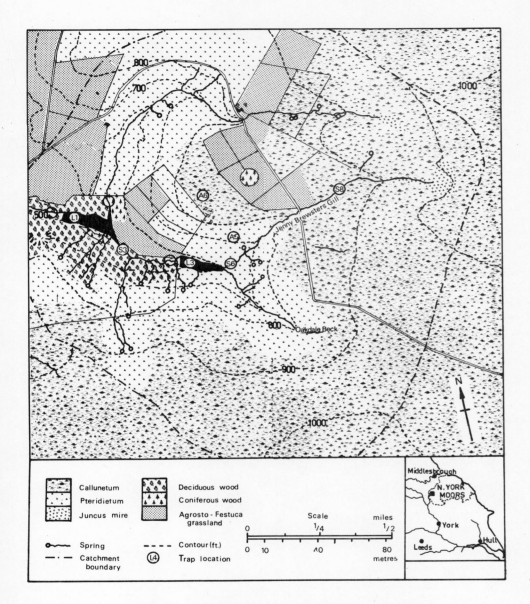

Figure 1. The Oakdale catchment, showing topography, vegetation, and trap locations. A_1–A_6—air traps. L_1–L_3—lake traps. S_3–S_8—stream traps.

cularly on the north hillslope between the reservoirs, and on the moor top near stream trap S_8.

The woodland areas represent a mixture of planted, managed, and semi-natural communities. Between the reservoirs, on the southern hillslopes, a species rich oakwood exists, with standards of *Quercus petraea* and *Acer*, with *Alnus glutinosa* and *Corylus avellana* in wetter areas. This woodland is rich in both herb and fern species, and *Pteridium aquilinum* dominates the understorey. A thin band of planted coniferous wood composed of *Larix laricina* and *Pinus sylvestris* is present on the south shore of the lower reservoir, and the north shore supports a mixed deciduous/coniferous wood. Deciduous species include *Acer pseudoplatanus*, *Fagus sylvatica*, *Fraxinus excelsior*, and *Quercus petraea*, with *Larix* and *Pinus* as the principal conifers. Understorey herbs are very sparse here, in contrast to the other woodland stands. Upslope, on the north-facing side above the conifers, *Betula pubescens* is actively colonising the Pteridietum.

Aquatic plants grow as a continuous fringe in the lower reservoir to a depth of 3 m, and include, in shallow water, *Littorella uniflora*, *Polygonum amphibium*, and *Alisma plantago-aquatica*. In greater depths *Elodea canadensis* and three species of *Potamogeton* occur. A *Typha latifolia* swamp is present at the eastern inlet, below stream trap S_3, with *Iris pseudacorus* and *Sparganium angustifolium*. The upper reservoir is less eutrophic, and aquatic plant growth is limited to the shallow water at the eastern end, where there is *Potamogeton natans* and *Sparganium erectum*.

HYDROLOGY AND LIMNOLOGY

The hydrology of the catchment is relatively simple. The streams rise on the moorland and are further fed by a strong line of springs from the base of the Ellerbeck limestone bed, which outcrops at approximately 210 m. Stream discharge responds quickly to rainfall, with the time to maximum rise occurring within 3–4 hours after rain, depending on catchment antecedent conditions. During the sampling year, stream discharge records were read at a British Standard (1964) design square notch weir at S_3, on the inflow into the lower reservoir, and at a 90° 'V' notch weir at S_6, flowing into the upper reservoir. For a period at the end of 1968 into 1969, and again during flood water surveys of 1970–71, a Munro continuous recording gauge was established at the latter site, upstream of the weir.

Both reservoirs have almost identical capacities, but the lower reservoir is shallower, reaching only 9 m maximum depth, but with greater surface dimensions: 226 m length × 58 m width. The upper reservoir has maximum depth of 11 m, maximum length of 194 m, and breadth 61 m. During the sampling year a check on stratification was made using a thermistor thermometer to record temperature pro-

Figure 2. Aerial photographs of the reservoir sites (photographs by J.K. St. Joseph, Cambridge University Collection). Left L_1, showing distribution of woodland round the margins. Right L^3, surrounded by Pteridietum. (Trap positions shown by arrows).

files for each lake in three separate months. The upper reservoir remained almost isothermal in both September and December, and demonstrated only a gradual temperature decrease of 3°C from top to bottom in July. In the lower reservoir, the July temperature change with depth was well marked, showing a total decrease 6°C, with two layers where gradients reached 1°C change per metre of depth. This lake was also isothermal in September, as expected in an autumn overturn, and the December readings, under ice cover, revealed a gradual rise of 1.5°C with increasing depth. The difference between the reservoirs was attributed to the increased exposure of the upper reservoir, which is at higher altitude and without tree wind-breaks (Fig. 2), with consequent wind-created circulation (Bye 1965, Langmuir 1938).

EQUIPMENT AND SAMPLING SCHEME

On the assumption that the characters of air and water flow are not dissimilar, perspex traps, modelled on the design of Tauber (1967), were used for sampling both media. In this manner, a cumulative sample was obtained, representing the catch at each site over one month. The traps used for stream sampling were heightened by x3, to increase the particle sedimentation to the base through a calm layer of water. These were staked into position in the stream centre, facing upstream of the supporting pole, and attached with removable brass collar. Subsequent research (Peck 1972) has shown that despite their low overall trapping efficiency (mean = 0.8% E), they catch a representative sample of the passing pollen spectrum, and cannot overload. An added advantage of the design is their inconspicuous nature once submerged.

For lake site samplings, air traps were mounted on polystyrene and plywood floats. Lids were approximately 45 cm from the water surface to be free of wave splash (Fig. 3). From these floats a line of traps were suspended, separated by one metre intervals, and each was secured to a metal platform mounting by wing nuts. To avoid lifting the weights, thus disturbing the bottom sediment at each trap change, the system in Fig. 3 was designed. The float and buoy are kept permanently apart, to avoid trap line and spare rope from tangling up, by a wooden pole the same length as the iron girder on the lake bottom. Spare rope is released from storage at the buoy (A). Trap lines are unhooked from the base of the float at D, and slowly raised with the spare rope travelling through the system of pulleys B and C. Traps are corked and removed on reaching the surface, and replacements put in. The line is lowered, and the spare rope pulled back, through the system to the buoy for storage. This system proved satisfactory in all sampling months, except under conditions of moving ice rafts, at the spring thaw, though even then twisted lines, caused by excessive drag, were unwound without disturbance to the suspended trap samples.

All samples were processed in the laboratory to yield numerical values of total catches, by tested methods previously described by Matthews (1969) and Davis (1965).

Pollen deposition at the reservoir sites

An assessment of the importance of the aerial and water transported pollen components to the final pollen influx to the reservoir sediment was made by an initial comparison of air/water catches during the sampling year, followed by a comparison of influx into suspended traps and reservoir sediment per annum. The second

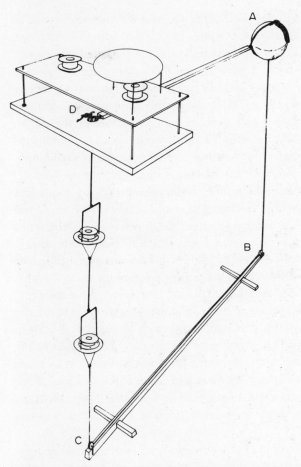

Figure 3. Diagrammatic representation of lake trapping equipment. Letters refer to text.

analysis was made to determine the presence of recycled pollen in the lake trap catches, which might invalidate the estimates of air and stream component significance if included in the calculation.

DEPOSITION THROUGH AIR AND WATER DURING THE MONITORING PERIOD

Figures for pollen deposition from the air to the lake surface are taken from the Tauber trap catches. This assumes that pollen influx into the trap is a true representation both of the composition and magnitude of pollen impinging on the lake surface. Unless this assumption is made, no estimate of aerial deposition is possible.

Preliminary results of wind tunnel tests by Tauber (personal communication), to determine trap efficiency, yield slightly lower values of 'p' (Deposition Coefficient) in comparison to other available outdoor experimental values determined by Gregory (1961) and Chamberlain (1966) for the flux of *Lycopodium* spores from a cloud to short grass. Making allowance for a water surface being more retentive of particles sedimenting or impacting on to it than grass blades, where rebound occurs (Chamberlain 1966) on impaction, the Tauber trap probably under-estimates the pollen reaching the lake surface, particularly in periods of high wind velocity. Air trapped pollen from the sheltered lower reservoir is therefore likely to have greater reliability, but so far it has not been possible to apply any correction factors.

Pollen figures in Table 2 are for selected months of the period 1968–69, displaying the number of grains deposited per cm² in air and suspended lake water traps, and the calculated proportion that the air numbers form of the deposition into the water in each month. The months which are tabulated display data for both the flowering and non-flowering seasons. May represents the highest production period for tree pollen, with *Quercus*, *Fraxinus*, and *Fagus* at a maximum. In consequence, the proportion of water trapped pollen (L_1 and L_3) explained by aerial deposition (A_1 and A_3) is high (20–26%). In the following month, June, with grasses flowering, the air catch in the upper lake accounts for the total herb pollen found that month in the upper reservoir (L_3). In September, pteridophyte spores are in evidence in the air traps but only explain 1% of the water pteridophyte total; however, in October this proportion rises to 10% at the upper reservoir site when *Pteridium* spores are airborne. In contrast, after October, without any new pollen production, air traps explain very little of the total in the water traps (see December and February values in Table 2).

Totalising the grain deposition per cm² for the monitored period, aerial deposition explains no more than 4% of the total deposition in the upper reservoir (L_3), and only 3% at L_1. The 'excess' of pollen in the lake water in most months can be attributed either to the action of pollen recycling within the waterbody, or to the presence of a stream-borne component in the total, or to a combination of both.

Table 2. Deposition (per cm²) at reservoir sites during the sampling period.

		LOWER RESERVOIR			UPPER RESERVOIR		
		ΣA_1	ΣL_1	$(\Sigma A_1/\Sigma L_1)\%$	ΣA_3	ΣL_3	$(\Sigma A_3/\Sigma L_3)\%$
MAY 1968	TDP	668	3,969	17.72	1,124	4,182	26.88
	tree	630	2,993	21.04	1,052	3,870	26.67
	herb	38	580	6.55	71	236	30.08
	fern	0	209	0	0	76	0
JUNE	TDP	1,129	7,614	14.83	1,652	1,446	100.00
	tree	127	3,176	4.00	97	764	12.70
	herb	997	4,115	24.23	1,553	771	100.00
	fern	0	339	0	0	119	0
SEPTEMBER	TDP	66	35,840	0.18	96	9,839	0.97
	tree	20	17,300	0.12	2	3,683	0.05
	herb	196	13,737	1.43	82	4,739	1.73
	fern	48	4,752	1.01	11	1,430	0.77
DECEMBER	TDP	35	4,553	0.79	15	2,705	0.55
	tree	28	1,717	1.63	8	1,038	0.77
	herb	7	2,163	0.32	6	1,171	0.51
	fern	0	671	0	1	52	0.11
FEBRUARY 1969	TDP	23	12,499	0.18	17	152	1.11
	tree	13	5,605	0.23	14	544	2.57
	herb	9	4,990	0.18	2	286	0.70
	fern	0	1,904	0	1	697	0.14
ANNUAL	TDP	3,705	158,419	2.34	4,542	146,883	3.09
	tree	2,016	77,722	2.59	2,208	67,945	3.23
	herb	1,650	58,142	2.84	2,287	58,524	3.91
	fern	39	21,923	0.18	47	20,046	0.23

A_1, A_3 Air traps
L_1, L_3 Lake traps
TDP = Total Determinable Pollen

RESERVOIR SEDIMENT CORES

Short cores from locations near the suspended traps were taken from each lake at the end of the sampling year, using a one metre Mackereth corer worked by a diver. Sediment thickness in the upper reservoir (constructed in 1911) was thin, as bedrock was struck at 49 cm. In the lower reservoir a 90 cm core was taken. Samples of known volume were removed for analysis, initially at 5 cm intervals, and later at closer intervals across detrital layers, with the object of determining the top of the drowned soil layer and thus the depth of real reservoir sediment deposited since construction. Sediment description is a modified Troels-Smith (1955) system, and organic fractions were calculated on all samples. Diatom analysis at critical levels was completed by E.Y. Haworth of the Freshwater Biological Association.

POLLEN CONCENTRATION AND INFLUX VALUES

In the upper reservoir a well marked detrital layer was found at 19–20 cm (Fig. 4). At this level there was a significant increase in total pollen concentration per cc, accompanied by a rise in the indeterminable pollen component. The most marked increase at this level is in herb pollen and pteridophyte spores, especially of Gramineae, *Calluna*, Rosaceae, and *Urtica* pollen among the herbs. This layer is immediately overlain by 2 cm of deposit poor in total pollen concentration, above which the concentration rises again and remains relatively stable for each taxon up to the sediment surface. It would appear that the layer with a high *Pteridium* spore concentration represents the old soil surface, with 2 cm of mineral sediment above resulting from redeposition of soil particles eroded and suspended during the rise of water level associated with the flooding of the valley.

In the lower reservoir the situation was more complex, with two detrital layers present (Fig. 4), one at 14–15 cm, the other at 28–29 cm. The occurrence of particles of coarse-sand size is limited to below 29 cm in the core, and the highest concentration of indeterminable grains is at 30–31 cm. Of the pollen and spore types *Sphagnum* spores reach high proportions at this level, along with a slight rise in *Pteridium* spores. Grains of *Myriophyllum*, normally a plant of ponds or slow-flowing water, only occur below this level. Sediment below 31 cm could have been part of a wet alluvial stream-terrace before drowning occurred. Between this level and 27 cm the marked fall in total concentration can be attributed to the same cause as for the upper reservoir, i.e. soil suspension, and above 27 cm, the total pollen concentration appears relatively stable. Small landslips on the Alum Shales are not uncommon round the lower reservoir, and the 14–15 cm detrital layer might be the result of such a disturbance. Diatom analysis of the 31–32 cm level showed a mixture of types indicative of some soil inwash, but in the 28–29 cm level, the proportion of lake types increased and includes species normally associated with eutrophic conditions. These must be indicative of the onset of true reservoir conditions.

Pollen influx figures (Table 3) are calculated on the assumption that 28 cm of mineral sediment accumulated in 76 years in the lower reservoir (constructed 1893), and 18 cm in 58 years in the upper reservoir. It is also assumed for the purposes of these calculations that vegetation change within the catchment since 1893 has been minimal.

THE RECYCLING COMPONENT

Mean pollen influx figures (grains/cm^2/year) for reservoir sediment and suspended traps, presented in Table 3, illustrate a close comparison between the two influx values in the lower reservoir, but a striking disparity between the trap and the sediment influx in the upper reservoir. In the latter case, the trap is overcatching pollen by a factor of three. In a short term study such as this, small discrepancies caused by

Table 3. Pollen influx (grains/cm^2/year) to the Oakdale reservoirs

	LOWER RESERVOIR (L_1)		UPPER RESERVOIR (L_3)	
	Mean figs. from 0–28 cm sediment core	Trap accumulation L_1 1.4.68–1.4.69	L_3	Mean figs. from 0–18 cm sediment core
Betula	16,780	30,053	33,871	7,559
Alnus	18,226	28,217	18,537	6,216
Corylus	10,474	6,955	7,161	4,652
Fraxinus	1,888	1,511	480	105
Pinus	1,836	2,471	829	359
Quercus	2,957	5,093	4,777	1,406
Salix	552	602	488	189
Ulmus	652	687	783	113
Others	348	2,133	1,019	289
Compositae	1,067	816	262	209
Ericaceae	15,130	20,632	28,974	8,524
Gramineae	18,188	31,737	23,406	6,814
Plantaginaceae	1,293	928	632	251
Rosaceae	1,230	1,401	1,326	647
Others	1,727	2,628	3,924	440
Dryopteris filix-mas type	993	1,267	555	478
Dryopteris dilatata type	1,866	2,686	405	71
Pteridium aquilinum	10,329	7,563	9,256	6,518
Polypodiaceae undiff.	8,848	7,785	7,757	3,343
Sphagnum	1,892	1,962	700	491
Others	767	660	1,373	233
Indeterminable	12,397	23,812	22,853	8,944
ΣTree	53,713	77,722	67,945	20,888
ΣHerb	38,635	58,142	58,524	16,885
ΣPteridophyte	24,695	21,923	20,046	11,134
ΣP	136,840	182,231	169,736	58,015

UPPER RESERVOIR (L₃)

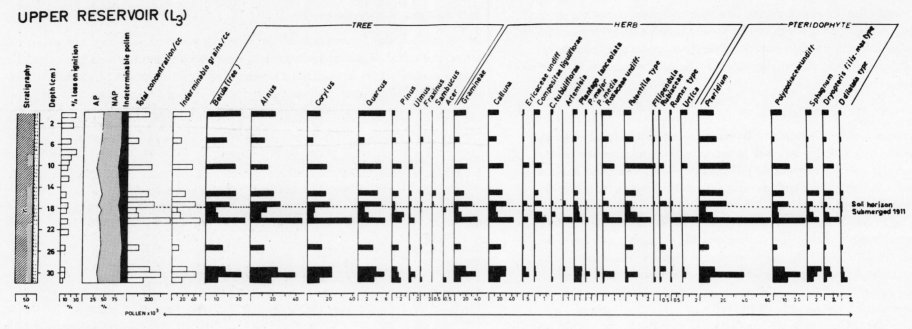

LOWER RESERVOIR (L₁)

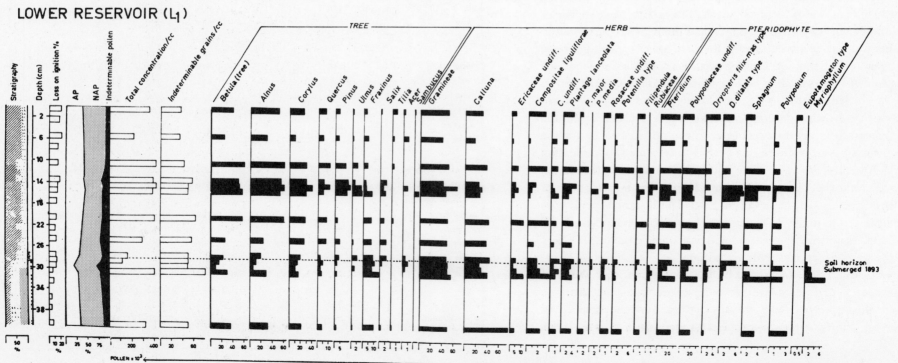

monitoring pollen production in a single season are to be expected. A partioned χ^2 test (Mosimann 1965) was applied to both sets of figures to determine the extent of similarity between pollen spectral composition of trap and sediment in each reservoir, and whether they could be considered representative of the same population. Though both results fell outside the 95% probability level, the χ^2 value was substantially lower for the upper reservoir than for the lower reservoir. An explanation for this result, which indicates a greater homogeneity between traps and sediment core spectra in L_3, particularly when taken in conjunction with the difference in influx values, must be that large-scale recycling of water causes a mixing of both old and new pollen input, resulting in a lowered final sedimentation.

This interpretation accords well with the lack of stratification detected in the upper reservoir, which eliminates seasonal overturn as a possible cause of redeposition (Davis 1968), and demonstrates the effect of wind-driven circulation on sedimentation in shallow exposed lakes. In contrast, recycling is less in evidence in the lower reservoir, which is at lower altitude, and well sheltered by trees. Water residence time is also longer in the upper reservoir.

REASSESSMENT OF AERIAL DEPOSITION

Using influx figures from the reservoir cores, thereby eliminating the recycling component monitored in the suspended traps, Table 4 presents a reassessment of the importance of the aerial component to the final pollen deposition per year into the sediment. There is a negligible increase in the percentage explained in the lower reservoir for total grains (cf. Table 2), but a rise from 4-9% for comparable figures in the upper reservoir.

Table 4. Reassessment of Aerial deposition component

ANNUAL TOTAL	LOWER RESERVOIR			UPPER RESERVOIR		
	ΣA_1	L_1 influx	$(\Sigma A_1/\text{influx})\%$	ΣA_3	L_3 influx	$(\Sigma A_3/\text{influx})\%$
T.D.P.	3,705	124,443	3.98	4,542	49,071	9.26
Σtree	2,016	53,713	3.75	2,208	20,888	10.57
herb	1,650	38,635	4.27	2,287	16,885	13.35
fern	39	24,695	0.16	47	11,134	0.42

Figure 4. Pollen diagrams from the Oakdale Reservoirs, North Yorkshire. All values are pollen concentration (grains/cm^3 × 10^3).

Water-transported pollen

SIGNIFICANCE AND CHARACTER OF THE WATER-BORNE COMPONENT IN OAKDALE

A tentative assessment for the proportion of new pollen supply brought to the site through stream transport in this sampling year, must, therefore, be 97% and 91% for the lower and upper reservoirs, respectively (Table 4). Of particular note is the very high proportion of water-borne pteridophyte spores. A subjective review is sufficient to demonstrate the high degree of correspondence between the pollen composition of the lake sediment and stream traps, indicating the importance of a water-borne component to the pollen finally incorporated in the lake sediment, and the low correspondence between the lake and air trap spectra. Both lake cores are surprisingly alike, and are characterised by high percentages of pteridophyte spores (19-20%), although they form only 1% of the total aerial deposition per year at each lake site. Tree pollen percentages are lower than those in the air, as are those for Gramineae pollen. In contrast there is six times the total Ericaceae proportion (13%) in the sediment as in the air. The stream traps show total pteridophyte values, for the year, of at least 15%, and Ericaceae totals of 8-13%, although tree percentages are more similar to those in the air. There seems little doubt that the pollen and spores of species that are poorly air-dispersed, such as *Calluna* and *Pteridium*, are almost entirely water-transported to the lake sites, and that the final spectrum of pollen types in the sediment is thereby widened to give a more realistic representation of the vegetation in the total catchment.

In contrast to the air trap catches (Fig 5), with marked seasonal fluctuations in grain numbers and spectral composition, the stream trap catches remain high throughout the year. Pollen in the air is negligible between October and April, whereas total water numbers peak outside the pollen production season, i.e. in November, January, and March. Maximum recorded air figures reach 40,000 in one month: in the same period maximum catches in the streams rise above 3 million grains. In any one month the total number of taxa is higher in the water than in the air traps, and it may be greater by a factor of three.

The composition of the air traps is formed mainly of tree pollen in April and May, herbs in June, July, and August, and with occurrences of pteridophyte spores in September. Stream trap spectra also show on close inspection some traces of seasonality. There is a larger tree component at the beginning of the year than in September, but although herb pollen is high in June and July, the largest values are recorded in January and March. Pteridophyte spores peak in November, but are also important in July, September, and March. Unless insect-borne pollen enters

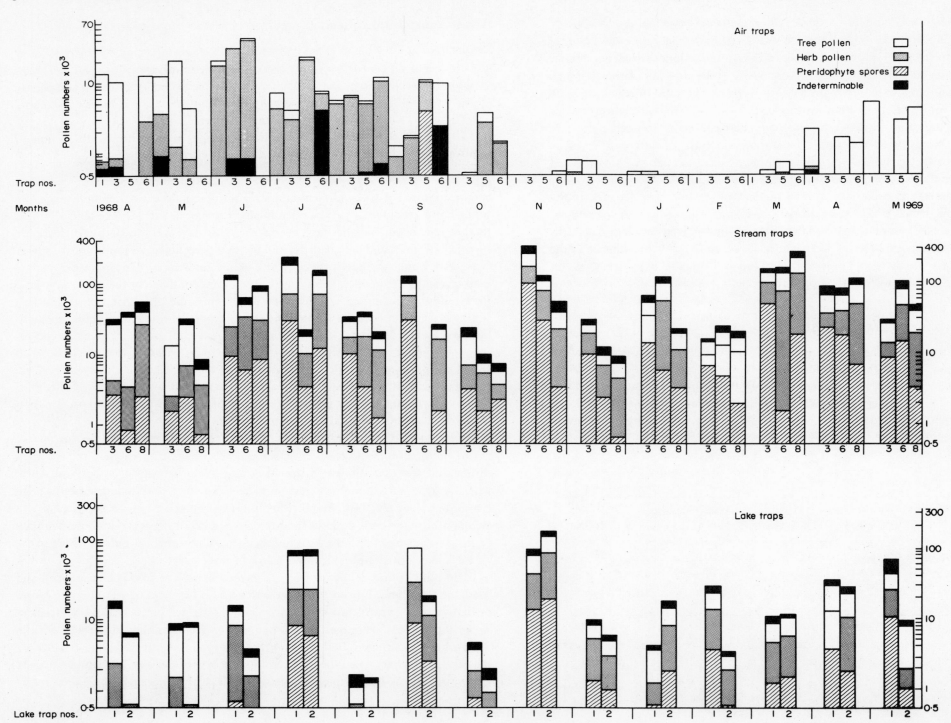

the air traps (as in July at A_6), the proportion of indeterminable grains caught remains low. In stream traps, this proportion can rise as high as 25% of total recorded pollen. Both air and water traps show differences in the same month according to the site within the catchment, though these are more marked for the air catches, e.g. the rise in traps A_5 and A_6 in June, July, August, and September is associated with a greater production of herb pollen at the upper end of the catchment in these months.

RELATION BETWEEN MINERAL PARTICLES AND POLLEN IN WATER

Previous evidence of the relationship between pollen and sediment has come from marine and brackish-water depositional studies. Stanley (1969) has suggested that the hydraulic equivalent of pollen would be fine or very-fine silt (diameter 4–16μm) rather than coarse/medium silt (16–62μm), the size range of which most pollen types occur. Muller (1959) estimated that for marine sites, pollen deposition occurs at a faster rate than is theoretically calculated from Stokes Law, and suggests an average settling velocity of 12 cm/hr, which compares with a settling velocity of quartz particles 4–6μm diameter, or the very fine silt fraction. Traverse and Ginsburg (1966) explained the distribution of pine pollen round the Bahama Bank in terms of sedimentation patterns; they mentioned the effects of water turbulence in keeping large quantities of pollen in suspension for long periods, and reported the highest pollen concentrations coincident with areas of fine grained mineral sediment. Groot (1966), in a study on the Delaware estuary, also indicated a relationship between total pollen and quantity of suspended mineral matter, and suggested a common derivation through water transport.

Figure 6 is a plot of total pollen numbers against weight of sediment less than 100μm diameter, trapped monthly during the sampling period at the S_3 and S_6 stream sites in Oakdale. Calculated regression lines are drawn and correlations are significant with a probability of 0.01. This is a clear demonstration that pollen, once wetted (Hopkins 1950), acts as any other fine clastic particle, and that within the body of water flow, is transported as part of the 'washload'. This is fine sediment, below 0.08 mm diameter, which is held constantly in suspension by upward turbulence; it thus becomes evenly distributed throughout the stream profile, and travels downstream at the velocity of water flow (Leopold, Wolman & Miller 1964, Colby 1963). With this assumption, it is possible to apply some concepts used in mineral erosion and transport studies to problems of pollen transport. However, since pollen numbers at the soil surface fluctuate both seasonally and yearly according to production, and decrease downward in the profile (Dimbleby 1957), the analogy with sediment will not necessarily be perfect.

Figure 5. Composition of air and water pollen trapped in Oakdale from April 1968 to May 1969.

FACTORS CONTROLLING FINE SEDIMENT SUPPLY TO STREAMS

Smith and Wishmeier (1962) have summarised the factors found by many authors that affect the supply of mineral particles to streams, as follows: (a) *rainfall character*, including its intensity and total; (b) *soil erodibility*, as it effects infiltration and percolation, and thereby runoff; (c) *topography*, including slope angle and slope length, and (d) *plant cover* and presence or absence of litter.

Particles are loosened and set in motion on slopes mainly through the action of raindrop impact (Ellison 1945, Ekern 1950, Emmett 1970). Subsequent removal under gravity through sheet or rill flow follows, and in consequence erosion and transport are most successful on bare surfaces (Banky 1959, Imeson 1971). There is a significant reduction of soil loss under water-absorbing litter or humus (Lowdermilk 1930, Copeland 1963), though there is also some evidence from Pierce (1967) at the

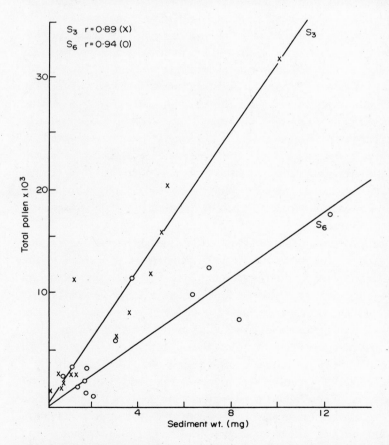

Figure 6. Graph of total pollen plotted against weight of fine sediment in stream traps S_3 and S_6. O S_6, $r = 0.885$, X S_3, $r = 0.940$.

Hubbard Brook Catchment to show that during intense rainstorms, leaves of hardwoods become compacted sufficiently to direct discrete runoff downslope above the soil horizons. Kittredge (1954) had previously demonstrated that pine needles behaved in a similar manner.

DERIVATION OF STREAM TRANSPORTED POLLEN

Stream-borne pollen can be derived from three major sources: (a) by direct fall from species growing on the bankside; (b) from bank erosion, and (c) through overland runoff. For the Oakdale catchment, with narrow watercourses flowing mainly over bedrock, except for certain large producers such as *Alnus* (Janssen 1959), the last source assumes overriding importance.

Pollen reaches the ground, to be available for transport, by both direct and indirect methods, depending on the plant cover. Chamberlain (1966) has shown in wind tunnel experiments using artificial grass, that *direct deposition* by sedimentation to the substrate can be as high as 37% of the total deposition at low wind speeds, though at wind velocities higher than 5m/sec, impaction on to the leaf blades accounts for a greater proportion of the total deposition. Tauber (1965) has shown theoretically for woodland areas that filtration by twigs is highly efficient, and has recovered large numbers of grains from willow twigs by washing (Tauber 1967). The amount of rainfall penetrating hardwood and softwood canopies has been investigated by Ovington (1954), Rutter (1963), and Trimble and Weitzman (1954). In high intensity rainstorms, once the interception capacity of the canopy is filled, leaf drip begins. Total rainfall reduction, due to through fall, may be as low as 20%. During low intensity storms stem flow is common, and thus filtered pollen may reach the ground surface *indirectly* by this means.

In shrub communities, measurements of through flow have been made on *Calluna vulgaris* by Rutter (1963) who showed a 79-80% penetration of rainfall to the ground, and Leyton, Reynolds and Thompson (1967), investigating this phenomenon for *Pteridium*, found increasing stem flow as foliage died down in November. As for mineral soil particles, the easiest transport of pollen should be away from bare surfaces. However, since this implies an absence of pollen-producing vegetation, the lower availability of pollen may result in low overall transport. In S_6 (Fig. 6), there is a higher weight of fine sediment recorded per 1,000 grains than in S_3, a partly wooded site. This could be related to the high mineral erosion from burnt-over swiddons on the moor, combined with the consequent low local pollen production in this part of the catchment. Andersen (1970) postulated that large numbers of grains should be deposited beneath tree canopies. Both soil and litter analyses in Oakdale, and Dimbleby's (1957) figures for other areas within and without woodland, substantiate the hypothesis that there is abundant pollen on the soil surface available for transport away, by overland flow, by through flow in the top soil horizons (Kirkby & Chorley 1967), or through the litter layer by the mechanism suggested by Pierce (1967).

ANALYSIS OF FACTORS EFFECTING POLLEN SUPPLY

In an attempt to explain pollen monthly variations in the stream flow, the matrix of data from the monitoring period was subjected to a multiple regression analysis against a series of parameters chosen to represent possible affecting factors.

Stream discharge expressed in the form of daily and monthly total flows, peak flow, and mean monthly flow, was chosen as the first variable, as it represents the end product of all runoff and water balance factors in the catchment.

Rainfall variables initially included in the analysis were monthly total, mean daily, and maximum 24 hour rainfall.

Pollen production factors presented difficulties of definition, as there are insufficient independent measures of plant productivity, and not all areas of the catchment were completely monitored. Finally an expression on a relative scale was utilised, taking deposition in each air trap in one month as a percentage of the highest deposition recorded at that site during the monitored period.

During the analysis factors not significantly improving the explained variance were rejected. Some caution in interpretation was necessary as runoff and rainfall variables tend to be interrelated. A significant explanation of variance was achieved in two stream sites, S_6 and S_3, using three variables in combination.

For station S_6, above the upper reservoir, a correlation of 0.775 ($N = 14$) was obtained between total pollen and mean monthly stream discharge. This was slightly improved by the inclusion of the production variable, to give a total correlation of 0.807, explaining 65.1% of the total variance. Figure 7 shows the plot of discharge against total pollen, total tree pollen, herb pollen, and pteridophyte spores. Though correlations with this variable are significant at a probability of $P = 0.01$, in all cases the points are well scattered. Pollen and spore types also found to correlate significantly with discharge included *Betula*, *Calluna*, Gramineae, total Compositae, Rosaceae, *Pteridium*, and undifferentiated Polypodiaceae spores. The relation with tree genera is notably weaker, which is a reflection on the relative scarcity of this pollen on the ground in this section of the catchment compared to herb species. Significant improvement in the regressions for *Pinus*, *Quercus*, *Fraxinus* (from 0.082 correlation to 0.917 total), *Alnus*, and Plantaginaceae was obtained on inclusion of the production variable (Table 5). Within the total stream catch therefore, some species do show seasonal fluctuations, showing a response in the water within the

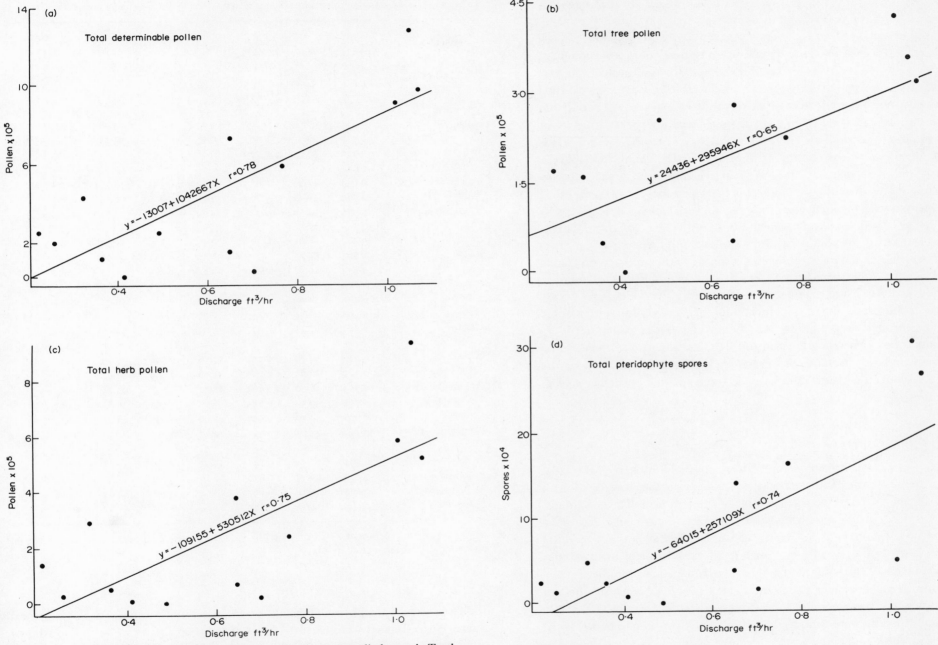

Figure 7. Graphs of pollen catch at stream site S_6 plotted against stream discharge. A. Total determinable pollen. B. Total tree pollen. C. Total herb pollen. D. Total pteridophyte spores.

same month to the highest intensities of terrestial flowering. At this site however, the behaviour of *Alnus* seemed unrelated to any factors included in the regression.

For the S_3 site, above the lower reservoir, the best overall correlation for a single factor with total water-borne pollen was for the maximum 24 hour recorded rain in each month ($r = 0.787$, $p = 0.01$, $N = 14$). Though initially unexpected, this may reflect the earlier discussed point that heavy rainfalls are able to penetrate hardwood canopies sufficiently well to cause surface runoff on the litter beneath. The correlation between maximum rain and total tree pollen again showed lower values than that for total herb and total pteridophyte spores ($r = 0.668$, $r = 0.737$, $r = 0.732$ respectively). Individual taxa also correlating significantly with this variable were *Alnus*, *Betula*, *Corylus*, *Sambucus*, *Calluna*, Gramineae, Rosaceae, and Plantaginaceae pollen and Polypodiaceae spores. Partial correlations for *Pteridium* spores and Compositae pollen were improved with the inclusion of the second variable, mean monthly discharge, and further significant improvement for *Ulmus*, *Acer*, *Pinus*, and Plantaginaceae pollen was made by adding the third variable, production.

From both stream sites genera with the best response to the production variable were the trees *Ulmus* and *Pinus*, with *Quercus*, *Fraxinus*, *Acer*, and *Plantago* following. Good correlation with discharge parameters must indicate the widespread occurrence of the genera concerned on the catchment, and their availability for transport all the year round through overland flow. Apart from *Betula* tree pollen in the S_6 catchment, all other taxa concerned grow in the area, and their pollen tends to be poorly dispersed by air, e.g. *Calluna* and *Pteridium*. In S_3, as a reflection on the wooded area of catchment, a higher number of tree taxa, as well as the herbs, correlate with the first variable.

POLLEN IN FLOOD DISCHARGE

As the above analysis was performed on cumulative pollen and discharge values, the importance of extremes in the data is removed or masked. Further investigation on the pollen and stream discharge relation was made through sampling flood water pollen during the winter of 1970–71, whilst aerial production was negligible.

A flood board modelled on the U.S. Geological Survey single stage sampler (1961) was erected, as shown in Fig. 8, with 0.25 inch (c 0.4 cm) diameter brass nozzles pointing upstream clear of the baffle shield. These instantaneously sample, at their respective heights, during the rising flood, and as long as vertical exhaust tubes are fitted to the bottles, the containers once filled, cannot refill. Attempts to sample falling floods automatically were unsuccessful. Four one-litre sample bottles were attached by rubber tubing, and the system changed once a week after floods. In addition, once a week, instantaneous depth integrated samples (U.S. Geological Survey 1952 Report) were taken by hand at the weir site, in order to furnish low water samples. Six separate flood events were analysed.

Table 5. Monthly variation in stream pollen at S_6 explained by three-variable regression

VARIABLES TAXA	MEAN DAILY DISCHARGE		+MEAN DAILY RAINFALL		+PRODUCTION		3 variable correlation significance
	C.C.	E.V.	C.C.	E.V.	C.C.	E.V.	'P'
Total Pollen	0.775	60.03	0.776	60.27	0.807	65.08	0.001
T. Determinable P.	0.777	60.47	0.781	60.97	0.809	65.52	0.001
Indeterminable P.	0.770	59.33	0.770	59.36	0.799	63.88	0.005
Total tree	0.654	42.78	0.673	45.29	0.753	56.73	0.005
T. herb	0.749	56.11	0.749	56.13	0.819	67.14	0.001
T. pteridophyte	0.737	54.39	0.737	54.39	0.739	54.61	0.005
Alnus	0.294	8.62	0.308	9.46	0.356	12.70	—
Betula	0.763	58.21	0.767	58.81	0.779	60.79	0.005
Corylus	0.453	20.54	0.456	20.82	0.465	21.68	0.200
Pinus	0.001	00.00	0.245	5.99	0.876	76.83	0.001
Quercus	0.461	21.22	0.609	37.11	0.753	56.67	0.005
Fraxinus	0.082	00.68	0.259	6.69	0.967	93.44	0.001
Ulmus	0.227	5.15	0.377	14.24	0.632	39.99	0.025
Calluna	0.740	54.77	0.757	57.28	0.760	57.791	0.005
Total Ericaceae	0.727	52.81	0.735	54.06	0.739	54.64	0.005
Gramineae	0.753	56.70	0.765	58.52	0.790	62.44	0.005
T. Compositae	0.711	50.52	0.753	56.72	0.753	56.84	0.005
Rosaceae	0.636	40.56	0.647	41.90	0.676	45.72	0.025
Plantaginaceae	0.490	24.02	0.495	24.53	0.815	66.50	0.001
Pteridium	0.731	53.44	0.748	55.93	0.759	57.73	0.005
Polypodium	0.500	25.05	0.838	70.28	—	—	—
T. Polypodiaceae	0.664	44.12	0.697	48.64	0.697	48.65	0.010
Sphagnum	0.585	34.22	0.641	41.10	0.665	41.67	0.025

C.C. = Correlation coefficient
E.V. = Explained variance

Figure 9 shows the increase in pollen numbers per litre of water during the passage of the floods. Samples labelled A_1 are from the base of the board, and represent the beginning of the flood rise, and samples A_4 are from the top. It is significant that the highest concentration of pollen does not always occur at the highest water flow. This is a phenomenon not uncommon in sediment sampling (Imeson 1970) and indicates an initial removal by rainfall of previously loosened particles from the surface, with subsequent rainfall causing clogging and compaction, resulting in a higher runoff but a lower sediment transport. This reasoning can also be applied to the pollen supply during floods. The significance of these events to the total pollen carriage by the streams cannot be overemphasised. Between flood water, samples averaged 315 grains/litre: at the height of flood, a maximum of 100,000 grains/litre were recorded. Table 6 gives pollen numbers and percentages recorded before and during the passage of the November flood. Both the numerical contrast and change in the proportions of the main taxa during the flood are clear. Particularly noticeable is the fall in proportion of tree pollen, and a compensating rise in Ericaceae pollen and *Pteridium* spores.

Similar changes in composition of flood pollen also occurred in successive peaks during the passage of winter. At the height of November floods (Table 6) *Pteridium*

Figure 8. Flood sampling board prototype at S_6. (Later models featured 4 litre containers).

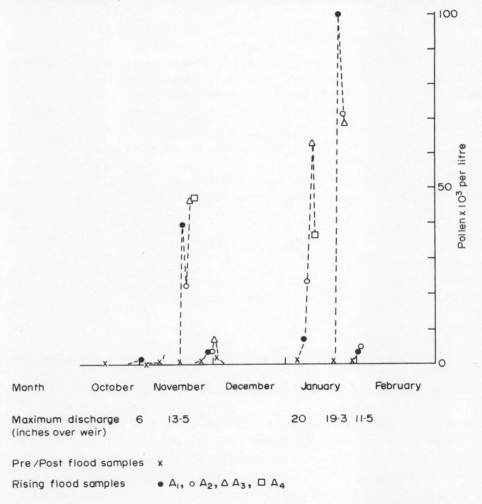

Figure 9. Pollen content of flood waters during the winter of 1970–71. X: Pre/Post flood samples. ●A_1, ○A_2, ▲A_3, □A_4 Rising flood samples.

Table 6. Pollen content of flood discharge

November 17th–18th 1970—Recorder Site—S6

	PRE-FLOOD		FLOOD RISE							
			A_1		A_2		A_3		A_4	
DISCHARGE l/sec	3.30	l/sec	4.60	l/sec	12.56	l/sec	34.53	l/sec	78.51	l/sec
POLLEN TYPES	Per litre nos.	%	Per litre nos.	%	Per litre nos.	%	Per litre nos.	%	Per litre nos.	%
Betula	15	10.80	2,695	6.70	1,815	8.08	3,054	6.41	3,553	7.43
Alnus	24	15.80	3,221	8.00	2,314	10.30	4,906	10.29	4,530	9.48
Quercus	3	1.70	460	1.20	499	2.22	555	1.17	335	0.78
Corylus	6	4.20	2,498	6.20	2,178	9.70	2,592	5.44	3,464	7.25
Pinus	1	0.80	394	1.00	0	0	185	0.39	266	0.56
Others	15	10.83	328	0.82	181	0.80	111	2.33	958	1.84
Gramineae	18	11.70	5,193	12.90	2,722	12.12	6,387	13.40	7,639	15.99
Total Ericaceae	4	2.60	4,011	10.81	2,722	12.12	7,220	15.50	8,261	17.28
Compositae	1	0.80	394	0.98	136	0.60	0	0	89	0.19
Plantaginaceae	3	2.50	394	0.98	136	0.60	278	0.58	267	0.56
Rosaceae	1	0.83	132	0.32	226	1.01	371	0.77	444	0.94
Other herbs	8	5.00	790	1.95	498	2.21	371	0.77	1,067	2.24
Pteridium	13	10.80	9,401	23.40	4,809	21.40	8,978	18.83	8,261	17.29
Sphagnum	0	0	394	0.98	136	0.61	463	0.97	178	0.37
Polypodiaceae undiff.	4	3.40	2,630	6.56	952	4.24	2,129	1.47	2,575	5.39
Other spores	0	0	0	0	45	0.20	43	0.19	0	0
TOTAL per litre	154		40,101		22,458		47,668		47,791	

spores formed between 18–23% of the total catch, as the flood followed directly on the production period. As winter progressed, the percentage representation of *Pteridium* spores became less, to a minimum of 10% in January, alongside rises in *Calluna*, Gramineae, and *Betula* pollen proportions. Figure 10 graphs all analysed flood samples of 1970–71, plotting both total pollen and weight of sediment against flood discharges. The best correlations (0.798 and 0.750 respectively), significant at the 0.1% level using Student's 't' distribution, were obtained using a log-log transformation of the data.

Environmental factors and lake water pollen variations

'Residual' lake pollen values (after subtraction of the air deposition component) were tested by multiple regression against the environmental factors previously defined for stream sites, in an attempt to discover the criteria controlling the monitored variations.

For the lower reservoir, a significant correlation with maximum 24 hour rainfall was found for total herb pollen, and total pteridophyte spores. Tree taxa were less well correlated, but *Fraxinus*, *Ulmus*, and *Acer* showed a marked improvement with the inclusion of the production variable, as seen in the stream sites. However, an attempt to show statistically, by use of the partitioned χ^2 test, that both lake and stream catches were derived from the same population (since they respond to identical factors) was unsuccessful. In the upper reservoir no significant correlation was found for any taxon with runoff or rainfall variables, though *Fraxinus*, *Pinus*, and *Ulmus* again showed a relation with the production variable. Inclusion of a wind-strength variable was also made to account for lake circulation, since a large recycling component was demonstrated, but the analysis was not improved.

Conclusions

It would be almost impossible for the pollen analyst to evaluate accurately the significance of every physical feature of the lake basin to his resulting fossil pollen spectrum. However, with a potentially large water-borne source of pollen, as demonstrated in Oakdale, there should be an increasing awareness not only of the effects of lake basin morphometry, but also of the surrounding catchment geography, as it controls the occurrence and magnitude of this component. Aspects of the geology and geomorphology affect topography, slope angle, soil structure, and thereby overland runoff: vegetation cover is equally important, particularly where it is a factor changing through the time covered by the fossil investigation, and climatic variables such as frost and snow occurrence, frequency and intensity of rain and its seasonal distribution, should also be considered in the final interpretation.

Thus the long distance transport of spruce pollen, monitored by Federova (1952) in the Volga, occurred solely because spring snow melt coincided with the spruce flowering period. On a smaller scale in Oakdale, the first winter floods of October and November following on the flowering of *Calluna* and *Pteridium*, are responsible for transporting large numbers of these grains and spores to the Reservoir sites, with important consequences on the composition of the final sediment pollen spectrum.

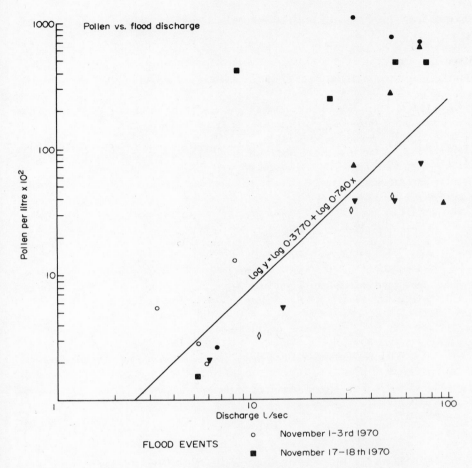
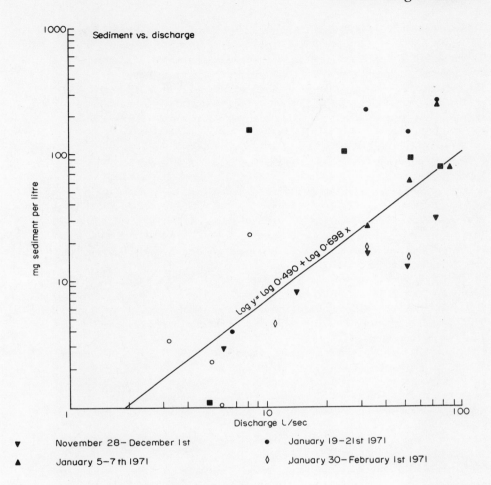

Figure 10. Graphs of pollen catch and sediment catch against flood discharge.

Acknowledgements

Research for this paper was completed in the Cambridge Botany School, with financial support from N.E.R.C. and the Cambridge Philosophical Society, to whom grateful acknowledgement is due. The Northallerton and Dales Waterboard kindly gave permission for work in Oakdale, and the active co-operation of the Osmotherley Filter Station Staff during the project is great appreciated. Thanks are also due to the Director of the Freshwater Biological Association, Ferry House, and his staff, particularly Miss E.Y. Haworth and Mrs W. Tutin, for loan of equipment and specialist analysis. I am particularly indebted to Drs H.J.B. and H.H. Birks and Mr J.I. Pitman for invaluable enthusiastic field help, to Dr R.G. West for critical supervision, and to Mr R.J. Chorley and Mr M.G. Anderson for much constructive discussion.

References

ANDERSEN S.TH. (1970) The relative pollen production and pollen representation of north European trees, and correction factors for tree pollen spectra. *Danm. geol. Unders.*, Ser. II, 96, 99p.

BANKY G.Y. (1959) Results of the erosion measuring station at Kisnana for the years 1956-58. *C.R. Ass. Int. Hydrologie Sci.* Hanover Symp. 1, 271.

BYE J.A.T. (1965) Wind Driven Circulation in unrestricted lakes. *Limnol. Oceanogr.*, 10, 451.

BRITISH STANDARDS INSTITUTION (1964) Methods of measurement of liquid flow in open channels: velocity area methods, *Br. Stand.* 3680(3).

CHAMBERLAIN A.C. (1966) Transport of *Lycopodium* spores and other small particles to rough surfaces. *Proc. R. Soc. A.* 296, 45.

COLBY B.R. (1963) Fluvial sediments—a summary of source, transportation, deposition and measurement of sediment discharge. *U.S. geol. Survey Bull.* 1181-A, 47p.

COPELAND O.L. (1963) Land use and ecological factors in relation to sediment yields. *U.S. Dept. Agr. Misc. Publ.* **970**, 72.

CROSS A.T., THOMPSON G.G. & ZAITEFF J.B. (1966) Source and distribution of palynomorphs in bottom sediments, southern part of Gulf of California. *Marine Geol.* **4**, 467-524.

DAVIS M.B. (1965) A method for determination of absolute pollen frequency. In *Handbook of Paleontological Techniques* (ed. by B. Kummel & D. Raup), pp. 674-6. W.H. Freeman, San Francisco.

DAVIS M.B. (1968) Pollen grains in lake sediments: redeposition caused by seasonal water circulation. *Science, N.Y.* **162**, 796-9.

DIMBLEBY G.W. (1957) Pollen analysis of terrestrial soils. *New Phytol.* **56**, 12-28.

EKERN P.C. (1950) Raindrop impact as the force initiating soil erosion. *Proc. Soil Sci. Soc. Am.* **15**, 7.

ELGEE F. (1914) The vegetation of the Eastern Moorlands of Yorkshire. *J. Ecol.* **2**, 1-18.

ELLISON W.D. (1945) Some effects of raindrops and surface flow on soil erosion. *Trans. Am. geophys. Un.* **26**, 415-30.

EMMETT W.W. (1970) The hydraulics of overland flow on hillslopes. *U.S. geol. Surv. Prof. Paper.* 662-A, 68p.

EVANS G.H. (1970) Pollen and diatom analysis of Late-Quaternary deposits in the Blelham Basin, N. Lancs. *New Phytol.* **69**, 821-74.

FEDEROVA R.V. (1952) Dissemination des pollen et spores par les eaux courantes. *Trans. Inst. Geogr., Acad. Sci. USSR.* **52**, 46-73.

GREGORY P.H. (1961) *The Microbiology of the Atmosphere.* Leonard Hill, London.

GRICHUK M.P. (1967) The Study of pollen spectra from Recent and ancient Alluvium. *Rev. Palaeobotan. Palynol.*, **4**, 107-112.

GROOT J.J. (1966) Some observations on pollen grains in suspension in the Estuary of the Delaware River. *Marine Geol.* **4**, 409-16.

HOPKINS J.S. (1950) Differential flotation and deposition of coniferous and deciduous tree pollen. *Ecology* **31**, 633-41.

HOROWITZ A. (1970) Palynological tracing of Saline water sources in Lake Kinneret-region (Israel). *J. Hydrology*, **10**, 177-84.

IMESON A.C. (1970) *Erosion in three East Yorkshire Catchments and Variations in Dissolved, Suspended and Bedload.* Unpublished Ph.D. thesis, University of Hull.

IMESON A.C. (1971) Heather burning and soil erosion on the North Yorkshire Moors. *J. appl. Ecol.* **8**, 537-42.

INGRAM H.A.P., ANDERSON M.C., ANDREWS S.M., CHINERY J.M., EVANS G.B. & RICHARDS C.M. (1959) Vegetational studies at Semerwater. *Naturalist, Hull*, 1959, 113-27.

JANSSEN C.R. (1959) Alnus as a disturbing factor in pollen diagrams. *Acta bot. neerl.*, **8**, 55-8.

KIRKBY M.J. & CHORLEY R.J. (1967) Through flow, overland flow and erosion. *Bull. int. Ass. scient. Hydrol.* **12**, 5-21.

KITTREDGE J. (1954) The influence of pine and grass on surface runoff and erosion. *J. Soil Wat. Conserv.* **9**, 179.

KORENEVA E.V. (1966) Marine palynological researches in the U.S.S.R. *Marine Geol.* **4**, 565-74.

LANGMUIR I. (1938) The surface motion of water induced by wind. *Science N.Y.* **87**, 119.

LEOPOLD L.B., WOLMAN M.G. & MILLER J.P. (1964) *Fluvial Processes in Geomorphology.* W.H. Freeman & Co., London.

LEYTON L., REYNOLDS E.R.C. & THOMPSON F.B. (1967). Rainfall interception in forest and moorland vegetation. In *Forest Hydrology* (ed. by W.E. Sopper & H.W. Lull), pp. 163-78. Pergamon Press, London.

LOWDERMILK W.C. (1930) Influence of forest litter on runoff, percolation, and erosion. *J. For.* **28**, 474.

MATTHEWS J. (1969) The assessment of a method for the determination of absolute pollen frequencies. *New Phytol.* **68**, 161-6.

MCVEAN D.N. & RATCLIFFE D.A. (1962) *Plant communities of the Scottish Highlands.* H.M.S.O. London.

MOSIMANN J.E. (1965) Statistical methods for the Pollen analyst: Multinomial and Negative multinomial techniques. In *Handbook of Paleontological Techniques* (ed. by B. Kummel & D. Raup), pp. 636-73. W.H. Freeman, San Francisco.

MULLER J. (1959) Palynology of recent Orinoco Delta and shelf sediments. *Micropaleontology* **5**, 1-32.

OVINGTON J.D. (1954) A comparison of rainfall in different woodlands. *Forestry* **27**, 41-53.

PECK R.M. (1972) Efficiency tests on the Tauber trap used as a pollen sampler in turbulent water flow. *New Phytol.* **71**, 187-98.

PIERCE R.S. (1967) Evidence of overland flow on forest watersheds. In *Forest Hydrology* (ed. by W.E. Sopper & H.W. Lull), pp. 247-53. Pergamon Press, London.

ROSSIGNOL M.L. (1961) Analyse Pollinique de sédiments marins quaternaires en Israel I: Sédiments récents. *Pollen Spores.* **3**, 303-24.

RUTTER A.J. (1963) Studies in the water relations of *Pinus sylvestris* in plantation conditions. I. Measurements of rainfall and interception. *J. Ecol.* **51**, 191-203.

SMITH D.D. & WISHMEIER W.H. (1962) Rainfall erosion. *Adv. Agron.* **14**, 109.

STANLEY E.A. (1969) Marine Palynology. *Oceanogr. Mar. Biol. Ann. Rev.*, **7**, 277-92.

TAUBER H. (1965) Differential pollen dispersion and the interpretation of pollen diagrams. *Danm. geol. Unders.* Ser. II, **89**, 69p.

TAUBER H. (1967) Investigations of the mode of pollen transfer in forested areas. *Rev. Palaeobotan. Palynol.* **3**, 277-86.

TRAVERSE A. & GINSBURG R.N. (1966) Palynology of the surface sediments of the Great Bahama Bank, as related to water movement and sedimentation. *Marine Geol.* **4**, 417-59.

TRIMBLE G.R. & WEITZMANN S. (1954) Effect of hardwood canopy on rainfall intensity. *Trans. Am. geophys. Un.* **35**, 226-34.

TROELS-SMITH J. (1955) Karakterisering af løse jordarter. *Danm. geol. Unders.* Ser. IV, **3**(10), 73p.

U.S. GEOLOGICAL SURVEY SUBCOMMITTEE ON SEDIMENTATION (1952) The design of improved types of suspended sediment samplers. Report No. 6. *Interagency committee on Water Resources Publication.*

U.S. GEOLOGICAL SURVEY, SUBCOMMITTEE ON SEDIMENTATION (1961) The single stage sampler for suspended sediment. Report No. 13. *Interagency committee on Water Resources Publications.*

VERSEY H.C. (1937) The Tertiary History of East Yorkshire. *Proc. Yorks. geol. Soc.*, **23**, 302-14.

Pollen in a small river basin: Wilton Creek, Ontario

ADÈLE A. CROWDER AND D.G. CUDDY *Department of Biology, Queen's University, Kingston, Ontario*

Introduction

Pollen analysis looks at past vegetation in a series of distorting mirrors; this paper is an attempt to look at the mirrors as they are currently being set up in a small physiographic unit, the basin of Wilton Creek, Ontario. This is a river which rises on Precambrian rock and flows some 35 km through sands and over limestone to Lake Ontario. The first quarter of the valley has fairly steep relief and the lower part is flat, with some underground drainage in the limestone. At Harrowsmith the creek drains a raised peat bog with a small lake beside it. The mouth of the river at Hay Bay is a winding inlet with wide marshes.

The present local vegetation has been sampled, and three 'mirrors' reflecting it have been investigated; these are moss cushions containing pollen rain from the last five to ten years, the surface peat of the bog at Harrowsmith, and the lake mud in Hay Bay. In order to evaluate the importance of water-borne pollen in the lake mud, and to see whether pollen is washed out of the bog, the pollen in the river water has been analysed each month for a year and a half.

Vegetation

The creek basin is contained in the area described in a *List of the Vascular Plants of the Kingston Region* (Beschel *et al.* 1970) which gives rough estimates of abundance. There are two studies within the basin—Garwood (1967) described the black spruce stand at Harrowsmith bog, and Jafri (1965) analysed the forest composition of the eastern half of the valley, relating it to topographic gradients. Jafri's data have been used, and his field method has been extended to the rest of the basin (Fig. 1).

The method takes the five kilometre grid on the 1:50,000 National Topographic maps as a base. Fourteen of these 5 × 5 km squares intersect Wilton Creek; a fifteenth square was added at the north end of the valley where the creek rises in an intermittent swamp which lies just south of the Ordovician limestone-Precambrian geological contact. This last square includes a granitic substrate.

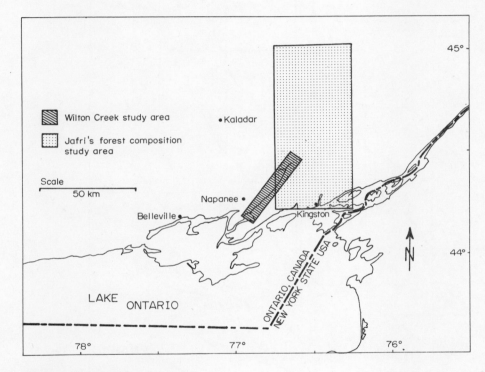

Figure 1. Location of study area.

The amount of woodland increases markedly in the north. This can be seen from air photographs, and from the green overlay on the topographic maps, which are in fact based on air photographs. The amount of forest cover was estimated by measuring the green overlay for each 5 km block, using a hundred point grid which reads off values as a percentage of the square. Fig. 2 shows the increase of forest cover from a minimum of 4% in the south to a maximum of 82% in the north. Swamp, marsh, cleared land, and open water were estimated from the maps and photographs using

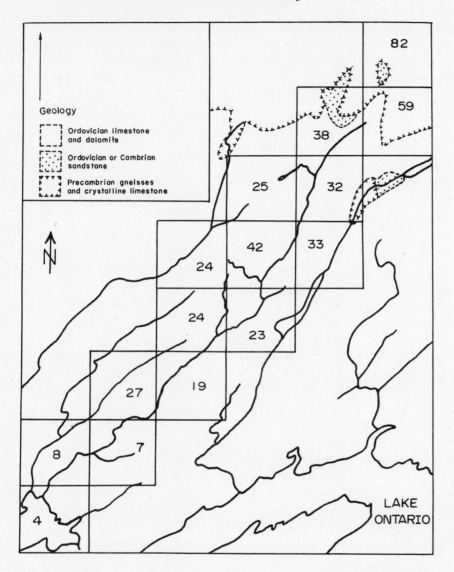

Figure 2. Forest cover. Numbers indicate percentage of 5 × 5 km square covered by forest.

the same method. Fig. 3 gives average values for these, dividing the basin into north, central and south zones.

The region was part of the Great Lakes—St. Lawrence forest until the eighteenth century; historical reasons explain the present lack of forest in the south. The Bay of Quinte, into which Hay Bay flows, was the site of a Sulpician mission to the Iroquois in 1669. This was shortlived, but by 1700 Mississauga Indians were culti-

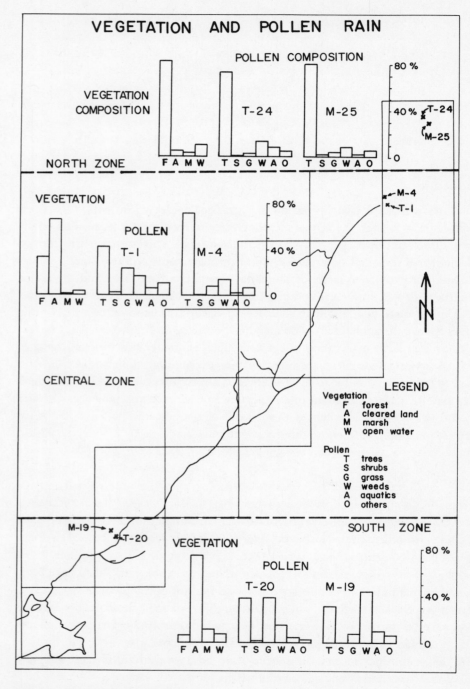

Figure 3. Vegetation and pollen rain.

vating some maize there. In 1783 an exploratory survey stated: 'the land is extraordinarily good for any kind of cultivation, the soil is deep and rich, the timber is beech, maple, elm, basswood, with some pine and white oak.' Following this the United Empire Loyalists settled the area, and sawmills were started (Blake 1957). The demand for timber was not just for the settlers, as the naval shipyard in Kingston used large quantities. Settlement increased steadily until the 1860s, together with more felling. An atlas of the area in 1878 (Meacham 1878) shows all the lots occupied, even in the northern part of the valley. After this the population began to decline, and farming changed gradually to dairying. Many fields have been abandoned and are in various stages of succession back to woodland; *Juniperus virginiana* is common on these old fields. Because settlement began on the shores of Lake Ontario, deforestation is greatest there.

Jafri's method of forest analysis uses six stands in each 5 km/square, a stand by definition having at least 50% cover of trees, each with at least 5 cm diameter at breast height. The six stands were each of different topographic aspect—north, south, east, and west slopes, well-drained flat ground, and wet flat ground. Stands were supposed to be at least 4 hectares in area, near a road if possible, and were selected using random co-ordinates. When all these prerequisites were met, the randomness was questionable, particularly in the south. In the western part of the basin there was often not a forest stand on all the six topographic aspects, so that only one or two stands could be sampled.

Within each stand two samples of twelve trees each were taken using a modified quarter method (Cottam & Curtis 1956). Circumference at breast height was recorded for each tree.

Jafri calculated importance values for each species of tree. He found it necessary to use the averages of nine squares to get the importance value for the species in the central square, in order to smooth out chance variation which was due to too small a sample size. We have used the same method of averaging squares but have calculated not importance value, but relative percentage basal area, which correlates better with pollen production. Hence a total of 46 squares, each 5 × 5 km, were used to obtain the relative percent basal areas for the tree species in the 15 squares through which Wilton Creek runs.

To make full use of the slope aspects the average percentage basal area of a species from the nine squares was multiplied by the frequency of that slope in the central square for which a value was being sought. The sum of these weighted basal area percentages for the six aspects gives the final percentage composition of the species in the forest, for that square.

For this preliminary study the species have been grouped into crude palynological entities quickly recognisable in the pollen counts, for instance, *Acer negundo*, *A. nigrum*, *A. rubrum*, *A. pennsylvanicum*, *A. spicatum*, *A. saccharinum*, and *A. saccharum* have been lumped together as *Acer*, *Pinus banksiana*, *P. strobus*, and *P. resinosa* have been lumped as *Pinus*, *Carya ovata* and *C. cordiformis* as *Carya*, whilst *Ostrya virginiana* and *Carpinus caroliniana* lose even their generic identity.

The quarter method, and Jafri's whole technique, will not work for hedgerows and patches of scrub vegetation which are most abundant in the south. It is intended to sample these, using transects, and to modify the values for the arboreal component accordingly. Herbaceous plants have been listed where moss cushions were collected, but without any attempt to quantify them.

The results of the forest sampling cannot be shown in their entirety. The relative abundance of some species remains the same throughout the basin. *Betula*, *Quercus* and perhaps *Acer* belong to this group. Others such as *Carya* are abundant in the south and rare in the north; *Juglans* on the other hand is rare on the southern limestone plain but is found in the north. *Pinus* tends to be moderately common on the deeper soils on limestone in the extreme south, rare in the central part where soils are shallower, and common again in the north on acid rock. *Picea* has a somewhat localised distribution, especially in the north and was not recorded in any of the 54 stands used to calculate the forest composition of the northern-most square. This is a defect of the sampling method which gives more accurate results for homogeneous stands.

Table 1 shows the average relative dominance (% basal area) for the major tree species of Jafri's study area. As can be seen in Fig. 1, his study area extends farther to the north and lies on mainly Precambrian rocks. Thus these average values are comparable to only the northern part of our basin. *Acer saccharum* is clearly the dominant species; it is followed by *Pinus strobus*, *Ulmus americana*, *Thuja occidentalis*, *Tsuga canadensis*, and *Quercus borealis*. Each of these contributes more than 5% to the total basal area. In the south as shown in Fig. 6 the positions are slightly changed. *Pinus strobus* is dominant with *Acer saccharum* close behind, followed by the two hickories (*C. cordiformis* and *C. ovata*) and *Quercus*. *Quercus macrocarpa* and *Q. alba* are more common than *Q. borealis* in the south. *Tsuga canadensis* and *Tilia americana* are less common in the south; usually they are found growing on north slopes there. The low value for *Ulmus* in the south is due to the recent spread of Dutch Elm disease. Only living trees were measured in the forest sampling.

Moss cushions and pollen rain

The ideal way in which to measure pollen-rain would be to catch it and analyse it over a period of several years; more than one year is necessary because the number of microspores shed by a species varies greatly from season to season, and so does the weather. Fig. 4 shows the magnitude of this fluctuation for one genus, *Ambrosia*, throughout four years, at several stations near Lake Ontario. (The numbers refer

Table 1. *Relative Dominance of Tree Species in the Study Area.* (Rel. dom. = % B.A.)

Most abundant species of the pollen type	% Basal Area	Pollen type	Number of Species included in type	% Basal Area
Abies balsamea (L.) Mill.	1.0	Abies	1	1.0
Acer saccharum Marsh	23.6	Acer	7	29.5
Betula papyrifera Marsh	1.8	Betula	3	2.4
Carya cordiformis Wang	1.0	Carya	2	1.5
Fagus grandifolia Ehrh.	4.2	Fagus	1	4.2
Fraxinus americana L.	1.9	Fraxinus	3	3.8
Juglans cinerea L.	0.3	Juglans	1	0.3
Larix laricina (DuRoi) K. Koch	0.1	Larix	1	0.1
Ostrya virginiana (Mill) K. Koch and Carpinus caroliniana Walt.	3.7	Ostrya	2	3.7
Picea glauca (Moench) Voss	0.3	Picea	2	0.6
Pinus strobus L.	10.4	Pinus	6	10.5
Populus tremuloides Michx.	0.7	Populus	6	1.2
Prunus serotina Ehrh.	0.2	Prunus	4	0.2
Quercus borealis Michx f.	5.9	Quercus	5	8.0
Salix amygdaloides Anderss.	0.1	Salix	12	0.1
Thuja occidentalis L. and Juniperus virginiana	6.7	Cupressaceae	2	6.7
Tilia americana L.	4.8	Tilia	1	4.8
Tsuga canadensis (L.) Carr	6.0	Tsuga	1	6.0
Ulmus americana L.	9.9	Ulmus	3	11.1

Other species which occur in the region but contribute less than 0.1% B.A.

Alnus rugosa (DuRoi) Spreng.
Amelanchier spp. (4)
Cornus alternifolia L. f.
Crataegus spp. (12)
Hamamelis virginiana L.
Morus rubra L.
Pyrus malus L.
Rhamnus catharticus L.
Rhus typhina L.
Robinia pseudoacacia L.
Sorbus spp. (2)
Syringa vulgaris L.
Viburnum lentago L.

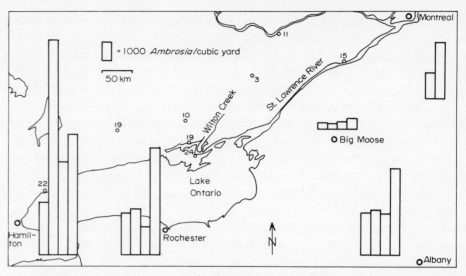

Figure 4. Fluctuation of *Ambrosia* pollen collected from the air at five sites around Lake Ontario during the years 1946-49 (Montreal 1948-49 only). Histograms show annual totals (from Durham 1950). Variation in the numbers of days with high *Ambrosia* counts during one summer is shown for eight sites north of the lake (from Bassett 1959).

to days with peak counts, see p. 72). The values are taken from Durham (1950) and Bassett (1959) and are based on daily counts.

Lacking such records the accepted method is to use the pollen caught in mosses; the age of these cushions collected at Wilton Creek was estimated at from five to fifteen years, (cf. Carrol 1943, Hansen 1949, Benninghoff 1960).

The area lies between several existing studies of pollen rain. King and Kapp (1963) sampled a series of cushions along a north-south line in Ontario about 150 km west, which covered the transition from forest to southern farmed land that the creek provides in a microcosm. To the north-east Terasmae and Mott's survey (1964) described bogs and lakes in the boreal forest, near Ottawa. Further afield there is the Davis and Goodlett (1964) study using Brownington Pond instead of moss cushions, and there are several papers relating surface samples and forest in Quebec (Potzger *et al.* 1956, Richard 1968).

Moss cushions were collected at approximately 3 km intervals along the valley and into its northern watershed. Paired samples of two species were taken, one from open country and the other from woodlands, to allow a comparison of the rain in the two, on the assumption that the regional rain would be found more evenly distributed in the open sites. The moss from the open clearings was *Tortula ruralis*, that from the woodlands was *Mnium cuspidatum*,—the *Tortula* usually grows on limestone or shallow calcareous soil, the *Mnium* on woodland litter. At the northernmost site there was no *Tortula*, so a moss of similar habit, *Orthotrichum anomalum*, was substituted. The paired samples were taken as close together as possible, usually from 100 to 500 m apart. Several pieces of moss were combined for each sample to decrease the variability. The local plants at each site were listed.

Several methods of extraction were tried to find one suitable both for the upright *Tortula*, which catches mineral particles, and the straggling *Mnium*. It was found that the *Mnium* retained different pollen at different levels in the cushion—this sieving effect is shown in Fig. 5. To avoid this hazard the mosses were cut off just at the surface of the substrate and about 250 ml were shaken in 500 ml of hot 5% KOH for 15 minutes. The moss fragments were removed by sieving through a single layer of cheese cloth and the pollen concentrated by centrifugation. The Faegri and Iversen (1964) technique was used to process it; all samples were treated alike, and all were subjected to hydrofluoric acid. The final suspension was mounted in silicone oil.

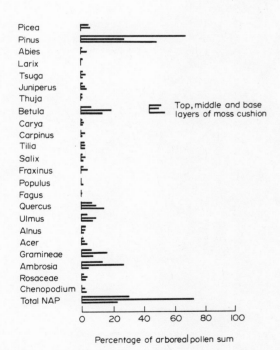

Figure 5. Sieving effects of *Mnium cuspidatum* cushions: results of counts at 3 different levels in a *Mnium* polster. The 3 bars represent 3 levels in the cushion, top, middle, and base. Differences could be caused by (a) size and shape of grains, (b) annual changes in pollen rain, (c) differential decay rates.

All the cushions contained ample pollen, usually several thousand grains. In this paper 6 samples have been used, three *Tortula*-type and three *Mnium*; these have been counted to 100 A.P. except for M4 in which the A.P. total is 624. The results of these counts, expressed as percentage A.P. and percentage P.S. (pollen sum = all pollen, but not spores) are given in Table 2. No value is given for significant variation as the number of samples is small.

The sites of the samples are shown in Fig. 3. Their localities are as follows:

T20 *Tortula ruralis* from a field, the southern-most sample taken near Hay Bay;

M19 *Mnium cuspidatum*, from a small maple-elm swamp, just north of T20;

T1   *T. ruralis* from limestone on the southwest side of a low hill in open pasture near the head of Wilton Creek;

M4   *M. cuspidatum*, from a small maple-elm swamp just northwest of T1;

M25 *M. cuspidatum* from wet ash woodland in a valley between ridges of granite, about 10 km NE of T1 and M4;

T24 *Orthotrichum anomalum*, from a small sloping natural clearing on the south shore of Otter Lake.

The pollen counts from the moss polsters generally reflect the expected shift from a low proportion of A.P. in the south to a very high one in the north. The forest cover of the basin as shown in Fig. 2 suggests three zones of density,—very low (south), moderate (central), and high (north); the histograms for each zone in Fig. 3 show the mean percentages of forest (F), cleared land (A), marsh (M), and open water (W). The pollen histograms for the same zones have been grouped as trees (T), shrubs (S), grasses (G), weeds (W), marsh and aquatic plants (A), and others (O). The units of the histograms are not directly comparable, but as Faegri (1966) has said 'studies on the relation between recent pollen rain and actual vegetation composition . . . should be used qualitatively only, and only for local comparison. It will hardly ever be possible to deduce general conversion factors from such—or any other—studies.'

The most marked correlation is that between the percentage forest cover (F) and the percentage A.P. (T). The effect of some two hundred years of farming in the south is quite clear, but the localised effect of clearing is shown by the fact that the northern A.P. percentage is so high.

The *Mnium* samples had the expected effect of amplifying the local pollen, and generally have higher A.P. proportions than do their paired *Tortula* cushions from open ground. M19 is an exception, but this was due to a local source of *Ambrosia*, a group of plants by a roadside, which produced 38% of the pollen sum. The greatest diversity of pollen was found in the *Tortula* sample T1, which came from a slope exposed to the prevailing southwesterly winds (cf. Putnam & Chapman 1938).

To compare the composition of the arboreal pollen and the tree distributions

Table 2. Results of preliminary pollen counts from moss polsters (based on approximately 100 A.P. with the exception of M-4 where 624 A.P. were counted).

	T-20		M-19		T-1		M-4		M-25		T-24		Average of 6 polsters	
	% A.P.	% P.S.	% A.P.	% P.S.	% A.P.	% P.S.	% A.P.	% P.S.	% A.P.	% P.S.	% A.P.	% P.S.	% A.P.	% P.S.
Trees														
Abies							2.1	1.5					1.1	0.6
Acer	3.1	1.2	2.8	0.9	0.9	0.4	1.1	0.7	2.5	2.5	0.9	0.4	1.5	0.9
Alnus					2.7	1.1	1.9	1.3	2.5	2.5			1.5	0.9
Betula	19.6	7.5	21.1	6.6	20.4	8.3	13.1	9.0	10.7	10.8	36.9	22.4	17.0	9.7
Carya	8.2	3.1	1.8	0.6	2.7	1.1	1.2	0.5	1.1	0.8	21.5	14.6	4.4	2.5
Castanea (cf.)					2.7	1.1							0.3	0.1
Cupressaceae	2.1	0.8	0.9	0.3	2.7	1.1	2.7	1.8					2.0	1.1
Fagus	1.0	0.4			0.9	0.4	0.2	0.1					0.3	0.1
Fraxinus	8.2	3.1	2.8	0.9	2.7	1.1	1.9	1.3	19.0	18.9	4.5	2.7	4.6	2.6
Juglans	1.0	0.4			0.9	0.4			6.6	6.7	1.8	1.0	1.0	0.6
Larix							0.5	0.3					0.3	0.1
Ostrya (+ Carpinus)	2.1	0.8	6.4	2.0	5.3	2.2	0.2	0.1	5.8	5.6	8.1	5.0	2.7	1.6
Picea	1.0	0.4	1.8	0.6	1.8	0.7	5.0	3.5	1.7	1.7			2.2	1.8
Pinus	20.6	7.8	7.3	2.3	21.2	8.6	49.8	35.5	20.7	20.4	19.8	12.1	34.7	19.8
Populus	4.1	1.6			0.9	0.4	0.3	0.2					0.6	0.3
Prunus					3.5	1.4							0.3	0.2
Quercus	25.8	9.8	36.7	11.6	20.4	8.3	9.3	6.4	3.3	3.2	18.0	11.3	14.5	8.2
Salix	2.1	0.8			2.7	1.1	1.2	0.8					1.1	0.6
Tilia					0.9	0.4	1.9	1.3	0.8	0.8	0.9	0.4	1.3	0.7
Tsuga			0.9	0.3	0.9	0.4	1.9	1.3	0.8	0.8			1.3	0.7
Ulmus	1.0	0.4	17.4	5.5	7.1	2.9	6.1	4.2	5.0	4.7	3.6	2.1	6.5	3.7
TOTALS		38.1		31.6		41.4		69.8		79.6		72.0		56.8
Shrubs														
Corylus	3.1	1.2			3.5	1.4	0.2	0.1	1.7	1.3	0.9	0.6	1.0	0.6
Ericaceae	1.0	0.4											0.1	+
Taxus					0.9	0.4							0.1	+
Vitis					0.9	0.4							0.1	+
TOTALS		1.6		0.0		2.2		0.1		1.3		0.6		0.6
Grasses														
Cerealia	4.1	1.6			16.8	6.8	0.3	0.2					2.1	1.2
Gramineae	96.9	36.9	24.8	7.8	39.8	16.1	10.6	7.3	5.0	3.9	3.6	2.6	20.6	11.7
TOTALS		38.5		7.8		22.9		7.5		3.9		2.6		12.9

Table 2. (continued)

	T-20		M-19		T-1		M-4		M-25		T-24		Average of 6 polsters	
	% A.P.	% P.S.	% A.P.	% P.S.	% A.P.	% P.S.	% A.P.	% P.S.	% A.P.	% P.S.	% A.P.	% P.S.	% A.P.	% P.S.
Weeds														
Ambrosia	27.8	10.6	121.0	38.2	40.0	10.6	17.7	12.2	7.4	5.9	10.8	7.8	27.7	15.8
Artemisia	1.0	0.4	0.9	0.3	0.9	0.4	0.2	0.1					0.3	0.2
Liguliflorae					2.7	1.1			0.8	0.7			0.3	0.2
Tubuliflorae	6.2	2.4	16.5	5.2	8.0	3.2					5.4	3.9	3.3	1.9
Chenopodiaceae	3.1	1.2			0.9	0.4	0.9	0.6	0.8	0.7	0.9	0.6	1.0	0.6
Plantago	1.0	0.4	1.8	0.6	0.9	0.4	0.2	0.1			0.9	0.6	0.8	0.4
Rumex							0.3	0.2	0.8	0.7			0.3	0.1
TOTALS		15.0		44.3		16.1		13.2		8.0		12.9		19.2
Aquatic and marsh plants														
Callitriche (cf.)					2.7	1.1							0.3	0.1
Cephalanthus			11.0	3.5									1.0	0.6
Cyperaceae	4.1	1.6	6.4	2.0	8.0	3.2	2.5	1.7			2.7	1.9	3.3	1.9
Juncaginaceae									0.8	0.7			0.1	+
Potamogeton			0.9	0.3									0.1	+
Sparganium					0.9	0.4							0.1	+
Typha	7.2	2.7	13.8	4.3	2.7	1.1			1.7	1.3	8.1	5.8	3.1	1.7
TOTALS		4.3		10.1		5.8		1.7		2.0		7.7		4.3
Others														
Campanulaceae									0.8	0.7			0.1	+
Caryophyllaceae			0.9	0.3	1.8	0.7	0.2	0.1	0.8	0.7			0.4	0.2
Cruciferae			1.8	0.6					1.7	1.3			0.3	0.2
Eupatorium			1.8	0.6									0.1	+
Hypericum			0.9	0.3					0.8	0.7			0.2	0.1
Impatiens			0.9	0.3									0.1	+
Labiatae					1.8	0.7							0.2	0.1
Leguminosae							0.6	0.4					0.3	0.2
Liliaceae			1.8	0.6	3.5	1.4	2.1	1.4					1.5	0.9
Ranunculaceae							0.8	0.5					0.4	0.2
Rosaceae	1.0	0.4	0.9	0.3	4.4	1.8	2.3	1.5					1.8	1.0
Rubiaceae					0.9	0.4							0.1	+
Scrophulariaceae	1.0	0.4	0.9	0.3	0.9	0.4	0.2	0.1			0.9	0.6	0.4	0.2
Solanaceae							2.1	1.4					0.1	+
Thalictrum			0.9	0.3							0.9	0.6	0.2	0.1

Table 2. (continued)

	T-20		M-19		T-1		M-4		M-25		T-24		Average of 6 polsters	
	% A.P.	% P.S.	% A.P.	% P.S.	% A.P.	% P.S.	% A.P.	% P.S.	% A.P.	% P.S.	% A.P.	% P.S.	% A.P.	% P.S.
Umbelliferae					1.8	0.7	0.2	0.1					0.3	0.1
Urticaceae			0.9	0.3			0.5	0.3	1.7	1.3	0.9	0.6	0.6	0.3
Incog.	5.2	2.0	7.3	2.3	9.7	4.0	1.1	0.7			3.6	2.6	2.4	1.4
TOTALS		2.8		6.2		10.1		6.5		4.7		4.4		5.0
Spores														
Moss		8.6		14.7		3.2		0.6		95.5		8.4	20.9	11.9
Liverwort						0.4				1.3			0.3	0.1
Equisetum				0.3									0.1	+
Lycopodium						0.4		0.2					0.3	0.1
Fern (monolete)		0.4		4.6				10.6		11.2		8.4	11.9	6.8
Fern (trilete)		0.4		1.2				0.3		2.0		1.3	0.9	0.5

Fungal—occur in all, but especially common in *Mnium* samples.
Protozoa—occur in all, especially common in *Mnium* samples.

the two *Tortula*-type cushions furthest apart were chosen. The A.P. of the ten most common tree types and their distribution in the north and south ends of the study area are shown in Fig. 6. As usual *Acer* is under-represented everywhere. Again as usual, *Betula*, *Quercus*, and *Pinus* are over-represented, but *Betula* to a much greater degree than *Quercus* and *Pinus*. The higher over-representation in the north could be due to south westerly winds picking up pollen, in addition to the long range grains they already contain coming in off Lake Ontario, and then dumping some of their load further up the valley where the relief is more marked. *Carya* pollen supports this interpretation, as the tree is rare in the north, but the pollen forms a significant part of the pollen sum there. The total absence of *Picea* in the north is misleading. The tree occurs in the north but not in stands extensive enough to be picked up by the sampling method.

The low *Ulmus* values in the south may be due to the spread of Dutch Elm disease being greater there. About 90% of the southern trees have been destroyed recently. The common species in both zones is *U. americana*.

Surface peat sample from Wolf Bog, Harrowsmith

The bog is unusual in this part of Ontario and extends into a swamp to the southwest with maples, elm, and ash. Much of the bog itself is overgrown with spruce, although it is said to have been burnt about 60 years ago. The spruce is mostly *Picea mariana*, but there is some *P. glauca*, and some hybrids are thought to occur. Open patches of *Sphagnum* alternate with dwarf shrubs, including *Kalmia angustifolia*, *K. polifolia*, *Chamaedaphne calyculata*, *Andromeda glaucophylla*, *Ledum groenlandicum*, and four species of *Vaccinium*. Two trees which are not common in the valley occur, *Larix laricina* and *Quercus bicolor* (Garwood 1967).

The pollen record of the bog was published by Terasmae (1969), and it has been a site of deposition for 10,000 years.

The bog does not have a well-defined edge, and its drainage seems to have been interfered with, as the 1878 atlas (Meachem 1878) shows a stream flowing out of the swamp to its southwest which still exists, and is part of the Napanee basin. The present drainage to Wilton Creek is not shown on that map.

The pollen in a surface sample was 64% arboreal. The most important trees were *Picea*, *Betula*, *Alnus*, *Quercus*, and *Ulmus* in that order. The count is shown in Fig. 9. The *Picea* reflects the large number of trees on the bog, but the *Betula* is over-represented.

The non-arboreal pollen failed to detect much pollen from the ericads, and instead has the regional *Ambrosia* and grasses as its major components, followed by other weeds (*Plantago*, *Chenopodium*, Urticaceae, etc.). The ericads, however, were more abundant than in any other site in the valley, and so was *Sphagnum*.

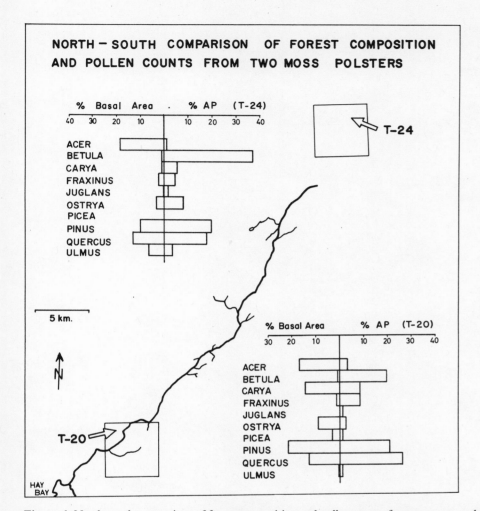

Figure 6. North-south comparison of forest composition and pollen counts from two moss polsters.

Hay Bay mud

The third site of pollen deposition to be looked at was the organic mud in the part of the creek bed where it opens out into the *Typha* marshes of Hay Bay. The channel is less than 2 m deep and joins the Bay of Quinte, which is fed by three much larger rivers. Borings were made through the ice in February; the sediment contained some sand and belongs to the shore types described by Lewis and McNeely (1967).

The Faegri and Iversen schedule for pollen preparation was used on 1 ml of fresh sediment, which yielded 9,030 grains. The shore sediments of Lake Ontario have less pollen than deep water deposits because they contain more large mineral particles (McAndrews & Power 1972), so this is a usual amount for such sediments.

The most abundant pollen was grass, then pine, oak, Cupressaceae, and birch, followed by spruce, maple, hornbeam, and blue beech. The counts are shown in Table 3. The second element of the non-arboreal pollen was *Typha*; both *T. latifolia* and *T. angustifolia* were present in the pollen count and in the marsh. The third most abundant herbaceous group was *Ambrosia* and then two groups, other weeds (*Chenopodium*, *Plantago*, etc.) and aquatics.

Because of the enormous amount of Gramineae, non-arboreal pollen was 80% of the total pollen sum. *Ambrosia* was correspondingly less important, being only 3% of the total sum. Unpublished counts for deeper water in Hay Bay made by Dr J.H. McAndrews and kindly made available by him show higher *Ambrosia* and less grass.

The Hay Bay pollen will be compared to that in deep water in Lake Ontario after a consideration of the role of the creek in transport.

River Water

Water samples were collected monthly, starting in September, 1970. The first four samples have been used only in the diagram of total pollen, as they were each only 10 l of water. To get pollen samples containing more than 250 A.P. at all times of the year the later samples were each 35 l. The samples were collected about 10 km up from the mouth of the stream, beside a flow gauge, using a depth-integrating sampler. In winter a hole was cut in the ice and the samples taken through this.

The water samples have a heavy load of both dissolved and dispersed sediment, and the clay proved difficult to remove. Centrifugation and separation with bromoform were tried, but the technique adopted was the removal of the smallest particles as the first step. This was done by sucking them through millipore filters with a pore size of 8 μm, using a vacuum system; when the filters get clogged they hold the finest particles, but enough of the colloidal inorganic sediment is removed to make it possible to continue with a Faegri and Iversen process. The KOH, in the initial stage of this, removed the millipore filters by dissolving them. Some samples required two hydrofluoric acid treatments to remove the silica; the amount and the state of breakdown of humus in the water was very variable.

Pollen and spores were found in the water at all seasons of the year. The lowest concentrations occurred in February, when the stream is battened down by 30 cm or more of ice. The highest concentration was found in July; the value for that month (17,000 grains/35 l/water) was slightly higher than Federova's (1952) figure for the Volga (15,000/35 l).

Table 3. Comparison of pollen counts from surface sample at Harrowsmith, samples of Wilton Creek water (monthly samples not weighted for water flow). Hay Bay surface mud sample and deep surface mud samples from Lake Ontario. Fungal and bryophyte spores and protozoa have been omitted. The N.A.P. figures for Lake Ontario are calculated on the A.P. sum with the individual N.A.P. count added; the Lake Ontario counts are from McAndrews and Power (1972). P.S. = total pollen; values less than 0.1% = +.

	Harrowsmith bog % A.P.	Harrowsmith bog % P.S.	Wilton Creek % A.P.	Wilton Creek % P.S.	Hay Bay % A.P.	Hay Bay % P.S.	Lake Ontario % A.P.
Trees							
Abies	8.0	5.0	0.9	0.2	1.7	0.4	1.1
Acer	3.4	2.3	4.3	1.0	5.4	1.0	5.5
Alnus	13.2	8.3	2.2	0.7	1.7	0.4	1.7
Betula	16.3	10.3	9.3	2.0	6.2	1.2	8.3
Carya	3.9	2.3	2.2	0.7	0.8	0.2	1.8
Celtis	0.9	0.5					0.8
Cupressaceae	0.4	0.3	6.9	1.7	8.7	1.6	6.3
Fagus			0.9	0.2	0.8	0.2	1.7
Fraxinus	0.4	0.3	4.9	1.1	1.2	0.2	2.5
Juglans			+	+	0.4	0.1	0.9
Larix			0.1	+			0.8
Ostrya (+ Carpinus)	3.0	1.9	1.5	0.3	3.7	0.7	1.9
Picea	19.0	12.0	6.4	1.6		1.1	3.3
Pinus	6.0	4.0	33.1	7.6	31.4	6.7	33.8
Populus			1.1	0.2	0.4	0.1	1.2
Prunus			0.1	+	0.8	0.2	
Quercus	9.5	6.1	18.3	4.1	16.5	3.1	23.7
Salix	1.1	0.8	2.2	0.7	2.1	0.4	2.5
Tilia	0.6	0.3	1.1	0.2	2.9	0.5	0.9
Tsuga			1.4	0.3	1.7	0.4	2.9
Ulmus	6.1	9.5	3.2	0.9	2.9	0.5	2.3
TOTAL TREES		63.9		23.5		19.0	
Shrubs							
Cornus			+	+			
Corylus	4.0	2.8	+	+	1.1	0.4	0.7
Ericaceae	2.9	2.0	+	+	1.1	0.4	0.8
Lonicera			+	+			
Myrica	0.9	0.5	+	+			0.8
Nemopanthus	0.9	0.5					
Ribes			+	+			
Rhamnus			+	+	0.7	0.4	
Rhus			+	+			
Sambucus			+	+			
Shepherdia			+	+			

Table 3. (continued)

	Harrowsmith bog % A.P.	Harrowsmith bog % P.S.	Wilton Creek % A.P.	Wilton Creek % P.S.	Hay Bay % A.P.	Hay Bay % P.S.	Lake Ontario % A.P.
Syringa			+	+			
Taxus			+	+			
Viburnum			+	+			
Zanthoxylum			+	+			
TOTAL SHRUBS		5.8				1.2	
Grasses							
Cerealia	2.0	1.1	2.5	0.7	4.8	0.9	
Gramineae	15.0	9.5	221.0	51.2	352.0	68.0	16.3
TOTAL GRASSES		10.6		51.9		68.9	
Weeds							
Ambrosia	12.0	7.5	31.3	16.3	16.8	3.2	29.9
Artemisia			3.4	0.1			1.6
Liguliflorae			10.5	0.8			0.9
Tubuliflorae			14.6	0.8	0.8	0.2	2.2
Chenopodiaceae	1.1	0.8	7.2	0.4	0.8	0.2	5.4
Plantago	4.0	2.3	2.1	0.5	2.4	0.5	1.2
Rumex			1.3	0.4			1.7
TOTAL WEEDS		10.6		19.3		4.1	
Aquatic & marsh plants							
Butomus			+	+			
Callitriche			+	+			
Cyperaceae	2.0	1.0	4.9	1.3	4.0	0.8	2.5
Epilobium					+	+	
Hydrocotyle			+	+			
Iris			0.3	0.1			
Lemna			+	+	0.4	0.1	
Myriophyllum			+	+	0.4	0.1	2.6
Nuphar			0.4	0.1	0.8	0.2	
Nymphaea			+	+	1.2	0.2	0.7
Potamogeton			2.9	0.8			0.7
Sagittaria			+	+			0.7
Sparganium			1.5	0.4	2.4	0.5	1.4
Typha			1.5	0.4	18.0	3.4	0.8
TOTAL AQUATICS		1.0		3.1		5.3	
Others							
Caryophyllaceae	0.4	0.3	0.7	0.1			0.7
Cruciferae	0.9	0.5	0.5	0.1	0.8	0.2	0.8

Table 3. (continued)

	Harrowsmith bog % A.P.	Harrowsmith bog % P.S.	Wilton Creek % A.P.	Wilton Creek % P.S.	Hay Bay % A.P.	Hay Bay % P.S.	Lake Ontario % A.P.
Geraniaceae	0.4	0.3	+	+			
Labiatae	0.4	0.3	0.3	0.1			0.7
Liliaceae	2.0	1.1	1.3	0.3			
Malvaceae	1.1	0.8					
Orchidaceae	0.4	0.3					
Ranunculaceae	2.0	1.1	1.1	0.4	+	+	
Rosaceae	1.1	0.8	2.9	0.8	2.0	0.8	1.2
Solanaceae	0.9	0.5	0.3	0.1	+	+	
Thalictrum			0.3	0.1			
Umbelliferae			1.4	0.4			0.8
Urticaceae	0.4	0.3	0.3	0.1			
Incog.	2.0	1.1	4.9	1.2	3.8	2.0	2.3
TOTAL OTHERS		7.4		3.7		3.0	
Ferns	+	+	3.4	0.9	8.0	1.6	1.7
Sphagnum	2.0	1.1	0.3	0.1	0.4	0.1	1.3

Figure 7 shows the pollen concentration in the river throughout 18 months, with the values for the flow through the gauge at the sampling point, and the dispersed sediment load at the same point. These values are not available for the full eighteen months. They have been calculated and made available by the courtesy of Dr Ongley of the Department of Geography at Queen's University, Kingston and the Ontario Water Resources Commission. They are less variable than the pollen counts, being based on weekly or daily samples.

The maximum pollen concentration was independent of the period of maximum flow and maximum sediment load which occurs when the snow melts. The maximum pollen concentration was in July when the daily flow rate varied from only 0.05 to 0.09 m^3/sec. The winter of 1970–71 had low counts throughout, but the second winter had surprisingly high counts in December and January. The first winter was cold, the stream froze in early December, and the snow remained on the ground, but in the second winter the stream did not freeze till December 21, the early snows melted, and the daily flow rate and sediment load increased. The mean flow rate in December was 1.52 m^3/sec.

One can conclude from this that in summer most of the pollen load could fall directly onto the water when it is moving slowly, but the fact that pollen is carried down river when there is none in the air suggests that it is also picked up from the soil all through the year. Federova (1952) assumed that the Volga system carried most pollen in spring, when the trees flowered and the rivers flooded, and that the pollen was carried to the river by surface flow. Possibly the time of maximum pollen concentration varies with the climatic regime. Wilton Creek has a precipitation fairly evenly distributed throughout the year. The concentration of pollen in the water is not the same as the load being carried down the river, which results from both concentration and the rate of flow. There is as much pollen carried down stream to Hay Bay in April and May as there is in July, because of the rapid water movement early in the year.

The composition of the monthly samples varied throughout the year. Fig. 8 shows four main groups of pollen-producers and the monthly distribution of their pollen in the river. The group 'other weeds' includes Compositae, *Chenopodium*, *Plantago*, *Polygonum*, *Artemisia*, and *Rumex*. Fig. 8 also shows the pattern of air-borne pollen at Toronto (from Collins-Williams & Best 1955).

There are several difficulties in comparing pollen rain from Toronto to pollen in Wilton Creek, in a different year. The total output of a species can vary greatly from year to year and from place to place (see Fig. 4). The time of anthesis can vary from year to year and from place to place; the difference in annual timing of flowering is greater in the spring—it can be from early March to mid-April for *Acer saccharinum*

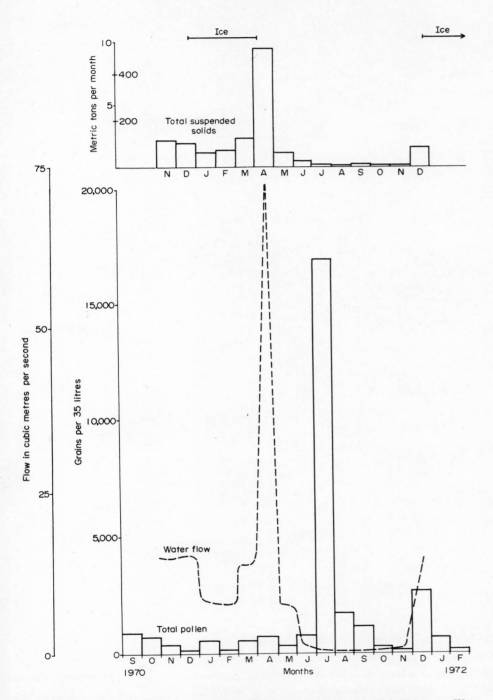

Figure 7. Monthly pollen counts from Wilton Creek, September 1970 to February 1972. Water flow and suspended sediment load are shown for 14 months. The period during which the river was covered with more than 2 cm of ice is indicated at the top of the diagrams.

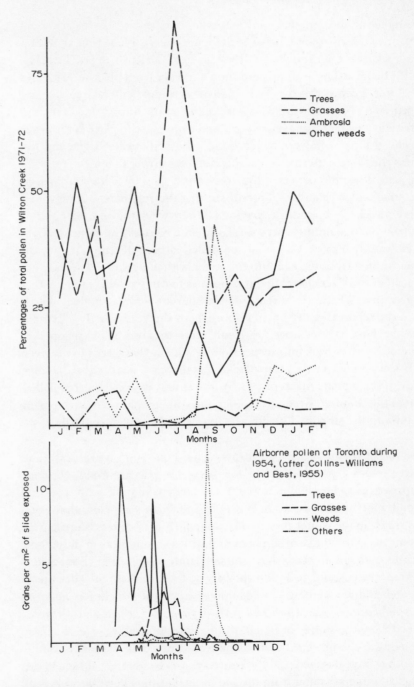

Figure 8. Monthly distribution of four groups of pollen in the water at Wilton Creek. The lower diagram shows monthly distribution of airborne pollen at Toronto. *Ambrosia* is the main component of Collins-Williams and Best's (1955) weed curve.

—than in late summer or autumn, when it is not more than two weeks for *Ambrosia* (cf. Minshall 1947; Durham 1950; Chafee & Settipane 1964). Flowering at Wilton Creek is about a week later than at Toronto (Beschel 1970; LaRush 1934). The peak time for the pollen being airborne and its duration also vary from place to place and from year to year. Fig. 4 shows this for *Ambrosia* in sites round Wilton Creek for one year (from Bassett 1959. Numbers are days with high counts, during one year).

What does remain constant from year to year and from place to place is the order in which the main groups of flowers bloom, and in both the waterborne and the airborne diagrams the same pattern is apparent; the trees flower in spring, they are followed by the peak flowering period of the grasses and then by *Ambrosia*. Wilton Creek has more grass pollen than does Toronto, so that the July maximum is much more marked in the creek curve. Because *Ambrosia* flowers when the time factor is not so variable the *Ambrosia* maximum is very similar in both media. The coincidence of the grass and *Ambrosia* peaks suggests that the pollen either falls directly into the creek or that it is washed from the soil into the water within a couple of weeks.

The winter load of the river shows that it continues to carry pollen into Hay Bay when the air is empty. There are three possible sources for this pollen, which includes much of the arboreal pollen carried throughout the year. Firstly, alternate snows and melts like those of December 1971 could scour the river bed so that pollen lying there would be brought back into suspension. It is likely that weeds in the river bed trap pollen in the summer, as they trap plankton of the same size range (Chandler 1937). The decay of the aquatic macrophytes by November would then free pollen trapped by them in the summer. Secondly, these snows and melts could wash pollen out of the soil and into the stream. Thirdly, the source of the pollen could be post-glacial sands.

The composition of the monthly load throws some light on this problem. The most common tree pollen types in the river for the whole year are *Pinus*, *Quercus*, *Betula*, and Cupressaceae. These all showed a summer maximum, with a second peak in the second winter. The less abundant pollen of *Picea* and *Alnus*, however, was more concentrated in the winter than in the summer (Fig. 9). Grass is present all year, overwhelming everything else in summer, and with a second peak in the second winter, and *Ambrosia* and 'other weeds' have the same time distribution. The aquatic plants form a significant percentage only if they are added together; they disappear altogether in winter and are highest at midsummer (the group included *Potamogeton*, *Nuphar*, *Nymphaea*, *Lemna*, *Callitriche*, *Myriophyllum*, *Caltha*, and *Typha*).

If the river bed were the source of much of the winter pollen one would expect pollen of the aquatics to be present because one would presume the plants would retain some until they have decayed. On the contrary, it is completely absent in the winter. The second source, washing from the soil in the basin, is however indicated

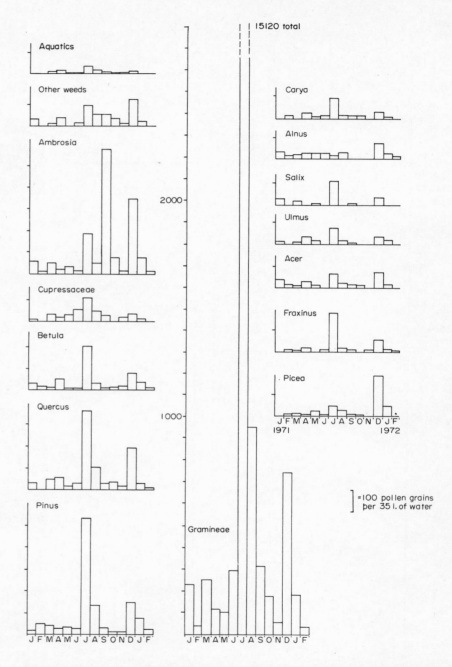

Figure 9. Major pollen types in Wilton Creek through the months January 1971 to February 1972. Each unit on the vertical scale is equivalent to 100 grains per 35 l sample of water.

by the winter peaks of *Picea* and *Alnus*. These two tree pollens are high at Harrowsmith, in the bog, so it seems likely that they are carried down in winter from there. This is backed up by the presence of the only grains of Ericaceae in the river water at the same time. On the whole, the bog retains its pollen like a sponge.

Pollen movement through the soil rather than on top of it is an unknown quantity. It could be abnormally high in this basin because of the underground drainage in the central zone, where the limestone has little soil cover.

It seems likely that pollen from all three sources is present all year in the river. The plants which are the biggest pollen producers such as pines, grass and ragweed shed so much at anthesis that they then have a seasonal peak in the water as well as the air; abundant plants such as maple which do not produce much pollen also tend to be present all year though in small amounts; plants which are patchily distributed such as spruce and alder may also have a patchy distribution in time.

Comparison of the river pollen with Hay Bay and Lake Ontario

The pollen in the Hay Bay mud has a spectrum matching the yearly load of the river very closely; grasses are the predominant pollen in both. The four most important trees make up the following percentages of the A.P.

	Hay Bay	Wilton Creek
Pinus	31	33
Quercus	16	18
Cupressaceae	8	7
Betula	6	9

Values from 80 bores in the deep water of Lake Ontario which are about to be published by McAndrews and Power (1972) show that the creek A.P. pollen matches them better than it does Hay Bay. The *Betula*, Cupressaceae, and *Pinus* values from the deep water and the creek are within one per cent. The values are shown in Table 3.

The positions of grass and ragweed in the creek water and in lake sediment are reversed, *Chenopodium* is not so abundant in the creek, but the Cyperaceae, Tubuliflorae, and ferns have the same importance. The low ragweed and high grass in the shallow deposit in Hay Bay could be an effect due to the difference in specific gravity between the grains. Grass pollen has a specific gravity of 0.9–1.01, whereas *Ambrosia* has values of 0.50–0.55 (Durham 1946). Grass pollen may thus settle before the *Ambrosia*, as the creek water slows in the inlet. Other possible explanations are the differential settling of airborne pollen and/or differential decay during the time of transport from the site of origin on land and the site of deposition in the lake mud, or erosion of the mud back to a time of low *Ambrosia* deposition.

The close match of the arboreal pollen suggests that river-borne pollen must be a major contributor to the lake muds. In calculating the annual pollen spectrum of the river no weighting for the rate of flow has been made; this should be done, but should be based on more than twelve pollen counts in the year.

Differential preservation

Sangster and Dale (1961) established that there are differential rates of decay both for different pollen types and for different types of deposit. Grass and juniper pollen are known to disappear in polsters (Davis & Goodlett 1960). In the moss cushions at Wilton Creek they do not accumulate in bulk, but they are abundant in the river and seem well preserved in the Hay Bay mud. Eroded exines were found in many types in the moss cushions, particularly *Carya* and *Ulmus*.

Populus does not occur in significant quantities anywhere in the system so that differences in its preservation were not noticed. In Hay Bay some of the *Ambrosia* grains were eroded, and it is possible that the unidentified grains included some *Ambrosia*.

Conclusions

Table 4 shows the abundance of the major types in the vegetation, the moss cushion pollen, the bog, the river mud, and Lake Ontario deposits.

Table 4. Comparison of the pollen types found in all the moss polsters with the amounts in the bog, the river, Hay Bay and L. Ontario surface samples (McAndrews & Power 1972). Values are percentages of arboreal pollen. Cupressaceae, *Typha*, and Ericaceae are local types, which are not part of the regional rain.

	Veg.	Moss	Bog	River	Hay Bay	L. Ont.
Acer	30	2	3	4	5	6
Betula	2	17	16	9	6	8
Carya	2	4	4	2	1	2
Fraxinus	4	5	+	5	1	3
Ostrya + Carpinus	4	3	3	2	8	2
Pinus	11	35	6	33	31	34
Quercus	8	15	10	18	17	24
Ulmus	11	7	10	3	3	2
Cupressaceae	7	2	+	7	5	6
Gramineae		21	15	221	352	16
Ambrosia		28	12	31	17	30
Typha		3	—	+	4	+
Ericaceae		+	3	+	+	+

The components of the pollen rain which were common to all the polsters were: *Acer*, *Betula*, *Carya*, *Fraxinus*, *Ostrya-Carpinus*, *Pinus*, *Quercus*, *Ulmus*, Gramineae, and *Ambrosia*. This highest common factor of ten types can be taken to be regional while the other plants in each spectrum are more local. The balance of the grass and weeds on the one hand and the trees on the other was found to match the forest cover closely. The most abundant tree pollens in the regional fraction were *Betula*, *Pinus*, and *Quercus* in that order. The glaring discrepancy with the vegetation is the under-representation of *Acer*. *Betula* is grossly over-represented, whereas *Pinus* and *Quercus* are not badly out of proportion to their basal areas.

The bog also received the highest common factor of the regional rain; again there is too little *Acer*, too much *Betula*, and again the balance of the open and forested ground is fairly represented. The trees present a localized picture because of the unusually high spruce and alder. There is a good diversity of non-arboreal pollen, perhaps reflecting good conditions of preservation.

The creek water and the pollen it deposits in Hay Bay proved to be a more accurate mirror than one would expect. The low A.P. percentage in both (25 and 19%) reflects the open farmed country in the southern zone. The trees are not so distorted as in the polsters or the bog, because while *Acer* is only 5% of the A.P., *Betula* drops to fourth place, and *Pinus* and *Quercus* are most abundant. A surprising aspect is the amount of pollen of Cupressaceae, which disappears from the other sites. *Thuja* trees are in fact a noticeable feature of the valley, and *Juniperus* is abundant in old fields and on rock outcrops, so the fluvial spectrum reflects an aspect of the landscape which is lost in the polsters and the bog.

All three distorting mirrors were found to distort in different directions, but it is interesting that not only do their local elements differ but also their spectra of the major trees. This difference should be considered in the interpretation of profiles from sites which change from marsh conditions at the base to bog peat higher up, as the differences in the tree pollen proportions coincident with the change may be due to a switch from waterborne to airborne pollen as the main influx.

The river has been shown to be an important contributor to the lake pollen, though one should perhaps quote Hutchinson (1963) who wrote 'the whole subject of rivers . . . appears as a marvellous foreign territory explored by workers whose audacity is admirable in view of the difficulty of getting a theoretical grasp of the subject.'

Acknowledgements

This research has been carried out, as part of a multi-disciplinary programme, with the support of the Ontario Department of University Affairs. We should like to thank the following for their help: the late Dr R.E. Beschel, Dr Ongley of the Department of Geography, Queen's University, the Ontario Water Resources Commission, and Dr J.H. McAndrews of the Royal Ontario Museum, for permission to use his values from Lake Ontario.

References

BASSETT I.J. (1959) Surveys of airborne regional pollen in Canada with reference to sites in Ontario. *Can. J. Pl. Sci.* **39**, 491-97.

BENNINGHOFF W.S. (1960) Pollen spectra from bryophytic polsters, Inverness Mud Lake bog, Michigan. *Mich. Acad. Sci. Arts & Letters* **45**, 41-60.

BESCHEL R.E. (1970) *Phenology of Ontario*. Paper read to the Canadian Botanical Association meeting at Laval, Quebec. Unpublished manuscript.

BESCHEL R.E., GARWOOD A.E., HAINAULT R., MACDONALD I.D., VANDERKLOET S.P. & ZAVITZ C.H. (1970) *List of the Vascular Plants of the Kingston Region*. Kingston.

BLAKE V.B. (1957) Historical section. In *Napanee Valley Conservation Report*. Dept. of Planning and Development, Toronto.

CARROL G. (1943) The use of bryophytic polsters and mats in the study of recent pollen deposition. *Am. J. Bot.* **30**, 361-66.

CHAFEE F.H. & SETTIPANE G.A. (1964) Atmospheric pollen and mold survey. *J. Allergy* **35**, 193-200.

CHANDLER D.C. (1937) Fate of typical lake plankton in streams. *Ecol. Monogr.* **7**, 445-79.

COLLINS-WILLIAMS C. & BEST C.A. (1955) Atmospheric pollen counts in Toronto, Canada. *J. Allergy* **26**, 461-67.

COTTAM G. & CURTIS J.T. (1956) The use of distance measures in phytosociological sampling. *Ecology* **37**, 451-60.

DAVIS M.B. & GOODLETT J.C. (1960) Comparison of the present vegetation with pollen-spectra in surface samples from Brownington Pond, Vermont. *Ecology* **41**, 346-57.

DURHAM O.C. (1946) The volumetric incidence of atmospheric allergens. III Rate of fall of pollen grains in still air. *J. Allergy* **17**, 70-8.

DURHAM O.C. (1950) Report of the pollen survey committee of the American Academy of Allergy for the season 1949. *J. Allergy*, **21**, 442-54.

FAEGRI K. & IVERSEN J. (1964) *Textbook of pollen analysis*. Munksgaard, Copenhagen.

FAEGRI K. (1966) Some problems of representativity in pollen analysis. *Palaeobotanist* **15**, 135-40.

FEDEROVA R.V. (1952) The spread of pollen and spores by water currents (in Russian). *Trans. Inst. Geogr., Acad. Sci. USSR* **52**, 46-73.

GARWOOD A.E. (1967) Some uncommon plants found near Kingston. *Blue Bill*, **14**, No. 4, 45.

HANSEN H.P. (1949) Pollen content of moss polsters in relation to forest composition. *Am. Midl. Nat.* **42**, 473-79.

HUTCHINSON G.E. (1963) The prospect before us. *Limnology in North America*. (ed. by D.G. Frey), pp. 683-60. University of Wisconsin Press, Madison.

JAFRI S. (1965) *Forest composition in the Frontenac axis region*. Unpublished M.Sc. thesis, Queen's University, Kingston.

KING J.E. & KAPP R.O. (1963) Modern pollen rain studies in eastern Ontario. *Can. J. Bot.* **41**, 243-52.

LEWIS C.F.M. & MCNEELY R.N. (1967) Survey of Lake Ontario bottom deposits. *Proc. 10th Conf.*

Great Lakes 1967, 133-42.

LaRush F. (1934) Pollen content of the air in Toronto, Canada, 1932. *J. Allergy* **5**, 306-17.

McAndrews J.H. & Power D.M. (1972) *Palynology of the Great Lakes: the surface sediments of Lake Ontario*. Unpublished manuscript.

Meacham J.H. (1878) *Illustrated historical atlas of the counties of Frontenac, Lennox and Addington*. Toronto.

Minshall W.H. (1947) First dates of anthesis for four trees at Ottawa, Ontario for the period 1936-1945. *Can. Fld. Nat.* **61**, 56-59.

Potzger J.E., Sylvio M. & Hueber F.M. (1957) Pollen from moss polsters on the Lac Shaw bog, Quebec, correlated with a forest survey. *Bull. Serv. Biogéogr. Univ. Montréal* **18**, 24-35.

Putnam D.F. & Chapman L.J. (1938) The climate of southern Ontario. *Sci. Agr.* **18**, 401-46.

Richard P. (1968) Un spectre pollinque type de la sapinière à bouleau blanc pour la forêt Montmorency. *Naturaliste can.* **95**, 565-76.

Sangster A.G. & Dale H.M. (1961) A preliminary study of differential pollen grain preservation. *Can. J. Bot.* **39**, 35-43.

Terasmae J. & Mott R. (1964) Pollen deposition in bogs and lakes near Ottawa, Canada. *Can. J. Bot.* **42**, 1355-63.

Terasmae J. (1969) A discussion of deglaciation and the boreal forest history in the northern Great Lakes region. *Proc. ent. Soc. Ont.* **99**, 31-43.

Absolute pollen frequencies in the sediments of lakes of different morphometry

WINIFRED PENNINGTON (Mrs T.G. Tutin) *Freshwater Biological Association and School of Biology, University of Leicester*

Introduction

The development and application of methods of absolute pollen analysis by M.B. Davis (Davis & Deevey 1964, Davis 1967, 1969, Davis *et al.* 1973), have provided a valuable new tool for the investigation of vegetation history at those horizons where progressive changes in abundance of one or more high pollen producers introduce complications into the interpretation of interdependent percentage curves. This may be of particular importance in comparison of a number of pollen diagrams in the study of regional vegetation history. This paper reviews some general findings from published data on absolute pollen analysis of lake sediments, and presents some new data on absolute pollen analysis which I have obtained from lake sediments from the uplands of N.W. Britain. It is appropriate in this volume dedicated to the late Dr Johs. Iversen to consider new data relevant to two problems, both involving the association of vegetation and soil changes, to which he contributed so much original data and stimulating ideas (Iversen 1954, 1958, 1960, 1964, 1969), namely problems of the late-glacial period and of the 'elm decline' in Northern Europe.

The late-glacial period and the 'elm decline' at the Atlantic-Sub-Boreal boundary are periods of vegetation history for which parallel changes in vegetation and soils have been demonstrated in lake sediment profiles from both N.W. England and N.W. Scotland (Mackereth 1965, 1966, Pennington 1970, Pennington *et al.* 1972). Multivariate statistical analysis of the data on sediment composition has shown a significant correlation between major changes in the composition of sediment, derived from soils of the catchments, and the pollen zone boundaries (cf. Figs. 1 & 7). This is interpreted as indicative of major environmental changes at these zone boundaries. It seemed possible that determination of absolute pollen frequency at these horizons would provide useful additional data on vegetation changes, and that pollen concentration and pollen deposition rates (Davis 1969) would provide a firmer basis than interdependent percentage curves for numerical treatment of these data aimed at an objective determination of pollen zone boundaries.

The determination of absolute pollen frequencies has several potential advantages in regional pollen studies. (1) The absolute pollen deposition rates for each taxon are advantageous over percentage figures as a basis for comparison of diagrams from many sites, when one or more high pollen producers vary from site to site. (2) The elimination of the need to use judgement in selecting a pollen sum as a basis for percentage calculations makes comparisons more objective. It should in theory be possible to determine, by comparison of many absolute pollen figures, the regional pollen component of Janssen (1966) or the long-distance component of West (1971)—this is, by definition, that component of the pollen rain which is constant in absolute terms throughout a region. In hilly regions the percentage value of the regional component varies in space—from site to site according to local pollen productivity—and it is recognised that in any one profile the percentage value of the regional (long-distance) component will vary in time, increasing during periods when pollen production by the local vegetation is low (Jessen 1949).

Before assessing the validity of such comparisons, however, it seems necessary to consider for each site what local influences, apart from pollen production by the vegetation represented, may affect pollen deposition rates. Davis (1967) suggested that 'the number of pollen grains entering a lake per unit area probably varies with lake size (Tauber, 1965)'. As yet no estimates are available for the numbers of grains which annually enter lakes of the characteristic types found in the glaciated uplands of N.W. Britain.

Absolute pollen frequencies in lake sediments

A regional study of the vegetation history of N.W. England based on open-water sediment from many lakes has revealed considerable differences between lakes with respect to pollen concentration in the post-glacial sediments, though results have been published only as percentage figures (Pennington 1947, 1964, 1970). Since the thickness of post-glacial sediment in these lakes varies only between 3 and 5 m, the fact that pollen slides prepared by a standard technique show such wide variations in pollen concentration indicates a real difference between lakes in annual pollen deposition rates, and argues against the hypothesis that 'it may become possible to read pollen deposition rate diagrams directly as a record of density of vegetation on the landscape' (Davis 1969). Furthermore, knowledge of the very uneven distribution of sediment on the bottom of some lakes had raised doubts about the meaning of deposition rates per unit area per year for pollen or any other variables. Published examples of uneven sediment accumulation include the organic late-glacial sediment in Windermere (Pennington 1943, 1947) and the post-glacial deposits of some lochs of northern Scotland, including Loch Ness, which are highly exposed to wind disturbance (Pennington et al. 1972). The results of absolute pollen counts at Rogers Lake (Davis & Deevey 1964, Davis 1967) stimulated me to attempt to quantify differences in pollen concentration in the deposits of British lakes.

Comparisons of some available estimates of pollen deposited per year in pollen traps in air with annual pollen deposition per similar unit area of lake mud indicate some degree of concentration within lakes (Table 1). Most estimates of pollen deposition rates in lakes appear to exceed average figures for deposition of pollen from the air into traps. Concentration of pollen in lake sediment results from two processes. One is recirculation by water movements of material once deposited on the bottom, followed by redeposition within a limited deep-water area; this has been demonstrated in lakes on several criteria (Lund 1954, Tutin 1955, Collins 1970) and shown by Davis (1968) to increase the annual pollen deposition rate in the deepest part of Frains Lake, Michigan. Secondly, in all except enclosed basins there is input of pollen from the lake's catchment in inflowing water; the presence of pollen in streams, especially after floods, has been demonstrated in hilly country (Peck 1973), and the effects have sometimes been recognised in percentage pollen diagrams (e.g. Pennington 1964, p. 40, Birks 1970). The results to be presented in this paper form part of an exploratory survey to find out whether the extent to which lakes have concentrated pollen in late-glacial and post-glacial times is a function of their morphometry, possibly including size. If it is, it will be necessary to standardise lake sites before their pollen deposition rates can be used to compare vegetation. Cushing (1967) has reminded us that 'the paleoecology of the site of deposition should be of as much concern to the pollen analyst as the paleoecology of the area from which the pollen came'.

Variation in pollen deposition rates between lakes, within what is shown by percentage pollen spectra to be the same vegetation region, is presumably the result of varying efficiency in the process of pollen concentration. The observed differences may be due either to different patterns of final settlement of material within the lake, or to differences in the amount of pollen recruited from the catchment via the inflows. Another factor is the extent to which pollen is lost from a lake's basin by rapid throughput of water. The effect of seasonal snow-melt and of flash floods in causing loss of microparticles down the outflows has been demonstrated for pollen (Maher 1963) and for bacteria (Collins 1970). Information about pollen deposition in lakes

Table 1. Fossil concentrations and deposition rates (influx) compared with contemporary pollen deposition rates.

A. Late-glacial

(i) North-west Europe	Concentration (grains/cm^3)	Influx grains/cm^2/yr.
Zone Ba (part of I)		
Blelham Bog	2.2×10^4	100–150
Low Wray Bay, Windermere	1.8×10^4	
Blea Tarn	2×10^4	
Zone Bb (part of I)		
Blelham Bog	8×10^4	$2-3 \times 10^3$
Low Wray Bay	1×10^5	
Blea Tarn	6 to 8×10^4	
Zone Bc (IIa)		
Blelham Bog	8×10^4	$2-3 \times 10^3$
Low Wray Bay	5×10^4	
Blea Tarn	8×10^4	
Zone Bd (IIb + c)		
Blelham Bog	$5-8 \times 10^4$	$1-2 \times 10^3$
Low Wray Bay	8×10^4	
Blea Tarn	1×10^5	
Zone Be + Bf (III)		
Blelham Bog	2×10^4	400
Low Wray Bay	less than 1×10^3	less than 100
Blea Tarn	4×10^4	1,000
Compare with:		
Leirvatn, at 1,300 m in S.W. Norway		
Surface mud	3×10^4 + 3.7×10^4 Pinus (from valleys)	
(ii) North America		
Rogers Lake[1]		
Zone T	2×10^4	600–900
c 11,500 B.P.	1×10^5	6×10^3
Compare with surface samples from:		
Arctic Canada[2]		
Sedge-moss tundra		22–65
Dwarf-shrub tundra		52–762
Forest tundra		361–2,372

Table 1. (continued)

B. Post-glacial

(i) North-west Europe	Concentration (grains/cm^3)	Influx grains/cm^2/yr.
c 5,000 B.P. (3000 B.C.)		
N.W. England—Mixed-oak forest region		
Windermere (South Basin)	1.5×10^5	1.6 to 2×10^3
Blea Tarn	3×10^5	1.5×10^4
N.W. Scotland—Pine-birch forest region		
Loch a'Chroisg	1×10^5	5×10^3
Loch Clair	3.3×10^5	1.5×10^4
Compare with figures for present:		
N.W. England		
Windermere surface mud	2.7×10^4	c 6.75×10^3
Blelham Tarn surface mud	6.0×10^4	4.8×10^4
cf. Vågåvatn (Table 2) (c 1,800 B.P.–present)	3.7×10^4	5×10^3
Pollen Traps		
Gantekrogsö[3]		6.6×10^3
Meathop Moss (excluding *Calluna*)[4]		6×10^3
Blelham Tarn, floating trap		3×10^3
(ii) North America		
Rogers Lake[5]		
10,000 B.P.	4×10^5	2×10^4
8,000 B.P.	5×10^5	5×10^4
5,000 B.P.	3×10^5	3.5×10^4
Present	2×10^5	2×10^4
Frains Lake[6]		
Present	$2.1 \times 10^4 \times 138$	2.1×10^4
Rutz Lake[7]		
10,000 B.P.		3.0×10^4
8,000 B.P.		1.5×10^4
5,000 B.P.		1.5×10^4
Riding Mountain, Manitoba[10]		
10,000 B.P.		2×10^3
8,000 B.P.		1.7×10^3
5,000 B.P.		1×10^3
Lake Louise[8]		
c 6,710 B.P.	2.6×10^5	2.2×10^4
c 8,510 B.P.	4.1×10^5	3.4×10^4
Lake of the Clouds[9]		
Pine-dominated zone		$2-4 \times 10^4$
Compare with pollen trap figures:[2]		
Northern conifer forest		1.16×10^4

[1] Davis (1967)
[2] Ritchie & Lichti-Federovich (1967)
[3] Tauber (1967)
[4] Oldfield, unpubl.
[5] Davis (1969)
[6] Davis (1968)
[7] Waddington (1969)
[8] Watts (1971)
[9] Craig (1972)
[10] Ritchie (1969)

now receiving seasonal melt-water from mountain snowfields and glaciers seemed relevant to the interpretation of late-glacial pollen deposition, and the nearest region of mountain glaciation, possibly comparable in some respects with the late-glacial environment of Highland Britain, is S.W. Norway. Pollen concentration figures will be given from surface cores (50 cm long) from two Norwegian lakes in which the sediment resembles lithologically the late-glacial clays of lakes of N.W. England.

Since one of the major arguments to be developed in this paper is that not all of the pollen entering a lake necessarily comes to permanent rest within the sediments, Davis's term 'pollen deposition rate' has been preferred to Cushing's 'pollen influx' (Davis 1969) to describe the annual addition of pollen to accumulating sediments.

Methods

FIELD AND LABORATORY METHODS

The new data included here were obtained from cores of lake sediments taken in open water using a Mackereth (1958) corer. After extrusion beside the lake into polythene-lined troughs, the sections were sealed in polythene sheeting and stored in a cold room until samples for ^{14}C dating and pollen analysis were withdrawn.

Surface cores from lakes in Norway were taken with a Gilson surface sampler, sealed in polythene, and stored in a cold room after return to Britain.

For absolute pollen analysis samples were withdrawn from the cores as soon as convenient, into small screw-topped, air-tight culture bottles, using a cork-borer of similar diameter to the bottle; the bottles were also stored in a cold room until preparations were made. From the sample in the bottle, 1 cm^3 or a fraction thereof was measured by displacement in water in a narrow-bore 10 ml graduated cylinder. The sample was then prepared for analysis by the methods of Faegri and Iversen (1964), and at the end of the preparation a measured weight of an assayed suspension of *Ailanthus* grains in glycerol was added, and the sample homogenised on a shaker (Bonny 1972). A total of 1,000 or 500 grains was then counted. Methods for calculation of the 90% confidence interval for the count have been given by Bonny (1972). Results of replicate samples will be given in the course of this paper.

Pollen counts were made from mounts in safranin-stained glycerine jelly. Criteria on which taxa were recognised were as given by Pennington *et al.* (1972).

Counts by Bonny for the site at Blelham Bog have been published (Pennington & Bonny 1970). Of the new counts published here, those for Windermere, South Basin were by A.P. Bonny, and the rest by the author.

INTERPRETATION OF A.P.F. DATA

Davis (1969) and Davis *et al.* (1973) have discussed the errors involved in both the processing and the selection and sampling of lakes. They suggest that 'changes in pollen deposition by a factor of five are probably the smallest attributable to variations in pollen input to the lake'.

The new data presented in this paper, like those from Blelham Bog, show a consistency which suggests that smaller differences may well be significant when this method is used on homogeneous, fine-textured open-water sediments from stony or rocky-shored lakes that have no marginal hydrosere to produce coarse plant detritus. These fine-textured muds form tractable material for pollen preparations for absolute frequencies, and no significant loss of pollen from the samples during preparation could be shown (Bonny 1972). Addition of exotic pollen at the end of the preparation removes any possibility of loss of these grains, and tests for homogeneity of the counts proved that the exotic pollen was adequately mixed by the use of a shaker. Duplicate assay of the exotic suspension using an inverted microscope (Bonny 1972) showed that *Ailanthus* pollen in suspension in glycerol can be adequately assayed by haemocytometer counts.

These new data represent the results of a preliminary exploration of sites to find out how absolute pollen frequency, estimated by the above method, differs (a) from level to level within a profile, and (b) from site to site at the same point in time. The amplitude of these differences, relative to the size of the confidence interval of the counts, will determine what problems can and cannot be solved by application of this method.

Absolute pollen frequency in late-glacial sediments

INTRODUCTION

The absolute pollen diagram from Rogers Lake (Davis 1967, 1969) shows the contrast between the lowest pollen zone (T, the Herb zone) of very low pollen deposition rate, and the overlying zone (A) in which not only are deposition rates of tree pollen much higher than in T, but pollen deposition rates from non-arboreal vegetation are also at least as high, if not higher, than in T. Within zone T, small differences in percentages of tree pollen from one sample to another 'represent statistical artifacts rather than significant vegetational changes' (Davis & Deevey 1964). The absolute diagram from Blelham Bog (Pennington & Bonny 1970) shows a similar contrast in pollen deposition rates at the boundary between the lowest zone, Ba, with high percentages of *Rumex* pollen, and the overlying zone Bb, with high percentages of *Juniperus*; within zone Bb deposition rates for pollen of *Rumex* and other herbs significantly exceed those of zone Ba. At both sites the authors have interpreted this significant increase in all pollen deposition rates as indicative of a major environmental change (a climatic amelioration) and consequent increase in pollen production. At Blelham Bog the significance of this change in the regional chronology of the Late-Weichselian period in N.W. Europe is that it both underlies in stratigraphic position and precedes in ^{14}C age the beginning of the classical Allerød period, the *Betula* pollen zone, at *c* 12,000 B.P. (cf. Fig. 4), but it is not separated from it by any evidence for climatic recession attributable to the Bölling/Allerød stade. The absolute pollen analytical evidence at Blelham Bog entirely confirms the reality of the post-Allerød climatic recession of Younger Dryas time, by showing significantly lower deposition rates for total pollen and of birch and juniper, and significantly higher rates for *Artemisia* pollen, within the deposits of the *Artemisia* percentage pollen zones, Be and Bf, correlated with zone III. In contrast absolute pollen data from North America have not confirmed the existence of the presumed climatic recession at the end of the Two Creeks interval (Davis & Deevey 1964).

SITES IN THE ENGLISH LAKE DISTRICT

Further evidence on the late-glacial sequence in N.W. England has now been sought by absolute pollen analysis of two other profiles for which percentage diagrams have been published. Within the 2,700 sq. miles (7,000 km²) of N.W. England there are four contrasted types of lakes, *viz*:

Type 1 Lowland lakes with lowland catchments, e.g. Esthwaite Water (Franks & Pennington 1961) and Helton Tarn (Smith 1958), and also the basins of former lakes now overgrown by bogs, e.g. Blelham Bog and Moorthwaite Moss (Walker 1966).

Type 2 Lowland lakes with highland catchments, e.g. Windermere (Pennington 1943, 1947, Godwin 1960), which include lakes of Type 4.

Type 3 Upland lakes lying in shallow basins and lacking fresh moraine in the catchments, e.g. Blea Tarn, Langdale (Tutin 1969, Pennington 1970, Pennington & Lishman 1971).

Type 4 Upland lakes in deep basins associated with fresh (Younger Dryas age) moraine, e.g. Small Water (Pennington 1970), and former lakes that are now valley mires, e.g. Langdale Combe (Walker 1965).

The three examples to be discussed in this paper, Blelham Bog, Low Wray Bay (Windermere), and Blea Tarn, represent respectively sites of Types 1, 2, and 3, but the history of Type 4 must be considered in interpreting Type 2. At sites of Types 1, 2, and 3 organic and polleniferous late-glacial sediments are present; at sites of Type 4 the only late-glacial sediment found is entirely minerogenic and barren in pollen, and is ascribed to deposition by active snowbeds or corrie glaciers during Younger Dryas time. As yet it has not been possible to penetrate this coarse minerogenic deposit to investigate what may lie beneath it; corers reach coarse gravel and react as to solid rock.

BLELHAM BOG AND LOW WRAY BAY, WINDERMERE

The series of pollen zones established for the dated profile at Blelham Bog (Type 1) (Pennington & Bonny 1970) is applied in this paper to the other re-investigated sites in the area.

Low Wray Bay, Windermere (Type 2) contains within its catchment many examples of Type 4 sites such as Langdale Combe and Red Tarn, Wrynose (Pennington 1964). The sequence of deposits in Low Wray Bay has been investigated in many cores and the results of stratigraphic and diatom analysis (Pennington 1943), and pollen and macrofossil analysis (Pennington 1947) have been published, together with later fuller pollen analysis and a ¹⁴C date (Godwin 1960). A similar sequence of deposits is present in other bays on the west and east shores of the North Basin of Windermere (Figs. 3 & 14a). As in all comparable large lakes of the English Lake

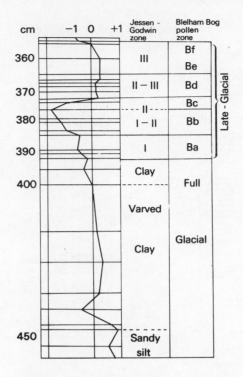

Figure 1. Blea Tarn late-glacial profile. First component of a Principal Component Analysis of the correlation matrix for 8 variables (chemical elements) in the analysis of sediment chemical composition, plotted stratigraphically, with pollen zone boundaries.

District (unpublished data) the brownish or greyish fossiliferous late-glacial silty clays and clay-silts lie between an upper and a lower varved clay. Both varved clays show strongly graded bedding (Pennington 1947). The lower clay in both composition (Mackereth 1966) and constancy of direction of its magnetic remanence (Mackereth 1971) shows evidence for very rapid input, confirming that the wide varves are approximately annual. This clay is entirely barren of microfossils in all cores examined. It is therefore assigned to the closing stages of the full-glacial period. The upper clay, above the organic late-glacial deposits, contains 400 to 500 pairs of laminations, and this, together with plant identifications from the underlying organic deposits, led to a correlation of the upper laminated clay with the Younger Dryas climatic recession (Pennington 1947), a correlation later confirmed by ¹⁴C dating (Godwin 1960; Mackereth 1971). The question of the length of time represented by the organic late-glacial deposits remained a problem until they were correlated by pollen analysis with the profile at Blelham Bog for which 11 ¹⁴C dates

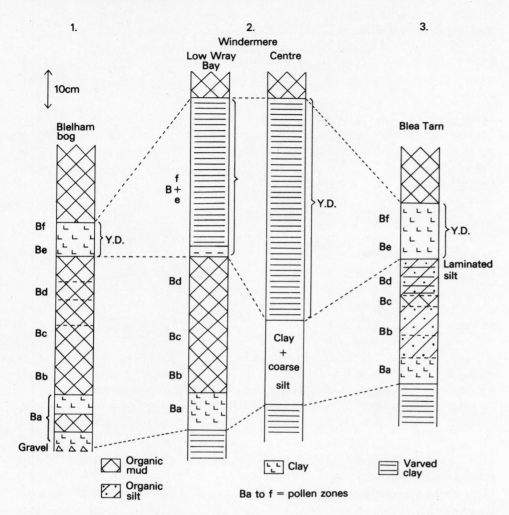

Figure 2. Late-glacial stratigraphy and pollen zones at Sites 1, 2, and 3. YD = Younger Dryas Time.

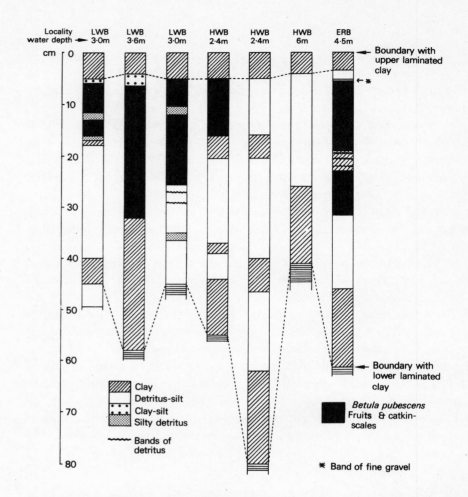

Figure 3. Site 2, Windermere. Variation in late-glacial stratigraphy at marginal sampling stations in the North Basin (LWB—Low Wray Bay, HWB—High Wray Bay, ERB—Ecclerigg Bay) showing the *Betula pubescens* macrofossil zone.

were obtained. These dates indicated that the deposits between the two varved clays of Windermere represented a period of 4 to 5 thousand years, and this has been confirmed by chemical evidence and changes in direction of remanent magnetisation (Mackereth 1971, p. 336).

Figure 3 shows the range of stratigraphy found in the many cores obtained from the marginal bays of Windermere during the 1940's, and the position of the zone in which fruits and catkin-scales of tree birches (*Betula pubescens*) were found in the organic silts; this zone constitutes only the upper portion of the organic sediment, and since the ^{14}C date of 9920 ± 120 B.C. (Godwin 1960) came from the base of the zone of *Betula* macrofossils it is therefore in good agreement with correlation of the zone of *Betula* macrofossils with the Alleröd period. Comparison of the pollen diagram published by Godwin (1960) with further unpublished analyses indicate a sequence of percentage pollen zones very similar to that found at Blelham Bog. The grey and brownish-grey silty clays and clay-silts in Windermere (Fig. 3) correspond in percentage pollen content with Blelham Bog zones Ba to Bd inclusive—that is, to

Absolute pollen frequencies in lake sediments

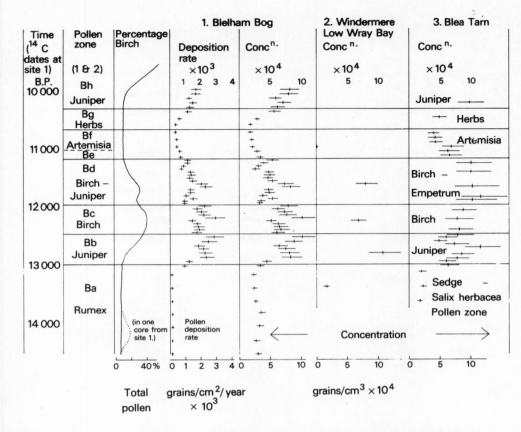

Figure 4. Late-glacial pollen deposition rates (influx) and concentration (total pollen) at Sites 1, 2, and 3.

the Jessen-Godwin zones I and II, but, as at Blelham Bog (Pennington 1970), the major increase in organic matter is found at the zone boundary Ba/Bb, namely the boundary between the *Rumex* percentage zone and the *Juniperus* zone.

Figure 2 compares the stratigraphic sequences of pollen zones Ba–Bd inclusive at Low Wray Bay and Blelham Bog. Figure 4 shows the new data for pollen concentration for one sample within each pollen zone at Low Wray Bay; comparison of the stratigraphy makes it clear that pollen deposition rates will not differ significantly from those at Blelham Bog within these zones. There is however a strong contrast between the two sites with respect to pollen zones Be and Bf, the deposits of Younger Dryas time. The concentration of pollen within the Windermere upper laminated clay is less than 1,000 grains per cm^3, whereas the unlaminated clay of pollen zones Be and Bf at Blelham Bog contains $c\ 2 \times 10^4$ grains/cm^3. Conversion to annual deposition rates (the length of Younger Dryas time is 500 years on both the ^{14}C time-scale from Blelham Bog and the number of presumed annual varves in Windermere) gives 500 grains/cm^2/year at Blelham Bog and less than 100 in Windermere—a difference highly significant in terms of the 90% confidence interval for the counts. Some factor was therefore operative in producing a significant difference in pollen deposition rates between these two neighbouring basins during Younger Dryas time, although during the preceding millennia of the Late-Weichselian period pollen deposition rates had been similar at the two sites, indicating that throughout the period of time represented by pollen zones Ba to Bd (I + II) the size of lake had not affected pollen deposition rates.

Further information about conditions of sedimentation in Windermere (and the other large lakes of the district) during the late-glacial period is provided by the contrast between deposits of deep and of shallow water (Pennington 1943, and Fig. 2). The organic muds and silts which represent pollen zones Ba to Bd in the marginal bays (Fig. 3) are not found in water depths greater than 30 m (Pennington 1947 Fig. 2, p. 143). In water deeper than 30 m the facies of late-glacial deposits is entirely minerogenic and barren of microfossils; clays and coarse silts are found between the lower and upper varved clays. The complete (North Basin of Windermere) or almost complete (South Basin of Windermere, Coniston Water) absence of pollen, diatoms and other microfossils from the entire late-glacial sequence in deep water in these large lakes indicates that no deposition of resuspended material from shallow water took place in late-glacial times. If organic material from shallow water was resuspended, it did not come to permanent rest in the deep-water areas but must have been permanently lost from the basin in outflowing water. Contrast in pollen deposition rates between Low Wray Bay and Blelham Bog during Younger Dryas time indicates very low pollen deposition rates even in marginal areas of the large lakes during this period.

Possible mechanisms for loss of pollen, operative (i) throughout the late-glacial period in deep-water areas of large lakes such as Windermere, and (ii) during Younger Dryas time in marginal areas of these lakes also, have been studied by estimation of pollen frequency in the surface deposits (0–50 cm) in Vågåvatn, S. Norway, an elongated (c 20 km) valley lake which receives the heavily silt-laden river Bövra from the mountain glaciers of Jotunheim. The steep valley sides around the lake are clothed in pine forest, interrupted by small areas of cleared ground. Surface cores from Vågåvatn show varves—paired laminations with a contrast in particle-size—below the top c 10 cm of unconsolidated deposit. These varves are presumably the result of seasonal melt of the mountain snowfields and glaciers. The Vågåvatn sediment is of clay and fine silt particle-size, the water content is c 35% and the organic content only 1% of the dry weight; in all these characteristics it resembles the late-

Table 2. Pollen deposition in Vågåvatn; S. Norway, compared with the pine-dominated zone in Scottish lakes.

	Concentration (grains/cm^3)		Pollen/ cm^2/year*
Vågåvatn			
0–1 cm	Arboreal pollen	18,462	?
	Pinus pollen	13,684	
13–14 cm below surface	Arboreal pollen	23,478	3913
	Pinus pollen	15,050	2508
31–32 cm below surface	Arboreal pollen	28,274	4039
	Pinus pollen	18,117	2588
Compare with pine-dominated zone in Scottish lakes:			
Loch Clair	Arboreal pollen	292,463	15,923
	Pinus pollen	94,824	5073
Loch a'Chroisg	Arboreal pollen	123,900	6422
	Pinus pollen	96,500	5000

*Assuming that one pair of laminations represents the deposition of one year.

glacial clays of Windermere (Mackereth 1971). Pollen concentration within the Vågåvatn sediments is significantly low in comparison with North American figures for lakes in pine-dominated zones (Table 1), and pollen deposition rates, obtained by assuming that the paired laminations are annual, are significantly lower than in pine-dominated zones of Scottish lakes or present annual deposition in pollen traps in coniferous forest (Tables 1 & 2). This implies that the rapid through-put of water during the summer melt in such a lake, receiving rivers from extensive mountain snowfields, leads to loss of pollen. It is suggested that similar losses, both of pollen and of organic sediment recirculated from marginal areas, took place in Windermere and similar lakes during the late-glacial and were particularly intense during Younger Dryas time, when there is both stratigraphic and geomorphological evidence for recrudescence of glaciation in the mountain corries of the Lake District (Pennington, 1964, 1970; Walker, 1965).

BLEA TARN (TYPE 3)

Comparison of absolute and percentage diagrams

This small lake (31 ha) occupies a shallow upland basin among steep and rocky mountains (Fig. 14b). Percentage pollen curves from its complete Late-Weichselian

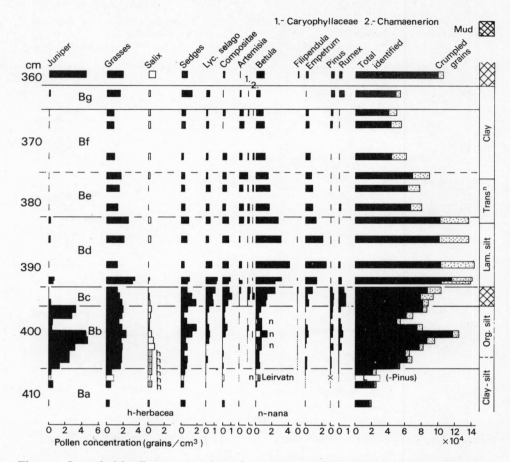

Figure 5. Late-glacial pollen concentrations of selected taxa at Site 3 (Blea Tarn) compared with the surface sample from Leirvatn. Division of (1) *Salix* histogram shows (i) undoubted *S. herbacea* grains, (ii) other grains (2) *Betula* histogram shows (i) undoubted *B. nana* grains (ii) other grains.

profile have been published (Tutin 1969, Pennington 1970) together with chemical data (Pennington & Lishman 1971) which have now been subjected to the multivariate statistical technique termed principal component analysis (P.C.A.) that seeks to extract the principal components (eigenvalues and eigenvectors) from the matrix of correlation coefficients of the chemical variables (Fig. 1). Pollen counts per cm^3 of wet sediment have been made on a new core obtained in 1971, and the diagrams shown in Figs 5 and 6 are for pollen concentration only. The carbon content of the late-glacial sediments is so low (at no level more than 5% dry weight) that critical ^{14}C dating is not possible. However, comparison of the pollen stratigraphy (Fig. 2)

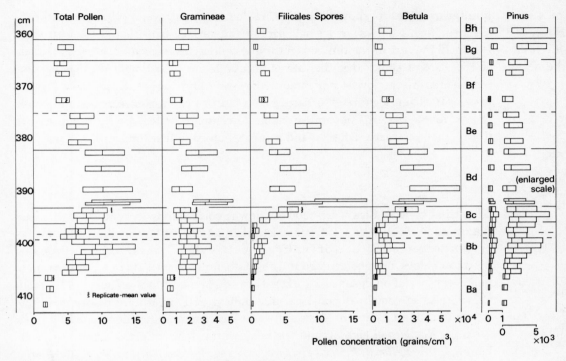
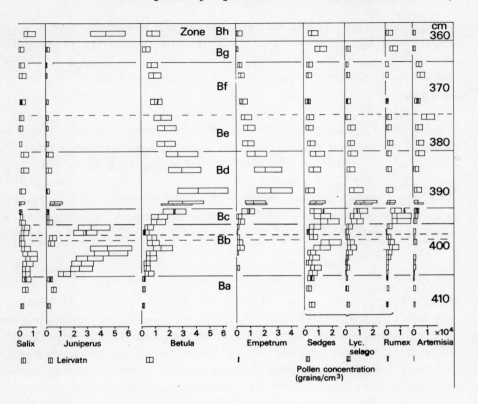

Figure 6. Late-glacial pollen concentration at Site 3 (Blea Tarn) showing the 90% confidence interval for the counts.
(a) Total pollen, Gramineae, *Betula*, fern spores and *Pinus*, (b) Taxa of significance in zoning the diagram.

shows that sediment accumulation rates are unlikely to differ from those of Blelham Bog by a factor of more than two. Conversion to pollen deposition rates would not change the information conveyed by the concentration diagram, except to accentuate the contrast between zone Ba (pre-interstadial) and the rest of the profile.

Problems of zonation of the percentage diagrams are discussed in the publications cited. The highest percentages of *Betula* pollen of tree birch type are found in minerogenic sediment above the late-glacial organic layers, so it was not immediately apparent where the boundaries of Jessen–Godwin zones I, II, and III should be placed. (High percentages of juniper within the lower organic deposits could have been interpreted as recording the interstadial vegetation at this upland site.) It seemed possible that the high percentages of *Betula* in minerogenic sediment resulted from secondary deposition within a period of low absolute local pollen production, so this deposit was at one time referred to a transition zone II/III in an attempt to apply the Jessen/Godwin zonation (see Haworth 1969), and to the 'post-interstadial' time of the Late-Weichselian (Pennington & Lishman 1971). Considerable difficulty was found in attempts to define a series of pollen assemblage zones (see Pennington & Lishman 1971, Fig. 7b). One object of absolute pollen analysis was to find out whether any method of numerical analysis of the data would provide a satisfactory objective pollen zonation, since advice received (P. Greig-Smith, personal communication) stated that there are some methods of multivariate analysis suitable for absolute pollen figures which cannot be used on interdependent percentage figures.

Mean values for pollen concentration are given in Fig. 5, and in Fig. 6 the diagram includes the 90% confidence intervals for the count of each taxon. Four samples were counted twice, the replicate preparation being made from a second sample from the same level in the core, using a different concentration of exotic pollen, to test the errors involved in the experimental procedures. Figure 6 shows that for most taxa the mean concentration in the replicate sample falls within the 90% confidence interval of the first. This is interpreted as showing that, in this type of sediment, experimental error does not materially increase the error inherent in the count. One sample (393-4 cm) showed larger than usual differences between replicates with respect to certain taxa; this is interpreted as evidence for non-homogeneity of the sediment with respect to these taxa at this horizon, which falls at a stratigraphic boundary.

Comparison of Fig. 6 with the published percentage diagrams shows a clearer division into zones in the concentration diagram. Zone boundaries can be drawn readily on inspection, on changes in concentration of both numerically dominant pollen types (Fig. 6a) and those taxa commonly used in zoning late-glacial diagrams (Fig. 6b). The series of zones correlates with that defined at Blelham Bog (Fig. 4) with differences which can readily be attributed, on ecological judgement, to the different altitude of the sites.

Comparison with Leirvatn

In an attempt to interpret late-glacial pollen spectra from the Lake District more objectively in terms of modern vegetation which includes similar taxa, surface samples from mountain lakes of similar size and morphometry in Jotunheim, S. Norway, have been examined (Tutin 1968a). Comparisons of percentage figures have been found to be complicated by the presence of much tree pollen in surface samples from these Norwegian lakes, which lie far above the tree-line in the lower and middle alpine regions (Faegri 1940, 1967) (cf. the findings of Faegri 1945, p. 110). In comparisons of percentage pollen spectra from these surface samples with late-glacial percentage spectra, conifer pollen and pollen of warmth-demanding trees in the surface samples can be discounted, but the genera *Betula* and *Salix* present problems when a pollen sum must be chosen for percentage calculations, because it is not possible to separate the pollen of these genera quantitatively into that of local and of distant plants. Figures for pollen concentration of each taxon, on the other hand, can be compared directly, and the amounts of non-local *Betula* and *Salix* pollen do not affect the figures for other taxa. As Faegri (1945) has pointed out, comparison of pollen concentrations means little if sedimentation rates have been different. Since the sediment of the Norwegian mountain lakes is as low in carbon content as the late-glacial sediments of Blea Tarn, there is no possibility of measuring sedimentation rates by ^{14}C dating. An attempt was made to measure the rate of accumulation by analysis for fall-out products, present in the atmosphere in quantity since 1954, a method which had given promising results when used for surface samples from British lakes (Pennington *et al.* 1973). Analysis (at A.E.R.E. Harwell) of samples of the top cm of sediment from the small lake Leirvatn at 1,350 m in Jotunheim (Plate 1c) showed no measurable Cs-137, suggesting that the post-1954 sediment forms only a very small part of the surface cm—that is, that 16 years is a small fraction of the time taken for one centimetre to accumulate. Comparison of Figs. 2 & 4 shows that the rate of accumulation of sediment during the time represented by the lowest pollen zone, Ba, was very slow at all the Lake District sites; at Blelham Bog three ^{14}C dates gave a rate of 1 cm in 190 years (Pennington & Bonny 1970). It seems not unreasonable to compare pollen concentration figures from Leirvatn surface sediment and from the lowest pollen zone at Blea Tarn.

Zone Ba

This basal zone at Blea Tarn (406 cm down) corresponds to zone Ba at Blelham Bog in its low pollen concentration (less than 25,000/cm^3) contributed mainly by grasses, sedges and *Salix*. It differs from zone Ba at Low Wray Bay and Blelham Bog in its low values, both percentage and absolute, for *Rumex* (Table 3). Comparison of pollen concentration in zone Ba at Blea Tarn and in the surface sediment of Leirvatn shows close agreement if non-local tree pollens at Leirvatn are discounted. The vegetation round Leirvatn consists of snow-bed communities dominated by *Salix herbacea*, *Ranunculus glacialis* and *Anthelia*, with patches of grass-sedge heath containing *Carex bigelowii*, *Festuca vivipara*, *F. ovina*, *Vahlodia atropurpurea* and *Juncus trifidus*. Faegri (1935, 1940, 1953) and Hafsten (1963) have described profiles from S.W. Norway where, in their view, snow-bed vegetation was dominant during the time represented by zone I of the late-glacial. I would now interpret the basal zone at Blea Tarn as indicative of the existence of snow-bed vegetation round that site, at the time when a pollen assemblage containing much more *Rumex* was being deposited at the lowland sites in the Lake District (cf. Table 3).

Zone Bb

The increase (two- to three-fold) in pollen concentration which defines the zone boundary Ba/Bb at 406 cm results mainly from equally large increases in juniper and grass pollen, but total *Betula* (including many grains of *B. nana* type) also increases. The sudden nature of the change in both juniper and total pollen concentration at this horizon (Figs. 4 & 5) supports the interpretation (Pennington & Bonny 1970) of a major climatic amelioration, synchronous at all sites, rather than a progressive and non-synchronous uphill movement of vegetation belts.

The division of the juniper percentage curve, which suggested (Pennington & Lishman 1971) the effects of a temporary climatic recession unfavourable to juniper

Table 3. N.W. England, Basal zone of very low pollen concentration, Ba.

	Percentage total pollen						Pollen concentration (grains/cm³)	
	Blelham Bog a&b		L.W. Bay	Burnmoor Tarn a&b		Blea Tarn Mean of 2 samples	Blea Tarn Mean of 2 samples	Leirvatn (surface mud)
Rumex	14.4	11.1	21.4	27.2	26.6	1.0	304	568
Juniperus	3.0	6.7	3.6	7.2		12.3	3,283	2,355
Betula	8.0	9.2	16.5	9.8	5.2	7.6	2,014	5,278
Salix	2.1	4.6	8.4	8.4	17.2	9.8	2,427	81
Salix cf. *herbacea*	—	—	5.0	0.6	4.0	10.3	2,565	2,842
Gramineae	16.7	15.4	19.5	26.6	22.6	27.4	7,215	9,906
Cyperaceae	4.2	4.0	6.3	6.6	9.2	18.3	4,703	2,274
Artemisia	2.5	1.5	5.0	4.0	4.6	0.5	138	244
Taraxacum type	0.3	0.2	0.5	0.6	2.6	1.0	304	650
Other Compositae	2.1	0.2	0.9	0.6	2.0	1.0	304	1,380
Empetrum	0.1	1.4	1.6	0.6	1.2	0.5	97	893
Rosaceae		0.1	0.5	0.6	0.6			
Filipendula	0.3	0.1						
Caryophyllaceae	0.2	0.1		1.2		0.5	138	81
Chenopodiaceae	0.1			0.6	1.2			81
Rubiaceae	2.0	0.7	3.8	2.6	0.6	1.5	373	
Thalictrum	5.4	6.8	1.4			1.5	415	244
Chamaenerion	0.2	0.1						
Umbelliferae	0.2	0.1						
Labiatae		0.1	0.2					81
Leguminosae	0.1		0.5	1.2	0.6	0.7	166	
Plantago sp.	0.1	0.2						81
Polygonaceae		0.1						81
Cruciferae						0.3	69	81
Ranunculaceae	0.9	0.7	0.5		0.6	0.7	166	81
Helianthemum	0.2		0.2					
Lycop. selago	0.1		0.2			2.35	566	162
Lycop. spp.			0.2					1,786
Selaginella	0.4				0.6	1.4	331	893
Filicales	2.8	1.3	3.0	6.6	4.0	10.8	2,772	12,424
Polypodium	0.1		0.2	0.6	0.6			81
Sphagnum	0.2				0.6			731
Other Bryophytes	205.0	62.0	54.0	79.0	100	22.2	7,936	19,730

Table 3. (continued)

	Percentage total pollen					Pollen concentration (grains/cm³)	
	Blelham Bog a&b	L.W. Bay	Burnmoor Tarn a&b	Blea Tarn Mean of 2 samples	Blea Tarn Mean of 2 samples	Leirvatn (surface mud)	
Corylus	0.1						
Pinus	0.9	0.4	0.2	1.95	884	37,596	
Hippophae		0.2		4.3	1,104		
						Alnus 812	
						Picea 406	
						Ulmus 81	
						Tilia 81	

*Pennington (1970)
a&b indicates two horizons within the zone
'Mean of 2 samples' indicates mean percentage of replicates

at this altitude, is confirmed by the concentration diagram (Fig. 5). Juniper and grass appear to have increased their pollen production simultaneously at the Ba/Bb boundary, appreciably earlier than the herbaceous community represented by *Rumex*, sedge and *Lycopodium selago* (cf. Fig. 6b). All these taxa show a temporary absolute decline at the horizon shown by broken lines in Fig. 6; above this, juniper and grass again increase absolutely before the herbaceous group.

Zone Bc

The zone boundary Bb/Bc is defined by a steep fall in both juniper percentages and concentration. It differs from the corresponding horizon at Blelham Bog in that *Betula*, instead of reaching maximum values immediately (cf. Pennington & Bonny 1970, Fig. 2) shows a steady increase in percentages and concentration throughout zone Bc at the higher site. At Blea Tarn the immediate increase in pollen concentration at the Bb/Bc boundary is from the herbaceous taxa, *Rumex*, sedges and composites, accompanied by *Lycopodium selago*. A rise in *Empetrum*, both in percentages and concentration, through this zone, is interpreted as the results of declining soil base-status (Haworth 1969, Pennington & Lishman 1971).

Within zone Bc, *Pinus* reaches its highest late-glacial pollen concentration; the numbers of grains are so low that changes barely exceed the 90% confidence interval, but the consistency of higher concentrations in zone Bc shows that the deposition rate of *Pinus* was not constant during late-glacial time. Disappearance of *Salix herbacea* pollen shows reduction of snow-beds.

Zone Bd

Maximum concentrations of *Betula*, *Empetrum*, grasses, fern spores and of total pollen are found within the laminated and predominantly mineral sediment previously referred to a II/III transition zone. These high concentrations of *Betula* show that it was a mistake to conclude (Pennington 1970) that the high percentages resulted from low local pollen production. Separate analysis of two contiguous colour laminations showed no significant difference in pollen content (Figs. 5 & 6, 391–2 cm). The numbers of crumpled grains are significantly higher in this zone than any other. *Artemisia* appears in the highest concentration yet found here, together with *Chamaenerion* and grains of *Cerastium* type.

These facts together are interpreted as evidence for the onset of severe winter soil erosion within this catchment (shown by lithology, abundance of crumpled grains, increase in *Artemisia*) during the second part of the period represented by the birch pollen zone (= Alleröd). This strengthens the already published suggestion (Pennington 1970 pp. 54-6) that the effects of falling temperatures can be seen in profiles from N.W. England during the later part of Alleröd time. The course of the first component of the P.C.A. analysis of sediment chemical composition (Fig. 1) indicates that the major lithological change is found at the base of this pollen zone.

Zones Be and Bf

Decreases in concentration of *Betula*, *Empetrum* and grass pollen draw a sharper change, at 382 cm, than is found in the percentage diagram. An increase in concentration of *Artemisia* parallels increased percentages, and total pollen concentration falls significantly. These features were used at Blelham Bog to define zones Be (of declining pollen influx) and Bf (low pollen influx), together correlated with zone III. The concentration of both total *Salix* and of *S. herbacea* type remains low—that is, there is no evidence for any return of snow-bed vegetation (cf. Hafsten's (1963) similar finding at Lista). The low concentration of *Pinus* in zones Be and Bf together with its low deposition rate at Blelham Bog, supports the view (Jessen 1949) that when high percentages of *Pinus* are found in zone III it reflects only low concentration of local pollen.

Figure 4 shows the total pollen concentration in this zone to be twice that at Blelham Bog, and Fig. 2 shows that accumulation of sediment during Younger Dryas time must have been more rapid than at that site. Therefore a significant difference appears between pollen deposition rates during Younger Dryas time at these two sites. A higher contribution of secondary pollen from redeposited soil at the higher site, Blea Tarn, would explain this. Chemical and diatom evidence (Pennington 1970, Haworth 1969) both indicate redeposition of mineral soils at Blea Tarn at this time.

Discussion

LATE-GLACIAL POLLEN DEPOSITION

The steep increase in annual rates of pollen deposition at Rogers Lake began about 12,000 years ago on that time-scale (^{14}C age minus 730 years, the apparent age of the surface sediment), and rose steadily from c 1,000 pollen grains/cm^2/year to a maximum of 50,000 grains about 9,000 years ago. At the three sites in N.W. England the rise of the curve for pollen deposition rates is interrupted, during the ninth millennium B.C., by the Younger Dryas recession. On the Blelham Bog time-scale the first rise begins just after 13,000 years ago; this is the uncorrected ^{14}C date, since in this filled-in kettlehole no correction is possible for apparent age of surface sediment. The rise begins from a lower figure than at Rogers Lake—c 200 grains/cm^2/year. Concentration figures and stratigraphy at the two newly analysed sites indicate pollen deposition rates not significantly different from Blelham Bog. It is not possible to guess whether the difference between 200 and 1,000 grains in pollen deposition rates represents higher pollen production by different vegetation in Connecticut or more efficient pollen concentration by Rogers Lake than any of the English sites.

Comparison of the three English sites shows that similar absolute pollen frequency figures are found in lakes (a) of contrasted morphometry (type 1—small kettlehole, 2—shallow-water area of large lake, 3—small lake 8 m deep), and (b) of different percentage pollen spectra, sites 1 and 2 in the lowlands having 14 to 20% *Rumex* in the grass-sedge-*Salix* assemblage, and site 3 at 200 m having 40% *Salix* (including much *S. herbacea* type) and little *Rumex*.

At the horizon of transition to the spruce zone at Rogers Lake the incoming of trees into the local vegetation did not depress the deposition rates of pollen of herbaceous taxa—grasses and sedges. Similarly at the transition to the juniper zone (Bb) at the English sites an increase is found in almost all taxa except those of snow-bed communities at the upland site (Blea Tarn). This is impossible to explain except by supposing there to have been a great increase in pollen productivity of both juniper and herbaceous taxa consequent on climatic amelioration. It is not possible to invoke any change in pattern of pollen deposition (independent of change in pollen production) which could have produced this effect simultaneously at three lakes so strongly contrasted as the three English sites. This major environmental change must surely be interpreted as the opening of the Late-Weichselian interstadial in N.W. England.

During the interstadial, represented by the deposits of the juniper and birch zones (Bb-Bd), annual pollen deposition rates at Blelham Bog reached 3,000 grains/cm^2/year; comparison of pollen concentration and stratigraphy at the other two sites shows they cannot have differed materially in deposition rates. Deposition rates for *Pinus* remain so low at the English sites that this tree cannot be supposed to have been present in N. England during the interstadial, but the absolute curves for this tree at Blelham Bog and Blea Tarn consistently suggest maximum input during the birch zone, and so correlate better with the northward movement of the pine forest during Alleröd time on the Continental mainland (Firbas 1949, Behre 1967) than do percentage curves for pine in N. Britain (Pennington 1970).

Since there is no pollen zone in N. American profiles which is comparable with the juniper and birch zones of the N.W. European interstadial, no further direct comparison between the English sites and Rogers Lake is profitable. Similarity in absolute pollen frequency between the three English sites of contrasted morphometry suggests that interstadial pollen deposition rates were more closely dependent on pollen production than on factors affecting dispersal and sedimentation in lakes. This being so, it may well be that further absolute pollen analyses of late-glacial sediments, in different regions of N.W. Europe, would be profitable in terms of assessment of regional differences in pollen production; it may for instance, become possible to define such terms as 'park-tundra' and 'birch copses' quantitatively in terms of pollen deposition rates.

The results from Blea Tarn, a late-glacial site hitherto impossible to zone satisfactorily, either on the Jessen–Godwin zonation or by comparison with Blelham Bog, show the value of absolute pollen analysis in producing independent curves for each taxon. Much work remains to be done on numerical analysis of the data from this site. Meanwhile comparison of pollen concentration figures with surface samples from Leirvatn, a lake surrounded by snowbed vegetation, has provided for the first time an interpretation of the vegetation represented by the lowest pollen zone at this lake.

Within the birch zone (Bc) correlated with the first part of the Allerӧd, concentration figures for Blea Tarn reinforce the suggestion of percentage diagrams; at Blea Tarn the steep fall in juniper deposition rates *preceded* the main rise of birch and was separated from it by an episode of high deposition rates for a herbaceous pollen assemblage. In deposits correlated with the second part of the Allerӧd, zone Bd, it is surprising that concentration of total pollen and of birch is consistently higher at Blea Tarn than at Blelham Bog. The laminations within the sediment of this zone at Blea Tarn suggest rapid input due to seasonal solifluction; organic laminae are present at the base of the zone (without significant difference in pollen concentration) and disappear upwards as the sediment becomes a silt of low organic content with colour laminations but no particle-size contrast. This suggestion of rapid input (15 laminations to 1 cm) implies high pollen deposition rates. But the presence of *Empetrum* suggests the presence of areas of open ground, so the high concentration of birch pollen cannot be attributed to high production from continuous birch woodland. It seems more likely that solifluction led to a higher concentration of the pollen from the catchment in the lake, in later Allerӧd time and in the succeeding Younger Dryas time. High concentrations of fern spores, associated with the *Betula* pollen, also suggest input from the catchment (Peck 1973).

Contrasts between pollen deposition rates at the three English sites within the minerogenic deposits of the *Artemisia* pollen zones (Younger Dryas) can be directly related to the palaeoecology of the site. Table 4 gives the estimated annual deposition

Table 4. Pollen deposition rates within deposits of Younger Dryas time. (Calculated from the concentration figures for Low Wray Bay and Blea Tarn by assuming that the deposits of the *Artemisia* pollen zones, Be and Bf, represent the 500 years of Younger Dryas time as they do at Blelham Bog on ^{14}C evidence.)

Pollen/cm²/year at:	Blelham Bog	Low Wray Bay	Blea Tarn
Total pollen	400	less than 100	1,200
Artemisia	71	less than 5	80
Gramineae	63	less than 10	260
Cyperaceae	66	less than 40	68
Pinus	17	—	17

rates for (a) total pollen, (b) *Artemisia*, grass and sedge pollen, representing local pollen, and (c) *Pinus*, representing the distant-transport component. Numbers are too small to compare the components with much statistical significance, but figures for total pollen show differences highly significant in relation to the 90% confidence interval of the count. These indicate loss of pollen from Windermere, with its assumed increase in throughput of water from seasonal melt of permanent snow and ice on its catchment (cf. Vågåvatn); and higher deposition rates at Blea Tarn than at Blelham Bog, attributed to more secondary deposition in the higher relief of the upland catchment. The exact numerical correspondence found in deposition rates for *Pinus* (the distant-transport component) at these two sites is, of course, largely due to chance, but suggests the possibility that the size of this distant-transport (regional) component could be identified reliably by larger counts.

RELEVANCE OF THE NEW APF DATA TO THE INTERPRETATION OF THE LATE-WEICHSELIAN RECORD IN N.W. ENGLAND

The APF data from two further sites in N.W. England have reinforced the conclusion first reached from percentage diagrams and two ^{14}C dates (Pennington 1970) and confirmed by absolute pollen analysis and 10 further ^{14}C dates from Blelham Bog (Pennington & Bonny 1970) that there is, in this region, evidence for only a single Late-Weichselian interstadial in profiles which cover the period of ^{14}C age *c* 15,000 to 10,000 years. This interstadial is defined by pollen concentrations and pollen deposition rates which are significantly higher than those found in underlying and overlying deposits. Deposits correlated with the Allerӧd (Jessen/Godwin zone II) on maximum pollen deposition rates for *Betula*, macroscopic remains of tree birches, and ^{14}C age, form the upper part of these interstadial deposits. The lower part is formed by the deposits of a juniper pollen zone, within which high pollen deposition rates for juniper and all herbaceous taxa represented indicate high pollen productivity of the vegetation compared with underlying deposits. On this evidence the major environmental change occurred at a horizon corresponding to the lower boundary of the juniper zone.

It has been suggested (Tutin 1968b, Pennington & Lishman 1971) that this sequence can be reconciled with the Continental sequence of two interstadials, Bölling and Alleröd (Iversen 1954, Van der Hammen *et al.* 1967) by correlating the base of the juniper zone with the onset of the Bölling amelioration, and by accepting that, within the oceanic climate of western Britain, there was no recession of climate or vegetation, at least in the lowlands, during the Bölling-Alleröd interval, termed zone Ic in Western Europe (cf. the ill-defined nature of zone Ic in S.W. Norway (Chanda 1965)). A succession of pollen assemblage sub-zones within the percentage juniper zone at the upland site, Blea Tarn, suggested (Pennington & Lishman 1971, Table 7) that it could record a minor climatic recession, not severe enough to cause solifluction (see analysis of sediment composition, Fig. 1) at a time which could well correlate with the ^{14}C dates for the Bölling-Alleröd recession (Fig. 4). Comparison of the absolute pollen figures for Blea Tarn supports this suggestion, for there is a significant temporary decrease in pollen deposition rates within the juniper zone, at the horizon where an increase in *Betula nana* type pollen suggests a climatic recession of small amplitude, comparable with the change in temperature and length of snow-lie which differentiates between communities of juniper and of *Betula nana* with increasing altitude on Norwegian mountain-sides. This is the only evidence found as yet for any division of the juniper zone in N.W. England.

Absolute diagrams from Blelham Bog and Blea Tarn, in which the amounts of juniper and of birch pollen are not interdependent, illustrate two different sequences. At Blelham Bog there is, at the zone boundary Bb/Bc, a steep fall in annual deposition rate of juniper as the *Betula* pollen reaches maximum deposition rates; this was attributed to interaction between the species, tree birches shading out juniper (Pennington & Bonny 1970). Moore's (1970) percentage diagram from the Elan valley in Mid-Wales, and Birks' (1964) diagram from Chat Moss, South Lancashire suggest that APF analysis there would give the same result. At Blea Tarn, by contrast, the absolute fall in juniper deposition precedes the rise of *Betula* to a maximum, and appears to be the result of some independent factor. The rise of the *Betula* deposition rate is at Blea Tarn delayed by contrast with Blelham Bog and Low Wray Bay.

It seems clear, however, that within the area of the British Isles, there was strong differentiation of the pollen-producing vegetation within Late-Weichselian time, and that regional differences may be more easily defined when absolute pollen analyses are more widely practised. No longer can we apply a simple interpretation of the northward movement of forest belts within Alleröd time. But it is my belief that it will prove possible to reconcile the sequence of late-glacial climatic changes in Britain with the sequence which has been widely demonstrated on the Continental mainland (Iversen 1954).

Absolute pollen frequency in lake sediments at c 5000 B.P. (3000 B.C.) in N.W. Britain

INTRODUCTION

Comprehensive reviews have been published both on the nature and interpretation of the 'elm decline' in northern Europe (Smith 1961) and on deductions as to climatic change which have been drawn from pollen diagrams (Iversen 1960, Troels-Smith 1960, Frenzel 1966). More recently Hibbert, Switsur & West (1971) have reviewed the ^{14}C-dating of this horizon in northern Europe, and have shown that the *Ulmus* decline is broadly synchronous.

In N.W. England, consistent changes in sediment composition were found in lake profiles at the horizon where percentages of elm pollen fall steeply (Mackereth 1965, 1966, Pennington 1964, 1970) and were attributed to accelerated soil erosion consequent on disturbance of primary mixed-oak forest by man. In this region human settlement is evidenced by the presence of artifacts ^{14}C-dated from 3010 B.C. onwards (Piggott 1954). Five samples from the elm decline radiocarbon dated in Copenhagen showed (i) that in three of the lakes of N.W. England the elm decline is synchronous and occurs just before 3000 B.C. (Tauber 1966, 1968), and (ii) that in Blea Tarn (Fig. 14b) dating of three contiguous samples indicated a uniform rate of sediment accumulation at, and just below, the elm decline—a promising situation for absolute pollen counts. Detailed chemical analysis by J.P. Lishman then showed (Tutin 1969) evidence for accelerated inwash and deposition rates just above the elm decline, so a new series of ^{14}C dates was obtained, including more of the profile above the elm decline, before absolute pollen frequencies were estimated. There was very good agreement between the two dating series. Absolute counts for this horizon have also been made by A.P. Bonny on a core from the South Basin of Windermere, and converted into pollen deposition rates by use of the accumulation rates provided by a series of ^{14}C dates on a parallel core (Mackereth 1971).

This horizon has now been investigated further by exploration of the sediments of lakes in the N.W. Highlands of Scotland, an area of base-poor Pre-Cambrian bedrock where percentage pollen analysis had shown that there had at no time in the post-glacial period been any significant expansion of pollen of deciduous trees of the mixed-oak forest. Comparison of chemical, pollen and diatom evidence from these lakes (Pennington *et al.* 1972) showed that leaching proceeded rapidly in this environment and that most soils were acid before the post-glacial immigration of birch forest at *c* 7000 B.C. This prevented any development of a 'mesocratic' stage (Iversen 1958) with deciduous forest. Pine-birch forest of the 'telocratic' stage succeeded birch-hazel (with low hazel percentages) at a date between 6000 and 5000 B.C. Retrogressive vegetational succession (Iversen 1964, 1969) then led to the development, accelerated growth, and swamping of mor soils which passed into blanket-bog (Iversen 1964, pp. 67–9) except in areas of high relief where free-draining soils persisted. Only in such areas did forest survive.

Since, in the region of mixed-oak forest including most of Britain at 3000 B.C., traces of forest clearance and land use by Early Neolithic man complicate the interpretation of clues to climatic change in the vegetation record, it seemed profitable to investigate what happened at this time in the different environment of N.W. Scotland, where there are few or no dated human artifacts of this period. Even if men were present, no economy based on exploitation of the elm could be envisaged in a country of pine-birch forest where no pollen diagram showed values for *Ulmus* higher than 4 to 5% of arboreal pollen, and most were 2–3%. Chemical analysis, and subsequent statistical analysis using P.C.A., revealed significant changes in sediment composition, indicative of increased waterlogging of soils and the first inwash of peat, at the horizon where these low percentages of *Ulmus* first fell to 1% or absence (Fig. 7); this horizon was ^{14}C dated to just before 3000 B.C. in two lakes (Pennington *et al.* 1972). It seemed probable that this percentage decline indicated a fall in *Ulmus* in the regional British pollen rain at this date, and that measurement of pollen deposition rates of *Ulmus* would give a measure of the regional component, which could then be used in attempts to apply the absolute pollen frequency technique to 'the problem of the areal extent of Neolithic forest clearance in western Europe' (Wright 1971).

Choice of lakes for this purpose was restricted to those where sediment appeared to have accumulated fairly evenly over the lake bottom. Variation in pollen concentration between lakes, similar to that observed in the Lake District (p. 79) became apparent, and one pollen-rich and one pollen-poor lake were studied—respectively Loch Clair (Fig. 15a), 33 m deep and 65 ha, and Loch a'Chroisg (Fig. 15b), 50 m deep and 259 ha. The two lakes are 10 km apart; percentage pollen analysis showed similar pine-birch-dominated pollen spectra until *c* 2250 B.C. after which taxa of bog communities replaced pine at Loch a'Chroisg.

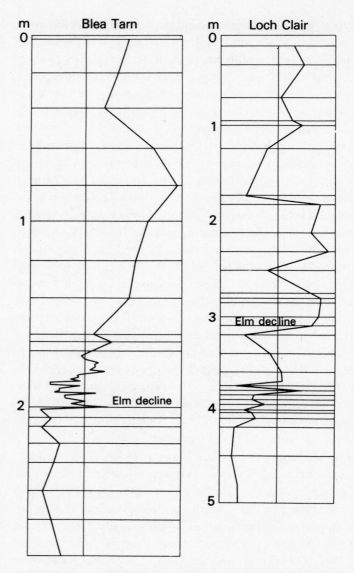

Figure 7. Post-glacial profiles: Blea Tarn and Loch Clair. First component of a Principal Component Analysis, as in Fig. 1—showing the horizon of the elm decline as determined by pollen analysis.

THE CHANGES IN POLLEN DEPOSITION AT THE ELM DECLINE AT THE FOUR SITES

Figure 8 compares annual deposition rates of *Ulmus*, other major trees, and total arboreal pollen at the four sites investigated, showing the 90% confidence interval for the counts; percentages of *Ulmus* of total arboreal pollen are also shown, with the 95% confidence interval given by Faegri and Iversen (1964, p. 126). Figure 12 compares size of lake and total pollen deposition rates c 3500 B.C. Figures 14a & 14b, 15a, 15b compare the catchments, and the percentage pollen spectra at c 3500 B.C. for Blea Tarn, Loch Clair, and Loch a'Chroisg are given in the pollen diagrams in Figs. 9, 10, and 11. Windermere and Blea Tarn lie within a region of present semi-natural mixed oak-birch-hazel-ash woods with a little *Ulmus glabra*; the catchment of Blea Tarn was cleared only within the last 100 years (Pennington 1964). Lochs Clair and a'Chroisg lie within a region where oak, elm and hazel have never, during the post-glacial period, reached percentages indicative of local presence, and the prevailingly base-poor soils of this region suggest that, within a radius of 25 km, there can never have been any sites for *Ulmus glabra*. Table 5 gives figures for deposition rates at the sites at c 3500 B.C. and after the elm decline.

The striking comparison is in the deposition rates for *Ulmus* pollen. In Windermere, where its percentage of A.P. below the elm decline is 10-15%, annual deposition rates are only 200-300 grains/cm^2/year. At Loch Clair, where *Ulmus* never exceeds and only rarely attains 5% A.P., which many interpreters would not regard as indicative of local presence, the deposition of total pollen is so much greater than in Windermere that 700 grains of *Ulmus* pollen were being deposited cm^2/year. In Loch a'Chroisg, near to Loch Clair, and with similar percentage A.P. spectra at c 3500 B.C., *Ulmus* made up 4% of A.P., and less than 100 grains/cm^2/year were being deposited. Maximum rates of deposition of *Ulmus* pollen at c 3500 B.C. (1700 grains/cm^2/year) are found in Blea Tarn, the pollen rich lake where *Ulmus* made up c 15% of the arboreal pollen.

The percentage fall in *Ulmus* pollen at the elm decline is at all four sites significant in terms of the 95% confidence interval (Fig. 8). At Blea Tarn this fall has been recorded in precisely similar form in three percentage diagrams, from replicate cores from sites as near to each other as it is possible to fix on water. ^{14}C dates, from two of these cores, for the sample of 10 cm thickness, the upper boundary of which is the steep fall in percentage *Ulmus*, gave dates of 3370 B.C. ± 120 (K-958) and 3285 B.C.

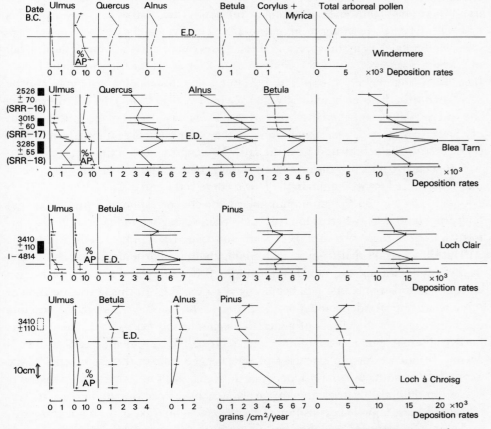

Figure 8. Pollen deposition rates at the elm decline (c 3000 B.C.) at four lake sites compared: Windermere, Blea Tarn, Loch Clair, Loch a'Chroisg. *Ulmus* also shown as percentage Arboreal Pollen.

Table 5. Pollen deposition rates at elm decline compared.

		Total pollen (pollen/cm^2/year)	*Ulmus* pollen
Before elm decline (c 3500 B.C.)			
Alder—oak—birch—elm zone,	Windermere (260 cm)	2,872	200
Ulmus 10-15% A.P.	Blea Tarn (160 cm)	23,215	1,716
Pine—birch—alder zone	Loch Clair (312 cm)	17,626	730
Ulmus 2-5% A.P.	Loch a'Chroisg (460 cm)	6,036	90
After elm decline (c 3000 B.C.)			
	Windermere	4,750	50
	Blea Tarn	23,400	350
	Loch Clair	22,000	170
	Loch a'Chroisg	3,830	24

±55 (SRR-18). The precision with which this horizon is recorded in Blea Tarn argues against appreciable disturbance by bottom fauna at the time.

The absolute fall in *Ulmus* pollen in terms of grains/cm^2/year is at all four sites significant in terms of the 90% confidence interval. At Loch Clair, concentration figures were counted on a replicate core, of stratigraphy identical with the dated core, and gave satisfactory results (see Fig. 10).

The Scottish data show that a numerically significant fall in absolute pollen deposition of *Ulmus* is found, just before 3000 B.C., in a region of Scotland where percentage arboreal figures for *Ulmus* do not exceed 5%. In the absence of suitable habitats for *Ulmus glabra* from an area of radius 25 km from each site, it is concluded that this records a fall in the regional British pollen rain of *Ulmus* at that date. The deposition rate of the regional component can be assessed from Loch a'Chroisg, a pollen-poor lake where the concentration of pollen is low, at just under 100 grains/cm^2/year of *Ulmus* pollen. In the nearby Loch Clair, concentration within the lake of the regional component produced a pollen deposition rate for *Ulmus* seven times as great.

Further work in Scotland will examine the validity of this conception of the regional component by measuring deposition rates of *Ulmus* pollen in other lakes, of different size and shape, at varying distances from outcrops of the Durness Limestone which give rise to areas of base-rich brown earths with base-demanding plants amid the general sea of peat and mor soils of N.W. Scotland. It seems possible that base-demanding trees, such as oak, elm, and hazel, may have grown in such habitats during the mid-post-glacial period.

Meanwhile, this comparison of these four sites with respect to *Ulmus* deposition rates must be taken as a warning against attempts to apply the technique of absolute pollen analysis 'to the problem of the areal extent of Neolithic forest clearance in western Europe' (Wright 1971), unless lake sites are carefully standardised for pollen-collecting efficiency. Figure 12 indicates a correlation between lake size and pollen deposition rates at c 3500 B.C. which bears out the suggestions of Davis (1967) and Tauber (1965) that more pollen grains per unit area enter small lakes than large ones. As yet, however, we have little information about the critical size of lake, and what other factors, such as throughput of water, affect its efficiency in concentrating pollen. The two pollen-poor lakes contrast with the pollen-rich lakes not only in size, but in shape. The pollen-poor lakes are 'river-lakes'—elongated valley troughs coincident with the course of a major stream. The pollen-rich lakes are more or less circular basins, one on the course of a major stream, but which does not flow through it in a direct course. Many more sites must be investigated.

CONTRIBUTION OF APF TO LOCAL VEGETATIONAL HISTORY AT EACH SITE

Blea Tarn

Changes in absolute pollen deposition rates are given in Fig. 9a and b for *Ulmus* and the other major taxa, and compared in detail with percentage values for the same counts.

Facts emerging from the APF analysis are:

1. There was a significant fall in deposition rate of *Ulmus* beginning at c 4424 B.C. associated with falls in *Alnus* and *Corylus*. No change in deposition rates of grasses or any other herbaceous taxon accompanied this fall. The decrease in *Alnus* and *Corylus* does not exceed the 90% confidence interval, but the consistency of the trends indicates significance; the same is true of total arboreal deposition rates. This first fall in *Ulmus*, reaching a minimum at 172 cm, is significant in terms of percentages of total A.P. in this core, and this horizon can be correlated with the first fall in *Ulmus* percentages within the Godwin zone VIIa at Lake District sites including Blea Tarn (Pennington 1964). The beginning of this fall in *Ulmus* may also be correlated with the zone boundary VIIa i/ii which Walker (1965) drew at the nearby site of Langdale Combe, where he expressed *Ulmus* as a percentage of the sum of dry land trees, *Ulmus*, *Pinus*, and *Quercus*, thereby producing an exaggerated percentage *Ulmus* fall compared with the fall when expressed as a percentage of total A.P. (which also included the absolutely falling *Alnus* in the pollen sum). It appears that Walker used the minimum in the *Ulmus* curve at the end of this fall as his VIIa/b boundary; on Fig. 9b this minimum falls at 170 cm corresponding with a date not long after 4000 B.C., whereas the true elm decline is higher and later. The total information from these two sites, Blea Tarn and Langdale Combe, confirmed by the APF at Blea Tarn, is that *Ulmus* was declining, both proportionately and absolutely, in these hills, during the fourth millennium B.C., and that no traces of human activity accompany the first fall, beginning at c 4400 B.C. In this fall, association of three relatively base-demanding species, *Ulmus*, *Alnus* and *Corylus*, in the absolute diagram, suggests the effects of generally declining soil base-status under the influence of leaching in a period of wet climate. (It should be remembered that there is chemical evidence for the extension of waterlogged soils in the catchments of the Scottish lakes studied, from c 4300 B.C.) It seems likely that Mesolithic man was present in the Lake District, though no artifacts have been found. There is no significant increase in pollen deposition of grasses, so no increase in the amount of open ground can be supposed, and this is borne out by the complete stability of the deposition rates of *Quercus* between 4400 and 3300 B.C.

2. The second fall in *Ulmus*, to an absolute minimum both absolutely and as per-

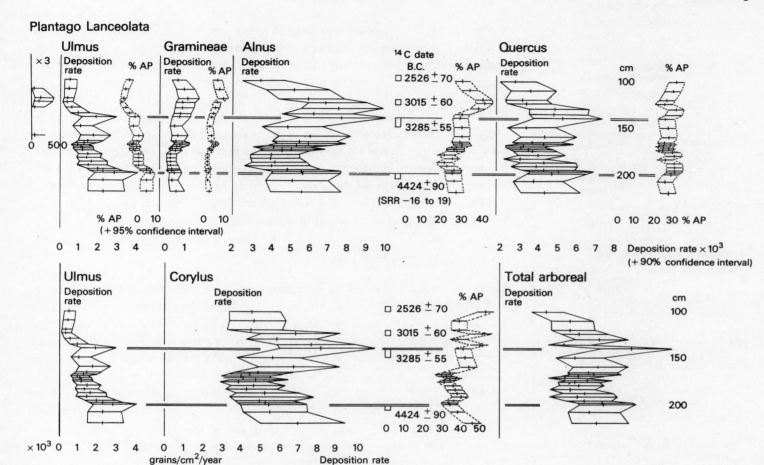

Figure 9a. Detail of pollen percentages and deposition rates at the elm decline at Blea Tarn; a. and b. ^{14}C dates from the same core.

centage total A.P. (Fig. 9b) at a date between 3285 ± 55 and 3015 ± 60 B.C., is accompanied not only by simultaneous increases in the deposition rates of grasses and *Plantago lanceolata*, but by falls in the deposition rates of *Pinus* and *Betula*, which latter are *not* apparent in the percentage diagrams. This is highly significant in the local context. Samples of wood charcoal associated with chipping floors, where stone (Cumbrian) axes appear to have been manufactured on a large scale on the hills surrounding the head of Great Langdale (Fig. 14b) have been dated recently at 2524 ± 52 B.C. and 2730 ± 135 B.C. (BM 676 & 281) (Clough 1973). This means that a decrease in pine and birch, trees which can be supposed to have been growing on the uplands (including the plateau at c 1700 ft (570 m) where the charcoal samples were found) occurred at the same time as the occupation of that plateau by the axe-makers, and coincides with the centuries immediately following the fall in *Ulmus* deposition rate, and with the increase in grasses and *Plantago lanceolata*, in the woods round Blea Tarn at c 600 ft (200 m).

These APF results do not therefore explain any further than hitherto the synchronous decline in elm pollen just before 3000 B.C., but show clearly that at this site the absolute decline in *Ulmus* took place in two stages, beginning at c 4400 B.C., and that both the first and the second stage were associated with absolute decreases in pollen deposition from other trees. This clear association makes it less likely that the fall in pollen production by the elm was the result of an isolated factor such as

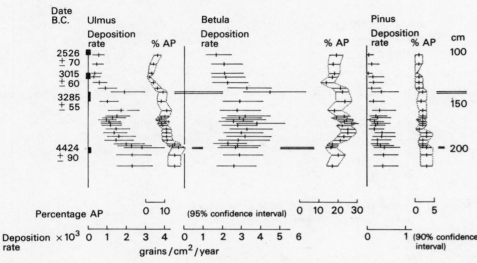

Figure 9b.

disease, affecting only the elm. The absolute increase in *Alnus* just above the elm decline is interpreted as the effect of the increased run-off which was demonstrated on chemical grounds.

Loch Clair

Figure 10 presents most of the APF information for this site. Pollen deposition rates were calculated from the time-scale (Fig. 13) provided by 6 ^{14}C dates. Percentage pollen analysis was carried out on closely spaced samples to examine a layer of distinct stratigraphic appearance at 300 cm which included coarse plant detritus and charcoal fragments. This layer was distinguished by high percentage values for *Calluna* pollen. The sample for dating and the samples for APF were chosen carefully to avoid this material of distinct composition suggesting inwash of soil material.

The diagrams for pollen deposition rates and concentration show that the absolute increases in *Calluna*, *Myrica* (+ some *Corylus*), and grasses which are found just above the elm decline do not disturb the pollen deposition rates for *Pinus* and *Betula*. This was therefore not a clearance of pine-birch forest, but an episode characterised by accelerated inwash of material containing both the coarse detritus and the pollen types to be expected in the soil of damp pine-woods. There was charcoal in this layer, but no significant fall in pine or birch suggestive of extensive forest fire is present. One grain of *Plantago lanceolata* was found, c 20 cm above the *Calluna* layer. It is not therefore possible to say with certainty whether the charcoal was naturally produced by lightning strikes or was the result of the presence of man, but if man was present, it must have been as a hunter and not to any extent as a pastoralist, for there is insufficient evidence to prove any anthropogenic effect on vegetation.

The replicate samples from a second core show that both above and below the elm decline the mean concentrations of pine, birch and elm pollen fall well within the 90% confidence interval for samples from the first core, whereas the concentration of *Alnus* pollen is significantly different in the replicate samples, indicating a difference between the cores (from closely neighbouring sites) with respect to this taxon. This suggests the possibility that *Alnus* pollen is not distributed uniformly over the bottom of the loch at this horizon in the sediment.

Loch a'Chroisg

Figure 11 shows the curves for the major taxa at this site, both as percentage A.P. and as pollen deposition rates, within that section of the core between c 4500 B.C. and 1010 B.C., with one higher sample. Three ^{14}C dates on this core give sediment accumulation rates for this section, and two lower dates have been transferred by correlation of the *Alnus* rise and the *Ulmus* fall from the neighbouring Loch Clair.

This core contributes little to our knowledge of the elm decline beyond the estimate of the deposition rate of *Ulmus* from the regional component, but its interest lies in its contribution to the history of the replacement of pine by the taxa of blanket-bog in N.W. Scotland. It is clear that at this site there has been a continuous decline in deposition rates of pine pollen since c 4500 B.C. This is correlated with the rise in water tables and development of peat on lake catchments of N.W. Scotland since c 4500 B.C., accelerated at 3000 B.C., which has been demonstrated on independent chemical grounds (Pennington et al. 1972). The elm decline is seen as occurring within a period of steeply declining deposition rate of *Pinus* pollen. A period of arrested decline or even recovery of *Pinus* between the dates of 2199 B.C. and 1633 B.C. is followed by a final steep fall. *Calluna*, *Myrica*, and grasses show uniformly increasing deposition rates as *Pinus* falls; these taxa are interpreted as produced by ground vegetation of increasingly waterlogged pine forests (cf. parts of the Coulin Forest today). As at some other sites in N.W. Scotland, including Loch Sionascaig, the final end of the pine forest corresponds both with the presence of continuous *Plantago lanceolata* pollen (at a time in agreement with the known occupation of part of this country by the people who built the chambered cairns (Daniel 1962)) and with the expansion of Cyperaceae which is interpreted as the final replacement of degenerating pine forest by open blanket-bog with

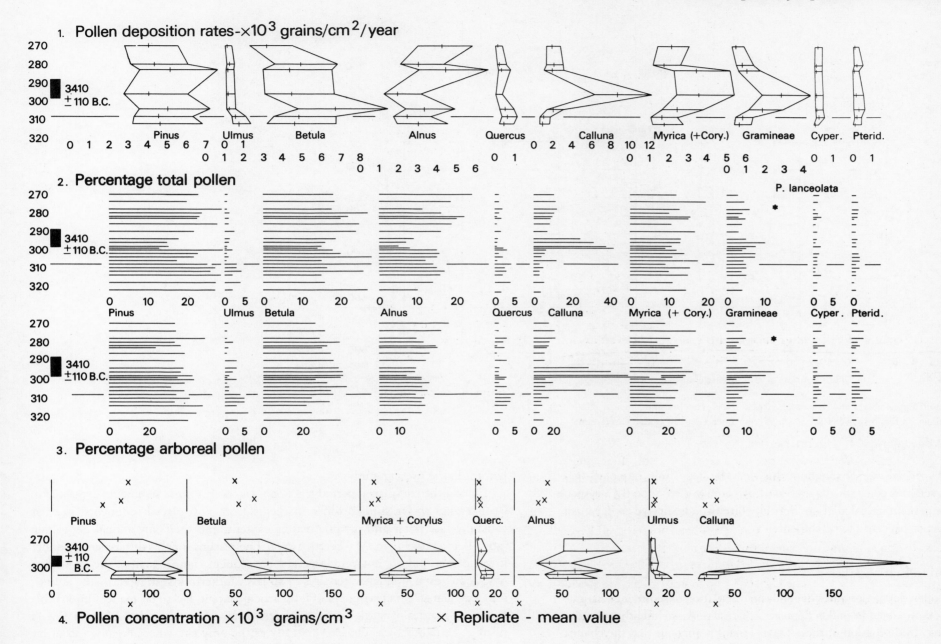

Figure 10. Detail of pollen percentages, concentration and deposition rates at the elm decline at Loch Clair, including replicate analyses on a second core. ^{14}C dates from the first core.

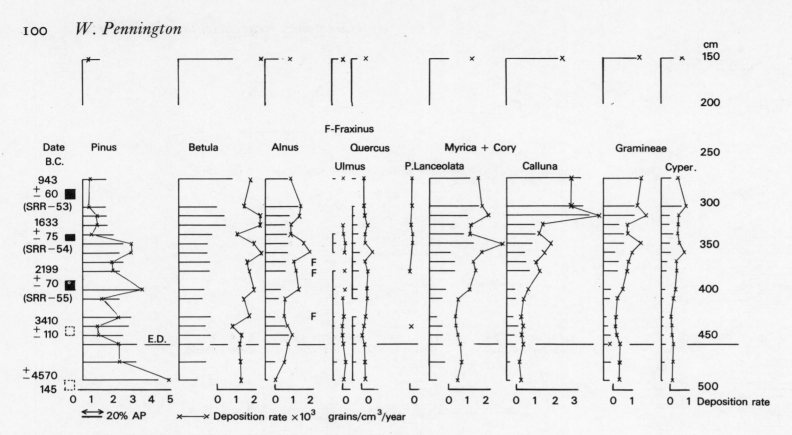

Figure 11. Detail of pollen percentages and deposition rates at the elm decline and above at Loch a'Chroisg. ^{14}C dates on this core in black, ^{14}C dates from Loch Clair, by pollen correlation, broken lines.

Trichophorum and *Eriophorum*. Throughout this time the deposition rate of *Betula* in Loch a'Chroisg remains constant; its increased percentage after 1700 B.C. is seen as an artifact of the percentage calculation. Scrubby birchwoods survive on dry flush habitats on the steeper parts of the catchment.

Pollen deposition in British lakes

Measurement of pollen deposition rates depends on calculation of sediment accumulation rates, so errors involved in pollen deposition estimations include the inherent errors of ^{14}C dating. In discussion of these Davis (1969) points out that the change in slope of the matrix accumulation curve at Rogers Lake at 2 m depth (2,500 years B.P.) is brought about by correction of older ^{14}C dates for changes in concentration of ^{14}C in the atmosphere: 'the difference in slope, however, is too small to be important in the present study'.

During the late-glacial period (c 15,000–10,000 B.P.) the sediments deposited in British lakes so far studied differ markedly from post-glacial sediments, in their generally low water and organic matter content, and their low annual deposition rates (Pennington 1970, Mackereth 1971). The reasons for this difference are under investigation, but are not yet clear. The consequent compression of the deposition of c 5,000 years, at open-water sites in all lakes hitherto investigated by the author, into less than 40 cm of sediment of low carbon content, has so far made it impossible to obtain a series of late-glacial ^{14}C dates for an open-water site. Blelham Bog, a filled-in kettlehole, was chosen for study as the only site where organic (20% dry weight = carbon) muds of late-glacial age, accessible to a wide-diameter peat-borer, were available. A wide-diameter aquatic sampler is under construction for use in Low Wray Bay, Windermere, in order to compare (i) the ^{14}C age of pollen zone

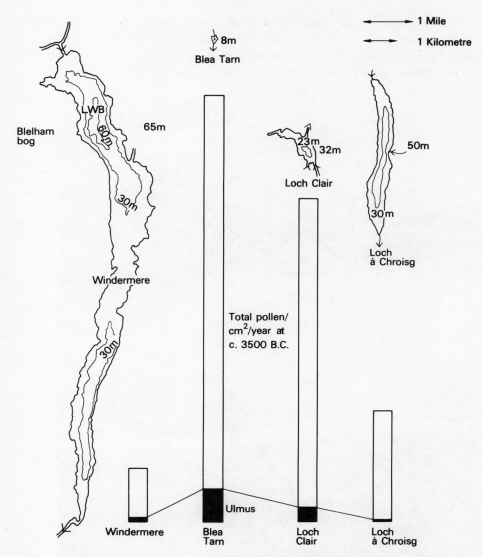

Figure 12. Comparative size of the four post-glacial sites (and Blelham Bog) together with histograms representing i. Total pollen and ii. *Ulmus* pollen/cm²/year, at *c* 3500 B.C. (before the elm decline).

boundaries, and (ii) pollen deposition rates, in these two neighbouring basins of such contrasted size. Results presented in this paper indicate that there is no significant difference in pollen deposition rates, except during Younger Dryas time, between lakes of different size. The spatially uneven deposition of organic sediment in large

Absolute pollen frequencies in lake sediments 101

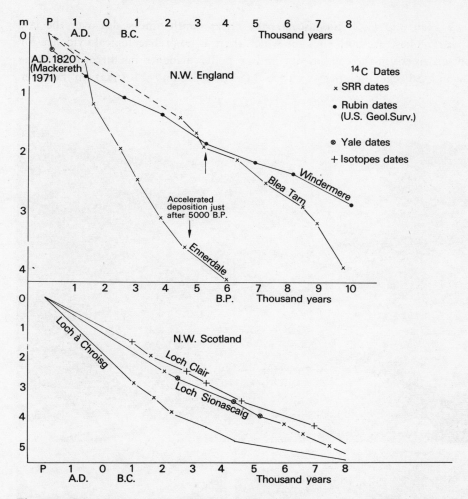

Figure 13. Post-glacial depth-time-scales at the sites. (For late-glacial depth-time-scale at Blelham Bog see Pennington & Bonny 1970, Fig. 1).

lakes with mountain catchments suggests the possibility of seasonal melt-water floods leading to loss of resuspended material down the outflows.

Figure 13 shows the post-glacial depth-time-scale of six British lakes; the dates for Windermere are corrected for changes in atmospheric ^{14}C (see Mackereth 1971) and the others are uncorrected. As yet the post-glacial dated samples from Britain are too few to provide anything but the rather crude approximation to sediment accumulation rates derived from a straight line between the means of neighbouring samples. Even so, comparison of these six depth-time-scales shows significant contrasts relevant to the palaeoecology of the sites.

Loch Clair and Loch Sionascaig have linear depth-time-scales over the last 9,500 years. These are both sites for which there is no archaeological evidence for human settlement in the catchments, and no pollen evidence for direct human influence on local vegetation before c 1500 B.C. (Pennington et al. 1972). In the other four lakes the sediment accumulation rate increases, steepening the slope of the depth-time-scale, at c 3000 B.C. At the three Lake District sites this increase is attributed to increased erosion of predominantly mineral soils (free-draining brown forest soils) from catchments which carried mixed-oak-forest. At Loch a'Chroisg

Figure 14. a. The southern Lake District from the air, including Blelham Bog & Low Wray Bay (arrowed) (C.H. Wood, Bradford, Ltd). b. Blea Tarn from the Langdale Pikes (M.A. Todd). c. Leirvatn (W. Pennington).

Figure 15. a. Loch Clair, showing Mackereth corer emerging (B. Walker). b. Loch a'Chroisg (B. Walker).

the increase is attributed to increased input of organic soils after 5,000 B.P. Similarly, organic soils (mors and peats) clearly developed on the catchments of Loch Clair and Loch Sionascaig during the last 5,000 years, but this did not lead to increased deposition rates in these lakes. All the evidence points to an environmental change at c 3000 B.C. leading to accelerated formation of blanket peat in N.W. Scotland and accelerated erosion of mineral soils in N.W. England. Thereafter an integration of local topography and human history appears to have determined whether or not these soil changes on the catchments led to increased deposition rates in the lakes. The influence of subsequent human history of the catchments is seen in the greatly steepened depth-time-scale for Ennerdale Water sediments between 0 and A.D. 600, a time of extensive forest clearance and cultivation of upland soils (Pennington 1970, Roberts, Turner & Ward 1973), and in the steep increase in the Windermere deposition rate since c A.D. 1820, attributed to cultural eutrophication (Pennington 1943, Mackereth 1971). In any consideration of pollen deposition rates in upland British lakes, soil history on the catchments and its effect on sediment accumulation rates must be taken into account, for in the prevailingly acid soils of the second part of the post-glacial period in upland Britain, pollen preservation has been good, and therefore increased soil erosion has accelerated the pollen deposition rates from the catchments.

For post-glacial sediments, analysis of the horizon at c 5,000 B.P. from four lakes of contrasted size has indicated a decrease in pollen deposition rates with increasing size of lake, and a possible correlation with other morphometric parameters. Comparison of absolute and percentage figures for *Ulmus* pollen at this horizon shows that, within this series of four lakes, the *Ulmus* deposition rate must have been determined by the relative efficiency of each lake as a pollen trap and not by the size of the local population of *Ulmus glabra*.

Acknowledgements

I am grateful to Dr Johs. Iversen for accepting for dating in Copenhagen the first samples from the elm decline in N.W. England. Most of the radiocarbon dates used in calculation of the pollen deposition rates given in this paper have been determined by Dr D.D. Harkness of the N.E.R.C. Dating Unit, Scottish Research Reactor Centre, to whom my thanks are due. I am also grateful to Prof P. Greig-Smith for the multivariate analysis of chemical data, Dr Margaret Davis for much helpful discussion of absolute pollen analysis, and Miss A.P. Bonny for the development of a laboratory method which gives consistent results. For help with field work I am much indebted to Miss E.Y. Haworth, the coring team of the Freshwater Biological Association (Mr B. Walker, Mr W. Askew and Mr P. Cubby) and the Brathay Exploration Group, particularly to Mr A.B. Ware, Principal of Brathay.

References

BEHRE K-E. (1967) The late glacial and early post glacial history of vegetation and climate in Northwestern Germany. *Rev. Palaeobotan. Palynol.* **4**, 149-61.

BIRKS H.J.B. (1964) Chat Moss, Lancashire. *Mem. Proc. Manchr. lit. phil. Soc.* **106**, 1-24.

BIRKS H.J.B. (1970) Inwashed pollen spectra at Loch Fada, Isle of Skye. *New Phytol.* **69**, 807-20.

BONNY A.P. (1972) A method for determining absolute pollen frequencies in lake sediments. *New Phytol.* **71**, 391-403.

CHANDA S. (1965) The history of vegetation of Bröndmyra: a late-glacial and early post-glacial deposit in Jaeren, south Norway. *Årbok Univ. Bergen (Mat.—Naturv. Ser.)* **1**, 1-17.

CLOUGH T.H.McK. (1973) Radiocarbon dates from a Langdale chipping site. *Antiquity* (in press).

COLLINS V.G. (1970) Recent studies of bacterial pathogens of freshwater fish. *Wat. Treat. Exam.* **19**, 3-31.

CRAIG A.J. (1972) Pollen influx to laminated sediments: a pollen diagram from northeastern Minnesota. *Ecology* **53**, 46-57.

CUSHING E.J. (1967) Evidence for differential pollen preservation in Late Quaternary sediments in Minnesota. *Rev. Palaeobotan. Palynol.* **4**, 87-101.

DANIEL G. (1962) The Megalithic Builders. In *The prehistoric peoples of Scotland* (ed. by S. Piggott), pp. 39-72. Routledge and Kegan Paul, London.

DAVIS M.B. (1967) Pollen accumulation rates at Rogers Lake, Connecticut, during late- and post-glacial time. *Rev. Palaeobotan. Palynol.* **2**, 219-30.

DAVIS M.B. (1968) Pollen grains in lake sediments: redeposition caused by seasonal water circulation. *Science* N.Y. **162**, 796-99.

DAVIS M.B. (1969) Climatic changes in southern Connecticut recorded by pollen deposition at Rogers Lake. *Ecology* **50**, 409-22.

DAVIS M.B., BRUBAKER L.B. & WEBB T. (1973) Calibration of absolute pollen influx. (this volume).

DAVIS M.B. & DEEVEY E.S. (1964) Pollen accumulation rates: estimates from late-glacial sediment of Rogers Lake. *Science* N.Y. **145**, 1293-5.

FAEGRI K. (1935) Quatärgeologische Untersuchungen im westlichen Norwegen 1. Uber zwei präboreale Klimaschwankungen in südwestlichsten Teil. *Bergens Mus. Årb. (Naturv. Ser.)* **8**, 1-38.

FAEGRI K. (1940) Quartärgeologische Untersuchungen in westlichen Norwegen 2. Zur spätquartären Geschichte Jaerens. *Bergens Mus. Årb. (Naturv. Ser.)* **7**, 1-201.

FAEGRI K. (1945) A pollen diagram from the sub-alpine region of central south Norway. *Norsk geol. Tidsskr.* **25**, 99-126.

FAEGRI K. (1953) On the peri-glacial flora of Jaeren. *Norsk geogr. Tidsskr.* **14**, 61-76.

FAEGRI K. (1967) *The plant world at Finse, Norway.* Univ. Bot. Mus., Bergen.

FAEGRI K. & IVERSEN J. (1964) *Textbook of pollen analysis* Blackwells, Oxford.

FIRBAS F. (1949) *Spät- und nacheiszeitliche Waldgeschichte Mitteleuropas nördlich der Alpen* Fischer, Jena.

FRANKS J.W. & PENNINGTON W. (Mrs T.G. Tutin) (1961) The late glacial and post-glacial deposits of the Esthwaite Basin, North Lancashire. *New Phytol.* **60**, 27-42.

FRENZEL B. (1966) Climatic change in the Atlantic/Sub-Boreal transition on the northern hemisphere: botanical evidence. In *World Climate from 8000 to 0 B.C.* (ed. by J.S. Sawyer), pp. 89-123. Royal Meteorological Society, London.

GODWIN H. (1960) Radiocarbon dating and Quaternary history in Britain. *Proc. R. Soc. B* **153**, 287-320.

HAFSTEN U. (1963) A late glacial pollen profile from Lista, South Norway. *Grana Palynol.* **4**(2), 326-37.
HAWORTH E.Y. (1969) The diatoms of a sediment core from Blea Tarn, Langdale. *J. Ecol.* **57**, 429-39.
HIBBERT F.A., SWITSUR V.R. & WEST R.G. (1971) Radiocarbon dating of Flandrian pollen zones at Red Moss, Lancashire. *Proc. R. Soc. B* **177**, 161-76.
IVERSEN J. (1954) The late-glacial flora of Denmark and its relation to climate and soil. *Danm. geol. Unders.* Ser. II, **80**, 87-119.
IVERSEN J. (1958) The bearing of glacial and interglacial epochs on the formation and extinction of plant taxa. *Uppsala Univ. Årsskr.* **6**, 210-15.
IVERSEN J. (1960) Problems of the early post-glacial forest development in Denmark. *Danm. geol. Unders.* Ser. IV, **4**(3), 32p.
IVERSEN J. (1964) Retrogressive vegetational succession in the post glacial. *J. Ecol.* **52** (Suppl.), 59-70.
IVERSEN J. (1969) Retrogressive development of a forest ecosystem demonstrated by pollen diagrams from fossil mor. *Oikos* Suppl. **12**, 35-49.
JANSSEN C.R. (1966) Recent pollen spectra from the deciduous and coniferous-deciduous forests of Northeastern Minnesota: a study in pollen dispersal. *Ecology* **47**, 804-25.
JESSEN K. (1949) Studies in Late Quaternary deposits and flora-history of Ireland. *Proc. R. Ir. Acad. B* **52**(6), 85-290.
LUND J.W.G. (1954) The seasonal cycle of the plankton diatom, *Melosira italica* (Ehr.) Kütz. subsp. *subarctica* Müll. *J. Ecol.* **42**, 151-79.
MACKERETH F.J.H. (1958) A portable core sampler for lake deposits. *Limnol. Oceanogr.* **3**, 181-91.
MACKERETH F.J.H. (1965) Chemical investigation of lake sediments and their interpretation. *Proc. R. Soc. B* **161**, 295-309.
MACKERETH F.J.H. (1966) Some chemical observations on post-glacial lake sediments. *Phil. Trans. R. Soc. B* **250**, 165-213.
MACKERETH F.J.H. (1971) On the variation in direction of the horizontal component of remanent magnetisation in lake sediments. *Earth and Planet. Sci. Letters* **12**, 332-8.
MAHER L.J. (1963) Pollen analyses of surface materials from the southern San Juan Mountains, Colorado. *Bull. geol. Soc. Am.* **74**, 1485-1504.
MOORE P.D. (1970) Studies of the vegetational history of Mid-Wales. II—The late glacial period in Cardiganshire. *New Phytol.* **69**, 363-75.
PECK R.M. (1973) Pollen budget studies in a small Yorkshire catchment. (this volume).
PENNINGTON W. (Mrs T.G. Tutin) (1943) Lake sediments: the bottom deposits of the north basin of Windermere, with special reference to the diatom succession. *New Phytol.* **42**, 1-27.
PENNINGTON W. (1947) Studies of the post-glacial history of British vegetation: VII Lake sediments. *Phil. Trans. R. Soc. B* **233**, 137-75.
PENNINGTON W. (1964) Pollen analyses from the deposits of six upland tarns in the Lake District. *Phil. Trans. R. Soc. B* **248**, 205-44.
PENNINGTON W. (1970) Vegetation history in the north-west of England: a regional synthesis. In *Studies in the vegetational history of the British Isles* (ed. by D. Walker & R.G. West), pp. 41-79. Cambridge University Press, London.
PENNINGTON W. & BONNY A.P. (1970) Absolute pollen diagram from the British late-glacial. *Nature, Lond.* **226**, 871-73.
PENNINGTON W., CAMBRAY R.S. & FISHER E.M. (1973) Observations on lake sediments using fallout Cs-137 as a tracer. *Nature, Lond.* **242**, 324-6.
PENNINGTON W. & LISHMAN J.P. (1971) Iodine in lake sediments in northern England and Scotland. *Biol. Rev.* **46**, 279-313.
PENNINGTON W., HAWORTH E.Y., BONNY A.P. & LISHMAN J.P. (1972) Lake sediments in northern Scotland. *Phil. Trans. R. Soc. B.* **264**, 191-294.
PIGGOTT S. (1954) *The Neolithic cultures of the British Isles* Cambridge University Press, London
RITCHIE J.C. (1969) Absolute pollen frequencies and carbon-14 age of a section of Holocene Lake sediment from the Riding Mountain area of Manitoba. *Can. J. Bot.* **47**, 1345-9.
RITCHIE J.C. & LICHTI-FEDEROVICH S. (1967) Pollen dispersal phenomena in arctic-subarctic Canada. *Rev. Palaeobotan. Palynol.* **3**, 255-66.
ROBERTS B.K., TURNER J. & WARD P.F. (1973) Recent forest history and land use in Weardale, Northern England. (this volume).
SMITH A.G. (1958) Two lacustrine deposits in the South of the English Lake District. *New Phytol.* **57**, 363-86.
SMITH A.G. (1961) The Atlantic-Sub-boreal transition. *Proc. Linn. Soc. Lond.* **172**, 38-49.
TAUBER H. (1965) Differential pollen dispersion and the interpretation of pollen diagrams. *Danm. geol. Unders.* Ser. II, **89**, 69p.
TAUBER H. (1966) Copenhagen radiocarbon dates VII. *Radiocarbon* **8**, 213-34.
TAUBER H. (1967) Investigations of the mode of pollen transfer in forested areas. *Rev. Palaeobotan. Palynol.* **3**, 277-86.
TAUBER H. (1968) Copenhagen radiocarbon dates IX. *Radiocarbon* **10**, 295-327.
TROELS-SMITH J. (1960) Ivy, Mistletoe and Elm. Climatic indicators—Fodder plants. *Danm. geol. Unders.* Ser IV, **4**(4), 32p.
TUTIN W. (1955) Preliminary observations on a year's cycle of sedimentation in Windermere, England. *Mem. Ist. Idrobiol.* suppl. **8**, 467-84.
TUTIN W. (*nee* Pennington) (1968a) In *Ann. Report Freshwater Biological Ass.*, 25-7.
TUTIN W. (*nee* Pennington) (1968b) Late-glacial soils and vegetation in north-west England. *J. Ecol.* **56**, 31-2P.
TUTIN W. (1969) The usefulness of pollen analysis in interpretation of stratigraphic horizons, both late-glacial and post-glacial. *Mitt. Int. Verein. Limnol.* **17**, 154-64.
VAN DER HAMMEN TH., MAARLEVELD G.C., VOGEL J.C. & ZAGWIJN W.H. (1967) Stratigraphy, climatic succession and radiocarbon dating of the last glacial in the Netherlands. *Geologie Mijnb* **46**, 79-95.
WADDINGTON J.C.B. (1969) A stratigraphic record of the pollen influx to a lake in the Big Woods of Minnesota. *Geol. Soc. Am. Special Paper* **123**, 263-82.
WALKER D. (1965) The post-glacial period in the Langdale Fells, English Lake District. *New Phytol.* **64**, 488-510.
WALKER D. (1966) The Late Quaternary history of the Cumberland Lowland. *Phil. Trans. R. Soc. B* **251**, 1-210.
WATTS W.A. (1971) Postglacial and interglacial vegetation history of southern Georgia and central Florida. *Ecology* **52**, 676-90.
WEST R.G. (1971) *Studying the past by pollen analysis.* Oxford University Press, Oxford.
WRIGHT H.E. (1971) Late Quaternary vegetational history of North America. In *The Late Cenozoic Glacial Ages* (ed. by K.K. Turekian), pp. 425-64. Yale University Press, New Haven and London.

Discussion on pollen dispersal and sedimentation

Recorded by Dr Hilary H. Birks,
Mr J.H.C. Davis, and Mr J. Dodd

Dr Janssen asked Dr Tutin whether the late-glacial influx data she presented showed the immigration of species, rather as Dr Davis had described, or whether the species were responding to a climatic change. Dr Tutin replied that *Betula* was not present during zone I, but immigrated during zone II. However, *Juniperus* and the herbs were present, and their increased pollen influx was due to an improvement in the climate. Dr Cushing wondered whether the difference in influx during the post-glacial between large and small lakes could be due to the flushing effect postulated for the pattern of sedimentation in lakes such as Windermere in the late-glacial. Dr Tutin replied that the flushing effect could not be invoked during the post-glacial, and that pollen influx was a function of lake morphometry. Elongated troughs are less efficient for depositing pollen than round deep lakes.

Professor Faegri considered that Dr Crowder's explanation of her *Betula* frequencies was too complicated. *Betula* may possibly flower abnormally. A similar situation was found for *Alnus* pollen in surface samples from Seattle, where abnormal wind transport could not be postulated. Here, *Alnus* pollen reached 90%, whereas the vegetation was dominated by hemlock and conifers. Dr Davis said that the increase of *Alnus* pollen in this region was culturally induced, and was equivalent to the *Ambrosia* pollen rise in the mid-West. Dr Crowder added that *Alnus* pollen does not decay so fast in moss cushions as it does in other surface sample materials. Dr Davis asked about the different *Ambrosia* pollen percentages at the mouth of Wilton Creek and in Lake Ontario. Dr Crowder said that one could only assume that *Ambrosia* pollen was washed further before settling than grass pollen. Dr Davis commented that in cases she had studied, *Ambrosia* and grass pollen behave similarly, but both sink relatively slowly compared with other pollen grains, and may be transported long distances by water before being deposited.

Dr H.J.B. Birks determined from Dr Janssen that, in his pollen diagram from Steven's Pond, the pollen curves were arranged according to their relative proportions and distribution in time, rather than on preconceived notions of the species' ecology. Dr Cushing then noted that, in Dr Janssen's transects across Myrtle Lake Bog, the values of the regional pollen types became much more irregular in forested areas round the islands. Dr Janssen explained that this was due to the much closer spacing of the samples in these parts, and no significant changes could be detected, which could be ascribed, for example, to selective filtration by the trees.

Section 3
Pollen representation

The differential pollen productivity of trees and its significance for the interpretation of a pollen diagram from a forested region

SVEND TH. ANDERSEN *The Geological Survey of Denmark, Copenhagen*

The differential pollen productivity of trees

Pollen analyses of surface samples from stands of mixed deciduous forest in Denmark have shown that the tree pollen spectra change within short distances and reflect the areal composition of the tree crown layer within small sample plots (Andersen 1970). If it is assumed that the amount of pollen of a tree species deposited at a sample point is proportional to its total crown area within the sample plot then

$$p = a \times P \qquad (1),$$

where p is the pollen deposition at the sample point, a is the crown area of the species within the same plot, and P is a constant which is high for the tree species with a high pollen production and low for the low pollen producers.

Equation (1) would be true if all of the pollen of the species were derived from the trees standing within the sample plot. If this is not the case equation (1) is modified into a regression equation:

$$p = a \times P + p_0 \qquad (2),$$

where $a \times P$ denotes the pollen derived from the trees within the sample plot and p_0 represents the pollen derived from trees standing outside the same plot, or local and extralocal pollen in the terms used by Janssen (1967).

The deposition of exotic pollen derived from plants growing outside the forest can be assumed to be constant within small forest stands (Andersen 1970). The relative pollen deposition of a tree species at various sample points can be found by dividing the number of pollen counted in the surface samples by the number of exotic pollen counted at the same time, and the results from the various sample points within a forest stand can be compared.

The relative tree pollen deposition is highly correlated with the tree-crown areas within circular sample plots of 30 m radius and linear regressions corresponding to equation (2) above can be calculated (Andersen 1970). It can be shown that a major portion of the tree pollen is deposited within 30 m and a small amount at larger distances.

P-values or pollen productivity factors are found from the calculated regression equations. The figures from various forest stands can be compared if normalised to the P-values of a common reference species. *Fagus silvatica* is favourable as a reference species because this tree is likely to flower equally well on various soils and because problems of species identification are avoided. Relative pollen productivity values (P_{rel}-values) calculated in 3 forest stands in southern Denmark are shown in Table 1. The species present are *Quercus robur*, *Betula pubescens*, *Alnus glutinosa*, *Carpinus betulus*, *Ulmus glabra*, *Fagus silvatica*, *Tilia cordata*, and *Fraxinus excelsior*.

The figures from the 3 forest stands in Table 1 do not differ greatly except for *Alnus glutinosa*. This species turned out to flower badly in section 386 in Draved forest presumably due to artificial drainage. The other figures show a high relative pollen productivity in *Quercus*, *Betula*, and *Alnus*, intermediate in *Carpinus* and *Ulmus*, and low values for *Tilia* and *Fraxinus*. They differ somewhat from the figures calculated by Pohl (1937) and found by Rempe (1937) but are based on more direct or comprehensive measurements. The figures calculated by Kabailiené (1969) resemble those of Pohl (1937). Kabailiené's figures were based on pollen spectra

Table 1. Relative pollen productivity (P_{rel}-values) in 3 Danish forest stands.

	Draved forest		Longelse forest
	Section 386	Section 370	
Quercus	5.8	6.0	5.5
Betula	4.4	4.2	–
Alnus	1.9	3.9	–
Carpinus	–	–	2.4
Ulmus	–	–	2.0
Fagus	1.0	1.0	1.0
Tilia	0.6	–	–
Fraxinus	–	0.4	0.4

from lakes and were recalculated in order to show relative pollen productivity by application of the formula for losses from dust clouds by ground deposition according to Chamberlain. Chamberlain's formula, however, is unlikely to apply to the losses from a pollen cloud passing over vegetation where filtration is likely to be the dominant cause for loss of pollen.

Pollen representation

The ratio pollen percentage: area percentage was called the R-value by Davis (1963). The R-values express the over- or under-representation of the species in individual pollen spectra. However, the R-values are not comparable from case to case, because they differ in various species combinations and vary with the species frequency (Davis 1963, Andersen 1970). Davis (1963) pointed out that the ratios of the R-values to the R-values of a reference species are constant.

Table 2. Relative pollen representation (R_{rel} values) in 3 Danish forest stands.

	Draved forest		Longelse forest
	Section 386	Section 370	
Quercus	4.6	3.8	3.3
Betula	4.8	4.6	–
Alnus	2.3	3.6	–
Carpinus	–	–	2.5
Ulmus	–	–	1.7
Fagus	1.0	1.0	1.0
Tilia	0.6	–	–
Fraxinus	–	0.4	0.5

The R_{rel}-values vary with the areal participation of the species if the pollen percentage of the species is not zero at zero area percentage (Andersen 1970). This is the case in the samples from forests mentioned above due to the presence of extra-local pollen dispersed from trees outside the sample plots. The R_{rel}-values of a species vary in these cases from infinite to about 1 when its area percentages in the sample plots vary from 0 to 100 and approach the P_{rel}-values only in cases when the area of the species is about 10–30 per cent.

R_{rel}-values were calculated from the average areal and pollen percentages from the 3 forest stands. The average area percentages of the various trees vary within the interval mentioned above, and the R_{rel}-values normalised with Fagus as a reference species are shown in Table 2.

It may be seen that the figures from the 3 forest stands are not unlike each other, and it may be noticed, that Alnus again is under-represented in section 386 from Draved forest for the reasons mentioned earlier.

Correction of tree pollen spectra

The relative pollen productivity and pollen representation values shown in Table 1 and 2 resemble each other. The largest difference occurs in Quercus, as the P_{rel}-values for this tree are slightly higher than the R_{rel}-values.

The P_{rel}- and R_{rel}-values can be used for a correction of the pollen spectra, and pollen percentages which resemble the area percentages can be calculated (Andersen 1970).

Correction factors thus were found for the most important tree species occurring spontaneously in Denmark. Correction factors for Pinus silvestris, Picea abies, and Abies alba were calculated from published data on tree composition and surface pollen spectra. The following correction factors were found (Andersen 1970):

Quercus, Betula, Alnus, Pinus	1 : 4
Carpinus	1 : 3
Ulmus, Picea	1 : 2
Fagus, Abies	1 : 1
Tilia, Fraxinus	1 × 2

These correction factors were calculated for pollen spectra within forests. It is more difficult to compare tree pollen spectra from open areas such as lakes or bogs with the forest composition because the pollen source areas are likely to be very large. Relative pollen representation values can be calculated in a few cases and are not unlike the ones found within the forests (Andersen 1970). Very high R-values occur in cases where the areal participation of the tree species is small (cf. Davis & Goodlett 1960, Janssen 1967, Livingstone 1968) due to pollen transported from a larger source area.

Iversen (1947, cf. Faegri & Iversen 1964) suggested correction factors for pollen spectra from lakes and bogs. Although Iversen's figures are arbitrary, they resemble the ones calculated above, but a few differences exist. The pollen productivity of Quercus turned out to be higher than assumed earlier, and the representation of this species in lake sediments cannot have been reduced by losses during dispersion, as the pollen grains of the species are rather small. Hence, Quercus is likely to be over-represented in most pollen diagrams, and the general opinions of the part played by oak in post-glacial and interglacial vegetational history must be revised.

Several authors suspected that *Tilia* is under-represented in lake pollen spectra, and Iversen (1960) maintained that *Tilia* was very important in the forests of Atlantic and Sub-Boreal time in spite of its low representation in pollen diagrams. The rather large *Tilia*-pollen grains are more likely to be retained during dispersion, a fact which will promote its under-representation in lake pollen spectra. The importance of *Fraxinus* certainly has been underestimated too, and this tree must have been more frequent in post-glacial and interglacial time than indicated by pollen diagrams.

The pollen productivity of Corylus avellana

A calculation of the pollen productivity of *Corylus avellana* presents special problems as this species may constitute an understorey in some cases and may form a canopy in other cases. The species grows under a tree canopy in one of the forest stands mentioned above. Its areal coverage corresponds rather well to pollen percentages based on the corrected tree pollen sum. This indicates that the pollen productivity of *Corylus* is similar to that of *Fagus* when the shrub is shaded. Its pollen productivity is presumably considerably larger in full illumination. Iversen (1947) suggested a correction factor 1:4 for such cases.

Application of the correction factors to a post-glacial pollen diagram

The correction factors mentioned above were calculated in forest stands where the trees are self-propagated and grow under at least semi-natural conditions. The stands occur within the natural distribution areas of the species considered. Hence, the correction factors may be fairly typical of the species in question and may be applied to pollen spectra from deposits formed under similar conditions.

SITE

A post-glacial pollen diagram was worked out in Eldrup forest in Djursland, eastern Jutland (Fig. 1).

The forest stand today consists of *Fagus silvatica* and *Quercus petraea*, and is a protected research area. It occurs on a morainal ridge at about 50 m above sea level, and the soil is morainic sand with a mor layer. *Quercus petraea* is frequent on the top of the ridge, and *Fagus silvatica* occurs particularly on the slopes.

The ridge contains several hollows in which gyttja and peat have accumulated. A pollen diagram was worked out from one of them, which is only 20 m across and contains about 90 cm of gyttja and peat. The hollow was drained by ditches in the last century, and the present vegetation on the peat is mainly *Molinia caerulea*.

The hollow is surrounded by the high-ground forest, mainly *Fagus silvatica*, and the tree canopy is nearly closed over it. Hence, the pollen diagram from the deposits can be assumed to record the history of the high-ground forest in the immediate vicinity according to the experiences from surface pollen analyses from forests mentioned above.

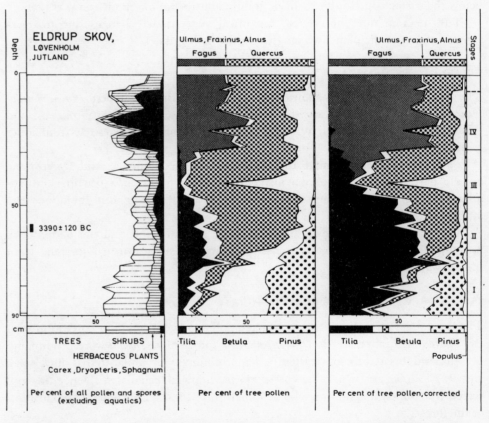

Figure 1. Eldrup forest, eastern Jutland, Denmark. Pollen diagrams from a small hollow in *Fagus silvatica-Quercus petraea* forest. Total pollen and spore diagram (left), tree pollen diagram (middle), and corrected tree pollen diagram (right). The tree pollen counts were corrected prior to the percentage calculation with the following factors: *Pinus*, *Quercus*, *Betula* and *Alnus* 1:4, *Ulmus* 1:2, *Fagus* and *Populus* 1 × 1, *Tilia* and *Fraxinus* 1 × 2.

TOTAL POLLEN DIAGRAM

The first diagram in Fig. 1 shows the frequencies of the trees, the shrubs, and the herbaceous plants. There is one radiocarbon date of an oak branch at 3390 ± 120 B.C. (K-1421).

Tree pollen dominates the sequence, and it may be concluded that the tree cover was fairly continuous throughout. Shrub pollen is important below the 70 cm level, and a local swamp vegetation of *Carex*, *Dryopteris*, and *Sphagnum* (shown separately) is important above 30 cm. The swamp vegetation disappeared abruptly at 6 cm below the surface due to the artificial draining of the hollow about 100 years ago.

The diagram does not reflect reciprocal vegetational changes due to over-representation of the plants which grew in the hollow itself.

TREE POLLEN DIAGRAM

The next diagram in Fig. 1 shows the frequencies of the tree pollen. *Pinus*, *Betula*, *Quercus*, *Tilia*, and *Fagus* are the main components, and *Ulmus*, *Fraxinus*, and *Alnus* are only represented by small amounts of pollen transported from other localities.

Pinus and *Betula* dominate the lower part of the sequence, and *Quercus* the middle part, which contains a conspicuous *Betula* peak. *Tilia* obtains low frequencies. *Fagus*, *Quercus*, and *Betula* are almost equally frequent in the upper part of the diagram.

The lower part of the diagram dates from Atlantic time, and the *Betula* peak presumably corresponds to an intensive cultural phase known in Djursland from the early Iron age.

THE CORRECTED TREE POLLEN DIAGRAM

The third diagram in Fig. 1 shows the tree pollen percentages after correction with the figures mentioned above. *Quercus petraea* and *Betula pendula* were presumably present, and it is assumed that their pollen productivity is similar to that of *Q. robur* and *B. pubescens*, for which the correction factors were calculated. Hence the corrected tree pollen diagram may be assumed to reflect the crown areal composition of the forest.

It now appears that *Tilia* and *Fagus* dominated the slopes around the hollow, and that *Betula* and *Pinus* only were important in an early stage and *Quercus* and *Betula* in an intermediate stage.

The slow expansion of *Tilia* in the early stage (I) is puzzling. The high frequency of *Pinus* and *Betula* and the presence of *Populus* indicate an open forest and conspicuous minima in the *Tilia* curve and peaks for *Betula* and *Pinus* suggest events which checked *Tilia*. The deposit is sandy and indicates an unstable surface, and it appears that the soil had not yet settled and that slides occurred which checked the *Tilia* forest and promoted the perseverance of the light-demanding trees.

The corrected pollen spectra show that *Tilia* was dominant for some time and *Quercus* was a subordinate member of the forest (stage II). Hence, we have here a picture of the composition of the forest on high ground in late Atlantic and Sub-Boreal time with *Tilia* as the dominant tree in the way predicted by Iversen (1960), and *Quercus* of restricted importance.

Tilia decreased violently in stage III, and *Quercus* and *Betula* became dominant. The *Betula* peaks indicate forest fires, and it seems obvious that human interference caused the destruction of the *Tilia* climax forest.

Fagus began to expand in stage III and became dominant in stage IV. A dense climax forest became established again, but *Fagus* now had replaced *Tilia*. *Quercus* was a subordinate member of the forest again. *Betula* remained present, and small peaks of *Betula* and *Quercus* indicate phases of restricted human interference. *Betula* disappeared suddenly. It was presumably removed by the modern foresters. A few old individuals are still present in the area.

It seems a coincidence that *Tilia* and *Fagus*, the shade trees, are low pollen producers and are under-represented in the distorted picture shown by the original tree pollen diagram. The revised pollen diagram shows a clearer picture, where the stable periods, the stages II and IV, are characterised by the development of climax forest of the shade trees. These stages stand out against periods of disturbance, the stages I and III, where the occurrence of the more light-demanding trees was promoted.

NON-TREE POLLEN AND SPORES

It is somewhat difficult to distinguish terrestrial and local non-tree vegetational components. Their pollen and spore productivity is not known very well, and the pollen and the spores probably were transferred in various manners. Hence it is advantageous to calculate the non-tree pollen and the spores outside the pollen total and this procedure may be justified in the present case because the tree cover was fairly continuous throughout the sequence.

The deposition of tree pollen can be assumed to have varied greatly because high and low pollen producers alternately dominated the forest. However, the corrected tree pollen sum constitutes a fair calculation basis because the variations in the tree pollen output are smoothed out in such a manner that the total tree pollen deposition at any level corresponds to that of a pure *Fagus* forest. The N.A.P.-frequencies thus become related to a rather constant factor. Their changes can be expected to reflect true changes in deposition but their relative frequencies cannot be assumed to express their areal importance directly.

Curves for the most important terrestrial non-tree components are shown in Fig. 2. They increase from right to left in order to emphasize their correlation with the light-requiring trees in the corrected tree pollen diagram.

It is rather difficult to decide how the *Corylus* pollen frequencies should be calculated. The shrub is likely to have flowered profusely if it grew without a tree canopy, and its pollen productivity was low if it formed an understorey. The frequencies shown in Fig. 2 were divided by 4 in the way suggested by Iversen (1947). They are fairly high in stage I and suggest that *Corylus* may have grown in patches without a tree cover at that time, and they are moderate in stage II and III, at which time the shrub probably formed an understorey. There is a small peak in stage III,

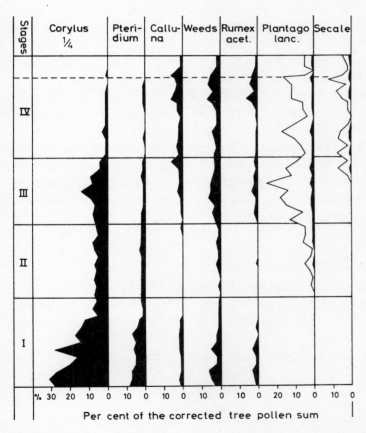

Figure 2. Eldrup forest. Frequencies of non-tree terrestrial vegetational components, in percentages of the corrected tree pollen sum. The *Corylus* frequencies were divided by 4.

which shows that *Corylus* reacted positively to the increased illumination. It vanished in stage IV.

The presence of *Pteridium* and *Calluna* in stage I indicates that there were small open patches in the forest, where the soil was stable and had become leached. They probably occurred on the hill tops near the hollow. *Pteridium* persisted with low frequencies in the later stages. *Calluna* disappeared in stage II and appeared in stage III and stage IV with rather low frequencies, which indicate that patches of heath vegetation were created.

The curve for 'weeds' includes the open-ground herbaceous plants. *Rumex acetosella* and *Plantago lanceolata* are shown separately. The curves indicate that patches of open ground occurred in stage I, dense forest in stage II and patches of open ground again in stage III and stage IV. The burnings in stage III thus did not result in deforestation but there was restricted grazing activity, which continued in stage IV. The open-ground plants decrease near the top because grazing was abandoned in the last century.

Plantago lanceolata pollen grains appear with low frequencies in stage II. The weed pollen frequencies are so low in this stage that no grazing activity is indicated, and the significance of the *Plantago lanceolata* grains is uncertain except for a suggestion of human activity somewhere in the neighbourhood. The presence of *Secale* pollen in the upper part of stage III shows that this level dates from the Iron age. The frequencies remain low and do not indicate fields at the site.

Curves for the most important local plants and for *Salix* and Gramineae are shown in Fig. 3.

The curves for *Salix* and Gramineae pollen are rather similar and a summation curve is shown in Fig. 3. Gramineae pollen with annulus diameters smaller and larger than 8 μm are shown separately. Nearly all the Gramineae pollen were crumpled and unsuitable for size measurements, but the diameter of the pore annulus could be measured in all cases. The pollen with the smaller annulus diameters belong to various wild grasses (cf. Leroi-Gourhan 1969), whereas grains with the larger annulus may belong to some wild grasses (Leroi-Gourhan 1969) or cereals (Faegri & Iversen 1964, Leroi-Gourhan 1969). Most of the grains with the larger annulus belong to *Glyceria* in the present case. A few cereal pollen grains may occur too, but they were difficult to distinguish except for the typical *Secale* pollen which was shown in Fig. 2.

Salix and wild grasses were common in stage I. *Salix* presumably grew around the hollow and an open tree canopy is suggested. The wild grass pollen also suggests good illumination. *Salix* and the grasses disappeared in stage II at which time a closed tree canopy is indicated. *Salix* and *Glyceria* responded to the forest fires in stage III with distinct maxima. The hollow clearly was well illuminated at that time

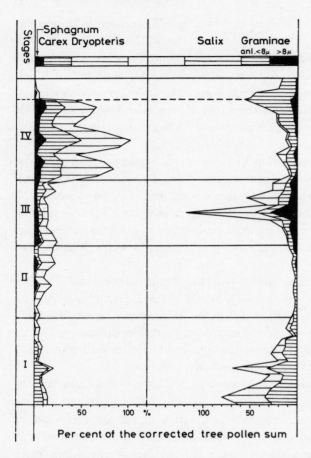

Figure 3. Eldrup forest. Frequencies of more or less local vegetational components, in percentages of the corrected tree pollen sum.

and a *Salix* scrub and a local *Glyceria* vegetation developed. They were suppressed in stage IV. A wild-grass maximum is seen near the top. It is unlikely that this peak is due to dry-land grasses, as the frequencies for the terrestrial non-tree pollen otherwise are low, and a local grass species, probably *Molinia coerulea*, is suggested.

The *Sphagnum*-, *Carex*-, and *Dryopteris*-curves in Fig. 3 are nearly parallel. *Carex* was present in stage I, *Sphagnum* and *Dryopteris* in stage II and stage III, and they increased in stage IV due to a swamp community in the hollow. The swamp vegetation disappeared as a result of artificial drainage, as mentioned earlier.

CONCLUSION

The calculations of the relative pollen productivity of the trees have made it possible to revise the pollen diagram from Eldrup forest and to establish a new picture of the composition of the local forest since early Atlantic time.

Tilia was the most important tree in Atlantic and Sub-Boreal time. The persistence of light-demanding trees such as *Pinus*, *Betula* and *Populus* nearly until the end of Atlantic time (stage I) is surprising. The N.A.P.-curves have shown that the forest contained patches of *Corylus* and herbaceous vegetation and the tree-crown cover allowed a *Salix* scrub to develop around the hollow.

The denseness of the forest in stage II is emphasized by scarceness of terrestrial non-tree components and suppression of the local vegetation at the hollow. *Quercus* became increasingly important in this stage.

The *Plantago lanceolata* curve shows that Neolithic man was present in the neighbourhood. One may wonder whether the increasing *Quercus* frequency could have been due to lopping of *Tilia* for leaf foddering.

The catastrophic destruction of the *Tilia* forest in late Sub-Boreal or early Sub-Atlantic time (stage III) is emphasized in the corrected pollen diagram. Burning and grazing presumably caused the expansion of *Betula* and *Quercus* and the moderate increase of *Corylus* and the terrestrial herbaceous plants. The *Salix* scrub and *Glyceria* vegetation at the hollow indicate an open tree canopy.

A vigorous expansion of *Fagus* (stage IV) was apparently due to decreased human activity. There was, however, moderate grazing activity, and the tree canopy was sufficiently open for a local *Carex-Dryopteris-Sphagnum* community. Abandonment of the grazing and the influence of modern forest management are traced in the uppermost part of the sequence.

Acknowledgements

The studies in Draved forest were supported by the Carlsberg Foundation and the Danish State Research Council. The research area in Eldrup forest was established in agreement with the Løvenholm Foundation and with financial support from the Carlsberg Foundation.

Mr H. Tauber, the C-14 dating laboratory, Copenhagen, provided the C-14 date K-1421.

References

ANDERSEN S.TH. (1970) The relative pollen productivity and pollen representation of north European trees, and correction factors for tree pollen spectra. *Danm. geol. Unders.* Ser. II, 96, 99p.

Davis M.B. (1963) On the theory of pollen analysis. *Am. J. Sci.* **261**, 897-912.

Davis M.B. & Goodlett J.C. (1960) Comparison of the present vegetation with pollen-spectra in surface samples from Brownington Pond, Vermont. *Ecology* **41**, 346-57.

Faegri K. & Iversen J. (1964) *Textbook of pollen analysis.* Munksgaard, Copenhagen.

Iversen J. (1947) Discussion in: Nordiskt kvartärgeologiskt möte den 5.-9. November 1945. *Geol. För. Stockh. Förh.* **69**, 205-6.

Iversen J. (1960) Problems of the Early Post-Glacial forest development in Denmark. *Danm. geol. Unders.* Ser. IV, 4(3), 32p.

Janssen C.R. (1967) A comparison between the recent regional pollen rain and the sub-recent vegetation in four major vegetation types in Minnesota (U.S.A.). *Rev. Palaeobotan. Palynol.* **2**, 331-42.

Kabailiené M. (1969) On formation of pollen spectra and restoration of vegetation. *Ministry of Geology of the USSR, Institute of Geology (Vilnius). Transactions,* 11.

Leroi-Gourhan A. (1969) Pollen grains of Gramineae and Cerealia from Shanidar and Zawi Chemi. In *Domestication and exploitation of plants and animals* (ed. by P.J. Ucko & G.W. Dimbleby), pp. 143-8. Duckworth, London.

Livingstone D.A. (1968) Some interstadial and postglacial pollen diagrams from Eastern Canada. *Ecol. Monogr.* **38**, 87-125.

Pohl F. (1937) Die Pollenerzeugung der Windblütler. *Beih. bot. Centralblatt.* **56A**, 365-470.

Rempe H. (1937) Untersuchungen über die Verbreitung des Blütenstaubes durch die Luftströmungen. *Planta* **27**, 93-147.

Pollen dispersal and deposition in an area of southeastern Sweden – some preliminary results

BJÖRN E. BERGLUND *Laboratory of Quaternary Biology, Institute of Geology, University of Lund, Sweden*

Introduction

The present investigation was inspired mainly by the work of three researchers:
(i) H. Tauber, who in 1965 published his well-known model of pollen transfer in a forested area with the following sub-divisions of the composite pollen deposition: (a) the trunk space component, Ct, (b) the canopy component, Cc, and (c) the rain-out component, Cr. Components a and b may be named the local component, and c the regional component. Later Tauber (1967) studied the pollen transport experimentally in a small forest in Zealand, using a specially designed pollen trap.
(ii) M.B. Davis, who in 1967 published results from an experimental study on pollen deposition in lakes. The lakes she studied were rather large, deep, and stratified.
(iii) S.Th. Andersen, who in 1970 published a monograph on the relative pollen productivity and pollen representation of North European trees, based on analyses of moss polsters in some Danish forests.

The study ought to give complementary information useful for a better understanding of the development of the human landscape in south Sweden. It ought to throw light on the following problems:
(i) The relation between local and regional pollen supply in basins of different sizes.
(ii) The pollen deposition in a vegetation transect from mixed coniferous forest region through a cultivated deciduous forest region, and into a deforested area, such as heathland.
(iii) The relation of pollen production to vegetation, especially in (a) the relics of deciduous forests with *Quercus*, *Tilia*, and *Acer*. (b) the old-fashioned 'park-meadow' with pollarded *Ulmus*, *Tilia*, and *Fraxinus*. (c) the partially forested human landscape as a whole.
(iv) The pollen deposition in a medium-sized shallow and non-stratified lake,

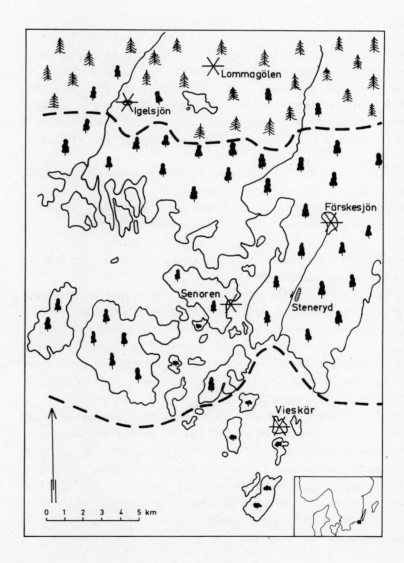

Figure 1. The investigation area of southeastern Blekinge. The sites on the transect from coniferous forest to deciduous forest, and the archipelago are indicated.

which ought to be representative for the main group of lakes now studied with palaeolimnological techniques.

As an investigation area the southeastern corner of Sweden (the eastern part of the Blekinge province) was selected for study. The vegetation and its history are well known due to previous studies (Berglund 1966), and there is a zonation of vegetation, human landscape, and lakes of the kind proposed above.

Methods

For collecting pollen in air the Tauber trap was used. This is now widely used and the values from different areas should be comparable. The circular opening of the trap is 5 cm diameter and the potential sedimentation area c 20 cm^2. Generally two traps, one open and one roofed, were used on each station to get an idea of the rain-out component. On land the traps were placed on poles with the trap opening situated 25 cm above the ground. On lakes they were placed on rafts with the mouth 35 cm above the water level. These traps were enlarged 10 cm to avoid spilling in stormy weather.

The traps were placed in the field before the spring flowering and were changed twice a year to get the following deposition periods:

(i) spring season (March-June) with the main flowering of the trees,

(ii) summer season (July-September) with *Tilia*, *Calluna*, herb, and graminid flowering, and

(iii) autumn season (October-December) a non-flowering period but with reflotation of pollen as described by Tauber (1967).

Preparation procedures followed the weighing method described by Jørgensen (1967). The absolute values obtained in this way are expressed as pollen content/sampler. It would be desirable to express them per unit area, but the efficiency of the traps is not known. It certainly varies with wind velocities and trap situation. According to wind tunnel investigations the efficiency relation for the traps when the wind velocity changes from 1 to 8 m/sec is about 10:1 for heavy and 4:1 for light pollen grains (H. Tauber, personal communication).

For collecting pollen in water, bottles with the same opening diameter as the air traps were used and placed on iron stands pushed into the bottom sediments. On each stand two bottles were placed, one with the mouth 20 cm and the other with the mouth 70 cm above the bottom surface. Preparation was performed and the pollen values expressed in the same manner as for the air traps. It must be noted that the efficiency for these traps is much greater than that for the air traps due to the slow water movements. These traps have been changed once each year—at the end of September with the assistance of a diver.

The vegetation mapping of the areas surrounding the pollen stations is, as yet, only superficial. Detailed maps based on special air photos will be made this year.

In the present paper results from three years, 1969-71, are presented. Sabotage and other accidents made the series from 1967 and 1968 discontinuous. The pollen production was very rich in 1969, very poor in 1970, and rather poor in 1971.

Investigation sites

1. Lommagölen

The lake is 100 × 100 m and is situated in a forested area with *Picea*, *Pinus*, and *Betula* dominant. 200 m northwest of the lake there are large deforested areas.

2. Igelsjön

The lake is 200 × 150 m and is situated in a mixed forest area where *Fagus* and *Quercus* are dominant in the west and south, and *Picea* and *Pinus* in the north and east. The *Picea* stand was, however, cut down in the winter of 1969. The border between the forest area and the cultivated valley plain is situated 200 m southwest of the lake.

3. Färskesjön

The lake is 500 × 1,000 m and is situated in a mosaic landscape, with *Calluna* heath with scattered stands of *Quercus*, *Betula*, and *Pinus* in the west, *Fagus* forest in the northwest, *Calluna* heath and arable land in the north, planted *Pinus* forest in the east, *Calluna* heath with *Quercus* and *Betula* and arable land in the southeast, and lastly *Carpinus* and *Fagus* in the southwest.

Three rafts are placed on this lake, one in the central part (named Central), one in the northwestern bay (named North), partly separated from the main basin by a small island, and one in the southern, narrow bay (named South).

Eight iron stands with bottles have been placed in this lake, four (A, B, C, D) of these in the southern bay forming a transect running SSW-NNE, one (E), in the central part and three (F, G, H) in the northwestern bay forming a transect running ESE-WNW. Stands C, E, and F are placed 20 m south of the rafts South, Central, and North respectively.

Water depths and the situation of rafts and stands are shown on Fig. 3. The lake has a small inlet brook in the northwestern bay. The outlet in the southern bay is dammed. During summer the water does not reach the overflow threshold.

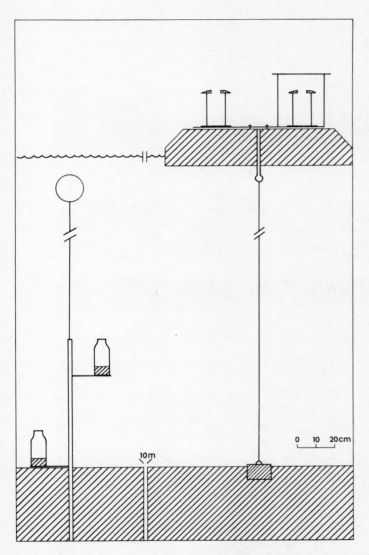

Figure 2. Equipment on three stations of Lake Färskesjön: raft with roofed and unroofed traps and iron stand with sediment traps.

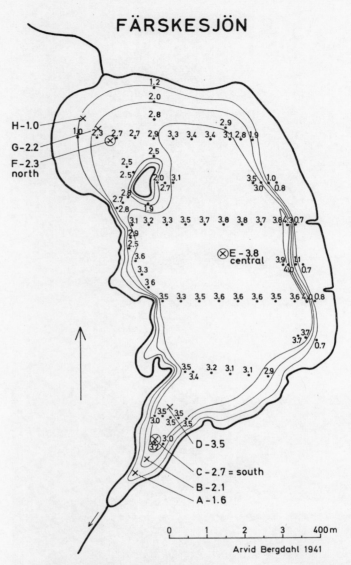

Figure 3. Lake Färskesjön with depth curves at metre intervals. Stations with sediment traps, A–H, and with rafts for air traps, South, Central and North, are indicated.

4. Senoren

This is an island in the inner archipelago belonging to the deciduous forest area. However, the forests are divided by heaths and arable fields. The actual site is the sea bay Sörviken exposed in the south to the sea. The raft is situated 50 m from the shore. The woods surrounding the bay are dominated by *Quercus* and *Tilia*, but *Acer platanoides* is also rather common. Beyond 400 m *Betula* woods become very common. In addition to the raft the local pollen deposition is measured at two forest stations and one heath station.

Figure 4. Lommagölen and Igelsjön.

Figure 5. Färskesjön: northern bay, central part and southern bay.

Figure 6. Sörevik and Vieskär (with the barren ridge surrounded by juniper shrubs in the back-ground).

5. Vieskär

This is an island in the outer archipelago with treeless, low-lying islands generally characterised by grass heaths of *Juniperus*. On the inhabited islands there are some planted trees, *i.a.* small *Pinus* stands. The raft is situated in a north-exposed bay. In addition three traps are placed on the island. The values from one of these, on a barren ridge, have been used here to represent the regional pollen deposition in the archipelago since the series from the raft was discontinuous.

Local and regional pollen supply

Data have been compiled in Table 1 to illustrate the relation of local to regional pollen supply. These terms are used in a general sense here. The local component corresponds to the dispersal from the trunk space and the canopy, and the regional component to the atmospheric fall-out and rain-out. For pollen-analytical work it is of great importance to know the 'effective area' of each pollen spectrum, defined by Tauber (1965) as the area from which 80% of all the pollen grains originate. It has been stated that the effective area is dependent on the size of the basin. Experimental studies by means of pollen trapping should give some indications of this relationship.

GEOGRAPHY OF THE SITES

According to their size the sites can be grouped into three broad classes:
(1) Lommagölen, Igelsjön, the bays of Färskesjön, and the bay of Senoren with an area of 0.5–5 ha and with a distance to the nearest forest edge of 50–150 m,
(2) Färskesjön with an area of 50 ha and with a distance to the nearest forest edge of 300 m,
(3) Vieskär in a woodless (treeless) archipelago with an area of c 25 km^2 and with a distance to the nearest forest edge of 600 m.

The forest density is highest around Lommagölen and Igelsjön, the areas around Färskesjön and on Senoren are mosiac, half-open landscapes, and the area around Vieskär is quite open.

THE TREE POLLEN SUM

Regarding the tree pollen sum the sites can be grouped into four classes, (1) high values of Lommagölen, Igelsjön, and Senoren, (2) lower at the bays of Färskesjön, (3) still lower in the central part of the latter lake, and (4) very low values at Vieskär. The relation is 4:3:2:1. The mean values/cm^2/year, if the effective area of the traps is 20 cm^2, are given in Table 1. However, due to the wind these values should be

Table 1. Local/regional pollen supply

Characters based on values 1969–71	Lommagölen	Igelsjön	Färskesjön N	Färskesjön C	Färskesjön S	Senoren	Vieskär
Distance to nearest forest edge (m)	50	75	150	300	70	50	600
Total area of lake or bay (ha)	0.5	2	5	50	1	2	2,500
Total sum of A.P. $\times 10^3$	219	200	140	93	158	205	56
Mean value/cm^2/year of A.P. $\times 10^3$	3.7	3.3	2.3	1.6	2.6	3.4	0.9
A.P. in % of total pollen sum	85	84	87	80	88	89	35
Reflotation, max. value in %	5	4	2	4	4	8	?
Relation roofed/unroofed trap, %	75	90	79	60	86	75	66
Local component %	75	72	60	40	65	75	0
Local → regional, numerical order	1	2	5	6	4	3	7
Local dominance I	×	×	—	—	—	×	—
Local dominance II	—	—	×	—	×	—	—
Regional dominance III	—	—	—	×	—	—	—
Extra regional dominance IV	—	—	—	—	—	—	×

corrected by multiplying by a factor in the range of 2 to 5.

The relation A.P./N.A.P. is approximately the same in the mainland and inner archipelago sites, i.e. the A.P.% is 80–89%. The value of 80% found at Färskesjön C may be due to the half-open landscape here and the more regional character of the pollen rain at this site. The Vieskär site differs completely with only 35% A.P. This is fully long-distance transported.

REFLOTATION

Tauber (1967) found that pollen was refloated during the autumn after capture on surrounding vegetation. He found a reflotation percentage of more than 50%. The local pollen producers were distinctly dominating. High reflotation is related to small basins and shrubby shore vegetation. At the sites investigated in Blekinge very low and equal reflotation percentages were found (always < 10%). The small differences cannot, with accuracy, be correlated to the amount of local pollen supply.

RELATION OF ROOFED TO UNROOFED TRAP

The difference between the content of the roofed and the unroofed traps ought to reflect, to some degree, vertical deposition, i.e. the magnitude of the atmospheric fall-out and rain-out elements, according to Tauber mainly the rain-out component. The relation of the roofed to unroofed samplers will thus give an idea of the relation between local and regional pollen supply. High percentages (75–90%) are quite well related to the small lakes and bays, and low values (c 60%) to Färskesjön C and Vieskär.

LOCAL COMPONENT

The Vieskär site may be used as a basis for the regional tree pollen deposition that characterise the whole investigation area. Using the absolute values from the Vieskär trap as the general regional component the values from the other sites have been

Table 2.

	Lake area	Lake radius	Eff. area radius
I. Local dominance	< 2 ha	50 m	< 500 m
II. Local dominance	5 ha	150 m	1,000 m
III. Regional dominance	50 ha	300 m	10,000 m
IV. Extra regional dominance	25 km^2	> 600 m	∞

reduced to show the local component at each site (Fig. 7). In Table 1 percentages for the local component are given, and show high values for Lommagölen, Igelsjön, and the bay at Senoren (c 75%), moderate values for the bays of Färskesjön (c 60%), and low values for the central part of Färskesjön (c 40%).

CLASSIFICATION

With regard to the relation of local to regional tree pollen supply the sites may be classified in the following manner. Approximate values for effective areas are also given. These figures, given in the Table 2, are approximate due to lack of vegetational mapping.

The relation of local to regional pollen supply is, as expected, related to the size of the basins and the distance to the nearest forest. Even the bays follow this general rule, the only exception being the southern bay of Färskesjön. The bay is small and narrow but exposed towards the lake centre. Besides this only the southwestern sector of the bay is covered with dense forest.

Until the vegetation of the areas has been mapped in detail it is not possible to analyse the local and regional pollen supply for different tree species.

At the moment the results seem to confirm Tauber's theoretic calculations. For small lakes of radius 50–100 m, he supposed the effective area to be c 300–1,000 m, for lakes with diameters of some kilometres he supposed the effective area to be c 30–100 km. Therefore the tables in Tauber (1965, pp. 24, 25), are useful in calculating effective areas.

The transect coniferous forest – deciduous forest – archipelago area

Pollen values along the transect from coniferous forest through the deciduous forest to the archipelago area are given in Table 3 and Fig. 7. The pollen spectra from the different sites reflect the vegetation zones rather poorly. This is due mainly to the over-representation of *Betula* and *Pinus*. The high values of these, especially the percentage, conceal the real differences between the coniferous and deciduous forest regions. *Quercus* seems to be distinctly under-represented in traps within the latter region, except at Senoren due to the local supply from the woods near the shore.

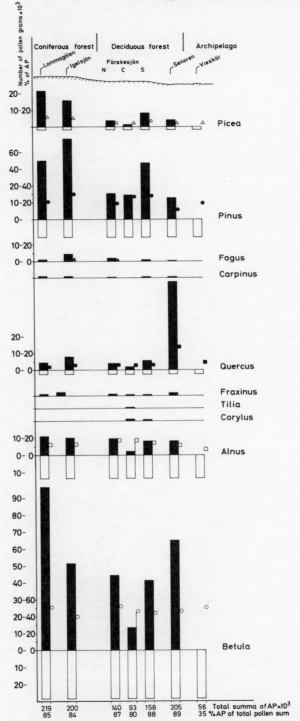

Figure 7. Diagram showing the absolute numbers of tree pollen in 1969–71 in the vegetation transect. Absolute values are illustrated by columns, relative values with ordinary pollen symbols. The values of the Vieskär trap are considered to represent the regional component within the whole area and therefore illustrated by open columns below a base line. In this way the black columns broadly illustrate the local component.

Table 3. Transect of pollen traps, 1969-1971

	Lommagölen		Igelsjön		Färskesjön North		Färskesjön Central		Färskesjön South		Senoren		Vieskär	
	Σ	%	Σ	%	Σ	%	Σ	%	Σ	%	Σ	%	Σ	%
Fagus	1,209	0.6	4,664	2.3	2,159	1.5	369	0.4	1,191	0.8	593	0.3	130	0.2
Quercus	7,134	3.3	10,546	5.3	6,625	4.7	4,507	4.8	8,237	5.2	55,469	27.1	2,636	4.5
Fraxinus	1,048	0.5	2,078	1.0	858	0.6	428	0.5	490	0.3	1,398	0.7	133	0.2
Ulmus	358	0.1	569	0.3	230	0.2	159	0.1	279	0.2	259	0.1	44	0.1
Tilia	180		118		34		389	0.4	149		88		31	
Acer	–		–		–		–		–		–		–	
Carpinus	1,092	0.5	861	0.4	422	0.3	134	0.1	652	0.4	648	0.3	197	0.3
Corylus	188		236	0.1	164	0.1	1,034	1.1	708	0.4	225	0.1	178	0.3
Betula	112,874	51.4	79,488	39.8	73,155	52.3	42,484	45.6	69,778	44.2	93,508	45.6	28,592	49.0
Alnus	24,224	11.0	23,233	11.6	22,804	16.3	15,191	16.3	21,020	13.3	21,022	10.3	13,489	23.1
Picea	24,088	11.0	17,711	8.9	5,721	4.1	2,798	3.0	9,966	6.3	6,302	3.1	1,823	3.1
Pinus	45,438	20.7	58,776	29.4	26,380	18.9	24,828	26.7	44,327	28.1	23,533	11.5	10,668	18.3
Other A.P.	1,575	0.7	1,437	0.7	1,235	0.9	810	0.8	924	0.6	1,842	0.9	489	0.8
Σ A.P.	219,408	99.8%	199,717	99.8%	139,787	99.9%	93,131	99.8%	157,721	99.8%	204,887	100%	58,410	99.9%
Juniperus	2,705	1.0	3,426	1.4	2,480	1.5	907	0.8	819	0.5	3,753	1.6	6,298	3.9
Calluna	299	0.1	107		160	0.1	128	0.1	587	0.3	207		180	
Gramineae	12,922	5.0	14,911	6.3	7,328	4.5	7,950	6.8	8,274	4.6	9,838	4.3	25,585	16.0
Cerealia	1,879	0.7	3,348	1.4	942	0.6	6,242	5.4	2,268	1.3	1,400	0.6	1,882	1.2
Rumex	4,264	1.6	2,749	1.2	2,652	1.6	1,448	1.2	2,758	1.5	4,516	2.0	46,616	29.1
Plantago	1,030	0.3	917	0.3	751	0.4	372	0.3	653	0.3	710	0.3	206	0.1
Other N.A.P.	16,272	6.3	11,608	4.9	7,139	4.4	5,901	5.1	6,750	3.8	4,795	2.1	20,615	12.9
Σ N.A.P.	39,371	15%	37,066	15.6%	21,452	13.1%	22,948	19.7%	22,109	12.3%	25,219	10.9%	101,382	63.2%
Σ P	258,779		236,883		161,239		116,079		179,830		230,106		159,792	

Note. In the table are given total values for pollen traps on rafts. However, Vieskär represents the trap of a rocky hill—the highest part of the island. Values from the autumn season 1971 are not included in this table.

Within the deciduous forest area *Picea* pollen occurs with percentages up to 7%, although only scattered trees grow there. The values of *Fagus* and *Carpinus* pollen do not exceed 1% although they are rather common in the area. Among the NAP *Calluna* and *Juniperus* are obviously under-represented. The pollen spectra are evidently too regional in character to show any important differences within such a short transect.

The relation between pollen production and local vegetation

FOREST VEGETATION

To acquire information on the relation between the pollen production and the real distribution of the dominant trees, analyses of moss polsters and pollen traps are planned at Senoren with *Betula*, *Quercus*, *Tilia*, and *Acer*, at Färskesjön with *Betula*, *Quercus*, *Fagus*, and *Carpinus*, and at Igelsjön with *Pinus*, *Picea*, *Betula*, *Quercus*, *Fagus*, and *Corylus*. Only some preliminary results are available from the Senoren woods. They indicate that the relation of 1:8 between *Quercus* and *Tilia-Acer* found by Andersen (1970) in Denmark may be changed to 1:10 for *Quercus:Tilia* and 1:20 for *Quercus:Acer*.

PARK-MEADOW

Changes in the pollen deposition in an overgrown park-meadow area named Steneryd is being studied during the restoration of the area. All *Ulmus*, *Tilia*, and *Fraxinus* trees will be pollarded and only *Quercus* trees will remain untouched. The restoration may reflect changes which occurred in the juvenile prehistoric forests with broad-leaved trees when they were cut down for leaf fodder. This is a long-time experiment which was started in 1967 and will continue for ten years.

HUMAN LANDSCAPE

The pollen production of an old-fashioned village landscape is being studied on the island of Senoren. The oak-lime forests mentioned above belong to the infield area and these are compared to the heath area of the outfields characterised by Gramineae-*Juniperus* and *Calluna* heaths with scattered *Betula* stands. It is obvious that *Betula* is over-represented and *Calluna* and *Juniperus* under-represented as found in the lake traps.

Pollen deposition in a shallow lake

The results from the sediment traps for each of the years 1969, 1970, and 1971 are shown in Fig. 8. A comparison between the content of bottles at 70 cm and 20 cm height respectively reveals much higher and more unregular absolute values of tree pollen in the 20 cm-bottles. This must be due to an additional amount of redeposited pollen grains from the lake bottom. Therefore the results of the 70 cm bottles are more reliable for studying differences within the lake and between the water and air sedimentation. The following discussion is based on these results.

REDEPOSITION

The number of tree pollen grains is much higher in the water traps than in the corresponding air traps. The relation is about 50-100:1. This is partly caused by the difference in efficiency. Due to this the values of the air traps may be multiplied by, at a maximum, 5 (see above). If this is correct at least 10-20 times more pollen grains are deposited per unit area on the lake bottom compared with the lake surface. The annual accumulation based on dated sediment cores has not been investigated yet but the air traps should reflect the yearly input to the sediments, too. In a 10 m deep lake Davis (1968) found 2-4 times greater deposition in sediment traps compared with the deposition measured in surface sediment cores. This indicates that the redeposition increases with diminishing water depth. Rising absolute pollen values from the central part to the shores of Färskesjön indicate the same feature.

Fig. 9 shows astonishingly similar percentage values for all sediment traps but fluctuating values for air traps. However, the mean values of the air traps for 1969-71 are similar to those of the sediment traps. But some species, for example *Fagus*, *Carpinus*, *Picea*, and *Calluna*, are distinctly under-represented even in these mean values. This is another indication that the sediment traps represent pollen deposition of several years (i.e. more than three years).

However, there is a good correlation between the changing yearly amounts of tree pollen in the water traps and those of the air traps. This indicates that the water traps partly reflect the annual deposition of pollen grains, i.e. primary deposition of material from water surface, *and* the secondary deposition of fresh bottom material.

ABSOLUTE AND RELATIVE VALUES

The diagram with absolute values for the 70 cm-traps shows very smooth curves with high values near the shore and minimum values in the central part of the lake. The relative values are also very similar from point to point. If they were identical this would indicate a total mixing of pollen due to bottom erosion. However, there are slight but important differences in the percentage values, namely consistently rising values of *Pinus* and falling values of *Betula* from north to south. This is also the most important difference in the relative values from the air traps. The cause of this is probably differences in the local vegetation, as there is more *Pinus* in the southwest and east of the southern bay, and more *Betula* surrounding the northern bay.

There is no indication of pollen enrichment in the near-shore water traps due to differential flotation. Generally this is shown by high *Pinus* values, both in absolute and relative terms, near the shores (Lundqvist 1927, Hopkins 1950), but this is not the case at the investigated stations. It is possible that this applies to the shores at still lower water levels.

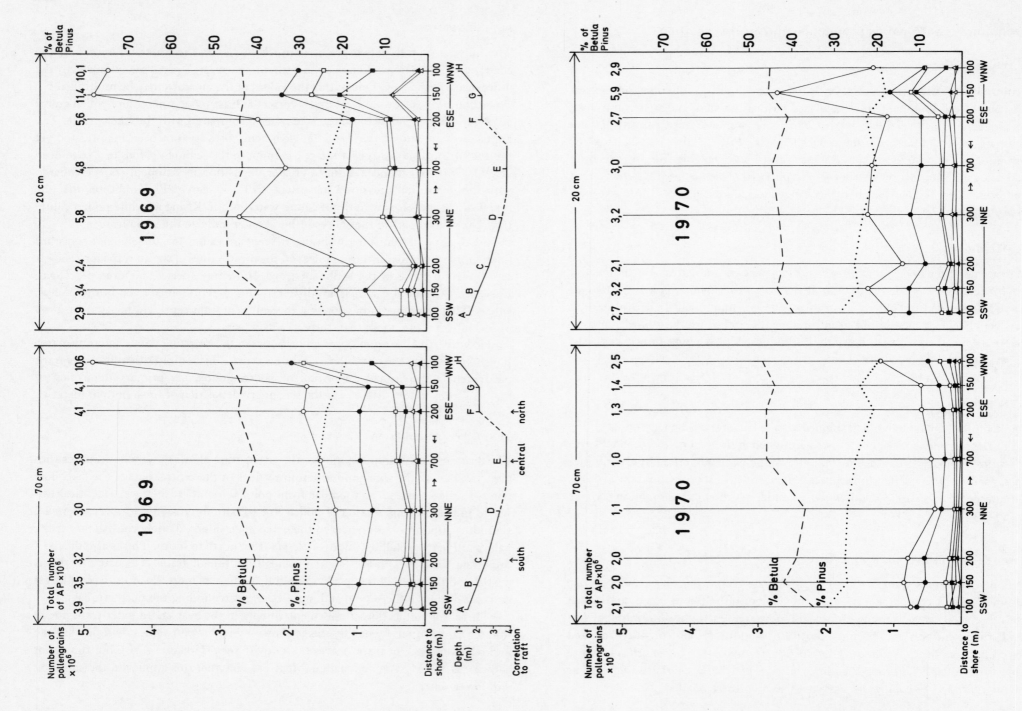

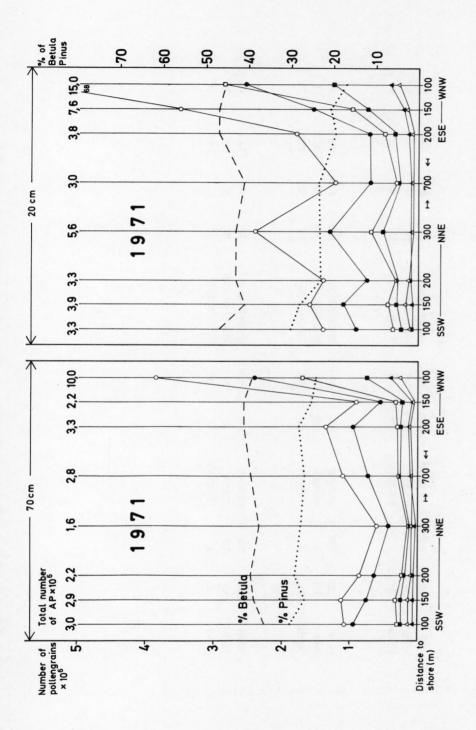

Figure 8. Absolute numbers of tree pollen in the sediment traps of Lake Färskesjön, 1969–71. The sites are arranged in a south-northern transect. Conventional pollen symbols are used, i.e. *Betula* = open circles, *Pinus* = filled circles, *Alnus* = open squares, *Quercus* = filled squares, *Picea* = open triangles, *Fagus* = filled triangles. Percentages are also given for the dominant trees, *Betula* and *Pinus*.

CORRELATION BETWEEN AIR TRAPS AND WATER TRAPS

In Fig. 9 the percentage values of the trees in the air traps and the water traps are compared. As stated above the percentage values are generally more constant in the water traps than in the air traps, both from year to year and from station to station. The reason for this can be two-fold; either erosion in older bottom material giving mean values for more than one year, or a mixture of the yearly deposition in the water. The first factor explains the constancy between the different years, the second factor the similarities between the different stations. Obviously pollen spectra from sediments in such a shallow lake as Färskesjön reveal mean values for the whole region independently of the sampling site. Possibly this applies to most non-stratified lakes with soft mud bottoms. Only air traps may be expected to reveal the local vegetation from the nearest land, as discussed below.

A correlation diagram based on the 1970-trapping is shown in Fig. 10. *Picea* is over-represented and *Quercus* under-represented in all air traps. This can be due to high and low pollen production respectively during 1970 and that the water traps contain a mixture of deposited material from more than one year. The herb pollen values show good correlations between air and water traps. However, *Calluna* is distinctly under-represented in the air traps. This must be due to its bad flowering during 1970 (the same is valid for 1969 and 1971) and that the water traps contain eroded material from more than this year.

The best correlation between air and water traps is found in the northern bay. The most probable cause of this seems to be that the local pollen supply from the surrounding woods dominates in the air as well as in the water. Due to the sheltered position erosion and mixing of bottom material may be of less importance than in the other parts of the lake.

CONCLUSION

In a shallow, non-stratified lake with an almost barren mud surface redeposition and lateral mixing of pollen are important. This has also been shown for deeper and larger lakes by Davis (1967). Relative pollen values of the mud are rather constant within the whole lake basin and reflect only slightly the local vegetation even when samples are taken in small bays. In the lake investigated here, bottom erosion has affected layers more than three years old.

It must also be noted that due to this erosion deposition values found experimentally cannot be compared with the accumulation rates found in mud cores. Sediment traps have revealed about 10 to 20 times higher annual pollen deposition than the estimated annual input to sediment. Corresponding figures for a deep lake

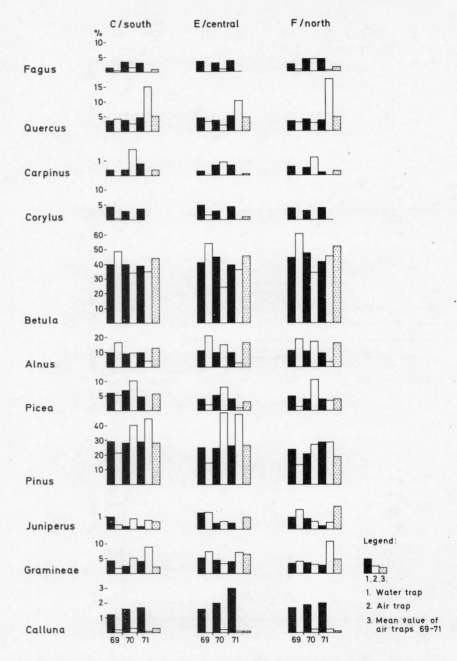

Figure 9. Comparison of relative pollen values from sediment traps and air traps at the three Färskesjön stations 1969–71.

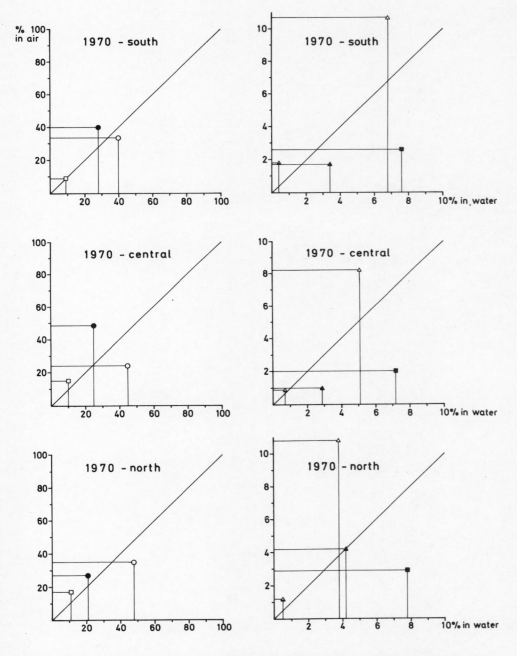

Figure 10. Correlation diagram for relative pollen values from sediment traps and air traps at the three Färskesjön stations 1970. Symbols as in Fig. 8. In addition *Carpinus* = half-filled triangles.

studied by Davis (1968) were 2 to 4 times. The redeposition is evidently much greater in shallow lakes. However, the main conclusion is a confirmation of the statement by Davis (1968): 'The process of redeposition may be largely responsible for the uniformity and consistency of pollen content that has made lake sediment such favorable material for paleoecological study'.

Future research

Based on this preliminary investigation the study will continue, if possible for a ten year period, but it will concentrate more on the deciduous forest region, i.e. on Senoren, in the Färskesjön region, and in the Steneryd park meadows. In an excursion report in 1970, Iversen considered these three areas to be of international interest for the understanding of vegetational development during the late post-glacial.

The heath area with oak, beech, and hornbeam woods west of Färskesjön is well suited for studies on present-day successions as well as Sub-Boreal and Sub-Atlantic successions based on pollen-analysis of raw-humus profiles. Experimental ecological studies as well as palaeoecological studies are planned so that the area may be a counterpart to the Draved forest research area from which Iversen had revealed many palaeoecological results of such fundamental importance (Iversen 1964, 1969).

References

ANDERSEN S.TH. (1970) The relative pollen productivity and pollen representation of North European trees, and correction factors for tree pollen spectra. *Danm. geol. Unders.* Ser. II, **96**, 99 p.

BERGLUND B.E. (1966) Late-Quaternary vegetation in eastern Blekinge, southeastern Sweden. A pollen-analytical study. II. Post-Glacial time. *Op. bot. Soc. bot. Lund* **12**(2), 190 p.

DAVIS M.B. (1967) Pollen deposition in lakes as measured by sediment traps. *Bull. geol. soc. Am.* **78**, 849–58.

DAVIS M.B. (1968) Pollen grains in lake sediments: Redeposition caused by seasonal water circulation. *Science N.Y.* **162**, 796–9.

HOPKINS J.S. (1950) Differential flotation and deposition of coniferous and deciduous tree pollen. *Ecology* **31**, 633–41.

IVERSEN J. (1964) Retrogressive vegetational succession in the Post-Glacial. *J. Ecol.* **52** (Suppl.), 59–70.

IVERSEN J. (1969) Retrogressive development of a forest ecosystem demonstrated by pollen diagrams from fossil mor. *Oikos* Suppl. **12**, 35–49.

JØRGENSEN S. (1967) A method of absolute pollen counting. *New Phytol.* **66**, 489–93.

LUNDQVIST G. (1927) Methoden zur Untersuchungen der Entwicklungsgeschichte der Seen. *Abderhalden, Handb. biol. Arb.-Meth.* 9(2), 427–462.

TAUBER H. (1965) Differential pollen dispersion and the interpretation of pollen diagrams. *Danm. geol. Unders.* Ser. II, **89**, 69 p.

TAUBER H. (1967) Investigations of the mode of pollen transfer in forested areas. *Rev. Palaeobotan. Palynol.* **3**, 277–86.

The use of modern pollen rain samples in the study of the vegetational history of tropical regions

J.R. FLENLEY *Geography Department, University of Hull*

Introduction

Pollen analysis was introduced by von Post (1916) as a geological tool, and it is therefore appropriate that it should adhere to geological principles. The greatest of these is Lyell's (1850) doctrine of uniformity—'The present is the key to the past'. In palynology this means that the pollen rain from existing plant communities should be used to interpret fossil pollen spectra in vegetational terms. The principle of uniformity must, however, be applied with care. In the first place it is not sufficient merely to show that a modern pollen spectrum resembles a fossil one, and thence to argue that both were produced by the same type of vegetation. In theory at least, it is necessary also to show that no other vegetation type produces a similar pollen spectrum. In practice this is impossible, but at least similar vegetation types should be eliminated. Secondly, the fossil spectrum may have come from a vegetation type no longer extant, either because species extinction has occurred or because species have become associated into new vegetation types.

Despite these difficulties, modern pollen rain samples (from moss polsters, water tanks, sticky slides, air samplers, rain gauges, pollen traps, or other devices) remain the most logical method for interpreting fossil results.

In the tropics (and I am confining my attention to the equatorial tropics, within c 10°north and south) palynology has developed late because it faces special difficulties, real or imagined. These are:

1 the legendary floristic diversity of the tropics has led to fears that determination of pollen types would be exceptionally difficult;
2 the great majority of tropical species are not wind pollinated (so far as is known), so that pollen production might be rather low (Faegri 1966);
3 tropical evergreen forests are frequently windless, so that it is possible no 'pollen rain' in the usual sense can form, and that pollen from each plant simply falls directly to the ground (Faegri 1966).

It is therefore understandable that almost all work to date has been in the tropical mountains. Here the diversity of species is reduced to a manageable level (although still higher than in many temperate regions), and the vegetation is often dominated by wind pollinated species (e.g. most Fagaceae) which presumably rely on anabatic/katabatic wind systems for pollination. In addition, the tropical mountains frequently bear the marks of glaciation to levels much below the present glacier limits, so that the search was soon on for corresponding palynological changes, and for modern pollen rain to interpret them.

Mountain pollen rain

SOUTH AMERICA

In general modern pollen spectra (Fig. 1) fit the existing vegetation (Cuatrecasas 1958) with surprising closeness. There is, however, one major exception to this: forest pollen can be carried well above the forest limit, presumably by up-valley winds. In the grassy paramo zone this has little effect on the spectra, but in the bare super-paramo forest pollen is a high proportion of the total. This is clearly an artefact of the percentage method, caused by the very low production of local vegetation. Even percentage spectra from the super-paramo can be recognised, however, by the presence of *Acaena* which appears to be a really good indicator of high mountain conditions; I understand (Th. van der Hammen, personal communication) there is a possibility that *Acaena* pollen is not separable from *Polylepis*, but the argument is unaffected since *Polylepis* is itself a shrub of the paramo zone (Schimper 1903).

Two examples will suffice to show the importance of modern pollen rain in interpreting South American diagrams. At Paramo de Palacio (3,500 m a.s.l.; van der Hammen & Gonzalez 1960) the lower spectra show high values of grass pollen, with *Acaena* sometimes present; these suggest paramo conditions. The upper spectra show higher values for forest pollen (although grass pollen is still present) and suggest that, at least on some occasions, forest vegetation surrounded the site.

From 1,000 m higher, the Valle de Lagunillas V diagram (Gonzalez, van der Hammen & Flint 1966) appears at first sight to suggest the reverse change; forest

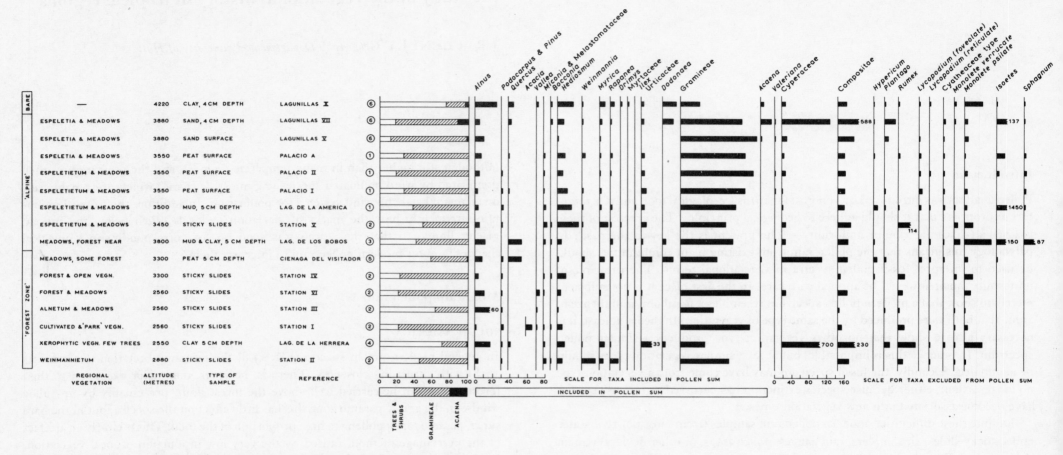

Figure 1. Modern pollen rain from the Colombian Andes: ① van der Hammen & Gonzalez 1960, ② van der Hammen 1961, ③ van der Hammen 1962, ④ van der Hammen & Gonzalez 1965a, ⑤ van der Hammen & Gonzalez 1965b, ⑥ Gonzalez, van der Hammen & Flint 1966.

pollen is more abundant in the lower spectra than in the upper ones. The presence of *Acaena* in the lower samples, however, combined with the authors' comment that pollen was sparse in these samples, gives away that these lower samples in fact indicate bare high mountain conditions, which were then followed by the grassy paramo. In general, the high altitude vegetation is indicated at the base of these diagrams, and I sometimes think that if von Post could have seen them he would have said that they exhibited 'invertence' and 'altitudinal parallelism'.

NEW GUINEA

The pollen rain (Fig. 2) again shows close parallels with the vegetation (Flenley 1969, Powell 1970), except that forest pollen, especially *Nothofagus*, is carried well outside the forest, and, in particular, well above the forest limit. Indeed I have yet to hear of a New Guinea pollen sample which does not contain *Nothofagus*; a fact which is hardly surprising since the genus holds the world record for long distance dispersal (S. America to Tristan de Cunha, about 4,000 km; Hafsten 1960). New Guinea workers are fortunate in having some good indicators of 'alpine' conditions, in particular *Astelia papuana*.

Figure 2. Modern pollen rain from the New Guinea Highlands. All data are from Powell (1970), except those marked * which are from Flenley (1967). Powell samples include *Ranunculus* in 'alpines', whereas Flenley samples include this in 'herbaceous non-forest'.

Modern pollen rain in tropical regions

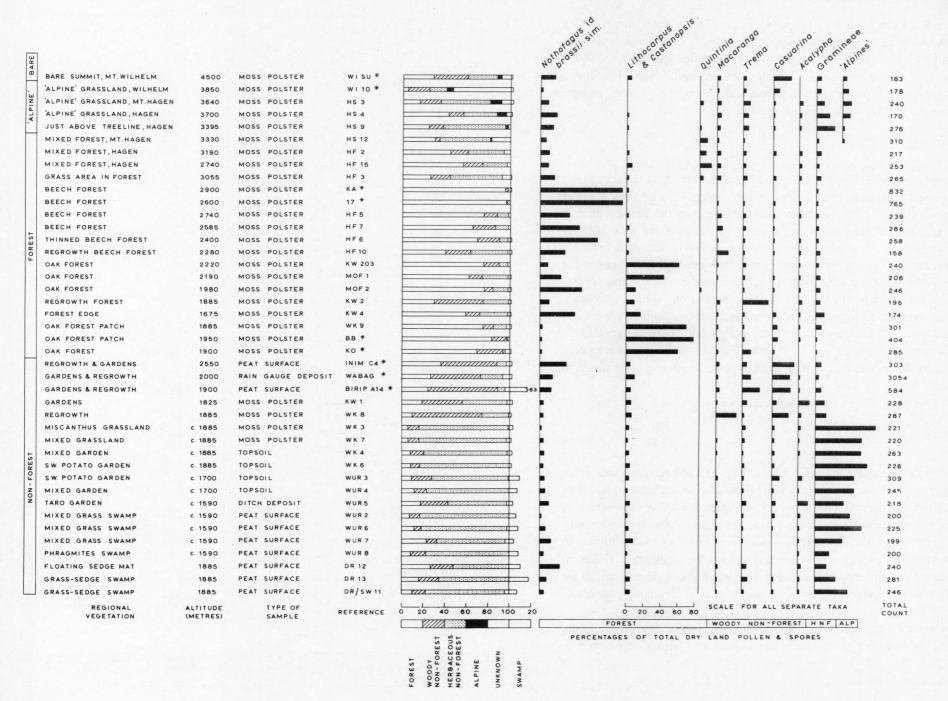

The diagram from Lake Inim C15 (2,550 m a.s.l.; Flenley 1967, 1972) shows lower spectra with about 50% *Nothofagus* pollen, about 20% Gramineae, and 20% 'alpines', especially *Astelia*. These are similar to modern pollen spectra from above the forest limit, except that the values for 'alpines' are unusually high. The upper spectra show no 'alpines', and increases in forest and 'woody non-forest' pollen; these are more like the spectra from forested areas today.

BORNEO

A series of modern pollen rain samples has been obtained from moss polsters in the relatively undisturbed vegetation (Stapf 1894, Gibbs 1914, Corner 1964 a, b) of Mt Kinabalu, Sabah (Fig. 3). The topmost sample came from the surface mud in the Sacrificial Pool near the summit. The results (Fig. 4) show the now familiar pattern; tree pollen is carried well above the forest limit. The chief tree involved is *Phyllocladus* in this case. The striking point here is the relatively low values for Gramineae. This, however, is not surprising when one considers the summit vegetation, for, most unusual among tropical mountains, there is no 'alpine' grassland. The trees give way directly to bare granite, with vegetation, including some Gramineae, largely restricted to rock crevices. In the pollen rain there are once again good 'alpine' indicators, however, e.g. *Trachymene*. Further striking changes occur at lower altitudes in the diagram, e.g. the change from Fagaceae dominance to conifer dominance.

We have as yet no fossil diagram from Kinabalu for comparison.

EAST AFRICA

The modern pollen rain diagram from E. Africa (Fig. 5) has been compiled from the work of numerous authors working on several isolated mountains whose flora is not uniform (Hamilton 1972). The zonation of vegetation differs considerably from one mountain to another (Hedberg 1951, Coe 1967), and it has been necessary to depart considerably from the altitudinal sequence in order to place samples in the vegetation zone or belt in which they are said to have been collected. Different authors used different pollen categories and different methods of calculation, so for uniformity all data were recalculated as percentages of total dry land pollen and spores.

The diagram shows once again the uphill drift of forest pollen, especially *Podocarpus*. In Africa, however, there is no clear indicator of 'alpine' conditions to

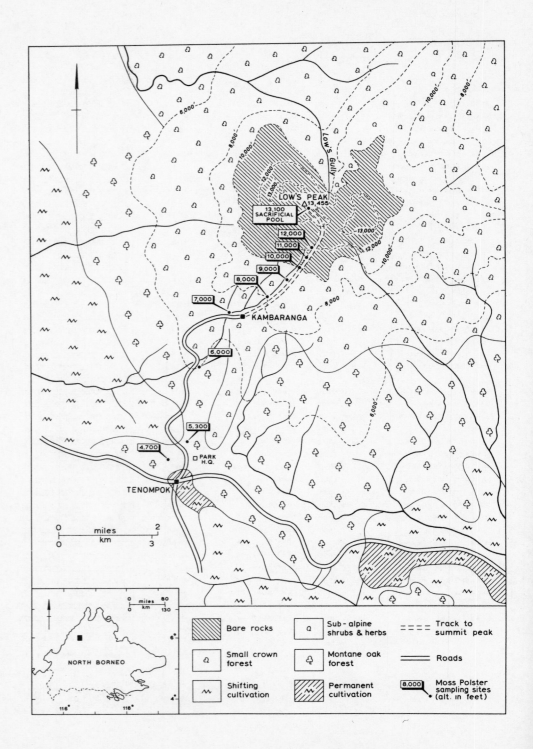

Figure 3. Mt Kinabalu, Sabah, Borneo.

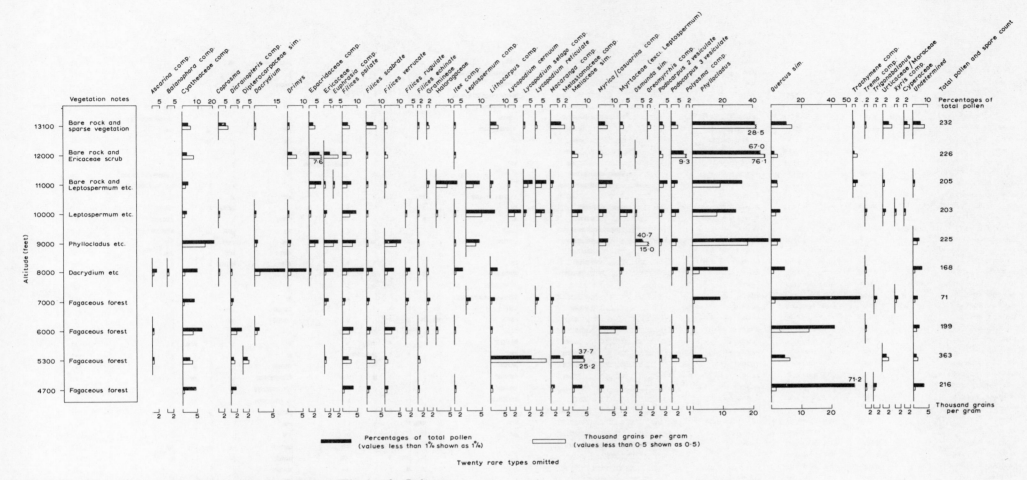

Figure 4. Modern pollen rain from Mt Kinabalu, Sabah, Borneo. The data for *Podocarpus* 2-vesiculate may include some *Pinus*.

redeem this situation. Several taxa are more abundant above the forest limit, e.g. *Alchemilla*, *Dendrosenecio* (where distinguished; most authors include it in Tubuliflorae), Gramineae—but all these can occur in significant quantities at lower altitudes. Only *Stoebe* offers some hope, but this is an indicator of the forest limit rather than of 'afro-alpine' conditions.

Despite this, there are now African diagrams which are difficult to interpret in terms other than altitudinal shift. Morrison (1968) has produced a diagram in which spectra suggesting 'afro-alpine' vegetation (base—950 cm), forest limit (950–700 cm), *Hagenia-Hypericum* zone (700–400 cm), and montane forest zone (400 cm—top) occur in sequence.

CONCLUSIONS ABOUT MOUNTAIN POLLEN RAIN

Absolute deposition rates

Since most workers have relied on moss polsters or surface organic deposits rather than pollen trapping over a known period, there are few absolute data available so far. In Colombia, however, van der Hammen (1961) used coated slides for pollen trapping and caught between 50 and over 500 grains/cm^2/annum. These are very low values. In New Guinea I collected sediment from a standard 5 inch (125 mm) rain gauge over 12 months and caught an estimated 144,000 grains. This gives a rate

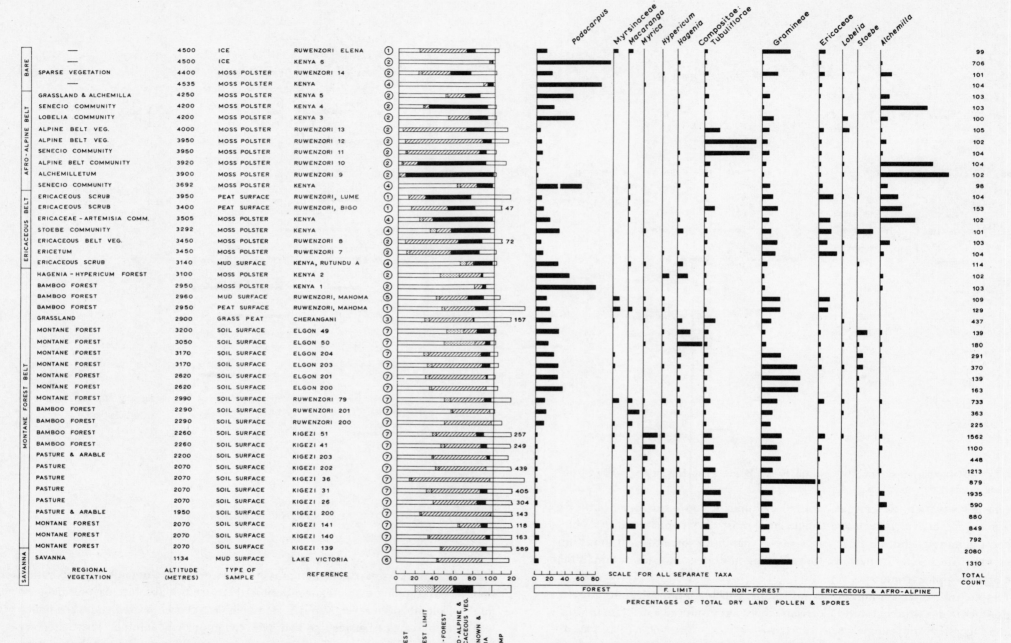

Figure 5. Modern pollen rain from East Africa: ① Osmaston 1958, ② Hedberg 1955, ③ van Zinderen Bakker 1964, ④ Coetzee 1967, ⑤ Livingstone 1967, ⑥ Kendall 1969, ⑦ Hamilton 1972.

of deposition of 1,130 grains/cm^2/annum which is not dissimilar from figures in Britain (Hyde 1952) but is much less than the rates of 20–25,000 grains/cm^2/annum quoted for North America during the post-glacial (Davis 1967).

Absolute dispersal

By this I mean quantitative absolute data on dispersal of individual pollen types. So far as I am aware, no such data are yet available for tropical species.

Relative production

Some ideas of relative production of different vegetation types can be obtained by examination of pollen spectra taken on the boundary between two vegetation types. In general, however, it is impossible to separate this from relative dispersal.

Relative dispersal

A classification of pollen types in terms of dispersal power has been introduced by Hamilton (1972). If a pollen type is present only in samples collected within vegetation in which it occurs, then he regards it as of low dispersal power. If it is present also in samples collected in adjacent vegetation it is termed of moderate dispersal power. Pollen types which occur in all or almost all samples, wherever collected, are designated as of high dispersal power. These are useful categories, but unfortunately titled. As long as all data from the tropical mountains are based on moss polsters or surface samples, and therefore on relative counts, it will be impossible to extract conclusions on relative dispersal, e.g. if in a grassland above the forest limit the pollen spectra contain large proportions of forest pollen, it is impossible to say whether this is the result of low relative *production* by the grassland, or high relative *dispersal* by the forest (or both). In such a case it might be best to speak of relative export.

Relative export

It is therefore suggested that Hamilton's terms be re-titled high, medium, and low relative export. In tropical montane regions, it is clear that pollen types of all three categories are present. The most conspicuous examples of high relative export are in the detection of forest pollen types well above the forest limit. In S. America *Alnus*, *Hedyosmum* etc., in Africa *Podocarpus*, in New Guinea *Nothofagus*, and on Mt. Kinabalu *Phyllocladus* all behave in this way. It is a surprising fact that this export is all upwards: it is extremely rare for pollen types from above the forest limit to be detected below it. Greater production by the forest (Hamilton 1972) is probably one reason for this, but meteorological conditions could also be important. Winds on tropical mountains are usually up-valley during the day and down-valley at night. Presumably most pollen is released during the day when lower humidity causes the anthers to open (Percival 1950). A gentle breeze of a few km per hour could carry pollen a long way uphill during the afternoon. Deposition would presumably occur in the afternoon or evening rainfall so common in the tropics, or in the still evening air. Thus little pollen would remain airborne to be carried down in the night-time wind. Upward pollen export is well known in temperate regions such as the Alps (Rudolph & Firbas 1926).

Pollen sum

Hamilton (1972) has argued that as well as omitting from the pollen sum the strictly local pollen (aquatics, etc.) one should also omit those types of high dispersal power (my high relative export). This is partially in line with Wright and Patten's (1963) view that the pollen sum should include those taxa which occur in the vegetation under investigation. Thus in those parts of the pollen diagram where montane grassland taxa dominate, the vegetation was probably a montane grassland and the trees are irrelevant and should be omitted. In most tropical montane diagrams, however, the upper part of the diagram is dominated by forest pollen; in that case the exclusion of trees would mean exclusion of constituent taxa of the likely vegetation. In any case there is a constant danger of circular argument. For complete interpretation of diagrams of this type, it is probably necessary to have the complete results calculated in several ways.

Interpretation of pollen diagrams

There has been a great deal of controversy in recent years as to whether the pollen diagrams from the montane tropics indicated changes in temperature or in the precipitation/evaporation ratio, or both. In Africa it has been particularly difficult to solve this problem because of the lack of good altitudinal indicators in the 'afro-alpine' zone. Hamilton's (1972) work lends more support to thermal than to hydrologic change in the mountains, but Kendall (1969) gives pollen and other evidence from Lake Victoria for marked hydrologic changes. Nor can these changes necessarily be explained by the simple view that lower temperatures cause lower evaporation and therefore lead to a wetter climate; indeed there is considerable evidence now for the reverse correlation (Livingstone & Kendall 1969, van Zinderen Bakker &

Coetzee 1972). A further discussion of these matters here would be inappropriate, but it is worth making two points. Firstly, the controversy has been particularly strong in relation to Africa, where the modern pollen rain is now shown to be lacking in good indicators of 'afro-alpine' conditions. Secondly, in all regions the modern pollen rain has been studied chiefly in humid areas, so that we have little idea of what pollen rain from 'dry' mountains is like. There is clearly room for research here, but in some regions (e.g. New Guinea) it would be impossible to find a 'dry' mountain on which to take samples.

Lowland pollen rain

MALAYSIA

If tropical montane pollen analysis is still at an immature stage, then tropical lowland studies in this field must be regarded as almost pre-natal. The reasons for this are not hard to find; in the lowlands the three great problems of pollen diversity, relative rarity of anemophily and relative windlessness are believed to reach their utmost culmination. These considerations led Faegri (1966) to the conclusion that pollen analysis in the usual sense might be impossible in the lowland tropics. Additional problems were the danger (for fossil work) that many pollen types might not preserve at the relatively high ambient temperatures, and (for modern pollen rain studies) the fact that moss polsters are surprisingly rare in the lowland forest. It is true that pollen does not preserve well in tropical lowland *soils*; Hamilton (1972) found little undamaged pollen in soils below 2,000 m a.s.l. But in anaerobic waterlogged conditions preservation is good (Muller 1965); many pre-Quaternary sediments rich in pollen and spores are assumed to have formed under tropical conditions.

Surface samples from lowland swamp forest in Borneo were analysed by Muller (1965) who found that lateral movement of pollen was indeed very restricted, so that samples tended to be dominated by the nearby trees.

To investigate the pollen rain of a lowland hill forest I have used a pollen trap which is a modification of Oldfield's (unpubl.) design (Fig. 6). The principle is that pollen is caught on a mass of de-oiled acetate yarn which is kept permanently moist by a water reservoir underneath. For use in the tropics it was judged impracticable to fit a reservoir large enough for a year's rainfall; instead a conical flask was used and excess water drains out of the air holes (Fig. 6). This trap is cheap in construction and has been tested in Britain. It was highly successful in catching pollen but in most cases the conical flask was shattered in Britain, presumably by freezing in winter; a polythene flask would clearly be more appropriate here. Three of these traps were placed on the floor of the rain forest within the plot recorded by McClure

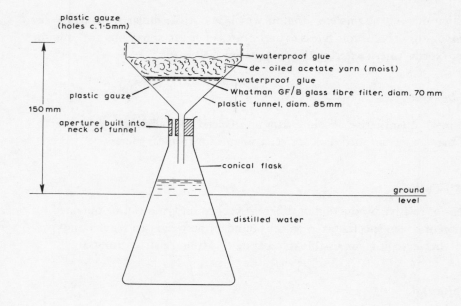

Figure 6. Modified Oldfield pollen trap.

(1966) and Wiedemann (1969) in the Ulu Gombak Virgin Jungle Reserve, Selangor, Malaysia, at intervals of 66 feet (20 m). A fourth trap was placed on a platform at a height of 43 m in a large tree of *Anisoptera laevis*. The ground traps were left out for one year (21 February 1969 to 1970), the trap in the tree from 4 May 1969 to approx. 4 May 1970. One of the ground traps was not recovered, having presumably been removed by one of the numerous primate species in the vicinity. From the recovered traps, the de-oiled acetate yarn was dissolved in acetone and the pollen prepared in the usual way.

The total grains and spores caught per year in a trap varied (Fig. 7) between 47,000 and 120,000, the lowest figure being for the canopy trap. Our trap has a surface area of c 59 cm^2, so that the catch varied between 800/cm^2/annum and 2,020/cm^2/annum. F. Oldfield (personal communication) had catches in four of his traps set out in Britain ranging from below 1,100 to 6,000 grains/cm^2/annum. It is, of course, difficult to compare different methods of catching, but Hyde (1950) caught c 1,600 to c 5,400 grains/cm^2/annum in Britain using slides coated in glycerine jelly. These figures are all of the same order, but are again much below those quoted by Davis (1967) for the post-glacial in North America.

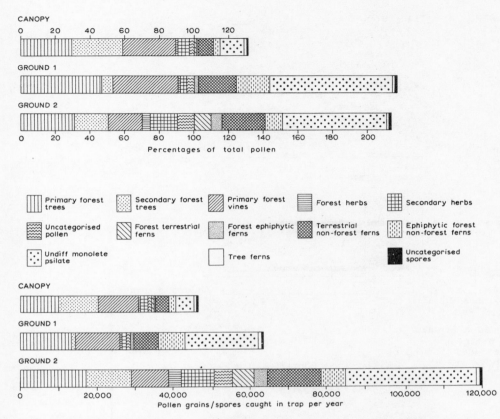

Figure 7. Summary diagram for modified Oldfield trap samples from Ulu Gombak Virgin Jungle Reserve, Selangor, Malaysia.

The composition of the spectra from these traps is also extremely interesting, although the paucity of data makes conclusions impossible. The quantity of *pollen* caught by the canopy trap was intermediate between the pollen catches of the two ground traps. On the other hand, the quantity of *fern spores* in the ground traps was at least twice that in the canopy trap. On a percentage basis (Fig. 8) these facts become even clearer. As percentages of total pollen, the ground traps each had about 50% psilate fern spores; the canopy trap had 10%. For *Asplenium* comp. the figures are 15% and 3% and for *Nephrolepis* comp. 15% and 6%. But the really interesting thing about these lowland forest samples is the sheer diversity of the catch. The mean number of types recognised (many still unidentified and capable of subdivision) was 60 in a count of 925 in the ground traps and 62 in a count of 779 in the canopy trap. In a detritus sample from mixed forest at 700 feet (213 m) a.s.l. near the foot of the Cameron Highlands road, Malaysia, 60 pollen and spore types were recognised in a count of 621 (including extra traverses examined for presence only). In this sample 91% of the pollen (excluding spores) was of one type (*Eugenia* comp.) and there was a tree of *Eugenia* sp. directly above, yet still the fantastic diversity was maintained.

Other workers have drawn attention to this diversity in relation to recent fossil material. Polak (1933) in the cores she studied in Java and Sumatra found a very large number of types (although few were identified). Muller (1965) lists 51 types found in peat in Sarawak (not all in one sample)—and all these are at the generic or family level, so that a large number of species could be represented, although the forest in the peat swamps is relatively undiverse (Anderson 1963).

SUGGESTIONS ABOUT THE LOWLAND MODERN POLLEN RAIN

It must be emphasised that suggestions are all that are being made; conclusions lie in the future.

1 The chief unusual feature of the pollen rain is that it is very diverse, although not impossibly so from the point of view of the pollen counter. Identifications frequently have to be only to the family or genus, which will make ecological interpretation of pollen diagrams very difficult.

2 The magnitude of the pollen rain is of the same order as that in at least some temperate regions.

3 Although there is a tendency for dominance by local trees, the diversity of a lowland sample is still so great as to make it recognisable, quite apart from the presence in small quantities of likely lowland forest indicators such as Dipterocarpaceae comp. Dominance by local trees is possibly less marked in hill forest, perhaps due to better development of winds there.

4 In terms of Tauber's (1965) model, with rain, canopy, and trunk-space components, the situation appears to be comparable with that in temperate forests. The trap in the tree presumably caught the rain and canopy components, the sum of which is thus shown to be a large proportion of the total. A considerable trunk-space component, consisting largely of fern spores, is also suggested by the additional catch of the ground traps.

General conclusions

1 There appears to be no fundamental difference yet demonstrated between tropical and temperate modern pollen rain. The differences are all of degree.

2 There is a great need for absolute studies on both modern and fossil pollen, in both the highland and lowland tropics.

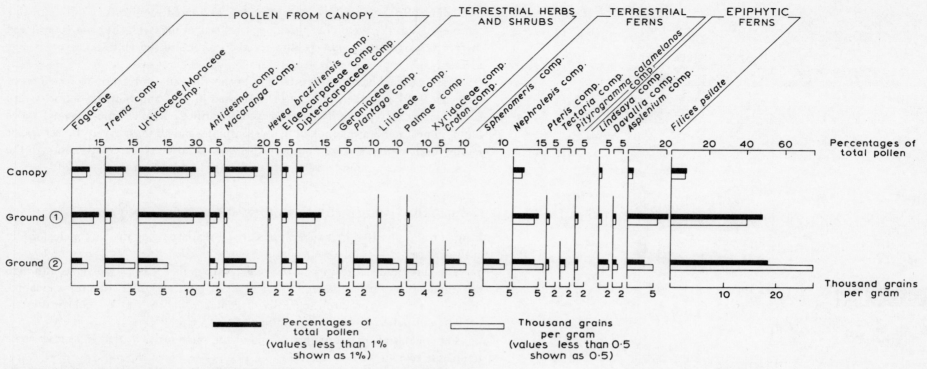

Figure 8. Selected pollen and spore types in modified Oldfield trap samples from Ulu Gombak Virgin Jungle Reserve, Selangor, Malaysia.

3 With the advent of computer assisted information retrieval in palynology (Walker et al. 1968, Germeraad & Muller 1970) no insuperable barrier is seen to the palynological investigation of even lowland tropical areas, as envisaged by von Post (1946).

Acknowledgements

I am particularly grateful to Dr J.M. Powell, Dr A.C. Hamilton, and Mr R.J. Morley for allowing me to use their unpublished results. I am also indebted to Professor F. Oldfield for supplying the yarn used in the pollen traps, to Dr E. Soepadmo for retrieving the ground traps, to Lord Medway for placing and retrieving the canopy trap, to Dr L.K. Wade for the moss polsters from Mt Wilhelm, and to Miss E. Spooner for help in the laboratory. The N.E.R.C. provided a research grant.

References

ANDERSON J.A.R. (1963) The flora of the peat swamp forests of Sarawak and Brunei, including a catalogue of all recorded species of flowering plants, ferns and fern allies. *Gdn's Bull., Singapore*, **20**, 131–228.

BAKKER E.M. VAN ZINDEREN (1964) A pollen diagram from Equatorial Africa: Cherangani, Kenya. *Geologie Mijnb.* **43**, 123–8.

BAKKER E.M. VAN ZINDEREN & COETZEE J.A. (1972) A re-appraisal of Late Quaternary Climatic Evidence from Tropical Africa. *Palaeoecology of Africa* **7** (in press).

COETZEE J.A. (1967) Pollen analytical studies in East and Southern Africa. *Palaeoecology of Africa* **3**, 1–146.

COE M.J. (1967) The ecology of the alpine zone of Mt Kenya. *Monographiae biologicae* **17**, 95p. The Hague.

CORNER E.J.H. (1964a) Royal Society Expedition to North Borneo 1961: Reports. *Proc. Linn. Soc. Lond.* **175**, 9–56.

CORNER E.J.H. (1964b) A discussion of the results of the Royal Society Expedition to North Borneo, 1961. *Proc. R. Soc. B.* **161**, 1–91.

CUATRECASAS J. (1958) Aspectos de la vegetation natural de Colombia. *Revta Acad. colomb. Cienc. exact. fis. nat.* 10, 221-64.

DAVIS M.B. (1967) Pollen accumulation rates at Rogers Lake, Connecticut, during Late- and Postglacial time. *Rev. Palaeobotan. Palynol.* 2, 219-30.

FAEGRI K. (1966) Some problems of representativity in pollen analysis. *Palaeobotanist* 15, 135-40.

FLENLEY J.R. (1967) *The present and former vegetation of the Wabag region of New Guinea.* Unpublished Ph.D. thesis, Australian National University.

FLENLEY J.R. (1969) The vegetation of the Wabag region, New Guinea Highlands: a numerical study. *J. Ecol.* 57, 465-90.

FLENLEY J.R. (1972) Evidence of Quaternary vegetational change in New Guinea. In *The Quaternary Era in Malesia* (ed. by P.S. Ashton), Proceedings of the Second Aberdeen-Hull Symposium on Malesian Ecology. *Hull University Geography Department, Miscellaneous Series* (in press).

GERMERAAD J.H. & MULLER J. (1970) A computer-based numerical coding system for the description of pollen grains and spores. *Rev. Palaeobotan. Palynol.* 10, 175-202.

GIBBS L.S. (1914) A contribution to the flora and plant-formations of Mount Kinabalu. *J. Linn. Soc., Bot.* 42, 1-240.

GONZALEZ E., HAMMEN TH. VAN DER & FLINT R.F. (1966) Late Quaternary glacial and vegetational sequence in Valle de Lagunillas, Sierra Nevada del Cocuy, Colombia. *Leid. geol. Meded.* 32, 157-82.

HAFSTEN U. (1960) Pleistocene development of vegetation and climate in Tristan da Cunha and Gough Island. *Årbok Univ. Bergen. (Mat.-Naturv. Ser.)* 20, 1-48.

HAMILTON A.C. (1972) The interpretation of pollen diagrams from Highland Uganda. *Palaeoecology of Africa* 7 (in press).

HAMMEN TH. VAN DER (1961) Deposicion reciente de polen atmosferico en la Sabana de Bogota y Alrededores. *Boletin Geologico (Servicio Geologico National de Colombia)* 7, No. 1-3, 183-94.

HAMMEN TH. VAN DER (1962) Palinologia de la Region de 'Laguna de los Bobos'. Historia de su clima, vegetacion y agricultura durante los ultimos 5,000 años. *Revta Acad. colomb. Cienc. exact. fis. nat.* 11, No. 44.

HAMMEN TH. VAN DER & GONZALEZ E. (1960) Holocene and Late Glacial climate and vegetation of Paramo de Palacio (E. Cordillera, Colombia, S. America). *Geologie Mijnb.* 39, 737-45.

HAMMEN TH. VAN DER & GONZALEZ E. (1965a) A pollen diagram from 'Laguna de la Herrera' (Sabana de Bogota). *Leid. geol. Meded.* 32, 183-91.

HAMMEN TH. VAN DER & GONZALEZ E. (1965b) A Late-glacial and Holocene pollen diagram from Cienaga del Visitador (Dept. Boyaca, Colombia). *Leid. geol. Meded.* 32, 193-201.

HEDBERG O. (1951) Vegetation belts of the East African Mountains. *Svensk bot. Tidskr.* 45, 140-202.

HEDBERG O. (1955) A pollen-analytical reconnaissance in Tropical East Africa. *Oikos* 5, 137-66.

HYDE H.A. (1950) Studies in atmospheric pollen. IV. Pollen deposition in Great Britain, 1943. *New Phytol.* 49, 398-420.

KENDALL R.L. (1969) An ecological history of the Lake Victoria Basin. *Ecol. Monogr.* 39, 121-76.

LIVINGSTONE D.A. (1967) Postglacial vegetation of the Ruwenzori Mountains in Equatorial Africa. *Ecol. Monogr.* 37, 25-52.

LIVINGSTONE D.A. & KENDALL R.L. (1969) Stratigraphic studies of East African Lakes. *Mitt. Int. Verein. Limnol.* 17, 147-53.

LYELL C. (1850) *Principles of Geology* (8th edition). London.

MCCLURE H.E. (1966) Flowering, fruiting and animals in the canopy of a tropical rain forest. *Malay. Forester* 29, 192-203.

MORRISON M.E.S. (1968) Vegetation and climate in the uplands of south-western Uganda during the later Pleistocene period. I. Muchoya Swamp, Kigezi District. *J. Ecol.* 56, 363-84.

MULLER J. (1965) Palynological study of Holocene peat in Sarawak. *Proc. Symposium on ecological research in humid tropics vegetation, Kuching, Sarawak, July 1963*, 147-56. UNESCO Science Co-operation Office for S.E. Asia.

OSMASTON H.A. (1958) *Pollen analysis in the study of the past vegetation and climate of Ruwenzori and its neighbourhood.* B.Sc. thesis, Oxford, and Uganda Forest Department.

PERCIVAL M. (1950) Pollen presentation and pollen collection. *New Phytol.* 49, 40-63.

POLAK E. (1933) Ueber Torf und Moor in niederlandisch Indien. *Verhand. Kon. Ned. Akad. Wet.* 30, 6-84.

POST L. VON (1916) Om skogstradpollen i sydsvenska torvmosselagerföljder. *Geol. För. Stockh. Forh.* 38, 384.

POST L. VON (1946) The prospect for pollen analysis in the study of the Earth's climatic history. *New Phytol.* 45, 193-217.

POWELL J.M. (1970) *The impact of man on the vegetation of the Mt Hagen region, New Guinea.* Unpublished Ph.D. thesis, Australian National University.

RUDOLPH K. & FIRBAS F. (1926) Pollenanalytische Untersuchung subalpiner Moores des Reisengebirges. *Ber. dt. bot. Ges.* 44, 227-38.

SCHIMPER A.F.W. (1903) *Plant-geography upon a physiological basis.* (English translation by W.R. Fisher), Clarendon Press, Oxford.

STAPF O. (1894) On the flora of Mount Kinabalu in North Borneo. *Trans. Linn. Soc. Lond. Bot.* 4, 69-263.

TAUBER H. (1965) Differential pollen dispersal and the interpretation of pollen diagrams, with a contribution to the interpretation of the elm fall. *Danm. geol. Unders.* Ser. 11, 89, 69p.

WALKER D., MILNE P., GUPPY J. & WILLIAMS J. (1968) The computer assisted storage and retrieval of pollen morphological data. *Pollen Spores* 10, 251-62.

WIEDEMANN A.M. (1969) A quadrat in the Ulu Gombak Jungle Reserve. *Malay. Nat. J.* 22, 159-63.

WRIGHT H.E. & PATTEN H.L. (1963) The pollen sum. *Pollen Spores* 5, 445-50.

Modern pollen rain studies in some arctic and alpine environments

H.J.B. BIRKS *The Botany School, University of Cambridge*

Introduction

Pollen analysis provides information relevant to the reconstruction of the past flora and vegetation of an area in the form of the numerical data resulting from the counts of the various pollen and spore types present in a stratigraphical series of sediment samples. The interpretation of such data in terms of past flora and vegetation has long been recognised as a complex problem, requiring not only a thorough knowledge of the present-day ecology of the taxa involved, but also information on the relationships between modern pollen rain and the vegetation from which it is derived.

Some functional relationship is usually presumed to exist between the number of pollen grains of a given taxon deposited in the sediment and the number of individuals of that taxon in the vegetation surrounding the site of deposition (Davis 1963). The function is undoubtedly complex and contains a large number of variables. Among these are the physiological and ecological factors affecting the flowering and pollen production of the individual plant, the abundance of the taxon within the vegetation, the structure of the community in which it occurs, the mode of pollen dispersal, the meteorological factors influencing pollen transportation, and the physical, chemical, and biological conditions that control pollen sedimentation and preservation at the site of deposition. The interactions of these factors are so complex (see Davis 1968, Tauber 1965, 1967) that few authors other than Livingstone and Estes (1967) and Livingstone (1968) have attempted any quantitative reconstructions of past vegetation.

For most purposes it is generally assumed that the pollen frequency of a given taxon is roughly proportional to the abundance of the taxon in the surrounding vegetation, and that gross changes in the numbers of pollen grains and spores (commonly expressed as relative frequencies) in a sediment are interpretable in terms of corresponding changes in the composition of the vegetation. Because pollen spectra give no direct information about the spatial composition and distribution of the plant communities that comprise the vegetation of an area, an inferential approach is required to reconstruct the past vegetation.

One approach to the problem of reconstructing plant communities from fossil pollen assemblages involves the use of numerical procedures. If an assemblage of fossils occurs consistently in a series of samples within and between stratigraphical sequences, interspecific associations or 'recurrent groups' can be derived statistically from the observed fossil data using an appropriate similarity coefficient (Fager 1957, Johnson 1962). Such an approach has been used in the reconstruction of past communities from fossil assemblages of marine invertebrates (Johnson 1962, Valentine & Mallory 1965), of conodonts (Kohut 1969), of microplankton (Brideaux 1971), and of Permian and Quaternary pollen and spores (Clapham 1970, Martin & Mosimann 1965, Harris & Norris 1972). In these instances, however, the unit of study, namely the fossil assemblage, is assumed to be closely related in space and time to the death assemblage or thanatocoenose, and thus to the life assemblage or biocoenose. In pollen analysis, much of the pollen rain that enters a site of deposition and thereby forms the death and subsequently the fossil assemblage, is derived from an undefinable source area and life assemblage. Recurrent groups of fossil pollen and spores can only refer, therefore, to 'associations' in time, whereas the Quaternary palaeoecologist is interested in the association of taxa both in time and space.

A second approach requires knowledge of both the modern pollen representation and the ecological tolerances and preferences of the *individual* taxa concerned. From the relationships between the pollen percentages of different taxa in modern spectra and the proportions of the same taxa in the surrounding vegetation, 'correction factors' or R-values can be calculated for the individual components of the pollen rain (Davis 1963). The theoretical composition of the former vegetation can then be inferred by assuming that the contemporary R-values for each taxon are applicable to the fossil pollen spectra. There are, however, many problems in the calculation and use of R-values (Davis 1969, Faegri 1966, Lichti-Federovich & Ritchie 1965). The most serious drawback is that the areal extent of the vegetation contributing pollen to a medium- or large-sized basin of deposition is generally not known with any certainty, and it is thus impossible to delimit accurately the size of plot to be sampled. It is only in specialised instances such as within a closed forest or in very small basins (Andersen 1970, 1973) that this problem can be overcome. A further problem to this approach is that the R-value for a given taxon depends not only on its abundance in a particular community, but on the structure and composition of the vegetation in which it grows. Janssen (1967a) and Comanor (1968) have demonstrated significant variations in R-values for the same tree taxa occurring in different forest types. In view of these variations from one area to another, contemporary R-values may not be generally applicable at anything less than order-of-magnitude level. Thus, as changes in environment, community structure, and pollen production and sedimentation may have occurred with time, R-values based on single points in space and time, may not be applied, with any great confidence, to fossil spectra derived from different depositional situations. Although R-values have been used in the interpretation of fossil pollen frequencies from medium- and large-sized basins (Davis 1963, Livingstone 1968) the quantitative significance of their conclusions is open to question.

A third approach, and the one discussed here, is to attempt to characterise a range of modern vegetation types by means of contemporary pollen spectra and then to compare the fossil pollen assemblages with modern pollen spectra (Davis 1967, Lichti-Federovich & Ritchie 1965, 1968, McAndrews 1966). If similarities in pollen content and proportions exist between the modern and the fossil spectra, it may be concluded that a similar vegetation produced the fossil assemblage, thereby providing some basis for the reconstruction of past vegetation and vegetation changes in terms of modern analogues (Wright 1967). If similar pollen assemblages can be recognised in a stratigraphic and thus a temporal sequence of fossil pollen spectra, the past changes in vegetation over time can thus be interpreted in terms of present differences in vegetation in space (see Davis 1969). Besides comparing modern and fossil pollen percentages, the occurrence of pollen and spore types of 'indicator species', here defined as morphologically distinctive pollen and spores of taxa that are characteristic or diagnostic of particular modern plant communities, in fossil spectra provide a further basis for reconstruction of past vegetation (see Janssen 1967b, 1970). If no match can be made with the fossil spectra, then in the absence of evidence for sediment disturbance and associated mixing of pollen and spores from different levels either modern analogues should be sought elsewhere, or it may be concluded that the vegetation of a particular time interval has no modern counterpart.

As there are floristic affinities between the Late-Weichselian ('late-glacial') pollen floras in Britain and elsewhere in north-west Europe and the present flora of arctic and alpine regions in northern Europe, a review of modern pollen spectra in relation to vegetation in arctic and alpine environments is presented to provide information for subsequent interpretations of Late-Weichselian pollen spectra in terms of modern vegetational analogues.

The plan of the paper is as follows. Section 1 discusses various approaches to the interpretation of fossil pollen spectra in terms of vegetation. Section 2 considers the general characteristics of the modern pollen rain in arctic and sub-arctic regions on the basis of studies in northern Europe and North America. Section 3 is concerned with the modern pollen rain of different communities within the arctic zone in Europe with reference to recent work in Greenland and Iceland. Section 4 discusses the results of a survey of modern pollen spectra produced by Scottish alpine and sub-alpine vegetation in an attempt to characterise contemporary communities in terms of modern pollen rain. Section 5 describes the application of various numerical procedures to problems of data-handling of the Scottish data, and Section 6 is a general discussion of the problems and limitations of the comparative approach.

No reference is made in this paper to work on modern pollen rain in alpine environments in North America (see, for example, Heusser 1969, Maher 1963, 1964) or in the Alps of Central Europe.

The use of the terms 'arctic' and 'alpine' follows Sjörs (1963, 1965) and Rune (1965). The alpine zone is here defined as the ground and its associated biota lying above the potential altitudinal limits of forest and tall scrub. No attempt is made to separate the alpine zone in Scotland into low-alpine, middle-alpine, and high-alpine belts which can be recognised in Scandinavia (Dahl 1956, Rune 1965), as such a division would result in an arbitrary separation between closely related vegetational types. Therefore the more general term 'alpine zone' is used (Birks 1973). The arctic zone is here defined as the ground and its associated biota lying north of the potential latitudinal limits of forest and tall scrub. A subdivision into low-arctic, mid-arctic, and high-arctic has been used in reference to the North American arctic zone, following Polunin (1951).

The terms sub-arctic and sub-alpine are used here to refer to the generally well-marked transitional zone of open woodland and tall scrub that occurs between the forest and the arctic or alpine zones in northern Europe and America today (see Löve 1970, Rune 1965, Sjörs 1963).

The botanical nomenclature follows that of Clapham, Tutin and Warburg (1962) for vascular plants; Warburg (1963) for mosses; and James (1965) for lichens. For taxa not considered by these authors, the relevant authorities are cited when the taxa are first mentioned in the text.

Modern pollen rain characteristics in arctic environments

The pioneer work on modern pollen rain in sub-arctic and arctic regions was done in Finnish Lapland by Auer (1927), Firbas (1934), and Aario (1940). Auer (1927) showed that on hills 400–500 m high, the modern pollen rain on the slopes covered by birch forest consisted primarily of *Betula* pollen with some *Pinus* pollen, whereas samples from the treeless summit areas showed a predominance of *Pinus* pollen and reduced *Betula* values. Firbas (1934) presented results of pollen analyses of a series of surface raw humus samples collected by Preuss in the sub-arctic forest–tundra transition in Finnish Lapland between 69° and 69°30′ N. In the samples within the pine forest region, pollen of *Pinus* predominated with moderate frequencies of *Betula* and *Picea* pollen, and low values of non-arboreal pollen (N.A.P.). In samples from the birch forest region the pollen percentages had at least 50% *Betula* pollen (expressed as percentages of total tree pollen) and less than 40% *Pinus* pollen. *Picea* pollen was rare or absent. N.A.P. values were generally higher than in the pine region to the south. As in the pine region the *Salix* pollen frequencies were very low, even in localities with dense willow scrub. In samples from the tundra, *Pinus* was generally the dominant tree pollen type with values of 40% or more. N.A.P. values were usually high but variable, ranging from 50 to 1,560% of total tree pollen.

Aario's (1940) data are rather similar to those of Firbas (Fig. 1). Aario showed that on a transect in northern Finland from spruce–pine forest, through the sub-arctic birch forest region and into the tundra the pine pollen percentages increased northwards from southern Finland, but that they decreased to 35% of total tree pollen in the birch forest region. North of the birch forest region the relative proportions of *Pinus* (as percentage total tree pollen) in samples from the tundra were similar to those observed in the pine forest region further south. Similarly *Picea* pollen had comparable percentages in the samples from the tundra to those in samples from spruce–pine forests in southern Finland. Some tundra samples contained *Tilia* and *Ulmus* pollen, presumably carried considerable distances by the wind from areas further south. Whereas the tree pollen in the tundra samples are mainly a result of long-distance transport, the N.A.P. reflect largely the composition of the tundra vegetation. According to Aario (1940) 30% of N.A.P. (expressed as percentages of total tree pollen) is enough to indicate the presence of tundra vegetation if no more than 50% of this is represented by only one pollen type. The mean number of N.A.P./100 Arboreal Pollen in the samples from the tundra is 122, whereas in the forested regions the corresponding values range from 9 to 34. The local influence of Ericaceae and Empetraceae pollen may increase this to 100 or more

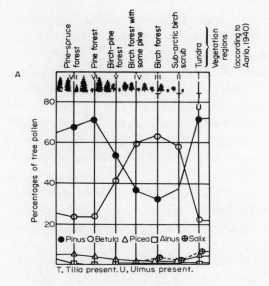
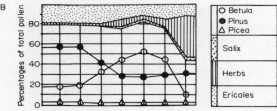
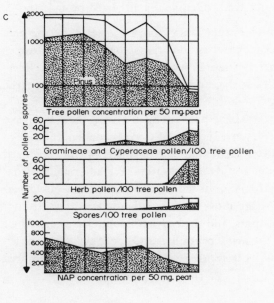

Figure 1. Pollen diagram of surface peat samples from different vegetational regions on a north–south transect in Finnish Lapland (from Aario 1940). Diagram A shows the percentage frequencies of the principal tree pollen types expressed as percentages of total tree pollen. Diagram B shows the percentage frequencies of the principal pollen types expressed as percentages of total pollen. Diagram C shows the pollen concentration of various pollen types expressed as number of grains in 50 mg of air dried peat.

in samples from heathy birchwoods or pinewoods. To permit ease of comparison with pollen spectra from 'late-glacial' sediments that are commonly expressed as percentages of total pollen, Aario's data are recalculated on this basis (Fig. 1b). Samples from the tundra are thus distinctive in their pollen content, with over 50% N.A.P. (mainly Gramineae and Cyperaceae) and with 30% *Pinus* pollen, whereas samples from the forested regions further south have low N.A.P. dominated by Ericaceae and *Empetrum* tetrads, and with low Gramineae and Cyperaceae pollen values. The principal forest regions are distinguished palynologically by the predominance of *Pinus* pollen in samples from the pine forest region and of *Betula* pollen in samples from the birch forests.

Aario's and Firbas' investigations dealt only with surface peat and humus, and the pollen values they obtained cannot be directly applied to or compared with fossil pollen spectra from limnic sediments. Despite this, their results indicate that there is often a striking difference in pollen rain between the tundra and the forested regions, particularly in terms of the relative frequencies of N.A.P. and its composition. Aario's (1940, 1942) data are particularly valuable in this context, for besides studying the relative pollen proportions, he determined the pollen concentration of the samples in terms of numbers of pollen per unit weight of sample (in his case 50 mg of air-dried peat) (see Fig. 1c). The mean values obtained were:

Samples from tundra	69 tree pollen/50 mg air-dried peat
Sub-arctic birch scrub	1,072–1,463 tree pollen/50 mg air-dried peat
Birch forest	1,679 tree pollen/50 mg air-dried peat
Pine forest	1,760–1,951 tree pollen/50 mg air-dried peat

The pollen concentration values for N.A.P. types vary from c 150 grains/50 mg peat in samples from the tundra to between c 500 and 700 grains/50 mg peat in samples from the forested regions. Unfortunately there is no means of assessing the likely sediment accumulation rate of Aario's surface peats and thus of determining the pollen influx in the different regions.

Ritchie and Lichti-Federovich (1967) have presented, as part of their extensive survey of modern pollen rain in the Western Interior of Canada, pollen influx data based on pollen catches in open Petri dish samplers enclosed within standard meteorological screens at 20 stations in arctic and sub-arctic Canada ranging from 74°41′ to 60°00′ N latitude. These stations occur within the high-arctic rock desert and fell-field regions, the mid-arctic sedge-moss tundra zone, the low-arctic dwarf-shrub tundra, the forest-tundra transitions, and the northern coniferous-forest belt (Fig. 2). The relative percentages of the principal pollen and spore types are shown in Fig. 2, and the estimated pollen influx in terms of grains deposited per unit area per year is also shown.

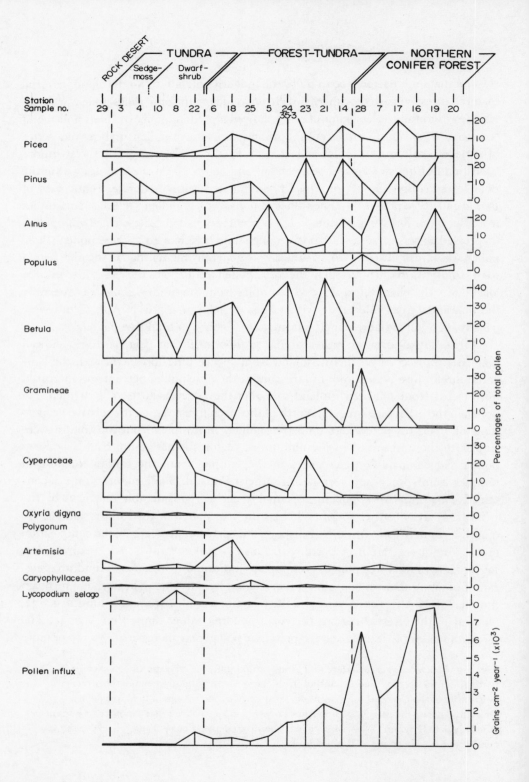

The modern pollen rain in the high- and mid-arctic zones (stations 29, 3, 4) show a high representation of *Pinus* and *Picea* pollen (Fig. 2). There do not appear to be any consistent features in terms of modern pollen composition or proportions that serve to distinguish the high- from the mid-arctic localities. The low-arctic dwarf-shrub tundra sites (10, 8, 22) are characterised by low relative frequencies of tree pollen types and by relatively high percentages of *Betula* (presumably *B. glandulosa* Michx.), Gramineae, and Cyperaceae pollen. It is difficult, however, to distinguish consistently these modern pollen assemblages in the tundra from spectra from the forest-tundra region (cf. Davis 1969), except by the higher relative frequencies of *Alnus* and *Picea* pollen in the modern rain in the forest-tundra region. The modern pollen rain in the northern coniferous forest is characterised by the preponderance of pollen of *Picea* and *Pinus* (see also Lichti-Federovich & Ritchie 1968, Davis 1967, Wright 1968).

The most distinctive feature of the modern pollen rain in the arctic tundra regions in Canada is the very low pollen influx (Fig. 2) with an estimated influx of 5 grains cm^{-2} year^{-1} in the high-arctic, a mean of 44 grains cm^{-2} year^{-1} (range = 22.5–65) in the mid-arctic, a mean of 335 grains cm^{-2} year^{-1} (range = 52.5–762.5) in the low-arctic, and a mean of 1,090 grains cm^{-2} year^{-1} (range = 275–2,372.5) in the forest-tundra zone. These low influx figures for the tundra contrast with the high values reported by Ritchie and Lichti-Federovich (1967) for pollen influx in the northern coniferous forests (mean = 5,004 grains cm^{-2} year^{-1}; range = 1,157–8,353). Fredskild (1969) reports very low values of modern pollen influx from high-arctic rock desert communities in Peary Land, North Greenland.

Lichti-Federovich and Ritchie (1968) have presented the results of relative pollen analyses of surficial mud samples from over one hundred lakes in the Western Interior of Canada on a transect from the low-arctic dwarf-shrub tundra of Keewatin to the grasslands and aspen parklands of southern Manitoba and Saskatchewan. Their study includes all the principal landform-vegetation types in this part of Canada. A summary of their results is shown in Fig. 3 as mean pollen percentages of the principal pollen types in all the samples from each vegetational type, as expressed as percentages of total pollen. The modern pollen assemblages being deposited in lakes in the low-arctic dwarf-shrub tundra and in the forest-tundra transition zone (Fig. 3) are broadly similar to those reported by Ritchie and Lichti-Federovich (1967) from open Petri dish samplers (Fig. 2), except for the consistently lower percentages of Gramineae pollen in the lake mud samples.

Figure 2. Results of pollen analyses of open Petri dish samples in various vegetational zones, Arctic and Sub-Arctic Canada (data from Ritchie & Lichti-Federovich 1967). Only the principal pollen and spore types are shown. Results are expressed as percentages of total pollen and spores. The estimated annual pollen influx in the various vegetational regions is also shown.

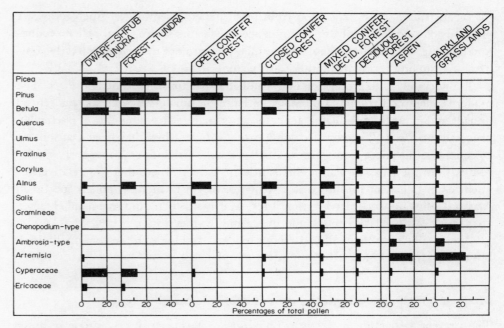

Figure 3. Summary pollen diagram of mean pollen percentages in surface lake muds of the principal pollen types in the main vegetational types of the Western Interior of Canada. All frequencies are expressed as percentages of total pollen (data from Lichti-Federovich & Ritchie 1968).

Although much of the modern pollen rain of the arctic zone has not been adequately studied, the following general features appear to characterise all the modern pollen spectra so far examined from treeless arctic environments (see also Bartley 1967, Colinvaux 1964, Fredskild 1967, 1969, Grichuk 1942, 1950, Iversen 1945, 1952–53, Livingstone 1955, Matthews 1970, Srodon 1960, Terasmae 1967, 1968):

1 A high proportion of N.A.P., particularly of Cyperaceae and Gramineae, and often low but consistent values of pollen of 'indicator species' such as *Oxyria digyna*. Although modern pollen assemblages from tundra regions are distinguishable from assemblages from forested regions, modern assemblages from the sub-arctic forest-tundra transition are rarely distinguishable from the modern pollen rain further south. As tundra plants appear to produce so little pollen relative to trees, a large proportion of the pollen rain may be blown long distances from the south. Within the forest region, large numbers of pollen of both trees and herbs are produced, with the result that ratio of A.P. to N.A.P. may be virtually identical in the forest-tundra zone and within the forest itself. The A.P./N.A.P. ratio has been used as a criterion

for reconstructing past vegetation from fossil pollen assemblages. Such ratios are primarily of local value only, being influenced by factors such as local pollen production (see Fig. 1 in Srodon 1960) and as such they are not necessarily directly comparable from one region to another (Livingstone 1955).

2 A moderately high proportion of tree pollen, particularly of *Pinus* and *Picea*. The modern pollen rain from arctic regions in large continental regions such as central Canada and in northern Finland, generally show high relative amounts of tree pollen primarily of *Pinus*, *Picea* and, to some extent, *Alnus* which has been transported considerable distances from sources further south. Grichuk (1950) reports similar results from the Russian plain. This is in contrast with the rather low relative frequencies of pollen of presumed long-distance origin in modern pollen spectra in Alaska (Colinvaux 1964, Livingstone 1955, Matthews 1970), in Greenland (Iversen 1945, 1952–53, Fredskild 1967), and in Iceland (Rymer 1972).

3 A generally low pollen influx. Reliable modern data for arctic regions are limited, but the observed range is from about 0.5×10^1 to 2.5×10^3 grains cm^{-2} year^{-1}.

Modern pollen rain characteristics of arctic vegetation types

The present vegetation in arctic regions does not consist of any single uniform vegetation type, but of a complex mosaic of communities including wind-blasted chionophobous 'fell-field' or 'fjaeldmark' communities of low plant cover, mildly chionophilous dwarf-shrub heaths, species—rich grasslands, and willow thickets, and markedly chionophilous snow-bed vegetation dominated by bryophytes and lichens. This section considers the available modern pollen rain data from the range of contemporary vegetation types in arctic environments.

GREENLAND

Iversen (1945) compared the relative pollen composition in surface mud samples from two lakes in the inland region of the Godthaab Fjord in western Greenland (64°20′ N, 51°50′ W) with the composition of the surrounding vegetation. Each lake lies in its own valley, and the composition of the surrounding vegetation was determined by means of a large number of small (0.1 m²) plots. The cover and relative frequency of the more important taxa were calculated for the two sites. One thousand plots were examined around lake I and two hundred around lake II.

The vegetation of the Godthaab Fjord area is described in detail by Trapnell (1931) and by Iversen (1954a). Briefly it consists of a mosaic of dense willow scrub (mainly *Salix glauca* L.) with *Alnus crispa* (Ait.) Pursh. along streams and in favourable, sheltered localities, alternating with *Betula nana*-dominated heaths, associated with ericaceous shrubs and dwarf willows in drier sites. *Juniperus communis* occurs more locally, particularly on south-facing slopes, whereas *Ledum groenlandicum* forms dense low scrub on west-facing areas. *Vaccinium*- and *Empetrum*-dominated heaths are common, especially at high altitudes and large wind-exposed areas are covered by lichen-rich or moss-rich heath with scattered herbs and dwarf shrubs.

Iversen's (1945) results are presented in Table 1. From these results he (in Faegri & Iversen 1964 p. 106) calculated *R*-values on the basis of the pollen percentages and the percentages of the corresponding taxa in the surrounding vegetation. They are:

Table 1. Comparison between the pollen content of surface muds in two lakes in Greenland and the composition of the surrounding vegetation (from Iversen 1945).

	Valley 1		Valley 2	
	Pollen analysis from surface mud	Composition of vegetation as % area covered	Pollen analysis from surface mud	Composition of vegetation as % area covered
Alnus crispa	13	3	10	5
Salix	6	13.5	6	12
Betula nana	56	19	58	33
Empetrum hermaphroditum	6	6	5	5
Ericaceae	5	26	4	18.5
Gramineae	3	11	5	13
Cyperaceae	10	22	12	17
Artemisia	1	×	0.5	×
Rumex acetosella	0.5	×	1	×
Σ Pollen	451	—	226	—
Sphagnum	2	5	2	10
Selaginella selaginoides	22	×	—	0
Lycopodium annotinum	23	1	4	1
L. complanatum L.	3	×	0.4	0
L. selago	0.4	×	—	0
Pinus + *Picea* (long-distance transport)	2	0	2	0

(Pollen expressed as percentages of total pollen; × = present in very low frequencies in the vegetation)

Modern pollen rain in arctic and alpine areas

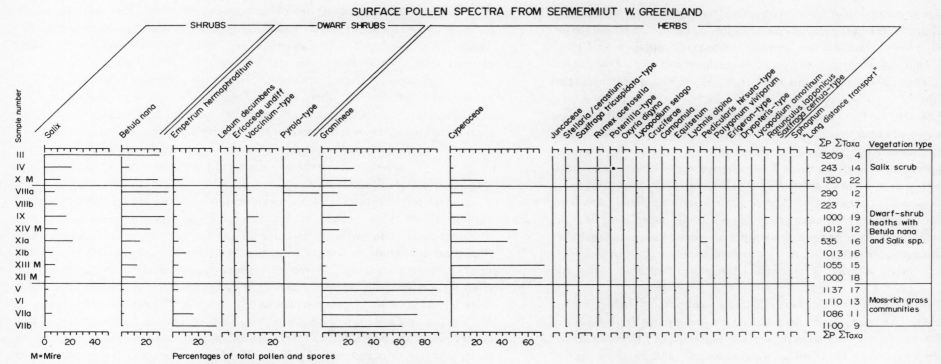

Figure 4. Pollen diagram for surface samples from Sermermiut, West Greenland (data from Fredskild 1967). Scale at base of diagram gives percentages for histograms. All histograms are plotted to the same scale, as percentages of total pollen and spores.
Abbreviations: anal. = analysed; undiff. = undifferentiated.

Fredskild (1967) presented results of pollen analyses of surface moss samples from a range of communities in the Sermermiut Valley near Jakobshavn in western Greenland (69° 12′ N, 51° 08′ W). A detailed account of the flora and vegetation of the area is given by Fredskild (1961). The results of the pollen analyses are shown in Fig. 4, and are expressed as percentages of total pollen and spores. The modern pollen spectra are grouped according to the vegetational characteristics of the plots from which the samples were collected. Three broad vegetational types are distinguishable: *Salix* scrub dominated by *S. glauca*; dwarf-shrub heath dominated by *Betula nana* with ericaceous and willow shrubs; and moss-rich grasslands characterised by *Poa arctica* R. Br. and *Alopecurus alpinus*. Samples XII, XIII, and XIV were collected from mires within dwarf-shrub heaths, and they are grouped on Fig. 4 according to their surrounding upland vegetation rather than by the local vegetation. Sample X is from a *Salix arctophila* Cockerell-dominated mire and it is grouped with samples from other *Salix*-dominated stands.

The pollen and spore types are grouped on Fig. 4 into three broad physiognomic categories of shrubs, dwarf-shrubs, and herbs (including pteridophytes), and individual histograms for all the pollen and spore types identified and included within

Alnus crispa	4.3	*Salix*	0.4
Betula nana	3.0	Gramineae	0.3
Empetrum hermaphroditum	1.0	Ericaceae*	0.2
Cyperaceae	0.5		

*(includes *Vaccinium*, *Ledum*, and *Rhododendron*).

These 'correction factors' were applied by Iversen (1952–53) to a post-glacial pollen diagram from the Godthaab Fjord.

From this study it appears that *Betula nana* and *Alnus crispa* are over-represented in the modern pollen rain compared with their abundance in the vegetation, whereas willows, grasses, sedges, and Ericaceae are grossly under-represented. A remarkable feature is the vast over-representation of *Lycopodium* spp., whereas *Sphagnum* is inconsistently under-represented.

the pollen sum are presented in order of occurrence from the bottom to the top of the diagram within the three physiognomic categories. *Sphagnum* spore values are plotted separately. The 'long distance transport' category comprises a total of 36 pollen grains of *Abies, Alnus, Ambrosia*-type, *Archangelica*-type, *Artemisia*, Chenopodiaceae, *Pinus, Plantago lanceolata*, and *Thalictrum*. All the histograms are drawn to a standard scale.

The modern pollen rain from the moss-rich grassland communities (Fig. 4) are characterised by an overwhelming dominance of Gramineae pollen. *Salix* and *Betula nana* pollen values range from 0.8 to 5.7% and 1.3 to 1.8% respectively, although neither taxon is present in any of the plots. Pollen frequencies of *Empetrum hermaphroditum* reach 16% or more in samples (VIIa, VIIb) from plots where this taxon grows locally. Besides the consistent but low representation of pollen of the 'long distance transport' component, there is a group of herbaceous pollen or spore types that occurs in low frequencies in samples from areas where the taxa concerned were not growing. This group includes *Stellaria/Cerastium, Rumex acetosella, Saxifraga tricuspidata*-type, *Oxyria digyna, Campanula, Equisetum, Potentilla*-type, and *Lychnis alpina*.

Modern pollen spectra from dwarf-shrub heaths are rather variable in their pollen composition. The spectra are characterised by consistently high values of *Betula nana* pollen and by the consistent occurrence of tetrads of *Empetrum hermaphroditum, Ledum decumbens* (Ait.) Small., *Vaccinium*-type, and Ericaceae undiff. *Salix* pollen values vary from 4.2 to 22%, and they generally show some relationship with the relative abundance of *Salix glauca* and *S. arctophila* in the vegetation. Samples from mire situations (XII, XIII, XIV) differ from those from upland sites in the high values of Cyperaceae pollen, presumably reflecting the local occurrence of species such as *Carex saxatilis, C. holostoma* Drej., and *Eriophorum scheuchzeri* Hoppe. A wide range of pollen types of herbaceous taxa occur in low frequencies such as *Stellaria/Cerastium Saxifraga tricuspidata*-type, *Potentilla*-type, *Rumex acetosella, Pedicularis hirsuta*-type, *Polygonum viviparum*, and *Ranunculus lapponicus* L. Pollen of some taxa, such as *Pedicularis* spp., *Pyrola grandiflora* Rad., and *Ranunculus lapponicus* occur in all the surface samples from plots where the relevant taxa occur in the vegetation, whereas others, such as *Polygonum viviparum, Equisetum arvense*, and *Stellaria* cf. *S. monantha* Hult., although present in several plots are only represented by pollen or spores in one or two surface samples. In contrast pollen of *Rumex acetosella, Potentilla*-type, and *Oxyria digyna* and spores of *Lycopodium selago* occur in several of the surface samples although none of the corresponding taxa occur in any of the plots.

Modern pollen spectra from *Salix*-dominated communities are characterised by high values of *Salix* pollen, by low values of *Empetrum hermaphroditum* and Ericaceae pollen, and by the absence of many of the pollen types of herbaceous taxa that occur commonly in spectra from the dwarf-shrub heaths. Gramineae pollen values vary from 0.4 to 25%, despite the consistent occurrence of *Poa arctica* in the plots. Cyperaceae pollen frequencies are low except in sample X (from a mire) where there is local growth of *Carex bigelowii* and *C. saxatilis*. The extremely high value of *Saxifraga tricuspidata*-type pollen in sample IV is striking, as *S. tricuspidata* Rottb. occurs within the plot from which the sample was collected. Similar discrepancies occur between the occurrence of pollen types of herbaceous taxa in the spectra from *Salix* scrub communities and the occurrence of the relevant taxa in the local vegetation, as were found in spectra from dwarf-shrub heaths and moss-rich grasslands.

The modern pollen rain data presented by Fredskild (1967) indicate that the three principal communities in this part of Greenland produce pollen spectra that are sufficiently distinctive in their pollen composition to permit the characterisation

Table 2. Summary of selected pollen percentages in surface samples at Sermermiut, western Greenland.

Vegetation type	Sample no.	*Salix*	*Betula nana*	*Empetrum hermaphroditum*	Ericaceae	Gramineae	Cyperaceae
Salix scrub	III, IV, X	12–99.6	0.03–28	0–7.3	0–3.7	0.4–25	0.03–26
		71.4	8.1	2.1	0.9	7.6	7.3
Dwarf-shrub heath	VIIIa, VIIIb, IX, XIV, XIa, XIb, XIII, XII	4.2–22 12.9	7.5–36 16.7	2.3–10 4.9	0.6–41.4 10.4	1.0–27 8.1	9.7–73 44.8
Moss-rich grassland	V, VI, VIIa, VIIb	0.8–5.7 2.7	1.3–1.8 1.6	0.7–34 13.4	0.1–0.9 0.4	62–94 79.9	0.3–0.9 0.6

Upper figures are range of percentages; lower figures are mean percentages. All percentages are based on sum of all determinable pollen and spores.

of the vegetation types in terms of modern pollen assemblages. The pollen percentages for selected taxa in the sixteen surface samples analysed by Fredskild are summarised as mean percentages and as percentage ranges in Table 2, for the three vegetational types.

Although the pollen sampling and vegetational description methods are different in the two Greenland studies (Iversen 1945, Fredskild 1967) it is clear that *Betula nana* is over-represented in the modern pollen rain when its pollen percentages are compared with its abundance in the vegetation. The low representation of members of the Ericaceae and of *Salix* spp. in the modern pollen rain is also shown by both studies. Fredskild's data indicate that *Empetrum hermaphroditum* is slightly over-represented and that Gramineae and Cyperaceae taxa are considerably over-represented. These conclusions contrast with the findings of Iversen. Such differences are probably an effect of different sampling procedures but in the absence of comparative data of spectra obtained from surface lake muds and moss polsters collected in the same area, it is difficult to evaluate the significance of the observed differences in terms of processes of dispersal, sedimentation, and preservation.

ICELAND

The only available data on the relationship between modern pollen rain and vegetation in Iceland come from the study by Rymer (1972). He examined surface moss polsters from six vegetational types around the Þingvellir region (64°15′ N, 21°10′ W) of southern Iceland and around the Akureyri area (65°40′ N, 18°5′ W) in the north. The results of the pollen analyses are shown in Fig. 5, and are expressed as percentages of total determinable pollen and spores. Unknown and indeterminable types are expressed as percentages of $\Sigma P + \Sigma Indet$.

As in the surface sample data from Greenland (Fig. 4) the Iceland spectra are grouped according to the vegetation of the plots from which the samples were collected. Six broad vegetational types are distinguishable: open 'fell-field' vegetation with *Dryas octopetala* and a few associated herbs; *Rhacomitrium* heath; grasslands dominated by *Festuca* spp. and *Agrostis* spp. but with willows, dwarf birch, and dwarf shrubs such as *Vaccinium uliginosum* and *Empetrum nigrum*; dwarf-shrub heath dominated by *Betula nana*, *Empetrum nigrum*, and *Vaccinium uliginosum*; mires within dwarf-shrub heaths characterised by an abundance of *Carex nigra* and *Eriophorum angustifolium*; and *Salix herbacea*-dominated slopes.

The pollen and spore types are grouped on Fig. 5 in the same way as the Greenland data (Fig. 4). All pollen and spore types identified and included within the sum are shown, including pollen and spores of presumed 'long distance transport' such as *Pinus*, *Betula* undiff., *Picea*, and *Fraxinus*.

The single spectrum examined from the *Dryas* 'fell-field' differs from the other spectra in Iceland by the absence of *Salix* and *Vaccinium*-type pollen and by the rather low *Betula nana* frequencies. The most distinctive features are the high values of *Alchemilla* and *Cerastium alpinum*-type pollen and of *Lycopodium alpinum* spores. The high *Lycopodium* spore percentages are remarkable in view of the absence of the plant in the surrounding vegetation (cf. Greenland).

The two spectra from *Rhacomitrium*-heath communities differ from the 'fell-field' spectrum in the high values of *Vaccinium*-type pollen and by the significant amounts of *Thalictrum* pollen. The occurrence of pollen of *Armeria maritima*, *Cerastium alpinum*-type, Rubiaceae, *Silene* cf. *S. acaulis*, and *Thymus*-type pollen and spores of *Selaginella selaginoides* reflect the local occurrence of the corresponding taxa in the plots. Besides the occasional grains of *Betula* undiff. and *Fraxinus*, there is a rather surprising group of pollen and spore types of herbaceous taxa present in the spectra but with the relevant taxa absent in the surrounding vegetation. This group includes *Botrychium lunaria*, Chenopodiaceae, *Ranunculus acris*-type, *Rumex acetosa*, and *Lycopodium alpinum*. The very high values of Polypodiaceae spores (mainly of *Thelypteris dryopteris*) are unexpected and remain unexplained at present.

Modern spectra from the grassland communities are characterised, as they are in Greenland, by a predominance of Gramineae pollen. *Salix*, *Betula nana*, *Empetrum nigrum*, *Calluna vulgaris*, and *Vaccinium*-type pollen values are variable, but they are generally lower than in spectra from *Rhacomitrium*-heath or from dwarf-shrub heath. They generally show some relationship with the relative abundance of the shrubs in the plots. A wide range of pollen and spore types of herbaceous taxa occur in the spectra, including *Cerastium alpinum*-type, Rubiaceae, *Ranunculus acris*-type, *Rumex acetosa*, *Thymus*-type, *Armeria maritima*, *Potentilla*-type, *Selaginella selaginoides*, and *Botrychium lunaria*. Although consistently present in the plots only occasional grains of *Alchemilla*, *Rhinanthus*, *Silene* cf. *S. acaulis*, and *Polygonum viviparum* and spores of *Equisetum* occur in the spectra, whereas spores of *Lycopodium* spp. and Polypodiaceae occur consistently but the plants are absent from the nearby vegetation.

Surface samples from dwarf-shrub heaths are characterised by high values of *Betula nana*, *Empetrum nigrum*, and *Vaccinium*-type pollen, by correspondingly reduced frequencies of Gramineae pollen, and by an absence of many pollen and spore types of herbaceous taxa present in the spectra from grasslands. Spectra from *Carex nigra* mires situated within dwarf-shrub heaths differ from those from upland sites in the high values of Cyperaceae pollen, no doubt reflecting the local occurrence of sedges.

Sample 15 is from a *Salix-herbacea*-dominated slope with a few associated herbs

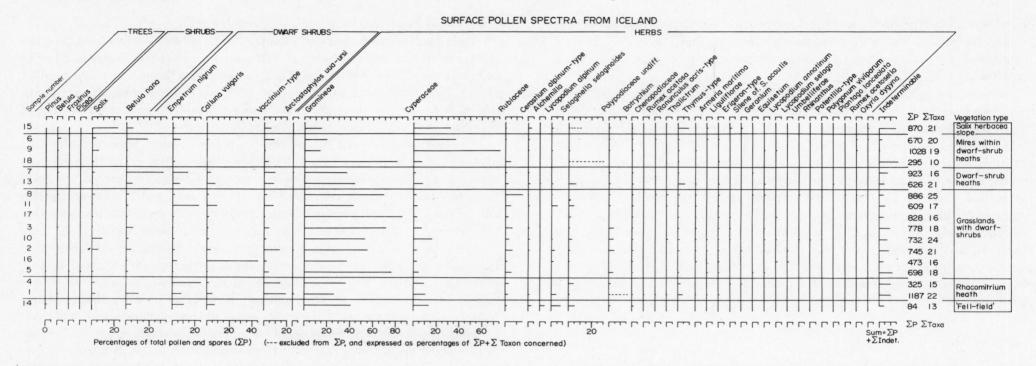

Figure 5. Pollen diagram for surface samples from Iceland (data from Rymer 1972). Scale at base of diagram gives percentages for histograms. All histograms are plotted to the same scale, as percentages of total determinable pollen and spores. If a pollen or spore type shows an unexpectedly high value in a particular sample, its frequency is calculated as above, but the other pollen and spore frequencies in the sample are expressed as a percentage of the pollen sum (ΣP) excluding that taxon (Wright & Patten 1963).
Abbreviations: anal. = analysed; indet. = indeterminable; undiff. = undifferentiated.

such as *Carex bigelowii*, *Alchemilla alpina*, and *Thalictrum alpinum*. The pollen rain is characterised by high values of *Salix*, Cyperaceae, and *Thalictrum* pollen.

Rymer's (1972) data indicate that the communities he examined in Iceland can be characterised, to some degree, by their modern pollen rain. The pollen percentages for selected taxa in the seventeen samples analysed are summarised as mean percentages and as percentage ranges in Table 3 for the six vegetational types sampled.

Comparisons with the vegetational data for the plots where the surface samples were collected (Rymer 1972) indicate that several taxa such as *Salix*, *Polygonum viviparum*, and *Equisetum arvense* are consistently under-represented in the modern pollen rain, as they are in Greenland. *Empetrum nigrum* appears to be slightly over-represented in Iceland, as it is in Sermermuit. This contrasts with its rather low representation in modern spectra in Scotland (Birks 1973). *Betula nana* is consistently over-represented in all the modern data from Greenland and Iceland. Several taxa that are well represented in the Iceland pollen rain such as Rubiaceae, *Thalictrum*, and *Lycopodium* spp. are similarly well represented in modern spectra from Scotland.

Modern pollen rain characteristics of alpine and sub-alpine vegetation types in Scotland

Very few studies on the relationships between modern pollen rain and contemporary vegetation as an aid in the interpretation of fossil pollen assemblages have been carried out in Britain or elsewhere in western Europe, possibly due to the assumption that the modern vegetation is so disturbed that such a study would be of little value in considering past vegetation. Although lowland vegetation in Britain is considerably disturbed, there is a variety of natural or semi-natural communities in the alpine and sub-alpine zones in Scotland today (McVean & Ratcliffe 1962, Birks 1973). In view of the strong floristic affinities between the late-glacial pollen flora of the Isle of Skye (Birks 1973) and the present Scottish montane flora, a survey of modern

Table 3. Summary of selected pollen percentages in surface samples from Iceland (from Rymer 1972).

Vegetation type	Sample no.	Salix	Betula nana	Empetrum nigrum	Calluna vulgaris	Ericaceae undiff.	Gramineae	Cyperaceae	Rubiaceae
Salix herbacea-slope	15	22	5.2	1.7	0	5.9	14.9	31.9	0.6
Mires within dwarf shrub heaths	6, 9, 18	0.7–5.6 3.8	0.7–18.2 6.9	0.5–7.3 2.9	0–0.4 0.3	0.5–8.7 3.6	14.3–81.5 21.8	7.3–75.3 55.9	0–3.9 0.4
Dwarf-shrub heath	7, 13	0.5–1.6 1.1	5.2–31.3 20.4	6.6–13.3 10.5	3.3–7.1 4.8	7.8–8.9 5.2	37.1–44.1 39.9	1.2–7.9 4.0	1.3–2.3 1.7
Grasslands with dwarf shrubs	8, 11, 17, 3, 10, 2, 16, 5	0–9.2 2.4	0.1–6.1 2.3	0.3–22.9 3.8	0.7–44.9 5.7	0.1–13.8 4.3	36.9–86.3 62.4	2.5–16.5 5.4	1.3–15.2 5.9
Rhacomitrium heath	1, 4	2.3–3.3 3.3	0–11.3 8.7	8.4–24.7 12.2	1.5–5.1 4.3	8.4–24.7 18.5	26.9–36.7 28.8	9.6–10.2 10.2	2.3–2.7 2.4
'Fell-field'	14	0	4	10.6	6.7	+	40.0	13.3	1.3

Upper figures are ranges of percentages; lower figures are mean percentages. All percentages are based on sum of all determinable pollen and spores (with selected exclusions).

pollen spectra from Scottish alpine and sub-alpine vegetation was undertaken in an attempt to find modern vegetational analogues to the fossil pollen assemblages. A detailed account of the field and laboratory methods, and of the palynological and vegetational data are presented elsewhere (Birks 1973). This section reviews the extent to which modern Scottish mountain vegetation types can be characterised by modern pollen spectra and it indicates pollen and spore types that may be used as 'indicator species' in vegetational reconstructions.

METHODS

Plots of uniform floristic composition and structure of at least 25 m^2 area were selected in the field from the range of alpine and sub-alpine vegetation types occurring in western Scotland today. All the taxa present within the plots were listed, and their cover and abundance were estimated and recorded on the 10-point Domin scale. Six to ten samples of moss (mainly *Rhacomitrium lanuginosum*) or surface soil were collected from different points within the plot and amalgamated and prepared for analysis as a single surface sample. Details of plot locations, sample preparation, and pollen counting and identifications are given by Birks (1973). A tabulation of the numbers of pollen grains and spores counted for each sample is available on request to the author.

The frequencies of all determinable local pollen and spore types are expressed on Fig. 6, as percentages relative to the sum (ΣP) of all such types. Local pollen and spores are here defined as pollen and spores morphologically indistinguishable from the pollen and spore types of the taxa occurring within or in close proximity to the vegetational plot sampled. As in the surface sample data from Greenland (Fig. 4) and Iceland (Fig. 5), the modern pollen spectra from Scotland (Fig. 6) are grouped according to the vegetation of the plots from which the samples were collected. Five principal vegetation types are distinguished here: alpine summit vegetation; sub-alpine grasslands, 'tall herb' communities, and *Salix* scrub; *Juniperus communis*—dominated vegetation; *Betula pubescens* woodland; and mixed *Betula pubescens*—*Corylus avellana* woodland. Nomenclature of vegetation types follows Birks (1973) and McVean and Ratcliffe (1962). Within each main vegetation type, the surface spectra are arranged according to the phytosociological affinities of the vegetation of the relevant plot, so that modern pollen spectra from the same community are grouped together. The pollen and spore types are grouped on Fig. 6 into four broad physiognomic categories—trees, shrubs, dwarf-shrubs, and herbs (including pteridophytes). All the histograms for the individual pollen and spore types identified and included within ΣP are drawn to a standard scale. The vegetational data for the plots examined are too extensive to present here, and the reader is referred to Plate 7 in Birks (1973).

ALPINE SUMMIT VEGETATION

The four principal types of alpine summit vegetation occurring in western Scotland today are the chionophobous Cariceto-Rhacomitretum lanuginosi Association ('*Rhacomitrium*-heath') and the wind-blasted *Festuca ovina-Luzula spicata* 'fell-

Figure 6. Pollen diagram for surface samples from Scotland (reproduced, with permission of Cambridge University Press, from Birks 1973). Scale at base of diagram gives percentages for histograms. All histograms are plotted to the same scale, as percentages of total determinable local pollen and spores (see text for further details). If a pollen or spore type shows an unexpectedly high value in a particular sample, its frequency is calculated as above, but the other pollen and spore frequencies in the sample are expressed as a percentage of the pollen sum (ΣP) excluding that taxon (Wright & Patten 1963). Abbreviations: anal. = analysed; indet. = indeterminable; Sph. = *Sphagnum*; undiff. = undifferentiated.

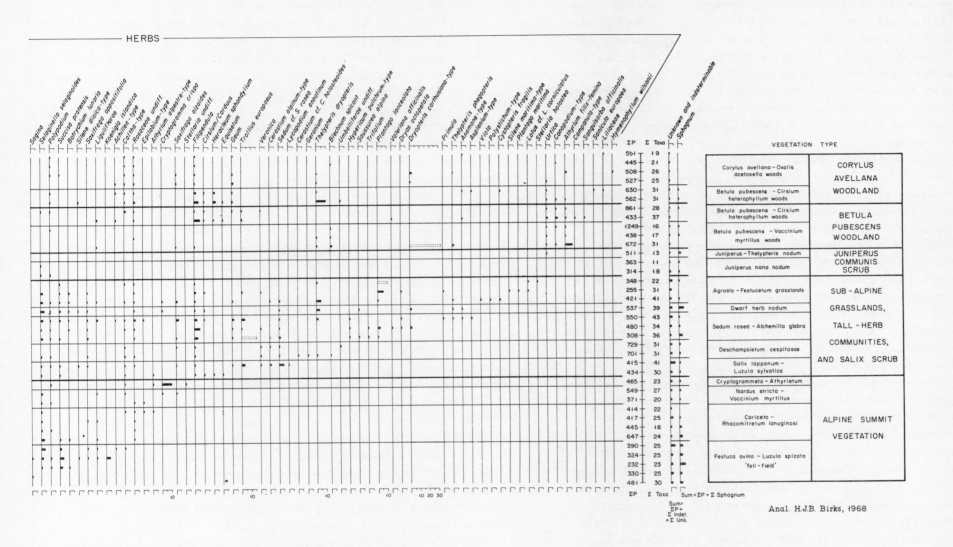

field' communities, and the chionophilous Cryptogrammeto-Athyrietum chionophilum Association and the *Nardus stricta-Vaccinium myrtillus* snow-patch communities. Details of the floristic composition, ecology, and inter-relationships of these communities are discussed by McVean and Ratcliffe (1962) and Birks (1973).

Although the four communities are varied in both their floristic composition and their present ecology, the main feature they share is the low cover of vascular plants, as a result either of bryophyte dominance, especially of *Rhacomitrium lanuginosum*, or of extreme openness of the vegetation. *Alchemilla alpina, Carex bigelowii, Festuca ovina* agg. (inc. *F. vivipara*), *Galium saxatile, Lycopodium selago*, and *Potentilla erecta* occur in 75% or more of the plots sampled. Montane taxa such as *Cherleria sedoides, Empetrum nigrum* agg., *Gnaphalium supinum, Lycopodium alpinum, Polygonum viviparum, Salix herbacea*, and *Sibbaldia procumbens* occur more rarely. *Artemisia norvegica, Deschampsia alpina, Juncus trifidus, Koenigia islandica*, and *Luzula spicata* are rare but characteristic species of the 'fell-field' plots.

The most prominent feature of the modern pollen rain in these alpine communities is the dominance of Cyperaceae pollen (Fig. 6). *Carex bigelowii* appears to have a higher pollen representation than any of the other taxa present, suggesting that high Cyperaceae pollen frequencies in fossil pollen spectra need not always indicate the local presence of sedge swamps (cf. Welten 1950). The second feature is the abundance of *Lycopodium selago* and *L. alpinum* spores despite the low cover of these species in the vegetation. *Lycopodium* spp. are similarly well represented in modern spectra from Greenland and Iceland (see above).

Gramineae pollen values are low in these spectra (Fig. 6) despite the abundance of grasses in several of the plots. These low pollen values may result from the reduced vitality and flowering in high-altitude chionophilous and chionophobous vegetation and from the predominance of viviparous grasses in alpine communities.

Several species have low cover in the vegetation but are well represented in the modern spectra. They include *Potentilla erecta* (*Potentilla*-type pollen), *Thymus drucei* (*Thymus*-type pollen), and *Galium saxatile* (Rubiaceae). In contrast some species have high cover values in the vegetation but very low pollen percentages in the modern spectra. They include *Alchemilla alpina, Cherleria sedoides* (*Arenaria*-type pollen) *Gnaphalium supinum* (*Antennaria*-type pollen), and *Vaccinium myrtillus* (*Vaccinium*-type pollen). Pollen of *Artemisia norvegica, Cryptogramma crispa, Polygonum viviparum*, and *Saxifraga stellaris* are only found in the spectra when the plants are present in the plots.

The pollen spectra from Scottish alpine vegetation are relatively homogeneous and distinctive from spectra produced by the other vegetational types examined. The dominance of Cyperaceae pollen and the consistent occurrence of *Lycopodium selago* and *L. alpinum* spores are characteristic. Pollen of *Antennaria*-type and *Saxifraga stellaris* are virtually exclusive to the alpine spectra. With the spectra from alpine vegetation as a whole, it does not appear possible to distinguish consistently the different vegetational types except by the occurrence of pollen or spores of 'indicator species' such as *Artemisia norvegica, Koenigia islandica*, and *Sagina* pollen in spectra from some 'fell-field' plots, and *Cryptogramma crispa* and *Athyrium alpestre* in spectra from the Cryptogrammeto-Athyrietum chionophilum Association. This homogeneity of pollen spectra is not surprising, as the modern alpine communities in Scotland have a rather similar vascular plant flora, the communities being differentiated floristically on the basis of the occurrence of species not represented in the pollen rain such as *Luzula spicata, Juncus trifidus, Rhacomitrium lanuginosum*, and *Cetraria islandica*.

SUB-ALPINE COMMUNITIES

Within the sub-alpine zone in western Scotland today there are five principal communities (Fig. 6). They are floristically and edaphically related, and all favour intermittently flushed brown-earth soils. The *Sedum rosea-Alchemilla glabra* Association rich in 'tall herbs' and the *Salix lapponum-Luzula sylvatica* Association characterised by a dominance of *S. lapponum* and, more rarely, *S. lanata* and by an abundance of 'tall herbs' generally occur as small fragmentary stands on north- or east-facing inaccessible ungrazed cliff ledges between 490 and 920 m. These communities are related floristically and ecologically to the more extensive and characteristic sub-alpine and low-alpine willow scrub of the Scandinavian mountains (Nordhagen 1943). On the grazed slopes below cliffs in Scotland species-rich grasslands are dominated by either *Deschampsia cespitosa* (Deschampsietum cespitosae alpinum), by *Agrostis* spp. and *Festuca ovina* (Agrosto-Festucetum), or by dwarf herbs such as *Silene acaulis, Alchemilla alpina*, and *Cherleria sedoides* (Dwarf herb *nodum*). Although these communities are dominated by grasses, 'tall herbs' such as *Rumex acetosa, Sedum rosea*, and *Angelica sylvestris* may occur as stunted plants within the grassy sward. These communities are almost certainly biotically derived from 'tall herb' stands as a result of extensive grazing by sheep and deer (Birks 1973).

Species occurring in 75% or more of the plots examined from the sub-alpine zone include *Alchemilla alpina, Anthoxanthum odoratum, Deschampsia cespitosa, Potentilla erecta, Ranunculus acris*, and *Viola riviniana*. 'Tall herbs' such as *Cirsium heterophyllum, Crepis paludosa, Filipendula ulmaria, Saussurea alpina*, and *Trollius europaeus* are locally abundant.

The first characteristic feature of the modern pollen spectra from sub-alpine communities is the dominance of Gramineae pollen (Fig. 6), reflecting the prolific pollen production of the dominant grasses *Anthoxanthum odoratum* and *Deschampsia cespitosa* (Pohl 1937). Secondly, pollen of *Rumex acetosa* is always relatively abundant, despite the low cover of the species in the plots. Other herbs such as *Filipendula ulmaria*, *Potentilla erecta* (*Potentilla*-type pollen), *Ranunculus acris* (*Ranunculus acris*-type pollen), *Saxifraga hypnoides*, and *Thalictrum alpinum* are also well represented in the modern spectra in spite of their often low cover in the vegetation.

In contrast, several taxa may be abundant in the vegetation but have very low pollen frequencies in the modern spectra. Willows such as *Salix lapponum* and *S. lanata* are particularly poorly represented. Similar under-representations of willows in modern spectra from arctic and sub-arctic regions have been shown in Lapland (Firbas 1934, van der Hammen 1951) and in Greenland and Iceland (see above) and from alpine regions in Switzerland (Welten 1950). Other poorly represented taxa in the Scottish spectra include *Vaccinium myrtillus* (*Vaccinium*-type pollen), *Angelica sylvestris* (*Angelica*-type pollen), *Cirsium heterophyllum* (*Cirsium/Carduus*), *Geum rivale*, *Sedum rosea*, *Trollius europaeus*, and *Valeriana officinalis*.

The modern spectra from sub-alpine vegetation in Scotland are characterised by their dominance of Gramineae pollen and by the consistent occurrence of *Ranunculus acris*-type, *Rumex acetosa*, and *Thalictrum* pollen. Pollen of *Filipendula ulmaria*, *Trollius europaeus*, *Alchemilla*, *Saxifraga aizoides*, *S. oppositifolia*, and *S. hypnoides* and spores of *Botrychium lunaria* and *Selaginella selaginoides* are more abundant in these sub-alpine spectra than in spectra from other vegetational types. Pollen of *Cerastium alpinum*-type, *Saussurea alpina*, *Sedum rosea*, and *Veronica* are exclusive to modern spectra from sub-alpine communities. It does not appear possible to distinguish the vegetational types within the sub-alpine zone on the basis of their modern pollen rain due to the poor pollen representation of many of the phytosociologically characteristic and differential species.

JUNIPERUS COMMUNIS SCRUB

Juniperus communis occurs in a range of growth forms and communities in Scotland today. In wind-exposed alpine and sub-alpine situations between 305 and 610 m it occurs as an extreme prostrate form (ssp. *nana*) that rarely flowers. Sample 48 (Fig. 6) is from such a stand. Patches of flowering juniper scrub dominated by small bushes intermediate in growth form between ssp. *nana* and ssp. *communis* occur locally in sub-alpine situations in western Scotland. Sample 24 is representative of such stands. Tall dense thickets of *J. communis* ssp. *communis* occur locally in the sub-alpine zone between 305 and 550 mm, often with a fern-rich understorey (*Juniperus-Thelypteris nodum*; McVean & Ratcliffe 1962). Sample 2 comes from one such stand. Mixed juniper-birch woods are very rare in Britain today, and sample 4 was collected from one of the few surviving examples.

There is considerable variation in the representation of juniper pollen in the modern spectra (Fig. 6): 2.5% in 48, 37% in 24, 49% in 2, and 7.1% in 4. It appears that its representation is influenced not only by the abundance of the plant but by its growth form and flowering (Iversen 1954b). The lowest pollen frequencies are found in samples from the alpine prostrate stand. It is of interest that the pollen rain from this stand closely resembles the modern spectra from other alpine communities in Scotland (Fig. 6). Intermediate well-grown juniper bushes are moderately well represented in the pollen rain. High pollen values are only produced, however, by tall dense scrub, whereas in mixed juniper-birch woodland the juniper pollen percentages are low, presumably because of its reduced flowering in shade and because of the high local pollen production of birch, tending to depress, on a percentage basis, the representation of the understorey juniper. Vasari and Vasari (1968) report similar modern data for juniper communities in northern Finland, with low pollen values in mixed juniper-birch-alder woods and high frequencies in treeless juniper-dominated situations. Iversen (1954b) records low pollen values in surface lake muds from 'a valley in Greenland in which *Juniperus* is frequent but dwarfed'.

BETULA PUBESCENS WOODLANDS

There are two principal types of woodland today in western Scotland that are dominated by *Betula pubescens* spp. *odorata*. Although these two woodland types are similar in terms of the dominant canopy trees and shrubs, they differ considerably in their understorey species. In the *Betula pubescens-Vaccinium myrtillus* woodlands the understorey is dominated by *Vaccinium myrtillus*, *Deschampsia flexuosa*, *Galium saxatile*, *Potentilla erecta*, and ferns. These woods generally occur on acidic podsols or poor brown-earth soils. In contrast the *Betula pubescens-Cirsium heterophyllum* woods have an abundance of 'tall herbs' such as *Cirsium heterophyllum*, *Crepis paludosa*, *Filipendula ulmaria*, and *Geum rivale* in the understorey. This woodland type only occurs on fertile irrigated brown earths in areas that are protected from grazing animals such as on steep blocky slopes and in rocky ravines.

All the pollen spectra from the woods are dominated by *Betula* pollen (Fig. 6). There is a significant correlation between pollen percentages and canopy area in the woods examined (Birks 1973). Associated trees and shrubs such as *Sorbus aucuparia*, *Populus tremula*, and *Salix* spp. are poorly represented. Although there are considerable floristic differences between the two woodland types, there are only minor differences in the composition of their pollen spectra due to the poor overall representation of the phytosociologically characteristic and differential understorey species. Within the spectra from the woods examined, pollen of the 'tall herbs' *Filipendula ulmaria*, *Cirsium heterophyllum*, *Geum rivale*, and *Trollius europaeus* are virtually exclusive to the spectra from stands of the *Betula pubescens-Cirsium heterophyllum* Association. In contrast the percentages of *Vaccinium*-type pollen and of fern spores (Polypodiaceae undiff., *Blechnum spicant*, and *Dryopteris filix-mas*-type) tend to be higher in spectra from the *Betula pubescens-Vaccinium myrtillus* woods than in spectra from the 'tall herb' woods. A further distinguishing feature is the higher diversity of pollen and spores in spectra from the 'tall herb' woods than in spectra from the *Betula pubescens-Vaccinium myrtillus* woods. Williams' (1964) α index of diversity (see below) for spectra 3 and 30 is 9.5 and 5.6 respectively compared with 6.8, 2.6, and 3.4 for the pollen spectra 4, 32, and 33 from the grazed woods.

BETULA PUBESCENS—CORYLUS AVELLANA WOODLANDS

Mixed woodlands dominated by birch and hazel are locally frequent today on fertile brown earth soils in western Scotland. Just as in the pure birch woods discussed above the two principal types of mixed birch-hazel woodland reflect differences in grazing pressures. In grazed situations the understorey is characterised by an abundance of *Conopodium majus*, *Endymion non-scriptus*, *Oxalis acetosella*, *Potentilla erecta*, and *Primula vulgaris* (*Corylus avellana—Oxalis acetosella* Association). In ungrazed rocky situations 'tall herbs' such as *Cirsium heterophyllum*, *Filipendula ulmaria*, *Heracleum sphondylium*, and *Silene dioica* dominate the understorey (*Betula pubescens-Cirsium heterophyllum* Association).

The modern pollen spectra from the mixed birch-hazel woods are dominated by *Betula* and *Corylus* pollen (Fig. 6). There is a significant correlation between pollen percentages and canopy cover for both taxa (Birks 1973) and although both trees are well represented, *Betula* appears to have a slightly higher pollen representation than hazel in these woods when comparing canopy percentages and pollen percentages. Associated trees and shrubs such as *Sorbus aucuparia* and *Salix* spp. are poorly represented throughout.

As in the modern spectra from the birch woods, there is a poor pollen representation of the understorey species, and it is thus difficult to distinguish the two types of birch-hazel woodland on pollen criteria alone. The main difference between the modern spectra from the two types is the consistent occurrence of pollen of *Angelica* type, *Cirsium/Carduus*, *Filipendula*, and *Heracleum sphondylium*, and the somewhat higher values of *Salix* pollen in spectra from the 'tall herb' woods, whereas in spectra from the *Corylus avellana—Oxalis acetosella* woods, many of these pollen types are absent, or, if present, their frequencies are markedly lower than in spectra from the 'tall herb' stands. Percentages of fern spores (Polypodiaceae undiff., *Dryopteris filix-mas*-type, *D. carthusiana*-type, and *Blechnum spicant*) tend to be higher in the spectra from the grazed woodlands than in spectra from the 'tall herb' woods. As in the spectra from the birch woods, there is a greater diversity of pollen and spore types in spectra from the woods rich in 'tall herbs' than in spectra from the grazed woodlands. The α index of diversity for pollen spectra 13 and 19 from 'tall herb' woods is 7.0 and 7.1 respectively in contrast to 5.9, 5.3, 3.8, and 4.5 for spectra 31, 34, 35, and 36 from the grazed birch-hazel woodlands.

Numerical analysis of the Scottish modern pollen rain data

The surface samples from Scotland were collected and analysed in order to see if a range of modern communities could be characterised by means of contemporary pollen spectra. In Fig. 6 the pollen spectra have been grouped, for the purposes of discussion, according to the vegetation of the plots from which the surface samples were collected. In this section the results are discussed of the application of various numerical procedures to the classification of the modern pollen spectra on the basis of the observed pollen percentages alone and then the comparison of this classification with the scheme based on the vegetation of the plots. Such an approach is potentially valuable as a means of testing the hypothesis that different communities produce distinctive modern pollen assemblages. If this hypothesis can be substantiated by a correspondence between the two classifications, then modern pollen spectra can be used in the vegetational interpretation of fossil pollen spectra with a greater degree of confidence than would otherwise be possible. If, on the other hand, no such correspondence is found, it is suggestive that the modern communities investigated here cannot be consistently characterised at the level of the association on the basis of modern pollen spectra. This conclusion would be of considerable significance in assessing the potential value of the approach of comparing fossil and modern pollen spectra as a means of interpreting fossil assemblages in terms of past vegetation.

CLUSTER ANALYSIS

Methods

Cluster analysis is a polythetic agglomerative classificatory procedure developed mainly by numerical taxonomists (Sokal & Sneath 1963) whereby objects are grouped on the basis of some measure of their overall similarity and each of the groups are in turn fused until the whole set of objects is accounted for. The purpose of any cluster analysis is thus to generate groups among objects $1, \ldots n$ on the basis of dissimilarity (or similarity) coefficients (DC) where each $DC(i, j)$ is a measure of the dissimilarity between objects i and j. The method of cluster analysis involves four major stages.

The first stage is the computation of a matrix of dissimilarity coefficients between all pairs of objects (in this case surface pollen spectra). Thus DC(i, j) gives a measure of how different spectrum i is from spectrum j. Let n be the total number of spectra we are considering, then i and j can take any integral value between 1 and n.

Dissimilarity coefficients must satisfy certain minimum conditions (Jardine & Sibson 1971) namely

(i) DC(i, j) ⩾ 0
(ii) DC(i, i) = 0
(iii) DC(i, j) = DC(j, i)

for $i, j = 1, 2, \ldots n$.

Two DC's which will be referred to later are defined here. Let P_{mn} be the proportion of the total pollen sum in spectrum n which is of the pollen type m, then DC1(i, j) = $\sum_{k=1}^{t} |P_{ki} - P_{kj}|$, where t is the number of pollen types considered. In this study DC1 was applied to all pairs of samples using all pollen and spore types included within the basic pollen sum on Fig. 6 that occur in six or more spectra. This DC has been used by Gordon and Birks (1972) in their numerical procedures for pollen zonation.

The second DC used in this study (DCJ) is DCJ(i, j) = $1 - S_J(i, j)$, where $S_J(i, j)$ is Jaccard's similarity coefficient (Sokal & Sneath 1963, Cheetham & Hazel 1969). Jaccard's coefficient of similarity is calculated as the number of pollen types common to two samples divided by the total number of different types present in both samples. Symbolically it takes the form

$$S_J(i, j) = \frac{a}{a + b + c}$$

where the positive and negative matches for two samples i and j are given by the following contingency table:

		sample i	
		present	absent
sample j	present	a	b
	absent	c	d

DCJ was calculated for all pairs of samples using the presence and absence of all pollen and spore types included within the basic pollen sum for the sample pair in question. A simple binary (presence-absence) dissimilarity coefficient was used in this study as a means of utilising the pollen and spore types of 'indicator species' that occur with very low frequencies in one or two samples only. The numerical contribution of these types to the pollen rain is slight, and they would thus have no significant effect on DC1 which is principally influenced by the numerically large pollen and spore taxa. Both approaches are potentially valuable in handling complex pollen analytical data.

An important feature of the Jaccard coefficient is that it omits negative matches from consideration (Sokal & Sneath 1963, Cheetham & Hazel 1969, Kaesler 1969). Thus if a pollen type occurs in one or more samples but not in samples i and j, its mutual absence from i and j would not contribute to the measure of similarity between i and j. Many other commonly used similarity coefficients do not have this feature. The simple matching coefficient (Sokal & Sneath 1963), for example, includes the number of negative matches in both the numerator and the denominator. The modern pollen rain data contain several pollen and spore types that occur in a few samples only. Use of the simple matching coefficient in such instances would essentially be a study of negative associations because most types occur in only a few of the samples. The use of the Jaccard coefficient avoids this difficulty (see Kaesler 1969).

The second stage is to represent the dissimilarity coefficients by an appropriate classificatory or clustering procedure. The wide range of clustering procedures currently available are reviewed by Sokal and Sneath (1963), Lance and Williams (1967), Cormack (1971), and Williams (1971). Jardine and Sibson (1971) show that only the single-link (nearest-neighbour) method satisfies certain obvious conditions, for example the requirement that continuous variation in the data should produce continuous variation in the resultant dendrogram. The standard objection (see, for example, Lance & Williams 1967) to the single-link method is that it groups together dissimilar objects at low levels that are linked by chains of intermediates, because of the likelihood that objects outside the group are more similar to one of the many objects inside the group, than to some object outside the group. Jardine and Sibson (1968, 1971) suggest that this is an inherent defect of hierarchical classification, and that the kinds of information concealed by chaining, for example information about the relative homogeneity of the groups recognised, can be revealed only by clustering procedures that lead to non-hierarchical systems. The various average-link methods attempt to avoid the defect of chaining, but are themselves unsatisfactory as they do not fulfil some of the basic requirements for clustering methods such as stability and continuity specified by Jardine and Sibson (1968, 1971).

In view of the theoretical justification for single-link cluster analysis, this procedure was used here. The method basically consists of clustering together those objects (in this case surface pollen spectra) with the lowest dissimilarity coefficients. The level of admission of objects is successively raised until all the objects have been amalgamated into a single group. The admission of an object or group into another group is by the criterion of single-linkage where the DC between groups is the DC between the most similar pair of objects, one in each group. If the groups concerned are G_1 and G_2, then $DC(G_1, G_2) = \min (DC(i,j)|i \in G_1, j \in G_2)$. Other clustering procedures such as average-link involve some averaging of the DC's of the groups G_1 and G_2 (see Sokal & Sneath 1963).

The third stage in cluster analysis is the geometrical representation of the clusterings. Such clusterings may be represented in a variety of ways. The clusterings of the surface sample data are shown as dendrograms in Figs. 7 and 8. The levels h at which objects cluster together in a numerically stratified hierarchic clustering satisfy the three conditions that define a DC, namely

$$h(i,j) \geqslant 0$$
$$h(i,i) = 0$$
$$h(i,j) = h(j,i)$$

The scale of cluster levels is shown alongside the dendrograms (Figs. 7 & 8). The levels at which each pair of objects fuse, the cluster levels, are determined by the DC from which the dendrogram is derived. A dendrogram also obeys the ultrametric condition where

$$h(i,k) \leqslant \text{ the larger of } h(i,j) \text{ and } h(j,k).$$

Dendrograms are characterised by ultrametric dissimilarity coefficients, but a dendrogram can rarely represent the dissimilarities between objects exactly, since these rarely satisfy the ultrametric condition.

The fourth step is thus a comparison of the output dissimilarities as shown by the dendrograms with those of the original input matrix of dissimilarity coefficients. This step is necessary because the clustering procedures involve distortion of the data in order to express the multidimensional DC matrix in a two-dimensional hierarchic relationship. Jardine and Sibson (1968, 1971) have developed methods of making such comparisons. In this study the following relationship proposed by Jardine and Sibson (1971) has been used as a guide to the accuracy with which a clustering obtained by a transformation D represents a DC d:

$$\hat{\Delta}_1 = \sum \frac{|d(i,j) - D(d)(i,j)|}{\sum d(i,j)}$$

This is a useful measure of the performance of the clustering methods and of the hierarchic classifiability of the modern pollen spectra. It is bounded by the value of 0 in case of perfect hierarchic classifiability and by the value 1 for totally imperfect hierarchic classifiability.

Results

Dendrograms of Figs. 7 and 8 show the results of the single-link clustering on the matrices of DCI and DCJ respectively. The surface sample numbers and the vegetation types from which the samples were collected are also shown. In Fig. 7 there is a broad correspondence between the classification of the modern pollen spectra on the

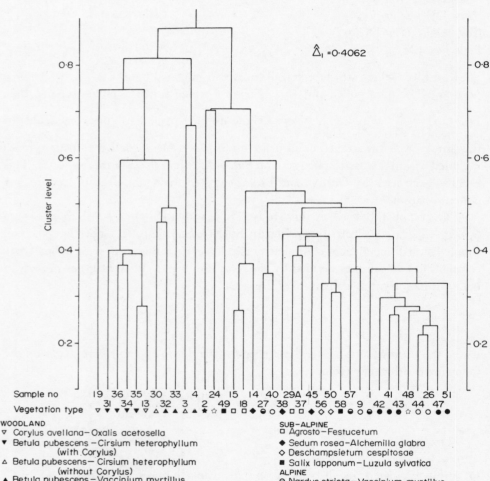

Figure 7. Dendrogram representing the hierarchic classification of Scottish surface pollen spectra obtained by a single-link cluster analysis of DCI. A key to the various vegetational types from which the surface samples were collected is also shown.

basis of pollen percentages (DCI) and the vegetation type from which the samples were derived. The clustered groups consist of samples 31, 36, 34, 35, and 13 (all from *Betula-Corylus* woods), of samples 30, 32, and 33 (all from *Betula* woods), of samples 3 and 4 (both from *Betula* woods), of samples 2 and 24 (both from juniper-dominated stands), of samples 15, 18, and 14 (all from sub-alpine communities), of samples 38, 29A, 37, 45, 56, 50, and 58 (all from sub-alpine communities), of samples 57, 9, 1, 42, 41, 43, 48, 44, 26, 47, and 51 (all from alpine or prostrate juniper communities), and of samples 27 and 40 (both from alpine communities). Samples 49 (sub-alpine willow scrub) and 19 (*Betula-Corylus* wood) are rather isolated and are amalgamated at a rather high level of dissimilarity. It is of interest that sample 48 collected from a wind-blasted dwarf juniper stand is classified with spectra from alpine communities, in view of the strong floristic and ecological affinities of these vegetation types. The $\hat{\Delta}_1$ value is 0.4062, suggesting that there is a reasonable but not perfect representation of the dissimilarity matrix in the two-dimensional hierarchic structure shown in Fig. 7.

In the dendrogram based on single-link clustering of the DCJ matrix (Fig. 8) only three clusters can be readily distinguished. One contains samples 44, 57, 9, 58, 40, 41, 27, 26, 42, 43, 1, 47, and 51. This group includes all the samples collected from the alpine zone plus sample 58 from a stand of sub-alpine willow scrub growing on cliff ledges. The pollen spectrum from this sample contains occasional pollen and spores of some taxa otherwise characteristic of alpine summit vegetation, such as *Lycopodium* spp. and *Artemisia norvegica* (Fig. 6). These grains were probably washed down onto the cliff ledge from the summit ground by flushing and by seepage of water onto the cliffs below. Samples 4, 3, 13, 19, 30, 31, 34, 35, 36, 32, 33, and 2 are grouped together on Fig. 8. These samples were collected from either birch-dominated woods, from mixed birch-hazel woods, or from a dense tall juniper thicket (2). Interestingly there is no distinction between the modern pollen spectra from the two major types of woodland sampled. In general the woodland pollen spectra are similar in their overall floristic composition as assessed in terms of presence or absence of pollen and spore types, differing only in the presence or absence of *Corylus avellana* pollen. When present *Corylus* attains high pollen percentages. In contrast spectra from the two woodland types are strikingly distinguished by DCI (Fig. 7). The third cluster of samples in Fig. 8 is rather heterogeneous and with a lot of chaining. It contains all the pollen spectra from the sub-alpine zone (except sample 58) and the two samples from the Juniperetum nanae (24, 48). The value of $\hat{\Delta}_1$ is 0.235, and this indicates that the clustering represented by Fig. 10 is a reasonable representation of the matrix of dissimilarity coefficients.

Conclusions

Single-link cluster analysis of either the matrix of DCI or of DCJ showed no consistent clustering of samples collected from the same vegetational community, such as all the samples collected from *Rhacomitrium*-heath within the alpine zone. This is not surprising in view of the minor differences detectable in either the pollen percentages or pollen composition in samples from different communities within the same zone, except for the rare occurrence of pollen or spores of a few 'indicator species'. The cluster analysis suggests that the observed differences in pollen percentages and composition in samples collected from the same plant community are as great as the differences in pollen percentages and composition in samples collected from different communities, but that the overall differences between spectra from alpine, sub-alpine, juniper, birch, and birch-hazel communities are sufficiently great to permit the consistent characterisation of these broad vegetational types in terms of modern pollen spectra.

Cluster analysis has several important disadvantages (see Kaesler 1970). One of the most important of these is the distortion introduced by clustering. The extent of the distortion can be measured, but it can rarely be corrected for. A more fundamental objection is that cluster analysis produces hierarchic clusters regardless of the structure of the original matrix of dissimilarity coefficients. Thus objects distributed uniformly in an hyperspace would be clustered into a hierarchy although

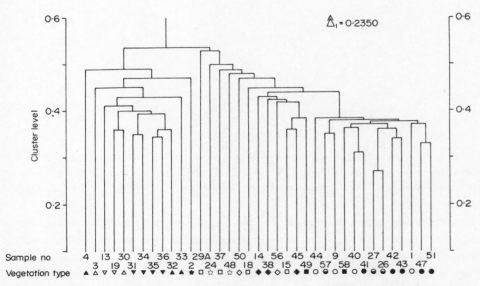

Figure 8. Dendrogram representing the hierarchic classification of Scottish surface pollen spectra obtained by a single-link cluster analysis of DCJ. For key to vegetational symbols see Fig. 7.

no such structure may exist in nature. The next section describes an alternative strategy involving a non-metric multidimensional scaling procedure as a non-hierarchical means of displaying the dissimilarities between samples.

NON-METRIC MULTIDIMENSIONAL SCALING

Methods

Given as data a matrix of dissimilarity coefficients between all pairs in a set of objects, the non-metric multidimensional scaling algorithm, devised by Kruskal (1964a, 1964b, 1971), may be used to obtain a configuration of the objects in a given number of dimensions, in which the distances between points representing the objects 'distort' the data as little as possible. The algorithm seeks an arrangement of points in a given number of Euclidean dimensions that minimises the stress measure S:

$$S = \sqrt{\frac{\Sigma\{d'(i,j) - \hat{d}(i,j)\}^2}{\Sigma\{d'(i,j)\}^2}}$$

where $d'(i,j)$ are the interpoint distances, and $\hat{d}(i,j)$ are numbers which minimise S, subject to the condition that they be monotonic with the values of the DC. In this case the objects are surface pollen spectra and the aim is to discover the extent to which the pattern of pollen dissimilarities, given by DC1 and DCJ, may be related to the vegetation type from which the samples were collected. The algorithm cannot be used to find the one-dimensional disposition of points which minimise the distortion measure, because in this case the optimisation procedure becomes trapped in local minima. The one-dimensional disposition has to be recovered by inspection of the disposition in the best-fitting two-dimensional representation.

The two-dimensional representation obtained for the modern pollen spectra based on DC1 and DCJ are shown in Figs. 9 and 10. The stress function, a measure of the goodness-of-fit, of the two-dimensional representation to the dissimilarity coefficients, is 9.6% and 14.9% for DC1 and DCJ respectively. These values can be evaluated verbally as 'fair' according to Kruskal (1964a). Non-metric multidimensional scaling differs from principal component analysis as a method of obtaining a representation of objects in a reduced number of dimensions in making no assumptions about linear regression. For that reason it is preferred in this case.

Results

In Fig. 9 the two-dimensional representation of the modern pollen spectra based on DC1 shows that four broad groups of spectra can be recognised. One contains all

Figure 9. Two dimensional representation of Scottish surface pollen spectra obtained by non-metric multidimensional scaling of DC1. S is the distortion imposed on DC1 by the interpoint distances. For key to vegetational symbols see Fig. 7.

the samples from alpine vegetation plus two samples from juniper-dominated stands (24, 48). The spectra from the sub-alpine communities are closely grouped within themselves and are well separated spatially from other spectra. The pollen spectra from woodlands are spatially distinct from other spectra, and there is a consistent pattern within the spectra, with the six samples from birch-hazel woods being closely grouped. The five spectra from birch woods are more widely scattered. Sample 2

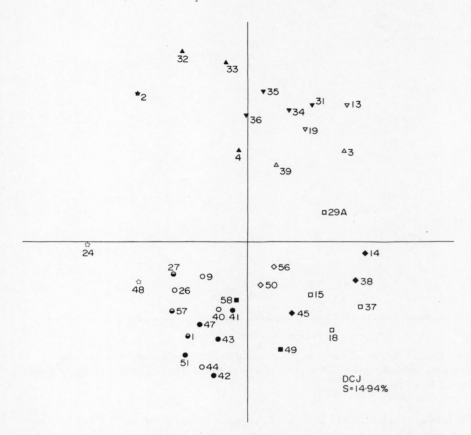

Figure 10. Two dimensional representation of Scottish surface pollen spectra obtained by non-metric multidimensional scaling of DCJ. S is the distortion imposed on DCJ by the interpoint distances. For key to vegetational symbols see Fig. 7.

(from tall juniper scrub) is positioned some distance from any of the other spectra.

The disposition of pollen spectra in two-dimensional space on the basis of DCJ (Fig. 10) shows a wider spatial scatter than is found with the scaling of the DC1 matrix. Spectra from alpine situations are closely grouped, and it is of interest that sample 58 (sub-alpine willow scrub) is positioned more closely to samples from the alpine zone than from the sub-alpine zone. A similar feature was found in the single-link cluster analysis (Fig. 8) of DCJ. The sub-alpine spectra are consistently positioned within themselves, whereas there is little spatial differentiation between spectra from the two principal woodland types, although the spectra from the woodlands as a whole are clearly separated from all the other pollen spectra considered. A similar feature emerged from the single-link cluster analysis. Spectra from juniper-dominated vegetation show no consistent grouping within themselves, as samples 24 and 48 are positioned nearest the spectra from alpine situations, whereas sample 2 is located nearest the woodland spectra.

Conclusions

As in the cluster analysis there is no consistent grouping of surface spectra from the same community. However, both numerical data in DC1 and presence/absence information for DCJ group spectra from the same broad vegetational type and altitudinal zone. Cluster analysis and non-metric multidimensional scaling are extremely consistent in their arrangement and classification of the data suggesting that the observed differences in pollen percentages and pollen composition in surface samples collected from the same community are as great as the differences in pollen percentages and composition between samples from different communities, but that the broad zonal vegetational types can be consistently characterised by modern pollen spectra.

DIVERSITY

Diversity is the relationship between the number of taxa in an assemblage and the number of individuals in that assemblage. This section discusses a preliminary attempt at assessing the diversity of pollen assemblages and examines the relationships between the diversity of modern pollen spectra and the diversity of the vegetation from which the pollen rain originates. If some relationship can be demonstrated between pollen and vegetational diversity at the present day, it may be possible to investigate temporal changes in vegetational diversity from changes in diversity of fossil pollen assemblages in a way analogous to that of Goulden's (1969) and Deevey's (1969) studies of diversity changes in assemblages of invertebrate fossils.

One of the simplest indices of diversity is Williams' (1964) α index of diversity. This concept of diversity is related to Fisher's logarithmic series. If n_i is the number of taxa represented by one individual only in a sample, then in the logarithmic series

$$n_i, \frac{n_i}{2}\chi, \frac{n_i}{3}\chi^2, \frac{n_i}{4}\chi^3, \ldots,$$

the second term gives the number of taxa with two individuals, the third the number with three individuals, etc. χ is a constant depending on the sample size, and n_i/χ is constant for the same population, irrespective of sample size. Williams (1964) has designated the constant n_i/χ, α the index of diversity.

If N is the total number of individuals in any sample and S is the number of taxa, it follows from the properties of the logarithmic series that

$$S = \alpha \ln\left(1 + \frac{N}{\alpha}\right)$$

A high value of α implies a great species diversity, and a low value indicates little diversity. α is only suitable as an index of diversity if the assemblage in question has many species and the species abundance forms a logarithmic series (Pielou 1969).

The relationship between the number of pollen and spore types and the number of individuals per type in a range of pollen assemblages examined by the author follow a logarithmic series, indicating that the use of α as an index of diversity is justified in this case. The form of the equation

$$S = \alpha \ln\left(1 + \frac{N}{\alpha}\right)$$

is such that a direct estimate of α from N and S is not straightforward. Two alternative approaches are available for the determination of α (see Williams 1964, Reyment 1971). One method is to use Williams' graph (1964, Fig. 126) of S against N giving approximate values of α. An alternative approach, and the one followed here, is to derive a value for χ from N/S by interpolation of the relationship between χ and N/S given by Williams (1964, Table 146). χ and α are related by the equation $\alpha = N(1 - \chi)/\chi$. The α indices of diversity for each modern pollen sample are tabulated in Table 4.

The α index of diversity for the vegetational types from which the surface samples were collected were calculated from the vegetational data presented in Birks (1973) and McVean and Ratcliffe (1962) using Dahl's (1956) formula:

$$\alpha = \frac{S_n - S_i}{\log n}$$

where S_n is the total number of taxa in the vegetation S_i is the mean number of taxa in the vegetation, and n is the number of floristic analyses.

In Fig. 11 the relationship between the mean index of diversity of the modern pollen assemblages and the diversity of the corresponding vegetation type from which the modern pollen rain is derived is shown graphically. There is a significant correlation between the two indices ($r = 0.91$, $P < 0.001$). That such a correlation exists is of interest for two reasons. Firstly, the diversity index of modern pollen spectra can provide a means of characterising spectra from different vegetational types. Secondly, the correlation demonstrates that the diversity of pollen spectra at the present day reflects the diversity of the vegetation from which the surface samples were collected. Thus observable changes in diversity of fossil pollen assemblages in pollen diagrams may be interpretable in terms of corresponding changes in vegeta-

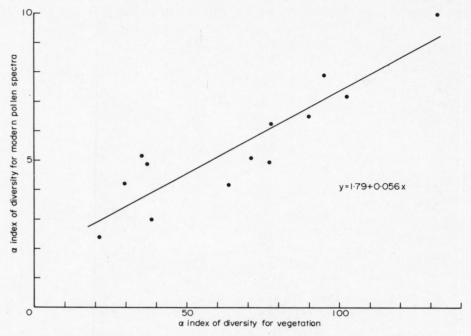

Figure 11. Plot of α index of diversity for modern pollen spectra and α index of diversity for vegetation from which the surface pollen samples were collected. The regression equation is shown ($r = 0.9091$, $N = 13$).

tional diversity. Such changes in diversity may aid in deducing the rates of development of past ecosystem diversity and community stability from fossil pollen sequences.

Problems in the comparison of modern and fossil pollen spectra

In general the available modern pollen rain data from arctic and alpine environments show consistent and distinctive relationships with the vegetation type from which the pollen rain is derived, at least when the vegetation type is considered at a rather broad sociological level. Thus if similarities in pollen content and proportions exist between modern spectra from known vegetation and fossil spectra, it may be suggested that a comparable vegetation type produced the fossil assemblage.

Some qualifications to this approach are, however, needed. A comparison between fossil and modern spectra from the same basin of deposition is moderately simple as it is only a comparison in time rather than one in time and space. However, past and present spectra from the same site are rarely, if ever, comparable, thereby posing the question—what type of plant community or range of communities does the fossil

Table 4. α Index of diversity values for modern pollen spectra from Scotland.

Sample No.	Vegetation type	α
Birch-hazel woods		
35	*Corylus avellana-Oxalis acetosella*	3.83
36	*Corylus avellana-Oxalis acetosella*	4.49
31	*Corylus avellana-Oxalis acetosella*	5.91
34	*Corylus avellana-Oxalis acetosella*	5.33
13	*Betula pubescens-Cirsium heterophyllum* (with *Corylus*)	7.01
19	*Betula pubescens-Cirsium heterophyllum* (with *Corylus*)	7.11
Birch woods		
30	*Betula pubescens-Cirsium heterophyllum* (without *Corylus*)	5.55
3	*Betula pubescens-Cirsium heterophyllum* (without *Corylus*)	9.51
32	*Betula pubescens-Vaccinium myrtillus*	2.63
33	*Betula pubescens-Vaccinium myrtillus*	3.44
4	*Betula pubescens-Vaccinium myrtillus*	6.79
Juniper scrub		
2	*Juniperus-Thelypteris* nodum	2.41
24	Juniperetum nanae	2.12
48	Juniperetum nanae	4.46
Sub-alpine vegetation		
29A	Agrosto-Festucetum (species-rich)	5.30
37	Agrosto-Festucetum (species-rich)	7.25
15	Agrosto-Festucetum (species-rich)	11.24
18	Dwarf herb nodum	9.56
14	*Sedum rosea-Alchemilla glabra*	11.22
38	*Sedum rosea-Alchemilla glabra*	8.05
45	*Sedum rosea-Alchemilla glabra*	10.51
50	Deschampsietum cespitosae alpinum	6.40
56	Deschampsietum cespitosae alpinum	6.58
49	*Salix lapponum-Luzula sylvatica*	10.08
58	*Salix lapponum-Luzula sylvatica*	6.06
Alpine vegetation		
57	Cryptogrammeto-Athyrietum chionophilum	5.17
1	*Nardus stricta-Vaccinium myrtillus*	5.11
27	*Nardus stricta-Vaccinium myrtillus*	4.51
9	Cariceto-Rhacomitretum lanuginosi	5.03
44	Cariceto-Rhacomitretum lanuginosi	5.71
26	Cariceto-Rhacomitretum lanuginosi	3.68
40	Cariceto-Rhacomitretum lanuginosi	4.82
41	*Festuca ovina-Luzula spicata*	5.94
42	*Festuca ovina-Luzula spicata*	4.28
43	*Festuca ovina-Luzula spicata*	5.19
47	*Festuca ovina-Luzula spicata*	4.22
51	*Festuca ovina-Luzula spicata*	5.83

spectrum resemble? Modern analogues are therefore sought elsewhere, thereby introducing a transfer in space as well as in time. The simplest situation is one in which the fossil spectrum can be matched in terms of pollen composition with modern spectra which (a) are distinctive from all other contemporary spectra, (b) refer to a characteristic and homogeneous vegetational type, and (c) are derived from comparable sites of deposition in topographically similar areas (Oldfield 1970).

A principal limitation of this approach is that much of the available modern pollen rain data is based on moss polsters or surface soils, whereas fossil spectra are generally based on sediments deposited in open water. There are, at present, no comparative data either to support or invalidate comparisons between the two depositional environments. A study of recent pollen spectra in surface lake muds in the alpine zone of Scotland or in northern Scandinavia, similar to those of Lichti-Federovich and Ritchie (1968) and McAndrews (1966) in North America, would be a particularly valuable project.

In theory it is possible for a particular pollen assemblage (expressed in relative frequencies) to be derived from two or more quite different vegetational types. Species with a very low pollen representation constitute 'blind spots' or 'silent areas' (*sensu* Davis 1963) in the vegetation that are barely registered in the pollen spectra. Communities that differ greatly in the extent of such 'silent areas', but that are similar in other respects, could be expected to produce similar relative pollen percentages (Davis 1963). To some degree, the Scottish alpine communities fall in this type, with a varying vegetational prominence of species not contributing to the pollen rain, such as *Luzula spicata*, *Juncus trifidus*, and *Rhacomitrium lanuginosum*.

An additional problem is the difficulty of evaluating the significance of the variation in pollen percentages between surface samples. For example, small variations in pollen percentages of taxa with a low pollen production may represent greater differences in terms of vegetational composition than large variations in percentages of taxa with a high pollen production. It is not possible to say that a distinct plant association always produces a distinct pollen spectrum. Cluster analysis and nonmetric multidimensional scaling of the Scottish data indicate that although the modern pollen rain from broad vegetational types is characteristic and distinctive, there appears to be as much variation in pollen composition in samples from one plant community as there is in samples from different but related communities. In attempting comparisons between modern and fossil pollen spectra, the variations in fossil pollen proportions from level to level and from site to site within the same pollen assemblage zone raise further problems. Averaging fossil spectra should be avoided, as trends may exist within a zone. More extensive studies of modern and fossil spectra and the development of numerical methods for comparison of spectra may ultimately provide some solution. In view of these problems it is preferable at present only to attempt comparisons between modern and fossil spectra on a rather broad basis, and not to seek an exact match between specific modern and fossil spectra (cf. Ogden 1969).

Despite these limitations and problems, the comparative approach to the reconstruction of past vegetation from fossil assemblages is considered the soundest basis currently available. If there is a close match between the modern and the fossil spectra a vegetational reconstruction can be made with some confidence. On the other hand if no modern pollen spectra can be found that are comparable with the fossil spectra, the conclusion is made that the past vegetation represented by the fossil spectra has no known modern counterpart. As it seems likely that the combination of edaphic, climatic, biotic, and successional conditions during Late-Weichselian times differed from any combination known today (Iversen 1954b), it may be misleading to assume that the same plant associations necessarily existed then as at present if the Gleasonian individualistic concept of vegetation is accepted as a working hypothesis. This concept recognises the uniqueness of plant communities in space and time, and stresses the behavioural rather than the floristic approach to vegetation. If no match can be made, some vegetational reconstruction can still be attempted using modern pollen rain data. The palaeofloristic data provide information as to the species formerly present and, from the surface pollen spectra, it is possible to assess the modern pollen representation of the taxa concerned, in general terms at least. Comparison between the modern values and the fossil percentages can then provide a crude estimate of the relative abundance of the taxa in the former vegetation.

In some cases the fossil pollen spectra may not only be dissimilar from present day spectra but they may be indicative of more than one plant community just as modern vegetation is rarely, if ever, a uniform homogeneous plant cover, it is likely that in the past there was also a complex mosaic of communities, each with its own floristic composition and structure and each associated with its own particular habitat. As fossil pollen spectra may represent an integrated pollen rain derived from several plant communities within the area reflected by the pollen rain, detailed comparisons with modern spectra from single uniform vegetational types may not be very profitable.

An alternative approach to vegetational reconstruction, proposed by Walker (1972), would require some estimate of the probability of various levels of pollen abundance, either in relative or absolute terms, occurring at various distances from a modern vegetational stand in which the parent plant occurred with a known measure of abundance in a sample plot of a particular areal extent. From such data it may be possible to construct, in probabilistic terms, synthetic fossil plant communities from fossil pollen assemblages. Such an approach would permit the modelling of vegetation changes implied in pollen diagrams, and may increase the resolution of pollen analysis as a palaeoecological technique.

Acknowledgements

I am grateful to Mr L. Rymer for permission to quote his unpublished data from Iceland, to Mr A.D. Gordon and Mrs M.E. Pettit for assistance with the numerical analyses, to Dr E.J. Cushing for stimulating discussions on several aspects of this work, and to Dr Hilary H. Birks for her critical reading of the manuscript.

References

AARIO L. (1940) Waldgrenzen und subrezenten Pollenspektren in Petsamo, Lappland. *Ann. Acad. Sci. Fenn. A* **54**(8), 1–120.

AARIO L. (1942) Turpeen siitepolyrunsauden käytösta metsätiheyden selvittäjäna. *Terra* **54**, 3–14.

ANDERSEN S.TH. (1970) The relative pollen productivity and pollen representation of north European trees, and correction factors for tree pollen spectra. *Danm. geol. Unders.* Ser. 11, **96**, 99p.

ANDERSEN S. TH. (1973) The differential pollen productivity of trees and its significance for the interpretation of a pollen diagram from a forested region. (this volume).

AUER V. (1927) Untersuchungen über die Waldgrenzen und Torfboden in Lappland. *Comm. Inst. Quaest. Forest. Finlandiae* **12**.

BARTLEY D.D. (1967) Pollen analysis of surface samples of vegetation from arctic Quebec. *Pollen Spores* **9**, 101–6.

BIRKS H.J.B. (1973) *The past and present vegetation of the Isle of Skye—a palaeoecological study.* Cambridge University Press, London.

BRIDEAUX W.W. (1971) Recurrent species groupings in fossil microplankton assemblages. *Palaeogeogr., Palaeoclimatol., Palaeoecol.* **9**, 101–22.

CHEETHAM A.H. & HAZEL J.E. (1969) Binary (presence-absence) similarity coefficients. *J. Paleontol.* **43**, 1130–36.

CLAPHAM A.R., TUTIN T.G. & WARBURG E.F. (1962) *Flora of the British Isles.* Cambridge University Press, London.

CLAPHAM W.B. (1970) Nature and paleogeography of Middle Permian floras of Oklahoma as inferred from their pollen record. *J. Geol.* **78**, 153–71.

COLINVAUX P.A. (1964) The environment of the Bering Land Bridge. *Ecol. Monogr.* **34**, 297–329.

COMANOR P.L. (1968) Forest vegetation and the pollen spectrum: an examination of the usefulness of the R value. *Bull. New Jers. Acad. Sci.* **13**, 7-19.

CORMACK R.M. (1971) A review of classification. *Jl. R. statist. Soc. A* **134**, 321-67.

DAHL E. (1956) Rondane. Mountain vegetation in South Norway and its relation to the environment. *Skr. norske Vidensk-Akad. 1 Mat.-Naturv. Klasse* 1956(3), 374p.

DAVIS M.B. (1963) On the theory of pollen analysis. *Am. J. Sci.* **261**, 897-912.

DAVIS M.B. (1967) Late-glacial climate in Northern United States: a comparison of New England and the Great Lakes Region. In *Quaternary Paleoecology* (ed. by E.J. Cushing & H.E. Wright), pp. 11-43. Yale University Press, New Haven and London.

DAVIS M.B. (1968) Pollen grains in lake sediments: redeposition caused by seasonal water circulation. *Science N.Y.* **162**, 796-99.

DAVIS M.B. (1969) Palynology and environmental history during the Quaternary period. *Am. Sci.* **57**, 317-32.

DEEVEY E.S. (1969) Specific diversity in fossil assemblages. *Brookhaven Symp. Biology* **22**, 224-41.

FAEGRI K. (1966) Some problems of representativity in pollen analysis. *Palaeobotanist* **15**, 135-40.

FAEGRI K. & IVERSEN J. (1964) *Textbook of pollen analysis.* Blackwell Scientific Publications, Oxford.

FAGER E.W. (1957) Determination and analysis of recurrent groups. *Ecology* **30**, 586-95.

FIRBAS F. (1934) Über die Bestimmung der Walddichte und der Vegetation Waldloser Gebiete mit Hilfe der Pollenanalyse. *Planta* **22**, 109-45.

FREDSKILD B. (1961) Floristic and ecological studies near Jakobshavn, West Greenland. *Meddr. Grønland* **163**(4), 82p.

FREDSKILD B. (1967) Palaeobotanical investigations at Sermermiut, Jakobshavn, West Greenland. *Meddr. Grønland* **178**(4), 54p.

FREDSKILD B. (1969) A postglacial standard pollen diagram from Peary Land, North Greenland. *Pollen Spores* **11**, 573-83.

GORDON A.D. & BIRKS H.J.B. (1972) Numerical methods in Quaternary palaeoecology. I. Zonation of pollen diagrams. *New Phytol.* **71**, 961-79.

GOULDEN C.E. (1969) Temporal changes in diversity. *Brookhaven Symp. Biology* **22**, 96-102.

GRICHUK V.P. (1942) An evaluation of the characteristic pollen composition of contemporary deposits of the different vegetation zones of the European part of the U.S.S.R. *Probl. Phys. Geography* **11**, 1-129 (in Russian).

GRICHUK V.P. (1950) The vegetation of the Russian plain in Early and Middle Quaternary time. *Tr. Inst. Geogr. Akad. Nauk S.S.S.R.* **46**, 5-202 (in Russian).

HARRIS W.F. & NORRIS G. (1972) Ecologic significance of recurrent groups of pollen and spores in Quaternary sequences from New Zealand. *Palaeogeogr., Palaeoclimatol., Palaeoecol.* **11**, 107-24.

HEUSSER C.J. (1969) Modern pollen spectra from the Olympic Peninsula, Washington. *Bull. Torrey bot. Club* **96**, 407-17.

IVERSEN J. (1945) Conditions of life for the large herbivorous mammals in the Late-glacial Period. In *The Bison in Denmark* by M. Degerbøl & J. Iversen. *Danm. geol. Unders.* Ser II, **73**, 62p.

IVERSEN J. (1952-3) Origin of the flora of western Greenland in the light of pollen analysis. *Oikos* **4**, 85-103.

IVERSEN J. (1954a) Über die Korrelationen zwischen den Pflanzenarten in einen Grönländischen Talgebiet. *Vegetatio* **5-6**, 238-46.

IVERSEN J. (1954b) The Late-glacial flora of Denmark and its relation to climate and soil. *Danm. geol. Unders.* Ser. II, **80**, 87-119.

JAMES P.W. (1965) A new check-list of British lichens. *Lichenologist* **3**, 95-153.

JANSSEN C.R. (1967a) A comparison between the regional pollen rain and the sub-recent vegetation in four major vegetation types in Minnesota (U.S.A.). *Rev. Palaeobotan. Palynol.* **2**, 331-42.

JANSSEN C.R. (1967b) A post-glacial pollen diagram from a small *Typha* swamp in northwestern Minnesota, interpreted from pollen indicators and surface samples. *Ecol. Monogr.* **37**, 145-72.

JANSSEN C.R. (1970) Problems in the recognition of plant communities in pollen diagrams. *Vegetatio* **20**, 187-98.

JARDINE N. & SIBSON R. (1968) The construction of hierarchic and non-hierarchic classifications. *Comput. J.* **11**, 177-84.

JARDINE N. & SIBSON R. (1971) *Mathematical taxonomy.* Wiley, New York.

JOHNSON R.G. (1962) Interspecific associations in Pennsylvanian fossil assemblages. *J. Geol.* **72**, 32-55.

KAESLER R.L. (1969) Aspects of quantitative distributional paleoecology. In *Computer applications in the Earth Sciences* (ed. by D.F. Merriam), pp. 99-120. Plenum Press, New York.

KAESLER R.L. (1970) Cophenetic correlation coefficient in paleoecology. *Bull. geol. Soc. Am.* **81**, 1261-66.

KOHUT J.J. (1969) Determination, statistical analysis, and interpretation of recurrent conodont groups in Middle and Upper Ordovician strata of the Cincinnati region (Ohio, Kentucky, and Indiana). *J. Paleont.* **43**, 392-412.

KRUSKAL J.B. (1964a) Multidimensional scaling by optimizing goodness-of-fit to a nonmetric hypothesis. *Psychometrika* **29**, 1-27.

KRUSKAL J.B. (1964b) Nonmetric multidimensional scaling: a numerical method. *Psychometrika* **29**, 115-29.

KRUSKAL J.B. (1971) Multi-dimensional scaling in archaeology: time is not the only dimension. In *Mathematics in the Archaeological and Historical Sciences* (ed. by F.R. Hodson, D.G. Kendall & P. Tăutu), pp. 119-32. Edinburgh University Press, Edinburgh.

LANCE G.N. & WILLIAMS W.T. (1967) A general theory of classificatory sorting strategies. I. Hierarchical systems. *Comput. J.* **9**, 373-80.

LICHTI-FEDEROVICH S. & RITCHIE J.C. (1965) Contemporary pollen spectra in Central Canada. II. The Forest-Grassland Transition in Manitoba. *Pollen Spores* **7**, 63-87.

LICHTI-FEDEROVICH S. & RITCHIE J.C. (1968) Recent pollen assemblages from the western interior of Canada. *Rev. Palaeobotan. Palynol.* **7**, 297-344.

LIVINGSTONE D.A. (1955) Some pollen profiles from Arctic Alaska. *Ecology* **36**, 587-600.

LIVINGSTONE D.A. (1968) Some interstadial and postglacial pollen diagrams from eastern Canada. *Ecol. Monogr.* **38**, 87-125.

LIVINGSTONE D.A. & ESTES A.H. (1967) A carbon-dated pollen diagram from the Cape Breton plateau, Nova Scotia. *Can. J. Bot.* **45**, 339-59.

LÖVE D. (1970) Subarctic and subalpine: where and what? *Arctic and Alpine Research* **2**, 63-73.

MAHER L.J. (1963) Pollen analyses of surface materials from the Southern San Juan mountains. *Bull. geol. Soc. Am.* **74**, 1485-1504.

MAHER L.J. (1964) Problems and possibilities of pollen analysis in mountainous regions. Unpublished manuscript.

MARTIN P.S. & MOSIMANN J.E. (1965) Geochronology of Pluvial Lake Cochise, Southern Arizona. III. Pollen statistics and Pleistocene metastability. *Am. J. Sci.* **263**, 313-58.

MATTHEWS J.V. (1970) Quaternary environmental history of interior Alaska: pollen samples from organic colluvium and peats. *Arctic and Alpine Research* **2**, 241-51.

McAndrews J.H. (1966) Postglacial history of Prairie, Savanna, and Forest in Northwestern Minnesota. *Mem. Torrey bot. Club* **22**(2), 72p.

McVean D.N. & Ratcliffe D.A. (1962) *Plant communities of the Scottish Highlands.* H.M.S.O. London.

Nordhagen R. (1943) Sikilsdalen og Norges Fjellbeiter. *Bergens Museum Skr.* **22**, 607p.

Ogden J.G. (1969) Correlation of contemporary and late Pleistocene pollen records in the reconstruction of postglacial environments in Northeastern North America. *Mitt. Int. Verein. Limnol.* **17**, 64–77.

Oldfield F. (1970) Some aspects of scale and complexity in pollen-analytically based palaeoecology. *Pollen Spores* **12**, 163–71.

Pielou E.C. (1969) *An introduction to mathematical ecology.* Wiley, New York.

Pohl F. (1937) Die polleneizeugung der Windblüter. *Beih. bot. Centralblatt.* **56A**, 365–470.

Polunin N. (1951) The Real Arctic: suggestions for its delimitation, subdivision and characterization. *J. Ecol.* **39**, 308–15.

Reyment R.A. (1971) *Introduction to quantitative paleoecology.* Elsevier, Amsterdam, London and New York.

Ritchie J.C. & Lichti-Federovich S. (1967) Pollen dispersal phenomena in Arctic-Subarctic Canada. *Rev. Palaeobotan. Palynol.* **3**, 255–66.

Rune O. (1965) The mountain regions of Lappland. In *The Plant Cover of Sweden. Acta Phytogeogr. suec.* **50**, 64–77.

Rymer L. (1972) Modern pollen rain studies in Iceland. Unpublished manuscript.

Sjörs H. (1963) Amphi-Atlantic Zonation, Nemoral to Boreal. In *North Atlantic Biota and their History* (ed. by A. Löve & D. Löve), pp. 109–25. Pergamon Press, London.

Sjörs H. (1965) Forest regions. In *The Plant Cover of Sweden. Acta Phytogeogr. suec.* **50**, 48–63.

Sokal R.R. & Sneath P.H.A. (1963) *Principles of numerical taxonomy.* W.H. Freeman, San Francisco and London.

Srodon A. (1960) Pollen spectra from Spitsbergen. *Folia quatern.* **3**, 1–17.

Tauber H. (1965) Differential pollen dispersion and the interpretation of pollen diagrams. *Danm. geol. Unders.* Ser. 11, **89**, 69p.

Tauber H. (1967) Investigations of the mode of pollen transfer in forested areas. *Rev. Palaeobotan. Palynol.* **3**, 227–86.

Terasmae J. (1967) Recent pollen deposition in the northeastern district of Mackenzie (Northwest Territories, Canada). *Palaeogeogr., Palaeoclimatol., Palaeoecol.* **3**, 17–27.

Terasmae J. (1968) Some problems of the Quaternary palynology in the western mainland region of the Canadian Arctic. *Geol. Surv. Pap. Can.* **68-23**, 26p.

Trapnell C.G. (1933) Vegetation types in Godthaab Fjord in relation to those in other parts of West Greenland, and with special reference to Isersiutilik. *J. Ecol.* **21**, 294–334.

Valentine J.W. & Mallory B. (1965) Recurrent groups of bonded species in mixed death assemblages. *J. Geol.* **73**, 683–701.

van der Hammen Th. (1951) Late-glacial flora and periglacial phenomena in the Netherlands. *Leid. geol. Meded.* **17**, 71–184.

Vasari Y. & Vasari A. (1968) Late- and post-glacial macrophytic vegetation in the lochs of Northern Scotland. *Acta bot. Fenn.* **80**, 120p.

Walker D. (1972) Quantification in historical plant ecology. *Proc. Ecol. Soc. Australia* **6**, 91–104.

Warburg E.F. (1963) *Census Catalogue of British Mosses* (3rd edition). Ipswich.

Welten M. (1950) Beobachtungen über den rezenten Pollenniederschlag in alpiner Vegetation. *Ber. geobot. Forsch. Inst. Rübel f.d. 1949* 48–57.

Williams C.B. (1964) *Patterns in the balance of nature and related problems in quantitative ecology.* Academic Press, London and New York.

Williams W.T. (1971) Principles of clustering. *Ann. Rev. Ecol. Sys.* **2**, 303–26.

Wright H.E. (1967) The use of surface samples in Quaternary pollen analysis. *Rev. Palaeobotan. Palynol.* **2**, 321–30.

Wright H.E. (1968) The role of pine and spruce in the forest history of Minnesota and adjacent areas. *Ecology* **49**, 937–55.

Wright H.E. & Patten H.L. (1963) The Pollen Sum. *Pollen Spores* **5**, 445–50.

Discussion on pollen representation

Recorded by Dr Hilary H. Birks,
Mr J.H.C. Davis, and Mr J. Dodd

Dr H.J.B. Birks asked Dr Andersen what correction factor he had used for *Populus* in the Eldrup pollen diagram. Dr Andersen replied that the numbers had not been corrected because no information was available for *Populus*. Sir Harry Godwin said that the replacement of *Tilia* by *Fagus* need not necessarily be due to burning, but may be due to its exploitation by prehistoric people for bast. He also asked what species of *Tilia* was involved. Dr Andersen replied that no *Tilia platyphyllos*-type pollen had been found, and some fruits of *T. cordata* had been recovered. During Neolithic times, the gradual *Tilia* replacement by *Quercus* could have been due to lopping of the *Tilia* branches. The catastrophic decline was undoubtedly due to burning. It was not certain why *Fagus* replaced *Tilia*, as *Tilia* was not completely eliminated. Perhaps mor humus started to form due to soil deterioration, and the soil became too poor for *Tilia*. Mr Tauber was surprised to see *Plantago lanceolata* pollen present in Stage II. This would be before the date of 3300 B.C., and would thus be the earliest record in Denmark, before the start of Neolithic agriculture. Its first appearance is usually at *c* 2600 B.C., with the landnam phases. Sir Harry Godwin commented that the absence of *Plantago lanceolata* pollen in these sites does not necessarily imply the absence of the plant. It was probably present in Denmark in other situations, similar to the way it survived since late-glacial times in the British Isles. Dr Andersen emphasised that *Plantago lanceolata* pollen was absent from Stage I, a time of relatively open conditions, but it had a continuous curve after its first appearance, and therefore the plant was probably absent during Stage I. There are no late-glacial records of *Plantago lanceolata* in Denmark.

Dr Rybniček asked if it was possible to transfer Dr Andersen's representation figures to other areas. It was his impression that, in Czechoslovakia, the representation of *Quercus* was at about the same level as *Fagus*, and thus required no correction. Dr Andersen saw no reason why *Quercus* should not flower so well in Czechoslovakia. In Denmark, the pollen values depended upon the situation of the sample. His figures for *Quercus* applied only to situations within the forest. In samples from lakes, low values of *Quercus* were recorded because of the high regional values for *Pinus*.

Professor Berglund stated that, in comparison with the Danish value of 8:1, the relative pollen representation of *Quercus* and *Tilia* in southeast Sweden was 10:1. The relative values of *Quercus* and *Acer* were 20:1. Thus *Acer* is extremely under-represented, and 1% of *Acer* pollen during Sub-Boreal times would indicate a high proportion of *Acer* in the forest. Dr Andersen thought that the differences in the values for *Tilia* were probably not significant. However, in one forest he had studied, the values for *Acer* were very similar to those for *Tilia*, although he admitted that he had relatively little data. Dr Cushing added that, in comparison with the European *Acer platanoides*, the North American *Acer saccharum* appeared to be more highly represented. The other American *Acer* species were probably represented at the same level as *Acer platanoides*.

Mr Tauber was impressed by the similarity between his and Professor Berglund's data on pollen trapping, considering that his lake was only 100 × 200 m, and the catching area was only a few hundred metres. However, he was struck by the high proportion, 10-40%, of the rain-out component, in comparison with his own value of about 12%. Professor Berglund thought that the minor climatic differences between Denmark and southeast Sweden could not account for this, although differences between the yearly catches may correlate with the annual rainfall. Mr Tauber added that, in dry periods, the roofed and unroofed traps caught the same amount of pollen, and that the unroofed trap gave a true estimate of the rain-out component, which was related to the total amount of precipitation, rather than seasonal effects. If, as Professor Berglund had suggested, the rain-out component was related to the size of the lake, such that in the presence of a large local component, the rain-out component was very small, this may account for Mr Tauber's relatively low values. Sir Harry Godwin suggested that Professor Berglund's high rain-out component at his coastal site may be related to heavier rainfall, but Professor Berglund said that the coast here was very dry, and the largest rain-out components were found inland.

Dr Seddon wondered if the correction factors derived from the relative pollen production of trees within the forest could be applied to samples from lake sediments, in view of the intervening processes of transport and deposition. Dr Andersen stressed that his pollen diagram was from an extremely small hollow, completely overhung by the tree canopy, and thus the transport factor was avoided. Professor Berglund added that he hoped his study would complement that of Dr Andersen, and that values from a lake could soon be compared with those from within a forest. However, suitable sites were hard to find in southern Sweden. Central and northern Sweden was more promising, but here the variety of deciduous trees was smaller, and conifers predominated. So far, his preliminary results had indicated that differences in correction factors between forests and relatively small lakes would not be very great. Mr Tauber thought, however, that, even in small lakes, transport factors may be overriding, and it was probably not possible to compare stands near the lake with those at a distance of more than 1,000 metres from the sampling spot. Dr Davis stressed the importance of defining the problem to be solved. Dr Andersen's investigation concerned local ecological factors, whereas studies of large lakes gave a regional picture. Correction factors were of no practical use in large lakes, as she had previously shown, as pollen from some species was derived from a large area, and others from a small area. However, there was no doubt about the usefulness of correction factors in local ecological studies, as a means of approaching a truer picture of the vegetation.

Professor Faegri ascertained from Dr Flenley that the 50% proportion of fern spores caught in his pollen traps contributed to the total of 8,000–20,000 grains per year. Sir Harry Godwin was surprised to hear that *Astelia* was characteristic of the upper mountain flora in New Guinea, as he had seen species of *Astelia* in New Zealand growing as epiphytes and ground plants in the forest. Dr Flenley informed him that there was only one species of *Astelia* in New Guinea, which was confined to mountain tops. The section to which it belonged contained no epiphytic species, even in New Zealand, and the *Astelia* pollen from New Guinea could be identified to this section.

Dr. Osmaston wondered whether it was wise to regard the lowland tropics as windless, as presumably the pollen rain would behave in much the same way as the ordinary rain. Dr Flenley agreed that fear of windlessness was largely unfounded. Dr Osmaston then suggested that bamboos would be good altitudinal indicators in East Africa, as they grew in a band at the top of the forest belt. However, he admitted that their pollen could not be readily distinguished from that of other grasses. Mr Mabberley commented that, in his experience, bamboos had a scattered occurrence in the East African mountains, and did not appear to be restricted to certain altitudinal belts, and thus could not be used as good altitudinal indicators. Dr Cushing inquired how many species of grass were found in the mountain grasslands, to which Dr Flenley replied that there were very many species, but only a few ever became dominant.

Professor Faegri was surprised that Dr H.J.B. Birks had found no *Betula* pollen in his surface samples from the alpine zone in Scotland. Dr Birks replied that the diagram showed local pollen types only, non-local pollen being excluded from the pollen sum. The local pollen percentages were used for the numerical analyses. Dr Sonesson asked if any differences had been found between surface samples from the same vegetation type taken from different materials. Dr Birks replied that although samples of *Rhacomitrium* were generally taken *Sphagnum* and soil were also occasionally used. He had noticed no differences between them, but the number of samples was too small to make a strict comparison. Dr Cushing noted that, although there was a degree of homogeneity within Dr Birks' numerically defined units, they were rather heterogeneous. Perhaps the lack of clustering between samples from the same vegetation community within the unit was due to this overall heterogeneity, and that clusters would be obtained if there were more samples from each community. Dr Birks agreed that this might be the case.

Section 4:
Plant macrofossil assemblages

Modern macrofossil assemblages in lake sediments in Minnesota*

HILARY H. BIRKS *Limnological Research Center,*
University of Minnesota, U.S.A., and The Botany School, University of Cambridge†

Introduction

At the present time, there is an increasing interest in plant macrofossils as a means of reconstructing past flora, vegetation, and environment. Macrofossils have been used to supplement pollen data, or to provide all the palaeobotanical information where pollen data are not available, for example in many 'full-glacial' and archaeological deposits. Detailed quantitative macrofossil recording in a stratigraphic context has only recently been developed as a palaeoecological technique; see, for example, Watts and Winter (1966), Watts and Bright (1968), Wright and Watts (1969), West (1964), Wasylikowa (1967), and Rybníček and Rybníčková (1968), although a pioneer study on numerical analysis of Tertiary macrofossils was carried out long ago by Chaney (1924).

For a satisfactory interpretation of quantitative macrofossil data, some information is necessary concerning the modern representation and dispersal of individual macrofossil taxa, and the representation of contemporary vegetation by macrofossil assemblages. Much information of this sort has accumulated for the interpretation of pollen diagrams by means of modern surface samples. However, there is very little such information concerning macrofossils. Chaney's (1924) study attempted to relate modern leaf assemblages collected from still water in a riverine environment to the surrounding forest, and then applied the results to the interpretation of a Tertiary leaf assemblage. McQueen (1969) studied the dispersal of modern macrofossils in a lake in New Zealand and attempted to ascertain the degree of transport from higher altitudes via the inflow streams. Ryvarden (1971) studied macrofossil dispersal onto recently deglaciated ground in Norway by means of traps, from which he was able to measure the annual influx of macrofossils and assess the efficiency of both terrestrial and water transport.

The present study investigates the relationships between modern macrofossils and the vegetation of small lakes, as palaeoecologists frequently choose limnic sediments for study. To ensure that the data are as comparable as possible, the study was restricted to relatively small lakes of similar morphometry, or enclosed bays of larger lakes, of about 300 m diameter. Minnesota, in north-central U.S.A. was chosen for such a study, because lakes of all kinds are abundant, and because several detailed quantitative macrofossil diagrams have been produced from the area. In addition, there are relatively sharp vegetational boundaries running across the state, between the prairie in the west, the intermediate deciduous forest, and the coniferous forest in the north and east (Fig. 1). Thus a wide range of lakes can be sampled from different vegetation formations within the same broad geographical area, and this range also covers the range of vegetation types that pollen analysis has shown to succeed one another during post-Wisconsin times (e.g. McAndrews 1966).

Methods

Surface samples of peat or mud were taken by hand, or with an Eckman dredge. The top few centimetres of the dredge samples were carefully removed with a spoon. For each sample, the position in the lake and water depth were recorded. The vegetation was described on a 5-point scale of cover-abundance. For aquatic plants that were not readily visible, five throws with a grapple were made in different directions around the boat. A water sample was taken from the centre of the lake to determine the ionic conductivity. Studies by Bright (1968) and E. Gorham (unpublished) have shown that conductivity is the most informative single parameter that characterises lake water chemistry in Minnesota. Several surface samples were collected from each lake, usually along a transect across vegetational zones, and an effort was made to sample the principal types of aquatic vegetation present, as well as deep-water areas where vegetation was absent.

In the laboratory, known volumes of sediment, usually about 50 cm^3, but more if necessary, were analysed for their macrofossil content. The macrofossils were identified with the aid of the fairly comprehensive macrofossil reference collection at the Limnological Research Center, University of Minnesota. The numbers of

*Contribution No. 114, Limnological Research Center
†Present address: The Botany School, University of Cambridge

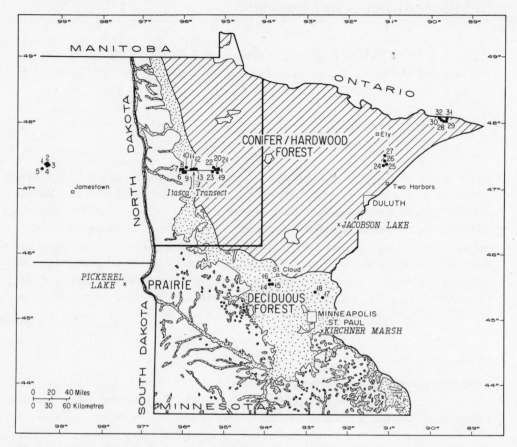

Figure 1. Map of Minnesota and North Dakota showing the locations of the lakes, numbered 1–32, used for modern surface samples, in relation to the major vegetation types. Sites of three stratigraphic macrofossil studies are also shown. The area of Fig. 2 is indicated.

each taxon were recorded, and the values standardised as number per 100 cm³ sediment. The sediment type was recorded according to the Troels-Smith system (Troels-Smith 1955).

Surface samples were analysed from 32 lakes (Fig. 1). They are arranged in a broad west-east transect, from the deep prairie of North Dakota, through the prairie border in Minnesota, the northern deciduous forest, the 'big-woods' deciduous forest near St. Cloud, the marginal coniferous forests round Itasca State Park, and an outlying area of coniferous forest north of Minneapolis, to the boreal coniferous forests of the Toimi drumlin field north of Two Harbors, and the Boundary Waters Canoe Area near Grand Marais.

For convenience, the lakes situated in the prairie will be termed 'prairie lakes', those in the deciduous forest 'deciduous lakes', and those in the coniferous forest 'coniferous lakes'. These terms imply nothing about the lake floras, but refer only to the upland vegetation type in which the lakes are situated.

Plant nomenclature follows Fernald (1950).

Present vegetation

In order to study the relationships of macrofossils to the vegetation that produced them, the vegetation had first to be studied. The lakes were chosen for their positions in the upland vegetation belts, and it was not known whether lakes from the different vegetation belts contained characteristic wetland and aquatic floras, and whether certain species assemblages were characteristic of certain lake types. As a pilot project, the vegetation from a restricted range of lakes was first studied. In northwest Minnesota, the vegetation boundaries are close together, and 18 lakes were selected along a west-east transect through the vegetation belts. It followed the line of the Itasca transect of McAndrews (1966), which within a distance of 30 miles (48 km) passes from the coniferous forest of the Itasca State Park through a narrow zone of deciduous forest into the prairie (Fig. 2). Not all the 18 lakes were used for macrofossil sampling, and the numbers do not correspond with the numbers on Fig. 1. From each lake, all the aquatic and wetland taxa seen were listed, and a water sample was taken for chemical analysis.

During the recording, it became apparent that some plants are characteristic of certain lakes. To test these suggestions, the data were subjected to a numerical ordination procedure similar to principal component analysis, which has been developed in France under the name 'Analyse factorielle des correspondences' (Escofier-Cordier 1969, M.O. Hill, in press and personal communication). The ordination of lakes on the basis of their contained species is plotted in Fig. 3. The points show that there is a continuum of variation in the data, with no discrete clusters detectable, except perhaps around lakes 1, 2, 4, and 6 and lakes 5, 8, 9, 10, and 11. The principal direction of variation appears to be the location of the lake in the upland vegetation type, for the lakes fall into three groups with no overlap, as indicated by the different symbols on Fig. 3. The lakes also fall along a gradient of conductivity values of the water, as shown in Fig. 4. Analyses of the major dissolved ions revealed a similar trend, with particularly high values of SO_4'' in lakes 1, 2 and 3. This suggests that water chemistry is a major environmental control on the wetland and aquatic flora, and that the water chemistry acts in the same direction as the gradient in upland vegetation.

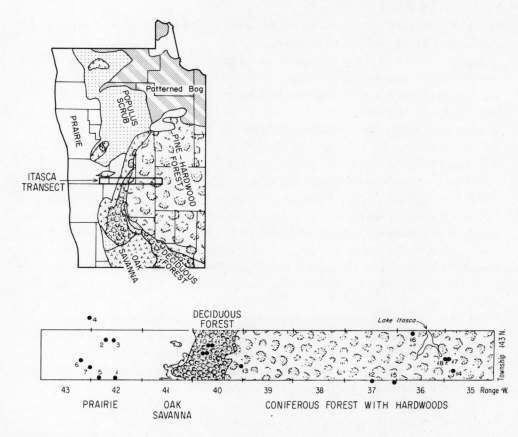

Figure 2. Map to show the Itasca transect and its relation to the major vegetation types. Lakes used for floristic recording are numbered 1-18.

When the same data were submitted to an agglomerative clustering technique (Orloci 1967) the lakes again fell into three main clusters, corresponding to their position in the upland vegetation belt, confirming that this is the major direction of variation.

When the species were ordinated on the basis of their occurrence in the lakes, the result was a continuum of points, arranged in a U-shaped pattern similar to the lake ordination. The major gradient again reflected the gradient in the upland vegetation. Several groups of species could be distinguished on the basis of their distribution in different lake types. Some of the commoner species characteristic of each group are given below.

Restricted to prairie lakes: *Juncus balticus*, *Scolochloa festucacea*, *Sonchus arvensis*, *Typha angustifolia*.

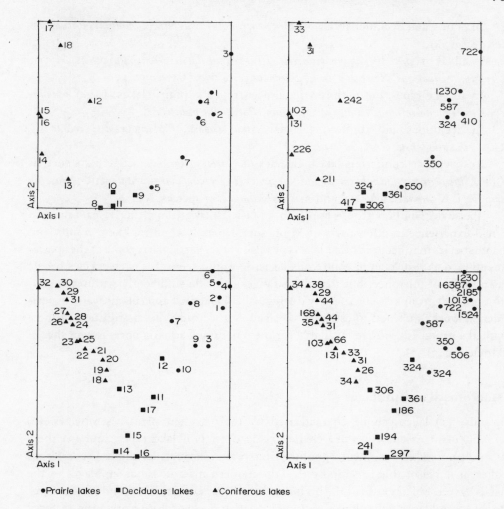

Figure 3. (Top left). Ordination plot of lakes along the Itasca transect, ordinated on the basis of their present flora. The lake numbers are those of Fig. 2.

Figure 4. (Top right). Conductivity values (μmhos at 25°C) of the lakes along the Itasca transect superimposed on the ordination plot of Fig. 3.

Figure 5. (Bottom left). Ordination plot of lakes used for surface sample studies, ordinated on the basis of their modern macrofossil content. The lake numbers are those of Fig. 1.

Figure 6. (Bottom right). Conductivity values (μmhos at 25°C) of the lakes used for surface samples superimposed on the ordination plot of Fig. 5.

In prairie and deciduous lakes: *Carex atherodes, Phragmites communis, Potamogeton pectinatus*.

In all lake types: *Eleocharis palustris, Lycopus uniflorus, Mentha arvensis, Myriophyllum exalbescens, Scirpus acutus, Sium suave, Utricularia vulgaris*.

In all lake types, but with restricted occurrence in prairie lakes: *Carex pseudocyperus, Lemna minor, Nuphar variegatum, Potentilla palustris*.

In deciduous and coniferous lakes: *Alnus rugosa, Calamagrostis canadensis, Glyceria canadensis, Sagittaria latifolia*.

Restricted to coniferous lakes: Species of *Carex lasiocarpa* sedge mats such as *Carex lasiocarpa, Dulichium arundinaceum, Hypericum virginicum*; and *Brasenia schreberi, Nymphaea odorata*, and many *Potamogeton* species.

These distribution groups apply only to the 18 lakes along the Itasca Transect. Field experience in other parts of Minnesota generally confirms these results, but some species may occur in other lake types elsewhere, particularly some of the species mentioned as restricted to coniferous lakes. It would seem that each species has its own range of tolerance, but that these ranges coincide sufficiently within the continuum that groups can be usefully distinguished. Detailed autecological and experimental work is needed to establish the factors governing the distribution of individual species. Little is yet available for the Minnesota aquatic flora, except that of Moyle (1945).

Macrofossil assemblages

Because (1) lakes can be characterised by their present plant assemblages, (2) certain plant assemblages are characteristic of certain lake types, and (3) there is a well marked environmental gradient from prairie through deciduous to coniferous lakes, it is reasonable to suppose that the modern macrofossil assemblages in the lakes have a similar gradient. If this is so, the comparison of modern and fossil macrofossil assemblages will allow a more informative ecological interpretation of fossil data.

To test whether this is the case, the modern macrofossil data were submitted to the same ordination procedure as the modern vegetation data. For each of the 32 lakes (Fig. 1) a list was compiled of all the modern macrofossils recovered from the surface samples within that lake, and the 93 taxa occurring in 3 or more lakes were used for the computation.

The ordination of the lakes on the basis of their macrofossil content (Fig. 5) shows a continuum, but with a very strong gradient correlated with upland vegetation type, similar to that found for the lakes along the Itasca transect. The values of water conductivity show a similar gradient (Fig. 6), with values of several thousand μmhos in the prairie lakes of North Dakota, to values in the region of 30 μmhos in the coniferous lakes of northeast Minnesota.

The ordination of the macrofossils on the basis of their occurrence in the lakes also shows a continuum, with no apparently distinct clusters (Fig. 7). However, the same geographical gradient is apparent, namely a trend from taxa characteristic of prairie lakes to taxa characteristic of coniferous lakes. Macrofossils of similar lake distribution, shown by symbols on the diagram, generally cluster together within the continuum, as did the plants of similar distribution along the Itasca transect. The macrofossils in the centre of the plot (Fig. 7) are widespread. If the fossil is identified to species, the macrofossils of that taxon are widespread, for example *Najas flexilis*. However, the majority of macrofossils in this group are not identifiable to species, or sometimes even genera, and thus may represent many species which altogether have a widespread distribution. This group includes *Calamagrostis, Carex* achenes (lenticular and trigonous), Compositae undiff., *Chara* oospores, *Eleocharis palustris* type, Gramineae undiff., *Lycopus uniflorus, Najas flexilis, Scirpus acutus/S. validus*, and *Typha*. Macrofossils characteristic of the other groups are given below.

Restricted to prairie lakes, or with a few occurrences in deciduous and coniferous lakes: *Alopecurus aequalis, Carex atherodes, Chenopodium album* type, *Chenopodium rubrum, Lemna, Mentha arvensis, Panicum capillare, Potamogeton pectinatus, Ranunculus sceleratus, Rumex maritimus* var. *fueginus, Scolochloa festucacea, Teucrium canadense, Utricularia vulgaris, Zannichellia palustris*.

In prairie and deciduous lakes: *Eleocharis calva* type, *Lycopus americanus, Myriophyllum exalbescens* type, *Polygonum lapathifolium, Sagittaria* cf. *S. cuneata, Taraxacum*.

Mainly in deciduous lakes, but with scattered occurrences elsewhere: *Carex pseudocyperus/C. comosa, Ceratophyllum demersum, Cyperus engelmanni, Eupatorium perfoliatum*.

In deciduous and coniferous lakes: *Betula papyrifera, Bidens cernua, Brasenia schreberi, Carex diandra, Carex synchnocephala* type, *Heteranthera dubia, Larix laricina* needles, *Nuphar variegatum, Nymphaea odorata, Potentilla norvegica, Sagittaria latifolia, Scirpus cyperinus*.

Mainly in coniferous lakes, but with scattered occurrences in deciduous lakes: *Najas gracillima, Potamogeton amplifolius, P. natans, Sagittaria rigida*.

Restricted to coniferous lakes: *Alnus rugosa, Carex lasiocarpa, Chamaedaphne calyculata, Diervilla lonicera, Dulichium arundinaceum, Eleocharis ovata, Hypericum virginicum, Myriophyllum farwellii, Nitella, Potamogeton epihydrus, P. spirillus, Potentilla palustris*.

If the mean conductivity value of the lakes in which each macrofossil taxon

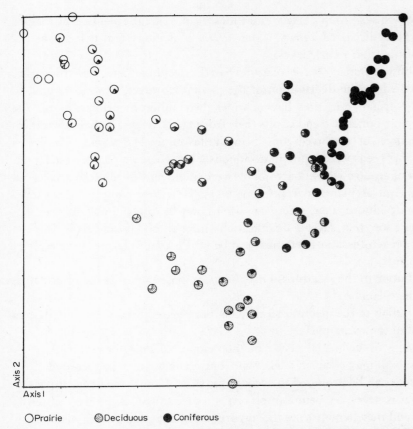

○ Prairie ◓ Deciduous ● Coniferous

Figure 7. Ordination plot of macrofossil taxa in surface samples, ordinated on the basis of their occurrence in lakes. The distribution of each taxon within the range of lakes is shown by 'clockfaces'.

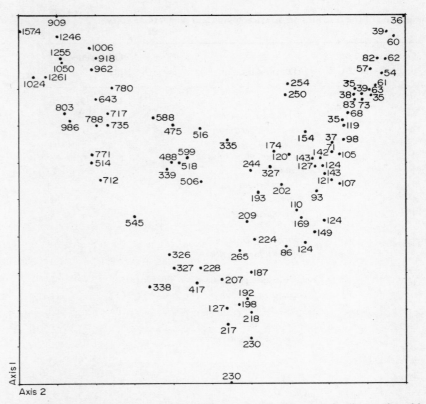

Figure 8. Mean conductivity values (μmhos at 25°C) for the macrofossil taxa plotted in Fig. 7. The mean value was calculated from the values of all the lakes in which each macrofossil taxon was found.

occurs is plotted on the macrofossil ordination (Fig. 8), a strong gradient is again apparent, from very high values for taxa characteristic of prairie lakes to low values for coniferous lake taxa. Thus the modern macrofossils appear to reflect this environmental parameter of lake water chemistry, as well as the upland vegetational setting.

Some examples of macrofossil taxa with characteristic distributions are compared in Table 1 with the distribution of the living plants. This latter information was taken from the vegetational notes made from all the 32 lakes, and differs in detail from the Itasca Transect data.

The taxa of group 1 on Table 1 have the same distribution of plants as of macrofossils. Group 2 taxa have a more widespread distribution towards the coniferous lakes as plants than their macrofossils, particularly *Sium suave* and *Utricularia vulgaris*. Better macrofossil representation towards the prairie is a common trend in Minnesota; it may be due to the warmer prairie climate, or to the higher concentration of dissolved ions, which may favour fruiting. The flowering of *Lemna* is certainly stimulated by warm conditions and stranding of the plants on exposed mud.

Sonchus arvensis and *Ranunculus sceleratus* in group 3 were only recorded as plants in prairie lakes, but their macrofossils were found throughout the range. The plants may have been missed during recording, or they may have an erratic appearance in deciduous and coniferous lakes, depending on local site conditions.

The macrofossils of group 4 taxa are found mainly in deciduous lakes. The plants tend to be more widespread, and there is a tendency for reduced macrofossil representation in prairie and coniferous lakes.

Table 1. Comparison of the distribution of macrofossils and plants in 32 lakes.

Group	Occurrence as macrofossils			Macrofossil taxon	Occurrence as plants		
	P	D	C		P	D	C
1	+			*Scolochloa festucacea*	+		
	+	+		*Lycopus americanus*	+	+	
2	+	O		*Potamogeton pectinatus*	+	+	
	+	+		*Lemna*	+	+	O
	+	+		*Myriophyllum exalbescens* type	+	+	O
	+	O		*Sium suave*	+	+	+
	+	O	O	*Utricularia vulgaris*	+	+	+
3	+	O	O	*Sonchus arvensis*	+		
	+	+	O	*Ranunculus sceleratus*	+		
4		+	O	*Eupatorium perfoliatum*	O	+	+
		+	O	*Sagittaria latifolia*	O	+	+
		+	O	*Carex pseudocyperus/C. comosa*	O	+	O
		+	+	*Bidens cernua*	O	+	+
		+		*Ceratophyllum demersum*		+	
5	+	+	+	*Eleocharis palustris* type	+	+	+
	+	+	+	*Typha*	+	+	+
	+	+	+	*Chara*	+	+	+
	+	+	+	*Lycopus uniflorus*	+	+	+
	+	+	+	*Najas flexilis*		+	+
6		+	+	*Nymphaea odorata*		+	+
		O	+	*Potamogeton amplifolius*		O	+
		O	+	*Potamogeton natans*		O	+
		O	+	*Sparganium fluctuans* type		O	+
		O	+	*Scutellaria*	+	+	+
		+	+	*Nuphar variegatum*	O	+	+
		+	+	*Brasenia schreberi*			+
7			+	*Dulichium arundinaceum*		O	+
			+	*Hypericum virginicum*		O	+
			+	*Glyceria canadensis*		O	+
			+	*Alnus rugosa*		O	+
			+	*Potentilla palustris*		O	+
			+	*Potamogeton epihydrus*			+
			+	*Chamaedaphne calyculata*			+

P = Prairie   D = Deciduous   C = Coniferous
+ = abundant occurrence   O = rare occurrence

The macrofossils of group 5 taxa are all widespread, as are the plants. The reasons for this have been discussed above. *Najas flexilis* is of interest in that it was not recorded as plants from prairie lakes.

The distribution of macrofossils of group 6 in the deciduous and coniferous lakes is closely matched by the distribution of the plants. However, *Scutellaria* and, to some degree, *Nuphar variegatum* have a wider distribution as plants towards the prairie. Perhaps an opposite trend affects their fruiting behaviour, high temperatures and concentrations of dissolved ions being unfavourable. *Brasenia schreberi* is anomalous in that seeds were found in deciduous lakes where no plants were found.

Of the taxa of group 7 whose macrofossils were found only in coniferous lakes, all are most abundant as plants in coniferous lakes. However, many have scattered occurrences in deciduous lakes, where conditions may be unfavourable for fruiting.

These data show that several qualifications have to be considered before the occurrence of a macrofossil in a sediment can be interpreted in terms of the ecology of the living plant.

1 The distribution of the macrofossil may be more biased towards the prairie than that of the living plant.
2 The distribution of the macrofossil may be more biased towards the coniferous lakes than that of the living plant.
3 The plant may occur in a lake type, but no macrofossils may be recorded.
4 The macrofossil may occur in a lake type, but no plants may be recorded.

Many reasons can be proposed to account for these anomalies. They emphasise the caution that is necessary before interpreting a macrofossil assemblage in terms of vegetation, and they demonstrate the importance of comparing fossil macrofossil assemblages with modern macrofossil assemblages, rather than directly with living plants.

Representation of macrofossils

The third point discussed above raises questions of representation and dispersal of macrofossils. This section will discuss representation.

There is little information on the seed production of obligate aquatic plants, although there are some figures for some wetland plants of reedswamps and muddy shores (Salisbury 1942, 1970). The majority of obligate aquatics are perennial, and they usually have efficient means of vegetative reproduction (Arber 1963). *Ceratophyllum demersum* is a good example of a plant that rarely fruits but relies on vegetative reproduction.

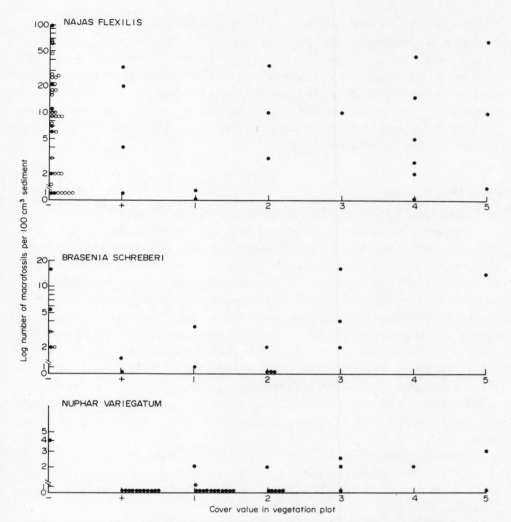

Figure 9. Representation plots for some obligate aquatic taxa: *Najas flexilis* (annual), *Brasenia schreberi* and *Nuphar variegatum* (perennials). The log scale is modified at the base to allow the depiction of zero values. For samples where there was zero cover value,
○ = sample where the plant was recorded as absent from the lake;
● = sample where the plant was growing elsewhere in the lake.

Of the few aquatic plants that are annuals, *Najas* spp. are the most notable examples in Minnesota. Such plants rely on seed production for over-wintering, and like most annual plants, they produce large numbers of seeds. *Najas* seeds are well preserved in sediments and can occur in high concentrations, often as broken halves, presumably as a result of germination of the embryo. Fig. 9 is a plot of the numbers of seeds of *Najas flexilis* per 100 cm^3 sediment in relation to the cover value of the plant in the vegetation. Wherever *Najas* plants occur, its seeds are found in some abundance. However, in many samples, seeds of *Najas* were found but no plants were recorded in the vegetation. Frequently, plants were found growing elsewhere in the lake (indicated by open circles) and their seeds may have been transported. However, some high numbers of seeds were found where *Najas* was recorded as absent from the lake. This absence may be an artifact of recording, as it was not possible to search every part of each lake. Alternatively, *Najas* may have formerly been present but recently disappeared, due to a change in conditions or to the chance fluctuations undergone by populations of annual species.

Fig. 9 also shows the representation of *Brasenia schreberi* and *Nuphar variegatum*, which are taken as good examples of the representation of perennial aquatic plants. In several cases, plants were present but no seeds were found, particularly with *Nuphar* but, in general, the higher the plant cover the more seeds were found.

Other perennial obligate aquatic species show similar patterns and levels of macrofossil representation. However, the plots are so irregular that no representation indices, such as the R-values used for pollen types, can be calculated, and only broad general conclusions can be made for each species. This does not necessarily invalidate quantitative macrofossil recording, as, for each species where enough data are available, the approximate number of macrofossils can be estimated which indicate a certain abundance of the plant. In a rather extreme example, 3 seeds per 100 cm^3 of *Brasenia schreberi* indicate that the plant had a moderate cover in the vegetation, whereas 3 seeds per 100 cm^3 of *Najas flexilis* indicate that the plant was probably very scarce in the vegetation, or that it was growing some distance from the sampling site. Thus some assessment of the former abundance of the plant may be made from a macrofossil count, as well as of the floristic composition of the vegetation.

The representation of wetland species in surface samples falls into two categories: (1) samples from the marginal vegetation where the plant was growing, and (2) samples from the lake itself, where the macrofossils can only be derived by transport from the shore a certain distance away. These latter samples provide information on dispersal, and are discussed in the next section.

Fig. 10 is a plot of numbers of macrofossils per 100 cm^3 sediment in relation to the cover value of the plant in the vegetation for two wetland taxa. In both cases, the numbers of seeds generally increase with increasing cover value. For *Typha*, very high numbers of macrofossils (in the hundreds per 100 cm^3) may be recorded where the plants are dominant, but relatively few where the plants are scattered. In contrast, the number of macrofossils produced by a *Carex lasiocarpa*-dominated sedge-mat are relatively few (about 10 per 100 cm^3). In general, the same qualifications hold true for the representation of wetland plants as for the obligate aquatic plants.

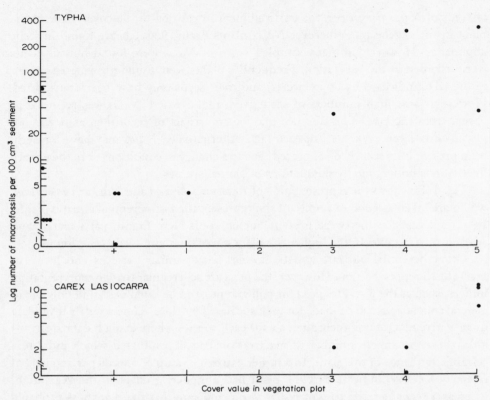

Figure 10. Representation plots for some wetland taxa, *Typha* spp. and *Carex lasiocarpa*. The log scale is modified to allow the depiction of zero.

Dispersal of macrofossils

Macrofossils can be derived from three broad categories of plants: upland plants, wetland plants, and obligate aquatic plants.

UPLAND PLANTS

The occurrence of fruits or seeds of upland plants in lake sediments is accidental as far as the plants are concerned. Many of the trees of the Minnesota forests have wind-dispersed fruits, and these are frequently blown into lakes. The commonest are the fruits of *Betula papyrifera*, which are produced in vast numbers and may float for a considerable time on water (Ridley 1930). They are often found in sediments in the centre of deep lakes where few other macrofossils are found. Winged conifer seeds are also wind-dispersed, and most commonly found are those of *Larix laricina*, which frequents boggy lake margins in the coniferous region.

Conifer needles are commonly found as macrofossils. Those of *Picea* spp. and *Pinus strobus* are usually found in low numbers near the lake shore, but *Larix laricina* needles are often exceedingly abundant and, with *Betula* fruits, are often the only macrofossils found in the centre of deep lakes. *Larix* is deciduous and the needles fall in great numbers in the winter, and may well be wind-transported over the ice and deposited in the sediments in spring.

WETLAND PLANTS

As for the upland plants, the occurrence of macrofossils of wetland plants is accidental as far as the plant is concerned, but in contrast many of these species utilise water dispersal. The seeds or fruits are often adapted for floating and are dispersed by water and ideally washed up on the shore in a position suitable for germination (Ridley 1930). Ridley gives data on the mechanisms of floating, and quotes data collected by Guppy and Praeger on the floating capacity of various species.

Fig. 11 illustrates the dispersal of several wetland taxa. Numbers of macrofossils per 100 cm^3 sediment are plotted in relation to the distance of the sample from the shore. *Chenopodium rubrum* and *Viola* are examples of species with no special dispersal adaptations. There is a sharp fall-off in seed numbers at 20 and 5 m respectively. The high values of *Chenopodium* seeds at 50, 90, and 150 m are from shallow prairie lakes. *C. rubrum* is a characteristic 'drawdown' species that colonises mud exposed during droughts. The high values probably reflect recent drying up of the lakes.

Ranunculus sceleratus, *Rumex maritimus* var. *fueginus*, and *Sagittaria latifolia* have seeds or fruits well adapted for floating. There is a fall-off of numbers near the shore, but macrofossils are consistently found considerable distances from the shore, up to a maximum of 150 m.

Fruits of *Bidens cernua*, *Dulichium arundinaceum*, and *Eleocharis palustris* type are adapted by means of spiny appendages for animal dispersal. There is a fall-off in numbers near the shore, but consistent occurrences are found away from the shore presumably as a result of transport by aquatic animals. *Bidens cernua* fruits also float well and seem to be fairly evenly dispersed in low numbers. There are no data on the floating capacity of *Dulichium* fruits, but they can reach 100 m in some abundance. *Eleocharis* fruits float only slightly and this is perhaps reflected in the concentration of *Eleocharis palustris* type fruits near the shore, with occasional large numbers at 90 and 100 m.

Figure 11. Dispersal plots for some wetland taxa. See text for explanation.

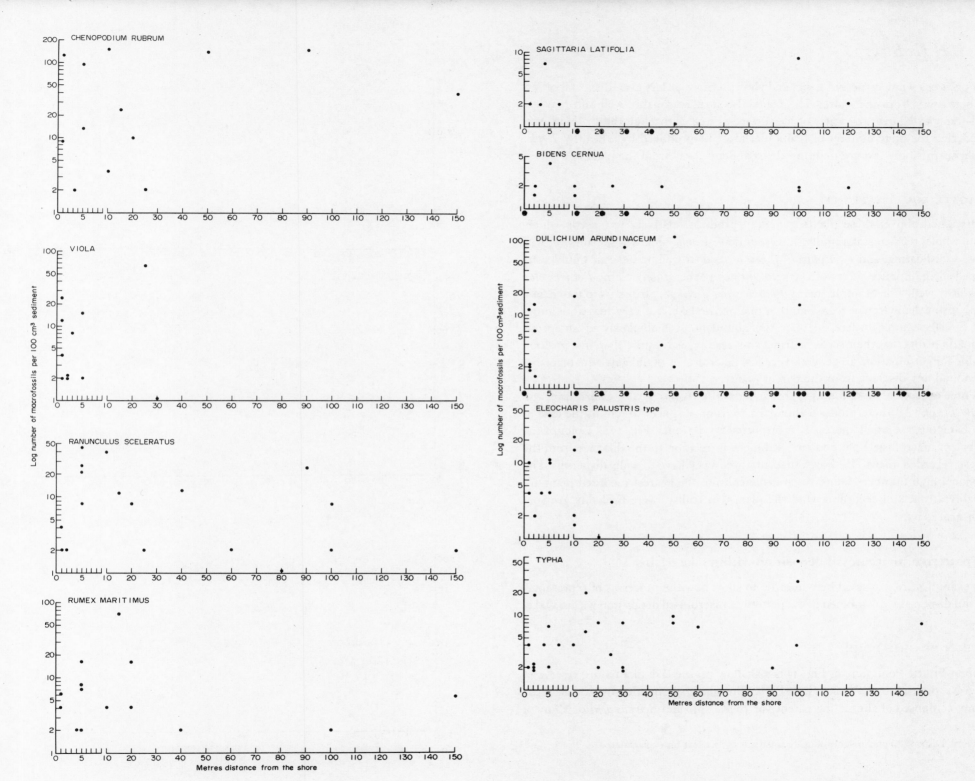

Typha has wind-dispersed fruits and shows a more or less even dispersal of low numbers away from the shore. The fruits also float reasonably well and may get wind-transported over the water or ice surface as well as through the air. However, considering the huge number of fruits that must be produced by a healthy *Typha* marsh, surprisingly few are found in the sediments away from the plants.

OBLIGATE AQUATIC PLANTS

As discussed above, these plants generally produce relatively few seeds which, on the whole, tend to sink rapidly. Exceptions are *Lemna* spp. whose fruits are well adapted for floating, and *Utricularia*, whose seeds float well because air bubbles are trapped in the minute ridges of the seed surface (Arber 1963). Some *Potamogeton* fruits float well. For example, many *Potamogeton pectinatus* fruits were found in a sample from the marginal reedswamp in one prairie lake, but very few were found from the lake samples where the plant was abundant. It is obviously advantageous to aquatic plants that their seeds or fruits sink readily, as they are then in a position suitable for germination. In general, very few macrofossils of obligate aquatic plants were found any distance from the parent plants.

Three exceptions were found to this general rule: *Najas flexilis*, *Chara* spp., and *Nitella flexilis*. All these produce abundant seeds or oospores which, in the case of the Characeae, are small and easily transported by currents. Fig. 12 is a plot of the number of macrofossils per 100 cm^3 sediment in relation to the distance from the nearest recorded plant. It shows that all three taxa have a wide dispersal. The occasional high numbers found some distance from the nearest recorded plant are probably artifacts of recording, and the plants, in reality, were probably growing much nearer by.

The pattern of macrofossil deposition within lake basins

Two examples are discussed here in order to show how the processes of representation and dispersal are combined in the pattern of macrofossil deposition within a lake.

WAUBUN PRAIRIE POND

Waubun Prairie Pond (site 6, Fig. 1) is a shallow prairie lake in a nature reserve of 'tall grass' prairie west of Itasca State Park. The lake is surrounded by a dense reedswamp, composed of alternating patches of *Typha* spp. and *Scirpus acutus*. Many of

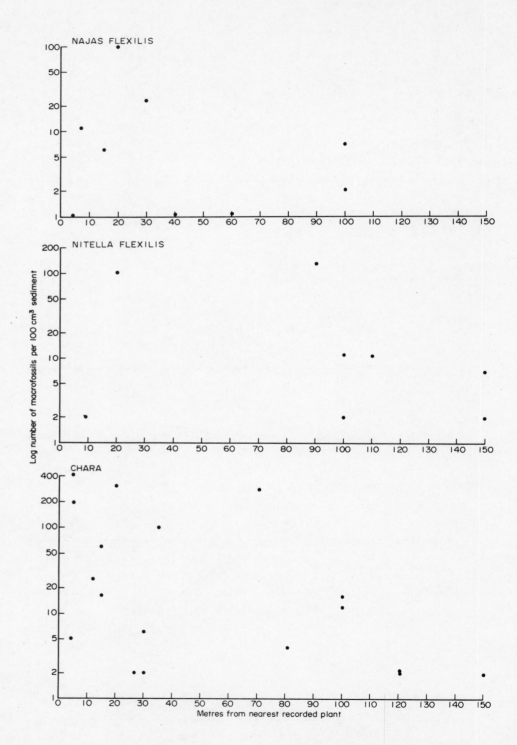

Figure 12. Dispersal plots for some obligate aquatic taxa. See text for explanation.

the wetland plants characteristic of prairie lakes grow in the marsh, for example *Juncus balticus*, *Lycopus americanus*, *Scolochloa festucacea*, *Sonchus arvensis*, and *Teucrium canadense*. In the lake the dominant aquatics are *Chara globularis* in the central area, and *Potamogeton pectinatus* with *Chara* in the marginal zone.

Three surface samples were taken: sample 1 is from the centre of the lake, sample 2 is one metre from the shore where it was dominated by *Scirpus acutus*, and sample 3 was 5 metres from the shore where it was dominated by *Typha*. For convenience, the three samples are represented in Fig. 13 as a linear transect, although their relative positions formed a flattened triangle. In Fig. 13 the cover values of the plants in the sample plots are shown as vertical bars, and the frequency of macrofossils per 100 cm^3 sediment by the silhouettes. The sediment composition is shown in the Troels-Smith (1955) notation, together with the loss on ignition. The increase in amount and coarseness of minerogenic material towards the shore is a common feature in prairie lakes, and it is due in part to windblown silt and sand from fields and roads being trapped by the reedswamp. The same principle is true for the distribution of macrofossils of upland plants in prairie lakes.

Chara globularis is present in all the samples and its oospores are correspondingly abundant. However, in common with most other *Potamogeton* species, fruits of *Potamogeton pectinatus* are only found where the plant is abundant and then only in small quantities. *Utricularia vulgaris* was recorded from sample 3, where no seeds were found. However, a seed was recovered from sample 1, suggesting the transport of the seeds from the margin.

The remaining macrofossils are all from upland or wetland plants. Very high numbers of *Scirpus acutus* fruits were found in the marginal samples, but relatively few in the centre, suggesting a low dispersal capacity. The fruits do not float well and are adapted for animal dispersal. *Typha* fruits are confined to the marginal samples where they occur in the low numbers characteristic of *Typha*. Macrofossils of the other wetland plants such as *Sonchus arvensis*, *Eleocharis palustris* type, *Rumex maritimus*, *Teucrium canadense*, *Mentha arvensis*, and *Lycopus* sp. show a similar pattern. *Carex lacustris* is exceptional in that its fruits reach the centre, but the fruits float well due to air becoming trapped within the utricle.

Macrofossils of a few upland plants, *Panicum capillare*, *Rubus* cf. *R. idaeus*, and *Cirsium* were found, but they were restricted to the marginal samples.

Chenopodium rubrum, *C. album* type, and *Ranunculus sceleratus* seeds belong to weedy annual taxa characteristic of exposed mud. The *Chenopodium* seeds are concentrated in large numbers near the lake shore. They are ill-adapted for floating and do not reach the lake centre. In contrast, seeds of *Ranunculus sceleratus* float well and a few were recovered from the lake centre.

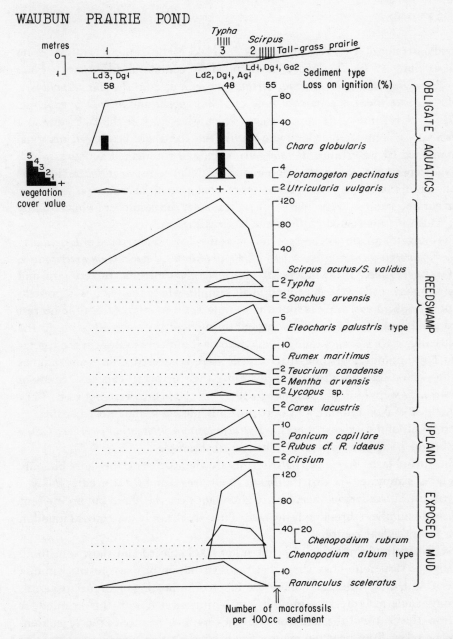

Figure 13. The pattern of macrofossil deposition in Waubun Prairie Pond. The vertical bars denote the cover value of the plant in the vegetation plots; the silhouettes denote the number of macrofossils per 100 cm^3 sediment.

MISSOURI POND

Missouri Pond (site 23, Fig. 1) is a relatively deep lake in the coniferous forest region near Itasca State Park. The lake is surrounded by coniferous forest, with *Pinus strobus*, *Picea glauca*, *Larix laricina*, *Betula papyrifera*, and *Populus tremuloides*. Towards the lake there is a narrow zone of *Alnus rugosa* and *Betula pumila* var. *glandulifera*, with some *Salix* spp. bushes, which gives way at the lake edge to a *Carex lasiocarpa*-dominated sedge-mat. Within the lake there is a broad marginal zone dominated by water lilies, *Potamogeton* spp., and submerged aquatics, which merges in deeper water into a zone dominated by *Potamogeton amplifolius*. There is no vegetation in the centre, where the water reaches a depth of 6 m.

Four surface samples were taken along a transect; the results are illustrated by Fig. 14, which is constructed on the same basis as Fig. 13.

Of the obligate aquatic taxa present, no macrofossils were recovered of *Utricularia vulgaris*, *Potamogeton gramineus*, *P. natans*, *P. amplifolius*, *Sparganium americanum*, or *Nuphar variegatum*. Of the others, *Scirpus subterminalis* fruits are very rare, and *Brasenia schreberi* seeds were not found in all samples where the plant was present. A seed of *Nymphaea odorata* was found in sample 2, seven metres from the nearest recorded plant, but *Nymphaea* was scattered throughout the waterlily zone. At the other extreme, *Chara globularis* and *Najas flexilis* have low cover values in the vegetation, but large numbers of their macrofossils were recovered even from samples where the plants were not growing, illustrating their dispersal capacity. Seeds of *Najas gracillima* were found in samples 1 and 2, but the plants were not seen. They resemble those of *Najas flexilis*, and were probably missed in the field recording.

Macrofossils of the sedge mat taxa, *Sagittaria latifolia*, *Potentilla palustris*, *Carex lasiocarpa*, and *Viola*, are restricted to the marginal sample.

Of the upland taxa, *Betula papyrifera* fruits are abundant near the shore but only reach as far as sample 3. The distribution of *Pinus strobus* and *Picea* needles follow a similar pattern. *Larix laricina* needles are abundant near the shore but are present in appreciable numbers throughout the lake. They are the only macrofossil found in sample 4 from the lake centre.

These two examples are typical of the other lakes studied and show something of the range of variation. They show that the distribution of macrofossils within a lake is very local, unless the macrofossils are adapted for floating or animal dispersal. Hence, any single macrofossil sample has to be interpreted with this in mind. A sample from the centre of the lake would give a very poor picture of the vegetation, whereas one taken from near the margin would give a much more comprehensive representation, and thus the maximum amount of information. In any vertical series of samples from a core, the appearance or disappearance or change in abundance of a

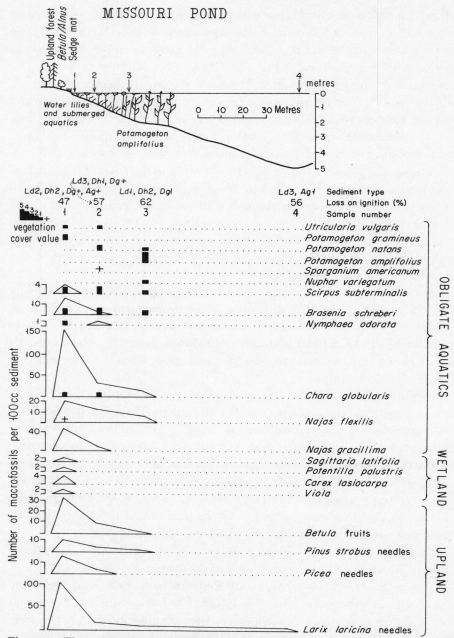

Figure 14. The pattern of macrofossil deposition in Missouri Pond. The vertical bars denote the cover value of the plant in the vegetation plots; the silhouettes denote the number of macrofossils per 100 cm³ sediment.

macrofossil taxon need not necessarily reflect changes in the vegetational composition of the lake, but merely a change in the vegetational pattern, or changes in the fruiting behaviour of the taxon in response to changing environmental conditions.

Some palaeoecological applications

Most of the major macrofossil diagrams from the Minnesota area have been produced by Watts (e.g. Watts & Winter 1966, Watts & Bright 1968, Wright & Watts 1969). The modern macrofossil studies confirm most of his conclusions and interpretations and these can hardly be improved upon here. However, in some cases, the new information allows more detailed interpretations to be made in terms of reconstruction of past communities and of past environmental conditions. A few such comments are made below.

RECONSTRUCTION OF COMMUNITIES

a. Late-Wisconsin assemblages

Watts (1967) has discussed the widespread occurrence of a distinctive assemblage of macrofossils of aquatic plants, which he terms 'pioneer aquatics', typically including *Hippuris vulgaris*, *Ranunculus* sect. *Batrachium*, and *Myriophyllum exalbescens* type, in Late-Wisconsin assemblages in Minnesota. No modern analogues of this assemblage were encountered during the present study, suggesting that the assemblage is rare or even absent in Minnesota lakes today. However, several of the species grow today in river habitats in the State. H.J.B. Birks (personal communication) has shown that the aquatic flora of newly formed lakes on recent glacial moraines in southwest Yukon closely resembles the Minnesota late-glacial assemblages, which may thus reflect the relatively unstable conditions in newly formed lakes after deglaciation.

The assemblage of macrofossils of upland plants characteristic of Late-Wisconsin sediments (Watts 1967) consists of needles of *Larix laricina* and *Picea*, associated with remains of understorey species of coniferous woodlands, such as *Cornus canadensis*, *Rubus* sp., and *Fragaria virginiana*. The assemblage is widespread, occurring from Woodworth Pond, North Dakota (McAndrews *et al.* 1967) to northeast Minnesota (Wright & Watts 1969). Similar modern assemblages were recovered from the coniferous forest region of northeast Minnesota, particularly from the Toimi area (sites 24, 25, 26, 27) and the Boundary Waters Canoe Area (sites 28, 29, 30, 31, 32). The floristic composition of the late-glacial spruce forest may have resembled that of the modern coniferous forest. However, pollen evidence (Cushing 1963, H.J.B. Birks, unpublished manuscript) indicates that the forests of this time may have been somewhat more diverse in their floristic composition than is suggested by the available macrofossil evidence.

b. Mid-post-glacial assemblages

The Late-Wisconsin spruce pollen zone is succeeded in most parts of Minnesota by a *Pinus* pollen assemblage zone. The associated macrofossils suggest the development of varied aquatic and reedswamp communities, which are difficult to reconcile with modern equivalents. At about 8,000 B.P., the pollen diagrams give evidence of the onset of more xeric climatic conditions, and the development of prairie in North and South Dakota and in the southern and western parts of Minnesota. The associated macrofossil assemblage of aquatic taxa at Pickerel Lake, in eastern South Dakota, is very similar to the modern assemblages recovered from prairie lakes in North Dakota (Fig. 7) except for the abundance of *Zannichellia palustris*. Very little modern information was obtained for this species, although it can occur abundantly in prairie lakes, and occasionally in deciduous lakes. The assemblage of wetland macrofossil taxa at Pickerel Lake also resembles modern assemblages from prairie lakes (Fig. 7), particularly in the occurrence of weedy annuals such as *Rumex maritimus* var. *fueginus* and *Chenopodium* spp. The abundance of these macrofossils at Pickerel Lake suggests that, due to the lowered water level, the plants grew abundantly on the exposed mud, and also that the sampling site was close to the new shoreline.

At Kirchner Marsh (Watts & Winter 1966), the same xeric climatic interval is shown by the appearance and increase of macrofossils of weedy annual plants, implying a fluctuating water level during periodic droughts. However, the assemblage differs in floristic composition from that of Pickerel Lake. The assemblage of macrofossils of obligate aquatic taxa most closely resembles the modern assemblages from deciduous lakes, or prairie lakes near the deciduous forest border (Fig. 7), with *Ceratophyllum demersum*, *Potamogeton praelongus*, *P. pusillus*, and *Heteranthera dubia*. Similarly, the wetland taxa and weedy annuals, with *Sagittaria latifolia*, *Lycopus americanus*, *Eupatorium perfoliatum*, *Cyperus engelmanni*, *Polygonum lapathifolium*, and *Bidens cernua*, are also more characteristic of such lakes, rather than true prairie lakes. The characteristic large peaks of *Cyperus erythrorhizos* and *C. engelmanni* macrofossils at Kirchner Marsh probably represent the abundant growth of these plants near the sampling site. No modern information was obtained for *C. erythrorhizos*, but limited information for *C. engelmanni* suggests that it is restricted to lakes within the deciduous forest (Fig. 7) and that it is well represented. Its fruits float well due to the attached aerenchymatous tissue of the bract. The values of *Chenopodium* spp. seeds at Kirchner Marsh are not as high as at Pickerel

Lake. If the sampling site was near the contemporary shore, this suggests that *Chenopodium* spp. were not growing abundantly at Kirchner Marsh. However, as they are rather poorly dispersed, it is possible that the sampling site was sufficiently far away from the shore for relatively few seeds to reach it.

At Jacobson Lake, the *Cyperus* spp. are present during this time interval in low numbers, together with small quantities of other weedy annuals, such as *Potentilla norvegica*. The significance of these occurrences is difficult to evaluate in terms of fluctuating water level, as the sampling site is in the middle of the basin. The main feature of this phase is the expansion of *Typha* fruits together with macrofossils of associated reedswamp plants characteristic of deciduous or coniferous lakes, such as *Eupatorium perfoliatum*, *Scirpus cyperinus*, *Epilobium glandulosum*, *Carex synchnocephala* type, *Sagittaria latifolia*, and *Polygonum lapathifolium*. Any fluctuations of lake level seem to have been sufficiently slight to encourage rather than destroy the growth of a dense *Typha* reedswamp.

c. *Post-glacial assemblages after 5,000 B.P.*

After about 5,000 B.P. the climate became progressively moister towards conditions of the present day, and the prairie gave way on its eastern border to deciduous forest (McAndrews 1966). The palaeobotanical evidence shows that Pickerel Lake remained in the prairie throughout, with only the localised development of riverine woodland beside the lake and in the valley floors (Watts & Bright 1968). The aquatic macrofossil assemblage changes little, with the exception of an increase in *Myriophyllum exalbescens* type, and the appearance of *Ruppia maritima*. However, the macrofossils of the weedy annuals decline, which Watts and Bright take as evidence of a more permanent, higher lake level. However, if the lake level were higher, the sampling site would be further from the contemporary shore and fewer macrofossils would be deposited there, even if there had been no decrease in the numbers of plants. Those taxa with good dispersal capacity, such as *Rumex maritimus* var. *fueginus* and *Ranunculus sceleratus*, are still present in small numbers, together with low numbers of *Chenopodium* spp. seeds.

At Kirchner Marsh, deciduous forest returned to the area after about 5,000 B.P. (Wright *et al.* 1963). Presumably the lake level became stabilised, as *Najas flexilis* becomes abundant, virtually to the exclusion of all other aquatic macrofossils. The macrofossils of the weedy annuals of exposed mud decrease, except for a small peak of *Cyperus engelmanni* and *Eleocharis acicularis* just before the macrofossils and the sediment indicate the overgrowth of the sampling site by reedswamp. The sampling site was then presumably near the shore, and the abundance of these drawdown plants indicates a brief period of dryness, or of disturbance, which exposed damp mud for colonisation. The sampling site became overgrown by a reedswamp, dominated at first by *Scirpus acutus* or *S. validus*, but joined later by *Typha*. *Brasenia schreberi* seeds become relatively abundant at this time, indicating a healthy growth of the plant. It can occur in deciduous lakes today (see Table 1), but has its optimum occurrence in coniferous lakes. Perhaps conditions were peculiarly favourable for it at Kirchner Marsh. Other wetland and aquatic taxa which appear at this time include *Nymphaea tuberosa*, *Menyanthes trifoliata*, *Dulichium arundinaceum*, *Carex comosa*, and *Lycopus uniflorus*. All these are characteristic of deciduous and coniferous lakes today (Fig. 7), although *Lycopus uniflorus* is widespread. Perhaps this change in the lake flora reflects the retreat of prairie influences, and the approach of the coniferous forest from the north.

At Jacobson Lake, at about 4,000 B.P., there is a marked change in the macrofossil flora, indicating the growth of a *Carex lasiocarpa* sedge mat and alder swamp, such as are characteristic of lakes in the coniferous region today. A decrease in base status during this development is detectable in the macrofossil assemblage. At first *Carex diandra* and *Eupatorium perfoliatum* are prominent components, but they are later replaced by a *Chamaedaphne* bog fringe, with *Scirpus purshianus*, *Hypericum boreale*, *Sarracenia purpurea*, *Vaccinium* spp., and *Eleocharis ovata*. *E. ovata*, together with *Bidens cernua*, appears to colonise exposed mud, for example along beaver trails beside lakes with a low conductivity. Other plants characteristic of lake margins in the coniferous region are represented by macrofossils of *Epilobium glandulosum*, *Carex synchnocephala* type, *Hypericum virginicum*, *Sagittaria latifolia*, *S. rigida*, *Carex rostrata*, *Lycopus uniflorus*, *Potentilla palustris*, *Scheuchzeria palustris*, *Viola pallens*, and *Larix laricina*. These macrofossils occur throughout the phase of sedge-mat development at Jacobson Lake, and tend to have a somewhat wider ecological tolerance today than the previously mentioned components.

The assemblage of macrofossils of obligate aquatics at Jacobson Lake during this time is similar to the modern assemblages recovered from low conductivity coniferous lakes in northeast Minnesota. *Potamogeton spirillus*, *Myriophyllum farwellii*, and *Najas gracillima* in particular are characteristic of coniferous lakes with a conductivity below 100 μmhos (Figs. 7 and 8), whilst *Potamogeton amplifolius*, *P. pusillus*, *Brasenia schreberi*, *Nymphaea tuberosa*, *Nuphar*, and *Najas flexilis* have a wider tolerance, but all occur regularly in low-conductivity coniferous lakes. The numbers of macrofossils are all relatively high, when the representation of the species is

Figure 15. Reconstruction of the environmental conditions in terms of conductivity in the correlative zones, C–b at Kirchner Marsh, and zone 3 at Pickerel Lake. The measured modern mean and range of conductivity values for the fossil taxa are plotted for each site, in comparison with the mean and range of values for the modern lakes studied.

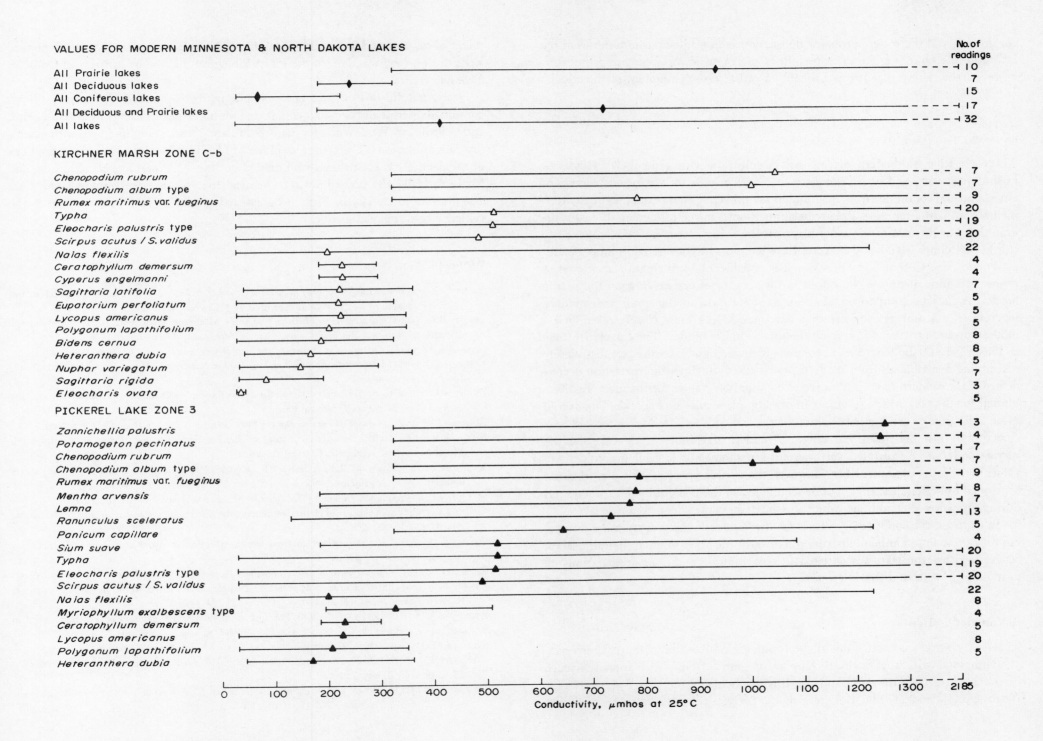

considered, and there were probably dense, flourishing aquatic communities in the lake near the sampling site. The quantities of *Najas flexilis* seeds are rather low, however, suggesting that it was probably a rather rare component of the aquatic vegetation at that time.

RECONSTRUCTION OF ENVIRONMENT

The correlation of modern macrofossil assemblages with present-day factors of upland vegetation and water chemistry provides a basis for the reconstruction of environmental conditions from macrofossil assemblages. Some examples of environmental reconstruction have already been mentioned in the discussion of past communities, but the case of the assemblages of the xeric 'prairie period' (8,000–5,000 B.P.) at Kirchner Marsh and Pickerel Lake will now be considered in more detail.

Pollen zone C-b at Kirchner Marsh is considered by Wright *et al.* (1963) to represent a time of prairie conditions in the area, or of oak savanna on the prairie border. A floristic comparison of the macrofossil taxa of this zone with modern macrofossil assemblages suggest that Kirchner Marsh most closely resembled a modern deciduous lake, or a lake on the forest-prairie border. The correlative zone at Pickerel Lake, pollen zone 3, contains a macrofossil assemblage most closely resembling assemblages from modern prairie lakes. Such an interpretation is confirmed if the modern mean and range of conductivity values for the macrofossils is plotted for the taxa where sufficient information is available (Fig. 15). The overall mean and range of conductivity values for the taxa at Kirchner Marsh most closely correspond with the mean and range of modern deciduous lakes. *Chenopodium rubrum* and *Rumex maritimus* var. *fueginus* are anomalous in having conductivity values in the range of prairie lakes. However, these species are probably not primarily controlled by water chemistry, but depend more on the availability of exposed mud during 'drawdown' conditions, which are today most prevalent in prairie lakes.

In contrast, the conductivity values for the Pickerel Lake taxa are more diverse, but they show a predominant distribution towards the range of modern prairie lakes, suggesting that the lake was situated in true prairie conditions at that time, as proposed by Watts and Bright (1968).

Acknowledgements

Laboratory work was carried out at the Limnological Research Center, University of Minnesota, with the financial support of grants from the National Science Foundation (Grant GB-7163) and the Environmental Protection Agency (Grant 16010 DXG), and at The Botany School, University of Cambridge, with the financial support of a Research Fellowship at Newnham College. Much of the field work during 1970 was done at the Itasca Biological Station, by courtesy of Dr W.H. Marshall (Director).

I am grateful to Professor H.E. Wright Jr. for supporting the project, for his continued interest and encouragement, and for his critical reading of the manuscript, and to Professor W.A. Watts for many stimulating discussions and help with macrofossil identifications. The assistance of Mr M.O. Hill, who carried out the ordinations of the data, Dr E.J. Cushing, who carried out the agglomerative classification, and Mr D.A. Hall, who carried out the chemical analyses, is greatly appreciated. I am also grateful to Dr H.J.B. Birks, who assisted in much of the field work, and who critically read the manuscript.

References

ARBER A. (1963) *Water Plants. A study of aquatic angiosperms*. Wheldon & Wesley, Ltd., Herts., and Hafner Publishing Co., New York, reprinted by J. Cramer, Weinheim.

BRIGHT R.C. (1968) Surface-water chemistry of some Minnesota lakes, with preliminary notes on diatoms. *Limnological Research Center, Univ. of Minnesota, Interim report* 3, 59p.

CHANEY R.W. (1924) Quantitative studies of the Bridge Creek flora. *Am. J. Sci.* 8, 127–144.

CUSHING E.J. (1963) *Late-Wisconsin pollen stratigraphy in east-central Minnesota*. Ph.D. thesis, University of Minnesota.

ESCOFIER-CORDIER B. (1969) L'analyse factorielle des correspondences. *Cahiers du Bureau Universitaire de Recherche Opérationelle* No. 13. Paris.

FERNALD M.P. (1950) *Gray's Manual of Botany* (8th edition). American Book Co., New York.

MCANDREWS J.H. (1966) Post-glacial history of Prairie, Savanna, and Forest in Northwestern Minnesota. *Mem. Torrey bot. Club* 22(2), 72p.

MCANDREWS J.H., STEWART R.E. & BRIGHT R.C. (1967) Paleoecology of a prairie pothole: a preliminary report. *North Dakota Geological Survey Miscellaneous Series* 30, 101–13.

MCQUEEN D.R. (1969) Macroscopic plant remains in recent lake sediments. *Tuatara* 17, 13–19.

MOYLE J.B. (1945) Some chemical factors influencing the distribution of aquatic plants in Minnesota. *Am. Midl. Nat.* 34, 402–20.

ORLOCI L. (1967) An agglomerative method for classification of plant communities. *J. Ecol.* 55, 193–206.

RIDLEY H.N. (1930) *The dispersal of plants throughout the world*. L. Reeve & Co. Ltd., Ashford, Kent.

RYVARDEN L. (1971) Studies in seed dispersal. 1. Trapping of diaspores in the alpine zone of Finse, Norway. *Norw. J. Bot.* 18, 215–26.

RYBNÍČEK K. & RYBNÍČKOVÁ E. (1968) The history of flora and vegetation on the Bláto Mire in southeastern Bohemia, Czechoslovakia. *Folia geobot. phytotax.* 3, 117–42.

SALISBURY E.J. (1942) *The reproductive capacity of plants*. G. Bell & Sons, London.

SALISBURY E.J. (1970) The pioneer vegetation of exposed muds and its biological features. *Phil. Trans. R. Soc. B.* 259, 207–55.

TROELS-SMITH J. (1955) Karakterisering af Løse jordarter. Characterisation of unconsolidated sediments. *Danm. geol. Unders.* Ser. IV, 3(10), 73p.

WASYLIKOWA K. (1967) Late Quaternary plant macrofossils from Lake Zeribar, western Iran. *Rev. Palaeobotan. Palynol.* **2**, 313-8.

WATTS W.A. (1967) Late-glacial plant macrofossils from Minnesota. In *Quaternary Paleoecology* (ed. by E.J. Cushing & H.E. Wright), pp. 89-97. Yale University Press, New Haven and London.

WATTS W.A. & BRIGHT R.C. (1968) Pollen, seed and mollusk analysis of a sediment core from Pickerel Lake, northeastern South Dakota. *Bull. geol. Soc. Am.* **79**, 855-76.

WATTS W.A. & WINTER T.C. (1966) Plant macrofossils from Kirchner Marsh, Minnesota—a paleoecological study. *Bull. geol. Soc. Am.* **77**, 1339-60.

WEST R.G. (1964) Inter-relationships of ecology and Quaternary palaeobotany. *J. Ecol.* **52** (Suppl.), 47-57.

WRIGHT H.E. & WATTS W.A. (1969) Glacial and vegetational history of northeastern Minnesota. *Minn. geol. Surv. Spec. Publ.* **11**, 59p.

WRIGHT H.E., WINTER T.C. & PATTEN H.L. (1963) Two pollen diagrams from southeastern Minnesota: problems in the regional late-glacial and post-glacial vegetational history. *Bull. geol. Soc. Am.* **74**, 1371-96.

Discussion on plant macrofossil assemblages

Recorded by Dr H.J.B. Birks,
Mr J.H.C. Davis, and Mr J. Dodd.

Professor Faegri noted that fruits may be transported long distances by ducks, particularly those of *Potamogeton* species. Dr H.H. Birks added that *Potamogeton* fruits had been proved to be resistant to digestion by ducks. *Potamogeton pectinatus* is a particularly important source of food for wildfowl, and its fruits may get transported many miles, and subsequently deposited in a habitat suitable for germination. Professor Watts asked about the proportions of macrofossils of aquatic to upland plants. Dr Birks replied that the numbers of macrofossils of obligate aquatics were usually less than those of wetland taxa. In prairie situations, the representation of upland taxa was poor, but in forested situations, macrofossils from upland trees, mainly *Betula* and conifers, were usually well represented.

Dr Cushing commented that the absence of a macrofossil in a surface sample may be related to the sample size. Dr Birks emphasised that the sample analysed was minute compared with the vegetation plot size of approximately $25\,m^2$, but not as small as a pollen sample! The number of taxa recovered would probably increase with increasing sample size, expecially if several samples within the plot were amalgamated. However, some samples were very rich, and an increased sample size may not yield much extra information. A discussion on the size of sample to be used for fossil and modern samples resulted in the general opinion that each sample was different, and that the sample size should be related to the concentration of macrofossils in it, the amount of time available, and the type of information required.

Dr Seddon was struck by the similarity of the ordination plots for modern vegetation and macrofossils. He found it difficult to distinguish clusters, and would interpret the results in terms of trends. Dr Birks agreed that the results showed a continuum, and that the direction of the trend had been brought out in terms of the upland vegetation and water chemistry. By this means, groups of closely similar lakes and plants could be delimited. For the ordinations the total species list from each lake had been treated as a single unit.

: Section 5:
Vegetational history and community development

Rates of change and stability in vegetation in the perspective of long periods of time*

W.A. WATTS *Department of Botany, Trinity College, Dublin*

Introduction

Geology and descriptive plant ecology allied with plant geography, have probably contributed most to the development of the interdisciplinary study of the vegetation of the past by pollen and macrofossil analysis. The kinds of investigation carried out, the methods used, and the form of analysis of the information obtained have tended to reflect the initial training of the investigator and the assumptions his discipline makes as to which kinds of information are significant. This may have the disadvantage that the full possibilities of the rich information contained in pollen and seed diagrams are not realised.

If the approach has been through geology, there will have been concern with the exact determination of fossils, regional stratigraphic correlation of fossil assemblages, and the establishment of an accurate time-scale. The geological approach means that the stratigraphic column at each locality must be divided into units for purposes of description and correlation, hence the use of pollen zones. Cushing (1967) has discussed the status of pollen zones from the viewpoint of stratigraphic geology. They are 'assemblage zones' which are defined by the American Commission on Stratigraphic Nomenclature (1961) as 'purely biostratigraphic units, which although highly discontinuous laterally can be recognised over a geographic and stratigraphic inverval sufficiently great to make them useful for correlation. They carry no ecologic or climatic implications.' The methodology and assumptions of stratigraphic geology have probably been the primary influence in determining how the data of pollen and macrofossil analysis are handled.

Descriptive plant ecology and plant geography provide another very important approach. Here the investigator's starting point is field experience of the taxonomic composition of different kinds of vegetation and of the occurrence of rare or disjunct species. The vegetation and plant distributions of the past are investigated, partly for their own interest and with a view to characterising them in detail, partly because their study illuminates the present. This disciplinary approach underlies attempts to describe the vegetation of the past by detailed comparative studies of modern and fossil pollen and seed rain (Davis 1967, Tauber 1967, Birks 1973). The questions it poses relate to the composition of the species assemblage and the geographical location of significant species at any one time. This essentially static and descriptive study generally leaves the more dynamic aspects of vegetation out of consideration.

It is possible to look at pollen and seed diagrams differently, from the point of view that they are a composite record of a number of plant populations, each of which might grow, remain stable, or decline with time. If one can isolate the record of any species, express it satisfactorily in numbers, and fit a time scale to it, a new kind of information may come to light. It relates to the population ecology of species seen in the perspective of long periods of time, far longer than the lifetime of one observer or than the greatest period for which we have historical documents. It is the thesis of this paper that an approach to pollen diagrams which draws on the discipline of population ecology for its ideas has not been tried, and that it is profitable to look at a number of published investigations in some detail to see whether such an approach might be of value.

It may be questioned whether pollen data are suitable for this kind of use, but it is also relevant to consider their advantages. As is well known, pollen and spores have such dispersal characteristics that they are effectively mixed in air, and often also in water (Davis 1968). A pollen count from sediment is generally a statistically valid sample of the contemporary pollen in the air. The mixing of pollen in air and water homogenises the separate records of the different plant communities which produced the pollen. The pollen curve for any species is, therefore, a record of that part of the population that is near enough to the site of sedimentation to contribute pollen, irrespective of the pattern in which individuals may be dispersed. The relation of the pollen representation of a species at any level to its population size in the field, expressed as numbers of individuals per unit area, poses well-known problems. The main difficulties lie in an oversimplified concept of 'pollen rain' (Tauber 1967) and in the likelihood that lakes of different sizes and shapes may have different characteristics as pollen traps (Davis 1967, Pennington 1973). However, at any site, level to level comparisons of the pollen representation of a species allow one to assess whether

*Contribution No. 115, Limnological Research Center

the trend for a population is growth, stability, or decline, irrespective of the actual numbers of individuals. The value of pollen diagrams for population studies is, therefore, that trends in population size can be established and an accurate time scale fitted to them. From this further conclusions of value may follow.

A number of technical advances in Quaternary palaeoecology favour the study of populations. The first is the provision of highly accurate time-scales. The fact that certain kinds of lake may deposit sediment in readily identifiable annual bands (Tippett 1964, Anthony 1971) means that an exact chronology is available at a few favoured sites. The only limitation is that the annual bands may be very thin, so that they cannot be sampled individually. At Lake of the Clouds, Minnesota (Craig 1970) annual bands were sampled for pollen analysis in segments representing 8–13 years of sedimentation (Table 1). A very valuable, but somewhat less accurate time scale is provided by closely spaced serial radiocarbon dates in homogeneous lake sediments (Davis 1967, Pennington & Bonny 1970, Stuiver 1971). From these the rate of sedimentation can be calculated and the annual increment of sediment estimated throughout a core. The second advance, which depends on accurate knowledge of sedimentation rates, is the study of annual pollen influx into lakes, expressed as influx/cm^2/annum for each taxon. This procedure, of which the work of Davis (1967) at Rogers Lake, Connecticut, is probably the best-known example, makes it possible to escape from one of the limitations of classical pollen analysis, in which the pollen representation of each taxon at any level is calculated as a percentage value of the sum of all taxa. This has the disadvantage that the numerical representation of each species is influenced by fluctuations in the values of others. The calculation of pollen influx values allows the investigator to follow the record of each taxon in isolation. In practice, pollen influx calculations are less important than might be thought. Throughout the post-glacial, sedimentation rates in lakes so far investigated and total pollen influx vary within narrow limits over long periods of time (Davis 1967, Stuiver 1971, Craig 1972). In these circumstances percentage pollen diagrams and pollen influx values provide closely similar records.

These considerations suggest that, given an accurate time scale provided by annual bands or serial radiocarbon dates, and given a pollen record based on pollen influx values, or less satisfactorily, in the post-glacial, on percentage pollen counts, it is possible to study the growth, stability, or decline of the population of certain species over long periods of time at certain favoured sites. This paper attempts to enlarge on this idea with reference to specific site investigations.

Lake of the Clouds

THE TIME SCALE OF AN INVASION

Lake of the Clouds lies in northeastern Minnesota near the border of the United States with Canada, some 40 miles (64 km) west of Lake Superior. The lake is in a wilderness area, the Boundary Waters Canoe Area, within the Superior National Forest. The regional vegetation is largely conifer forest with species of *Pinus*, *Picea*, *Abies*, *Thuja*, and *Larix* together with northern hardwood genera such as *Betula*, *Populus*, and *Alnus*. The map in Küchler (1964) shows the vegetation of the region to be a mosaic of 'Great Lakes Pine Forest' and 'Great Lakes Spruce-Fir Forest' (*Picea-Abies*). There are local conifer bogs.

The lake is remarkable in that its sediments contain a long sequence of annually deposited laminae each of which consists of a couplet of a light-coloured and a dark-coloured band. The lake provides an example of iron meromixis of the general type described by Ruttner (1963). The formation of light coloured layers occurs in autumn or spring when circulation of water in the lake introduces oxygen into the otherwise anaerobic iron-rich bottom waters and causes precipitation of ferric hydroxide. The darker layers may represent summer sedimentation or organic detritus. This is probably a very over-simplified explanation. The limnology of the site has been studied by Anthony (1971), who discusses the several forms in which iron occurs in the sediment and the possible processes by which they were formed. The physical conditions for sedimentation of annual bands are that the lake is deep (31 m at the sampling point), is relatively sheltered from wind, and has a very small catchment area with no inflowing streams. There is very little wave or current action to cause turbulence or to resuspend sediments, as is common in more exposed, shallow lakes (Davis 1968). There is little activity by organisms in the deep water at the mud-water interface to cause disturbance. This type of situation may be more common in nature than has been supposed. Tippett (1964) reported annual laminae from deep lakes in eastern Ontario. For obvious reasons deep lakes are less likely to be sampled than shallow ones and few have been investigated.

Craig (1972) counted some 9,400 laminae from a core from Lake of the Clouds, which covers nearly all of post-glacial time. He also prepared a relative pollen diagram and studied the pollen influx to the site. A long series of radiocarbon dates (Stuiver 1971) is also available. The dates amply confirm the assumption made on limnological grounds that the laminae were annual. Conversely, the laminae provide a most important check on the variation in C^{14} in the atmosphere during the last 9,000 years.

At Lake of the Clouds, Craig's data show that in pollen zones LC-4 and LC-5,

Table 1. Pollen percentage and influx data from Lake of the Clouds (from Craig 1970 with some further calculations). The bracketed figure after age gives vertical number of laminae (years) in pollen sample. Age is given as age of middle lamina of sample. (*Anomalously high results)

Pollen zone	Sample number	Age (lamina number)	Pollen sum	*Picea* %	*Pinus banksiana*/ *P. resinosa* %	*Pinus strobus* %	Σ *Pinus* %	Pollen influx (10³ grains/ cm²/year)	*Pinus banksiana*/ *P. resinosa* influx	*Pinus strobus* influx
	9	3168 (12)	347	2.9	27.1	39.5	66.6	28.6	7.7	11.3
	10	3624 (10)	308	1.0	32.5	37.6	70.1	23.2	7.5	8.7
	11	4102 (12)	339	0.6	31.6	33.0	64.6	30.0	9.5	9.9
LC-5	12	4518 (13)	392	0.8	22.7	46.2	68.9	55.9*	12.7*	25.8*
	13	5022 (13)	342	0.6	32.5	29.8	62.3	19.7	6.1	5.9
	14	5513 (13)	344	0.6	25.0	40.4	65.4	20.7	5.2	8.4
	15	5983 (13)	327	0.3	29.7	34.2	63.9	23.8	7.1	7.8
	16	6444 (10)	348	0.9	40.0	21.2	61.2	40.9	16.3	8.7
	17	6699 (8)	343	0.3	51.6	7.0	58.6	31.0	16.0	2.2
	18	7005 (6)	376	3.2	50.6	10.6	61.2	38.1	19.3	4.1
LC-4	19	7362 (12)	397	2.3	59.4	1.8	61.2	21.9	13.0	0.4
	20	7721 (12)	404	5.9	60.4	1.7	62.1	20.3	12.3	0.3
	21	8121 (9)	326	2.6	55.2	3.4	58.6	31.0	17.1	1.1
	22	8491 (10)	322	7.5	61.8	3.7	65.5	45.1	27.9	1.4
LC-3	23	8872 (10)	333	5.1	65.5	—	65.5	42.5	27.8	—
	24	9131 (17)	348	10.6	53.4	3.2	56.6	16.2	8.7	0.5
LC-2	25	9375 (13)	431	19.5	24.9	1.6	26.5	27.5	7.3	0.4
	26	not available	302	33.1	18.5	—	18.5	not available		

extending from about 8,300 to 3,000 years ago, *Pinus* is the most abundant pollen type with 58% to 70% of the pollen sum at all levels. *Betula* always makes up about 10% of the pollen and *Alnus* is somewhat less important with about 5% of the pollen. The remaining taxa which occur at all levels but in relatively small numbers are *Picea*, *Fraxinus nigra*, *Populus*, Cupressaceae, *Ulmus*, *Quercus*, and wind-pollinated herbs. The data from the Lake of the Clouds pollen diagram which are essential to this paper are presented in Table 1. The table is based on data from Craig (1970) and further calculations based on those data.

Three species of *Pinus* occur in Minnesota. One of these, *P. strobus* (white pine) is a five-needle pine in the section haploxylon. It can be identified separately by its pollen morphology. *P. banksiana* (jack pine) and *P. resinosa* (red pine) are diploxylon pines, two-needle species with pollen of closely similar size and morphology. It has not yet been possible to distinguish between these two in pollen analysis. From the earliest post-glacial, pollen of *P. banksiana*/*P. resinosa* was abundant. *P. strobus* was absent at first and later was present in small traces with up to 3% of the pollen sum at some levels in zone LC-4. *P. strobus* became as abundant as *P. banksiana*/*P. resinosa* in LC-5. In percentage terms *P. strobus* contributed 30–40% of the pollen sum in LC-5 and *P. banksiana*/*P. resinosa* about 30%. With this information, the time scale for the transition from LC-4 to LC-5 and the interpretation of the transition can be considered.

At the 7362 lamina (Table 1) *Pinus strobus* had 1.8% of the pollen sum. At the 7005 lamina a value of 10.6% had been attained, 21.2% at 6444, and 34% at 5983 by which time the percentage values for *P. strobus* were stabilised between 30% and 40% of the pollen sum. The pollen percentages of *P. banksiana*/*P. resinosa* fell steadily from a plateau of about 60% of the pollen sum at and below the 7362 lamina to a new plateau of 30% at lamina 5983. The interpretation is that a growing *P. strobus* population began to increase some time between 7,362 and 7,005 years ago and continued to expand its population until some time after 6,444 and before 5,983 years ago, i.e. the population expansion occupied at least 700 and perhaps as much as 1,000 years. There are only five pollen counts from the time when *P. strobus* was present as a trace to the time when it had built up a large and apparently stable population. It is clearly desirable to have more closely spaced pollen samples so that the course of the transition can be followed in more detail, but the basic picture is unlikely to change. The transition from the apparently rather homogeneous vegetation of zone LC-4 (duration about 1,800 years) to a new position of stability in zone LC-5 (duration about 3,000 years) lasted about 1,000 years. At the same site the expansion of the *P. banksiana/P. resinosa* population at the zone LC-2/LC-3 boundary cannot have occupied less than about 500 years (Table 1). Even in a 10,000 year time scale periods of 1,000 and 500 years are substantial portions of the whole and they should be looked at in detail. The arguments presented are based on percentage pollen counts. The pollen influx values have also been calculated and are presented in Table 1. The same general picture emerges but, as is discussed later, quite large level to level fluctuations in the pollen influx counts make closer sampling desirable so that the significance of the fluctuations can be tested.

The Lake of the Clouds situation, in which a zone to zone transition of long duration can be established, is probably very widespread in pollen diagrams, but can rarely be quantified. It emphasises that the drawing of pollen zone boundaries may be an arbitrary and crude procedure, determined by the greater interest of the investigator in periods of stability than in periods of change.

The present interest in defining pollen zones by objective statistical techniques (Gordon & Birks 1972, Cushing 1973) draws attention to this problem. Clustering techniques define blocks of relatively homogeneous pollen spectra and, by definition, will fail to place intermediate spectra satisfactorily. They underline the general character of the vegetational record contained in pollen diagrams, namely that periods of stability alternate with relatively long periods of change caused by invasions. A simple zonation system is a crude tool to deal with this relatively complex situation.

Smith (1965) is one of the few investigators who have drawn attention to the importance of transitions between zones. Burges (1960) places the subject in its ecological context when he points out how little attention has been paid to the time-scale of successional change, even when information was clearly available. Burges was mainly concerned with the time-scale of soil profile development, but his observation is also valid for palaeoecology. Pollen analysts have a sufficiently accurate time-scale to hand at many sites for the study of succession or invasion processes taking place at a rate too slow for the individual human observer.

SOME CHARACTERISTICS OF THE PINUS STROBUS INVASION

The invasion of a new region by a species such as *Pinus strobus* may have any of three causes. A climatic change may have begun to take place which permits it to grow in formerly unsuitable areas; it may be favoured by a stage reached in a soil-development process which itself was of long duration, or it may have become extinct over a wide area because of some such phenomenon as glaciation, and is now re-immigrating from a distant region. Clearly all three factors may interact, but the last seems to have been the most important for *Pinus strobus* (Wright 1968). Craig (1969) showed that it was present in the Appalachians of Virginia about 12,000 years ago. This may have been the main dispersal centre for its subsequent migrations into New England and the Mid-West. The advance of the pine through climax forest seems to have been checked only by the dry climate of the prairie-forest border (Wright 1968, 1971). It can be assumed that much of the more humid forested region of the Upper Mid-West was open to invasion as soon as it could arrive. If there were no environmental brakes on the invasion and growth of the population, the biology of the species itself will have imposed some limits on its capacity to invade, as must be true of any migrating species.

In trying to visualise and interpret the behaviour of white pine in Minnesota and the reaction of the other pine populations to its invasion it is useful to have information about their biology and, in particular, to know the average time from establishment to maturity for each of the three species. Fortunately, the three Minnesota pines are of economic importance and their autecology is well known (Fowells 1965).

Pinus banksiana flowers at 10–25 years of age, depending on whether it grows in open or closed stands. It does not reach a very large size and may not live for more than 100 years. It is fire-adapted with serotinous cones which open in the heat of forest fires. The seed germinates in ash after fire. As a result, *P. banksiana* is often found in pure stands which mark the site of former fires. It tends to occur on coarse sandy soils such as glacial outwash plains. *P. strobus* and *P. resinosa* are large and valuable timber species. They reach full flowering and fruiting in closed stands at 50 years and may live to over 200 years, after which their growth rate falls off. If the invasion occupied 700 to 1,000 years, the first *P. strobus* to arrive in the neighbourhood of Lake of the Clouds must have grown to maturity and died long before the population reached its stable size. This suggests that the immigration took the form of early establishment of a small population, grown from relatively small numbers of seeds which were far-dispersed from seed parents of the invader which were growing outside the area being invaded. As newly established trees matured they would have acted as seed parents both for the local increase in size of the population and for further invasion outside the local area. This process continued over several generations until a stable population size was achieved. It seems possible that plant invasions may normally follow this pattern. It is sometimes thought that invasions take the form of rapid saturation of all possible sites as a narrow migrating front advances through a region. This seems improbable. It assumes that the seed parents can produce a very large rain of propagules, although striking data are available for the *Pinus* species (Fowells 1965) as to the extent to which seeds fail to ripen, or are not dispersed from their cones, or are predated by mammals, birds, and insects. Furthermore, the migrating species is in competition with species already occupying the available sites. By their very presence, these species limit the capacity of the invader to become established. The resistance of existing climax vegetation to change is characterised by Pearsall (1959) as 'inertia'. Pearsall speaks of 500 to 1,000 years as the time that may be required to bring about an appreciable change in climax forest by wholly natural means. As the white pine invasion at Lake of the Clouds spread over 700 to 1,000 years, natural fires and wind-throws would have been sufficiently frequent for seedlings to become established in small or large light gaps, in which they would be in competition with other seedlings only. The chief factor controlling rate of population growth appears to be the time needed to colonise intermittently available light gaps by a small surplus of propagules.

In discussing the population build-up in *P. strobus* it is important to distinguish between several kinds of migrating population. There is invasion of bare land, such as a recently exposed moraine, or abandoned agricultural land, by forest. The trees have no competitors. Pure stands of birch forests may readily be established on young moraines after a brief period of soil development, just as the invasion of derelict farm land in the southeastern United States by pines is a familiar process. The invasion by *P. strobus* was different in character for it involved the insertion of a new species into a climax forest. We have no good modern analogy for this process because, in Europe at least, the activity of man has transformed the problem of dispersal (Webb 1966). Probably the best analogue for *P. strobus* is the migration history of *Picea abies* in Scandinavia (Faegri 1949, Tallantire 1972). Tallantire's map shows that spruce, which was present in eastern Finland about 3000 B.C. had reached southwest Finland by 1500 B.C. and had spread extensively into central Sweden by 1000–400 B.C. Further expansion in southern and northern Sweden had taken place by the early centuries A.D., and the spruce finally arrived in western Norway in the period 1100 to 1400 A.D. Faegri's pollen analytical data also show that in much of western Norway spruce is a very recent immigrant. Faegri's description (1949, p. 13) of the field occurrence of one spruce stand is of interest. 'Along the eastern banks of Lygnavatn the spruce occurs irregularly; in some places there is much of it, in other places practically nothing. The clumps of spruce consist of specimens of greatly varying size, but on the whole there are few big trees. The whole community gives the impression of a species expanding its area. In the vicinity of the bog where we collected our samples, there was a small clump of spruce, a few hundred trees in all, most of which were very young and hardly flowering yet.' Faegri also (p. 24) emphasises the importance of the establishment of scattered individuals as seed parents for a much larger population expansion. The present behaviour of spruce in western Norway may provide a valuable analogue for the behaviour of *P. strobus* in the early part of its population expansion. It is interesting to note the implication in both cases that the rate of population growth was not restricted by lack of suitable sites, but by the need for time to disperse a sufficient number of surplus propagules into light gaps as they become available, and to mature a sufficient number of seed parents.

THE GROWTH CURVE OF THE PINUS STROBUS POPULATION

One of the difficulties in studying rates of population increase from pollen and seed data is that one needs to know both the time and the numbers of individuals concerned. Annual laminae give a time scale which can clearly be refined enormously by closer sampling. There is no possibility of finding out numbers of individuals, and plainly pollen can be produced in varying quantity by trees of various sizes, ages, and distribution patterns about which we can gain no information, but it is reasonable to assume that, in a broad sense, the rate of increase in the population of a species is proportionate to the rate of increase in the number of pollen grains of the species sedimented per unit area in unit time. There is evidence (Pennington 1973) that each basin of sedimentation may have its own characteristics as a pollen trap, so that the absolute number of pollen grains/cm^2/annum may vary greatly from basin to basin and prevent one from establishing any relationship between pollen influx and upland population size. In spite of this, the growth curve for the population can be studied within one basin and may prove to have the same form and duration for all basins within a limited region, reflecting the behaviour of the species in relatively uniform upland vegetation.

The pollen curve for *Pinus strobus* at Lake of the Clouds, and at Jacobson Lake, Minnesota, where a time scale can also be fitted (Wright & Watts 1969, Stuiver 1971) takes the form of a rather long 'tail' with low percentage or influx values at first, then a steady rise which appears steep and brief in the pollen diagram, but occupies 700–1,000 years at Lake of the Clouds and can be calculated as a minimum of 500–600 years at Jacobson Lake, then a cessation of further growth with a flattening of the curve. The growth curve seems to have been 'S-shaped', as is common for biological populations which find it easy to exploit an initially uncompetitive situation. Examples of such curves are quoted in many ecological textbooks, for example by Kormondy (1969). The general pattern is an initial slow rate of growth followed by an increase in rate to a maximum, then a flattening of the curve as the population stabilises itself with relationship to its environment. The behaviour of the *Pinus strobus* population seems to have followed this pattern and may be visualised as follows. The initial slow growth rate corresponded to an early period of seedling establishment and maturation in small numbers. As soon as the first trees within the population were mature, they began to pollinate and seed freely. At this point a rapid increase in the growth rate of the population was possible, because the local seed rain within the forest was much intensified and the chance of establishment greatly increased. Finally, the population growth curve flattened out as the white pine began to meet environmental resistance in the form of competition from other species for resources, or, perhaps more probably, from intra-specific competition, expressing itself as density-determining self-thinning (Harper 1967). The 'S-shaped' curve as a model for population growth is largely based on laboratory experiments with single species in culture. In these cases the flattening of the curve after the phase of rapid expansion must be due to intra-specific competition for space and resources. The growth of the *Pinus strobus* population took place in the presence of other tree species. If there is an analogy to the laboratory experiments, then the white pine invasion was not effectively controlled by competition from other tree species at first. The control on numbers seems to have been intra-specific, a self-thinning effect which became effective as the white-pine density on the available sites reached an optimum.

At this point it is desirable to refer to some of the experimental difficulties which may make deductions of this type hazardous. We lack sufficiently closely spaced samples at Lake of the Clouds to make the assumption of the 'S-shaped' curve more than a plausible hypothesis. Obviously it is possible and desirable to increase the number of samples and the number of pollen grains counted at each level to improve the statistical reliability. In particular, the considerable level to level variation in pollen influx values (Table 1) requires attention. They may arise from experimental error inherent in the procedure used. Alternatively, there may be real and quite large fluctuations around a mean pollen influx value. In either case, a much larger number of samples with larger pollen sums will permit a better assessment of the data. With this information available, it should be possible to state the form of the curve more accurately and to test its consistency with existing models for population growth. This might permit a better evaluation of the extent to which the white pine encountered competition from other species during its phase of maximum growth rate.

There is a technical difficulty in distinguishing between the pollen of *Pinus strobus* and of other pine species. Well-preserved specimens present no difficulties, but broken or crumpled pollen grains, which may be the majority, may be indeterminable. The procedure used by Craig (1972) and by Wright and Watts (1969) is to count as many perfect pollen grains as is practicable (Table 2) and to calculate on the assumption that the proportion between pine species found in the perfect specimens will be repeated in the damaged ones. This is not necessarily true, and a particular problem arises at the beginning of the growth curve for *Pinus strobus* where the presence of only one or two of its rather conspicuous pollen grains may be given a spurious importance by the procedure adopted in calculating percentage and influx values. This problem can also be controlled more adequately by larger pollen counts. There is also a real difficulty in the early stage of establishment when the local pollen rain of the invader is small and it is supplemented by pollen of the same species transported from distant sources. Even if one equates pollen influx with population, it is valuable to be able to say at what point local establishment began, so as to avoid an excessively long 'tail' at the beginning of the growth curve. The only critical evidence which can be brought to bear on this is the presence of macrofossils. At Jacobson Lake (Wright & Watts 1969) macroscopic remains of *Pinus strobus*, both needle fragments and seeds, are present in a core taken near the shore of the lake from the middle point of rapid growth of the population, but not earlier (Table 2). It is interesting that *P. resinosa* is present macroscopically as needle fragments at the level of initial expansion of *P. strobus* (Table 2). This is the only critical evidence of which the author is aware from the whole Mid-West which identifies a pine species prior to the expansion of white pine. It is, of course, possible that *P. banksiana* was also present but positive evidence is lacking.

Table 2. Jacobson Lake. Near-shore core. Ratios of *Pinus strobus* pollen to *P. resinosa*/*P. banksiana* in 50 grains scored. (M)—macrofossils present. Data from Wright & Watts (1969).

Depth (cm)	*P. strobus*	*P. resinosa*/*P. banksiana*
480	44 (M)	6 (M)—*P. resinosa*
490	40 (M)	10 (M)—*P. resinosa*
500	39 (M)	11 (M)—*P. resinosa*
510	41 (M)	9
520	38 (M)	12
530	28 (M)	22
540	12	38
550	5	45 (M)—*P. resinosa*
560	4	46 (M)—*P. resinosa*
570	1	49
580	0	50
590	1	49
600	0	50
610	0	50
620	0	50

COMPETITION WITH EXISTING SPECIES

It is notable that, at Lake of the Clouds, *Pinus banksiana/P. resinosa* suffered most reduction in numbers as *P. strobus* invaded. Both the pollen percentage and influx figures (Table 1) make this clear. The pollen representation of other species was little changed (Craig 1972). The pines, particularly *P. strobus* and *P. resinosa*, both occupy the same kind of rather coarse, sandy, nutrient-poor, podsolised soils whereas *P. banksiana* has a more specialised ecology, as already outlined. It seems likely that what one is seeing at the LC-4/LC-5 boundary is the mutual numbers adjustment of two species with similar requirements. It is not possible to demonstrate this with any satisfying precision, although the macrofossil evidence from Jacobson Lake (Table 2) is consistent with this interpretation. It is worth noting that experiments have been reported in which two species have been grown mixed in various numerical proportions. Harper (1967) comments that 'the essential criterion for a balanced "co-occurrence" of two species is that the minority component should always be favoured'. In other words, given a *P. resinosa/P. strobus* mixture, both capable of occupying the same sites, the two species tended to stabilise their numbers after a period in which the less common species increased at the expense of the more common. This may have been the historical situation at Lake of the Clouds and Jacobson Lake. The assumption would be that *P. strobus* could increase its population essentially because it was not hindered by intra-specific competition, the main factor which stabilised the numbers of *P. resinosa*. While its population was small any seedling would have a higher chance of establishment in a light gap than any *P. resinosa* seedling. As its numbers grew it would be subject to the same constraint so that its advantage would decrease. As the *P. strobus* invasion progressed, the two species would adjust their seedling establishment strategies relative to one another by some form of niche specialisation to enable them to co-exist, a process in which natural selection would play its part. Examples of this type of process are reported in the ecological literature (Harper 1967, Kormondy 1969).

In this discussion it is assumed that the correlation of the immigration of *P. strobus* with climatic change is unnecessary, or that it is of only minor importance. The insertion of the species into the existing climax communities is an example of immigration followed by a new equilibrium. The immigration might have been favoured by climatic change, but need not have been so.

Faulenseemoos

At the Faulenseemoos, a drained swamp near Spiez beside the Lake of Thun in Switzerland, Welten (1944) reported the occurrence of annually banded sediments which occupy much of late-glacial and post-glacial time. At this site, annual couplets consist of light-coloured carbonate-rich bands, the result of the lake's summer productivity, and dark winter bands rich in fine organic detritus and clay.

Welten's most detailed pollen diagram (his Profile XI, Fig. 15) is accompanied by a chronology based on the annual laminae. Irrespective of its absolute accuracy, it provides a time scale for measuring rates of change. Data from a portion of the pollen diagram are abstracted in Table 3. The diagram excludes herb pollen from the pollen sum, but as the amounts of herb pollen at the relevant levels are small, and consist almost entirely of grass pollen and fern spores, they can be ignored in the present discussion. The pollen sum used, although it is not stated explicitly, appears to have been small, and may have been as low as 100 at many levels. A larger pollen sum would have been desirable to give greater statistical accuracy, but the events being studied are so strongly expressed that they emerge clearly in small counts.

The Faulenseemoos profile documents the invasion of *Corylus avellana* in the early post-glacial, a very striking feature of pollen diagrams from much of Europe north and west of the Alps. *Corylus* pollen is present in small percentages from 1,285 to 1,280 cm (Table 3). Above 1,280 cm the *Corylus* curve begins to rise steeply and continues to rise to 1,267.5 cm where it contributes 83% of the pollen sum. Above that level its percentage contribution is stabilised and falls back slightly. The period from the beginning of the *Corylus* rise to its maximum percentage value (based on lamina counts) is about 320 years.

Corylus probably flowers and fruits freely while as little as 25 years old. As with *Pinus strobus* at Lake of the Clouds, the evidence is that the growth of the hazel population, expressed by its pollen percentage curve, occupied the life time of several, perhaps as many as ten, generations of the shrub. Successful establishment of a small number of seedlings to act as seed parents for the second generation of invaders is all that was required, given the time available, to permit a major population expansion.

Corylus is included in the pollen sum in Table 3. Many authors, including Welten, follow the practice of excluding *Corylus* from the pollen sum on the grounds that it is a shrub and that the pollen sum should comprise canopy-forming trees only. Welten sets out the arguments in favour of this procedure. If it is followed, the expansion of the population of *Ulmus* + *Tilia* + *Quercus* extends from between 1,277.5 and 1,280 cm where the first traces appear, to 1,265 cm where the percentage values for the three trees stabilise at about 80%, a time scale of 400 years. This is a less useful figure than the figure for *Corylus*, because it assumes the concept of 'mixed oak forest', a common convention in pollen diagrams from continental Europe, whereas the species of the three genera must behave independently as populations.

Marks Tey

Turner (1970) reports annual laminae from an interglacial site at Marks Tey, in

Table 3. Pollen percentage data from Faulenseemoos, Profile XI. Abstracted and partially recalculated from Fig. 15 in Welten (1944). *Corylus* is included in the pollen sum. Infrequent pollen types, especially *Hippophae*, omitted. Tr.—Trace

Depth (cm)	Salix %	Betula %	Pinus %	Quercus + Ulmus + Tilia %	Corylus %	Timescale in annual laminae
1260	Tr.	1	2	20	77	
1262.5	Tr.	Tr.	2	21	77	
1265	2	—	3	16	79	
1267.5	—	2	5	10	83	
1270	—	2	7	12	79	c
1272.5	—	3	18	14	65	320 years
1275	—	18	39	15	28	
1277.5	1	27	53	3	17	
1280	2	17	77	Tr.	2	
1282.5	4	17	77		2	c
1285	6	13	79		2	1,400 years
1287.5	6	10	82		c	
1290	3	5	90			600 years
1292.5	4	13	82			
1295	6	8	83			
1297.5	3	11	83			
1300	3	6	89			
1302.5	3	19	77		Tr.	
1305	6	16	78		Tr.	
1307.5	2	33	64			
1310	1	27	69			

Essex, some 50 miles (80 km) north-east of London. The laminae are composed of couplets of light-coloured layers of diatom frustules, an almost pure population of *Stephanodiscus astraea* var *minutula*, associated with a thin layer of calcite crystals, followed by darker bands with an increased proportion of organic material. It is argued that the diatoms and calcite represent the early summer and summer flush of growth and photosynthesis in the Marks Tey lake and that organic material was increasingly precipitated at the end of the growing season and into winter. Turner's paper establishes that the duration of pollen zones in the Hoxnian interglacial period is quite comparable to that for similar events in the post-glacial. The very interesting phase of high percentages of herb pollen in zone Ho11c, attributed to local forest destruction by large browsing mammals, has a duration of at least 500 years, assuming a uniform thickness of the laminae. Unfortunately, the time scale provided by the laminae is not closely linked with the pollen diagram, and closer sampling would be necessary to study the detailed course of population establishment for any one species. A 25 cm sampling interval in Marks Tey zones Ho11c and Ho111a represents 250 years! As at Lake of the Clouds and Faulenseemoos, the annual bands at Marks Tey can provide a time scale for the minute study of transitions.

Jacobson Lake

STABILITY

During an invasion the population of the immigrant species grows vigorously for a time and then becomes stabilised as it meets increased environmental resistance. The other species of the climax forest adjust their numbers to admit the newcomer. After this period of stability follows in which the component species of the climax forest each reach an equilibrium level and their numbers may fluctuate around a mean with little amplitude to the fluctuations. This can be seen in the data from Faulenseemoos (Table 3) between 1,305 and 1,280 cm, a period of some 600 years in which a pine-birch-willow forest was present. This example is not altogether satisfactory because of the limited diversity of tree taxa. Table 4 presents raw pollen data for twelve important tree taxa for a period of 1,200 years in the later post-glacial at Jacobson Lake which has a much richer flora. The full pollen diagram is published in Wright and Watts (1969). The pollen spectra in Table 4 show a high degree of uniformity from level to level. From the statistical point of view the data are highly homogeneous. It is assumed that such a homogeneous series of pollen spectra is the expression of a stable climax forest which did not change significantly in the period under consideration. The plant associations which made up the vegetation cover were, no doubt, quite the same at the end of the period as they were at the beginning. This emphasises the view already expressed, that pollen diagrams record alternating periods of vegetational stability which may be of long duration with periods of invasion and re-adjustment which may also be protracted. The pollen zone concept is inadequate to express these two different kinds of situation.

Although the evidence for stability is on a sounder statistical basis in pollen diagrams than with macrofossils, fossil fruits, seeds, and leaves may also exhibit a high consistency in species composition from level to level. This is well illustrated in plant macrofossil assemblages from the late-glacial and early post-glacial of Minnesota (Watts 1967a). The same conclusion may be drawn, that the fossils reflect the persistence over long periods of time of well-defined upland and aquatic plant associations which were subject to change by invasion at the termination of a period of stability.

Table 4. Jacobson Lake. Off-shore profile Zone JL-4. Raw pollen data for 12 most important tree taxa. Pollen sum 300 at all levels. Time span about 1,200 years. Pollen diagram in Wright and Watts (1969).

Depth (cm)	Picea	Larix	Fraxinus nigra-type	Betula	Alnus	Abies	Pinus	Pinus strobus/ P. resinosa-type proportion in 50 grains	Ulmus	Quercus	Ostrya/ Carpinus	Tilia	Corylus
210	4	1	—	46	4	3	194	41: 9	—	10	2	1	—
220	10	1	2	48	10	4	177	40:10	1	17	3	—	—
230	9	2	3	49	11	5	169	38:12	2	19	3	—	1
240	4	1	1	24	4	7	238	44: 6	—	3	1	2	—
250	3	2	1	50	8	6	187	46: 4	1	16	5	—	1
260	1	7	1	45	5	3	218	39:11	—	9	2	—	1
270	3	3	1	50	10	2	200	44: 6	—	16	—	—	—
280	3	3	3	24	5	3	222	34:16	—	15	2	1	1
290	4	3	—	28	7	3	213	35:15	—	15	2	—	1
300	3	—	—	32	9	4	215	40:10	2	12	—	1	2
310	2	5	1	37	5	2	194	43: 7	2	18	5	—	1
320	—	2	—	37	12	2	203	43: 7	2	17	6	—	1
330	3	5	2	25	4	3	236	46: 4	1	11	2	—	—
340	2	2	2	38	6	5	210	39:11	1	17	1	—	—
350	2	—	1	28	10	3	217	47: 3	3	18	—	—	—

Species composition of stable communities

The modern plant sociologist who studies the forest flora of a species-poor region such as north-west Europe can distinguish a limited number of well-characterised plant associations with success. With considerable field experience of plant associations it becomes difficult to accept that their component species might occur in quite different combinations and numerical proportions, but the fossil evidence makes it clear that this was the case. Self-evidently invasion of the mid-western forests of North America by *Pinus strobus* created a new phytosociological situation, and there are well-known difficulties in relating in detail what we know about late-glacial floras to modern plant associations (Cushing 1967). Some of the most striking evidence comes from interglacial floras (West 1964). West states that 'our present plant communities have no long history in the Quaternary but are merely temporary aggregations under given conditions of climate, other environmental factors, and historical factors.' If one takes the post-glacial forest sequence in northwest Europe as a standard, it is possible to categorise the ways in which the interglacial successions are distinctively different. First, there is no consistent order of immigration of tree species. As an example, *Hedera* in the Gortian interglacial period in Ireland (Watts 1967b) immigrated with pine and birch before oak or elm, with which it is closely associated in the post-glacial. Species also vary greatly in their numerical representation in each interglacial. *Ulmus* is present only as traces in the Gortian deposits referred to, whereas it has higher percentage values in the pollen rain in post-glacial deposits in Ireland than anywhere else in western Europe. Assemblages in interglacial floras may differ from any now known, as the abundance of *Acer monspessulanum*, now a south European species, at some sites of Ipswichian interglacial age in Britain makes clear. Finally, apparently discordant elements can occur in forest assemblages. A good example is the occurrence of *Picea abies* in a flora in which *Abies alba*, *Rhododendron ponticum*, *Ilex aquifolium*, *Taxus baccata*, and hymenophyllaceous ferns are conspicuous components late in the Gortian interglacial in Ireland (Watts 1967b), a species normally found in areas with a severe winter climate in association with species that require winter warmth and an oceanic climate.

West's observation that plant communities are 'merely temporary aggregates' seems fully substantiated. It seems possible that forest assemblages owe their composition far more to purely historical factors such as the location of a surviving population after a period of unfavourable climate, and its capacity to expand and migrate subsequently, than to any necessity for particular groups of species to be associated together. Clearly species cannot associate together quite freely because, within broad limits, there are controls exercised by climate and other environmental factors such as soil characteristics, but within broad climatic and environmental limits forest associations appear to have formed readily in the past from the locally available species. What becomes of particular interest, is to define the characteristics or circumstances which enable a species to begin to act as an aggressive migrant.

Macrofossils

In an earlier paper (Watts & Winter 1965) the author made a case that the study of plant macrofossils would be on a better basis if fruits and seeds were extracted fully from standard volumes of sediment and the quantitative results expressed as a 'seed diagram'. This allows one to establish exact relationships between pollen diagrams and macrofossil records, so that the precise location in a core of a particular seed or fruit can be related to the events recorded in a pollen diagram. It was also pointed out that pollen grains and seeds had some common characteristics as fossils, and that some of the concepts of pollen-analysis usefully apply to seed-analysis if results are presented in a quantitative and systematic manner.

For the purposes of the present study macrofossils must play a secondary role to pollen analysis, because the study of population growth requires a better statistical control than is possible with macrofossils which may fluctuate greatly in number from level to level, and which are not effectively mixed in the way pollen is in the air, so that the macrofossil assemblage is hardly a valid sample of the 'seed rain'. However, there are two points where macrofossil studies are particularly useful in relation to population investigations. Pollen analysis cannot define critically the first moment at which an invading species is present locally. A species which has a large macrofossil and a large pollen production, such as *Pinus strobus* (Table 2) is very suitable for study, because there is a good chance that macrofossils may be found at critical levels which greatly assist interpretation. Second, the very fact that many pollen taxa cannot be identified beyond generic level means that one depends on macrofossil analysis for species determination, and for increasing the species list of an assemblage by adding species which may be unrepresented in the pollen rain. The consistent occurrence of certain macrofossil assemblages (West 1964, Watts 1967a, 1967b) provides some of the most critical evidence that plant associations are ephemeral and that quite different, and often surprising, associations commonly occurred in the past.

Conclusions

This paper began by stating the opinion that pollen and seed diagrams are rich sources of information about vegetation, but because interest has centred on stratigraphic correlation and the characterisation of vegetation at particular times in the past, the information has not been fully used. In particular, the possibilities of studying processes in vegetation which occupy long periods of time, and for which an exact time scale is available, have generally been overlooked. The emphasis of the paper has been to select a number of sites where a particularly accurate time-scale is available and, taking a number of species in isolation, to see how they behaved as populations over long periods. The question must arise whether the results reported have some general applicability, or whether they represent exceptional special cases. This can only be resolved in a completely satisfactory manner by studying a wider range of sites but it is considered that the pollen diagrams studied do not show exceptional features and that it is permissible to generalise from the results. The following generalisations are made:

1 Pollen diagrams, and, to a lesser extent, seed diagrams are a record of the alternation of periods of vegetational stability with periods of transition and change caused by plant invasions. Invasion phases may occupy as much as 500 to 1,000 years at sites where an accurate time scale is available. The traditional method of drawing pollen zone boundaries ignores transitions between stable phases. In this respect pollen zones are inadequate to express the information available.

2 Invasion phases are long enough to span several to many generations of the invading species. Invasion appears to be a gradual opportunist process, the success of which depends on high competitive ability in seedling establishment of the invader while its population has a low density.

3 The population growth curve of an invading species is 'S-shaped' and is reminiscent of growth curves in classic studies in population ecology. The exact form of the curve and its comparison with mathematical models of population growth requires more detailed study. The flattening of the growth curve when the population reaches its equilibrium level may be an effect of intra-specific competition and self-thinning.

4 After an invasion is complete, stability results which persists until a further invasion takes place and population numbers are adjusted once more to accommodate the invader. Stable populations offer little initial resistance to invasion. The ultimate control on the numbers of the invader's population appears to be intra- rather than inter-specific competition.

5 The evidence of invasions and the record of interglacial plant successions confirm that plant associations are not persistent in time. Forest species in north temperate regions, within broad climatic and environmental limits, can occur in a very wide variety of assemblages, many of which are unfamiliar to contemporary plant sociology. The main factor determining the species assemblage and numerical relations between species appears to be location of populations at critical times historically.

6 Macrofossils provide valuable subsidiary information to pollen diagrams. They permit accurate assessment of first local occurrence of invading species. They also allow a large range of species determinations to be made which are not possible with pollen. This permits more accurate and detailed statements to be made about species assemblages in the past than is possible by pollen analysis alone. However, macrofossils are much less suitable for statistical treatment than pollen influx data. In population studies, macrofossil analysis can only play a subsidiary role to pollen analysis.

This paper surveyed the existing information in published investigations which were not prepared with the possibility of detailed population studies in mind. Sites such as Lake of the Clouds, where the chronological control is very good, invite more detailed study of invasion processes by close-interval sampling and large pollen counts. No doubt other similar sites can be found when their potential value is fully appreciated. Some of the other issues discussed, such as competition and the limitations on ultimate population size for any one species are more speculative and it is doubtful whether much more progress can be made to illuminate these problems by fossil studies. However, the value of a rigorous attempt to visualise the processes recorded in palaeoecological records is asserted, and for this purpose some knowledge of the ideas of population ecology is clearly relevant to students of the Quaternary.

Acknowledgement

Much of the practical work referred to in this paper, particularly at the Jacobson Lake site, was carried out at the Limnological Research Center, University of Minnesota. I would like to record my thanks to Professor H.E. Wright, Jr., for the facilities made available to me at Minnesota and for valuable and stimulating discussion.

References

American Commission on Stratigraphic Nomenclature (1961) Code of Stratigraphic Nomenclature. *Bull. Am. Ass. Petrol. Geol.* **45**, 645–65.
Anthony R.S. (1971) *The mechanism of varve formation in Lake of the Clouds, Lake County, Minnesota.* M.S. thesis, University of Minnesota.
Birks H.H. (1973) Modern macrofossil assemblages in lake sediments in Minnesota. (this volume).
Burges A. (1960) Time and size as factors in ecology. *J. Ecol.* **48**, 273–85.
Craig A.J. (1969) Vegetational history of the Shenandoah Valley, Virginia. *Geol. Soc. Am. Special Paper* **123**, 283–96.
Craig A.J. (1970) *Absolute pollen analysis of laminated sediments: a pollen diagram from northeastern Minnesota.* M.S. thesis, University of Minnesota.
Craig A.J. (1972) Pollen influx to laminated sediments: a pollen diagram from northeastern Minnesota. *Ecology* **53**, 46–57.
Cushing E.J. (1967) Late-Wisconsin pollen stratigraphy and the glacial sequence in Minnesota. In *Quaternary Paleoecology* (ed. by E.J. Cushing & H.E. Wright), pp. 59–88. Yale University Press, New Haven and London.
Cushing E.J. (1973) Multivariate techniques applied to palynological problems: a review. (this volume).
Davis M.B. (1967) Pollen accumulation rates at Rogers Lake, Connecticut, during late- and post-glacial time. *Rev. Palaeobotan. Palynol.* **2**, 219–30.
Davis M.B. (1968) Pollen grains in lake sediments: redeposition caused by seasonal water circulation. *Science, N.Y.* **162**, 796–9.
Faegri K. (1949) Studies on the Pleistocene of Western Norway IV. On the immigration of *Picea abies* (L.) Karst. *Årbok Univ. Bergen (Naturv. Ser.)* **1**, 1–52.
Fowells H.A. (1965) ed. *Silvics of Forest Trees of the United States.* U.S. Dept. Agr. Handb. 271, 762p. Washington D.C.
Gordon A. & Birks H.J.B. (1972) Numerical methods in Quaternary Palaeoecology I. Zonation of pollen diagrams. *New Phytol.* **71**, 961–79.
Harper J. (1967) A Darwinian approach to plant ecology. *J. Ecol.* **55**, 242–70.
Kormondy E.J. (1969) *Concepts of Ecology.* Prentice-Hall, Englewood Cliffs, New Jersey.
Küchler A.W. (1964) Potential natural vegetation of the conterminous United States. *Am. Geogr. Soc. Spec. Publ.* **36**, 116p.
Pearsall W.H. (1959) The ecology of invasion: ecological stability and instability. *New Biology* **29**, 95–101.
Pennington W. (1973) Absolute pollen frequencies in the sediments of lakes of different morphometry. (this volume).
Pennington W. & Bonny A.P. (1970) Absolute pollen-diagram from the British Late-glacial. *Nature Lond.* **226**, 871–3.
Ruttner F. (1963) *Fundamentals of limnology.* University of Toronto Press, Toronto.
Smith A.G. (1965) Problems of inertia and threshold related to post-glacial habitat changes. *Proc. R. Soc. B* **161**, 331–42.
Stuiver M. (1971) Evidence for the variation of atmospheric C^{14} content in the Quaternary. In *The Late Cenozoic Glacial Ages* (ed. by K.K. Turekian), pp. 57–70. Yale University Press, New Haven and London.
Tallantire P.A. (1972) The regional spread of spruce (*Picea abies* (L.) Karst.) within Fennoscandia: a reassessment. *Norw. J. Bot.* **19**, 1–16.
Tauber H. (1967) Differential pollen dispersion and filtration. In *Quaternary Paleoecology* (ed. by E.J. Cushing & H.E. Wright), pp. 131–41. Yale University Press, New Haven and London.
Tippett R. (1964) An investigation into the nature of the layering of deep-water sediments in two Eastern Ontario lakes. *Can. J. Bot.* **42**, 1693–709.
Turner C. (1970) The Middle Pleistocene deposits at Marks Tey, Essex. *Phil. Trans. R. Soc. B* **257**, 373–440.
Watts W.A. (1967a) Late-glacial plant macrofossils from Minnesota. In *Quaternary Paleoecology* (ed. by E.J. Cushing & H.E. Wright), pp. 89–97. Yale University Press, New Haven and London.
Watts W.A. (1967b) Interglacial deposits in Kildromin Townland, near Herbertstown, Co. Limerick. *Proc. R. Ir. Acad. B* **65**, 339–48.
Watts W.A. & Winter T.C. (1965) Plant macrofossils from Kirchner Marsh, Minnesota—a paleoecological study. *Bull. geol. Soc. Am.* **77**, 1339–60.

WEBB D.A. (1966) Dispersal and establishment: what do we really know? In *Reproductive Biology and Taxonomy of Vascular Plants* (ed. by J.G. Hawkes), pp. 93-102, Pergamon Press, London.

WELTEN M. (1944) Pollenanalytische, stratigraphische and geochronologische Untersuchungen aus dem Faulenseemoos bei Spiez. *Veröff. geobot. Inst., Zürich* 21, 201p.

WEST R.G. (1964) Inter-relations of ecology and Quaternary palaeobotany. *J. Ecol.* 52 (Suppl.), 47-57.

WRIGHT H.E. (1968) The roles of pine and spruce in the forest history of Minnesota and adjacent areas. *Ecology* 49, 937-55.

WRIGHT H.E. (1971) Late Quaternary vegetational history of North America. In *The Late Cenozoic Glacial Ages* (ed. by K.K. Turekian), pp. 425-64. Yale University Press, New Haven and London.

WRIGHT H.E. & WATTS W.A. (1969) Glacial and vegetational history of northeastern Minnesota. *Minn. geol. Surv. Spec. Publ.* 11, 59p.

Recent forest history and land use in Weardale, Northern England

BRIAN K. ROBERTS, JUDITH TURNER AND
PAMELA F. WARD *University of Durham*

Introduction

This study of recent forest history in Weardale is based upon two small peat bogs, occurring near the margins of human settlement, and the object of this paper is to relate the story as revealed by the pollen record to the available documentary evidence. Our interest was first stimulated by the juxtaposition of one bog, that at Steward Shield Meadow, with deserted field banks of medieval origin, and the second bog, at Bollihope, was discovered while deliberately extending the scope of our enquiry to obtain comparative material.

The salient characteristics of the farming of the Weardale area of west Durham were outlined in 1880 by William Morley Egglestone who wrote, 'We find along the margin of the river Wear excellent meadow land, on the brow of the hills we have summer pasturage for cattle, and on the mountain tops extensive heather clad sheep-walks. . . . There is some tillage land about Stanhope and westward to Eastgate, but the largest portion of cultivated land is meadow.' Farming, however, was not the only form of economic activity, and he reminds us that 'the valley was the great iron and lead mining field, and hunting ground of the Bishops of Durham, and here were forests and forest castles, deer and deer parks'.

Steward Shield Meadow

Steward Shield Meadow lies to the north of the main Wear valley (Fig. 1) and the bog is just outside the modern enclosure wall but within the area enclosed by the medieval field banks (Fig. 10). It is approximately 60 m long and 15 m wide and covered with *Juncus effusus*, *Nardus stricta*, *Molinia coerulea*, and *Festuca ovina*. Associated species include *Potentilla erecta*, *Ranunculus flammula*, *R. repens*, *Cardamine pratensis*, *Montia fontana*, *Sagina nodosa*, *Luzula campestris*, *Carex nigra*, *C. panicea*, and *Equisetum palustre*. There are also a number of mosses including *Polytrichum commune*, *Pohlia nutans*, *Acrocladium cuspidatum*, and *Mnium punctatum*.

The stratigraphy of the bog was investigated by means of three lines of borings parallel with the boundary wall at 6 m, 12 m and 32 m respectively from it. A soil

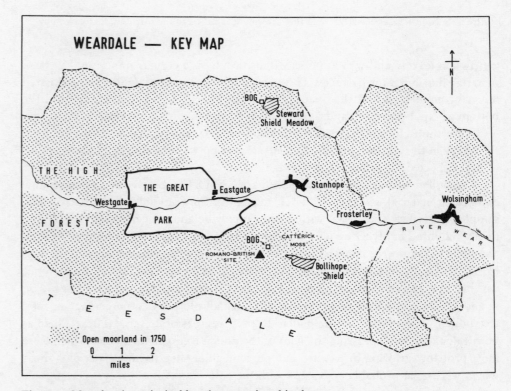

Figure 1. Map showing principal locations mentioned in the text.

auger was used. The results of the transect 32 m from the boundary wall, which crossed the deepest part of the bog, are shown in Fig. 2.

The local bedrock is covered with a grey boulder clay, presumably deposited during the last glaciation, and above the boulder clay there is a layer of hill wash, a brown-orange gleyed clay with sandstone fragments embedded in it. This latter layer thins out from the upslope side of the bog. The peat has developed on the flat bench on top of the boulder clay and hill wash. It is dark brown in colour, homo-

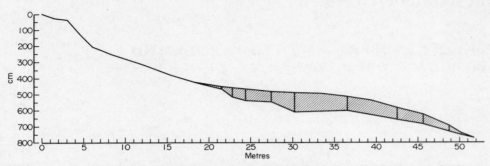

Figure 2. Section across Steward Shield Bog showing its relation to the surrounding topography.

geneous in texture and well humified. It also contains a certain amount of hill wash material, fine inorganic particles. The amount of this at various depths was estimated on a 0–5 point scale and the results are shown in Fig. 4. It is concentrated in the bottom 20 cm, between 43 and 61 cm, and in a band higher up between 6 and 21 cm.

The samples for pollen analysis were collected with a Russian peat sampler from the centre of the bog (see Fig. 2). At a later date a pit was dug nearby and a monolith of peat collected for radiocarbon dating.

The pollen samples were treated with 10% NaOH and sieved to remove humic and coarse material and then with HF to remove the fine inorganic particles. The samples were then acetolysed with acetic anhydride and concentrated H_2SO_4 and finally mounted in glycerine jelly stained with safranin ready for counting. At least 500 pollen grains were counted from each level except for samples between 45 and 61 cm where there was very little pollen and it was only possible to count totals varying between 152 and 425 grains. The results of the analyses are given in Fig. 3, 4, and 5. The individual pollen frequencies are all plotted as a percentage of total tree pollen. Also plotted on the pollen diagram is a pastoral/arable index. This has been calculated by expressing the sum of the pollen frequencies of taxa which were most probably growing in pasture as a percentage of such pastoral taxa plus those which were most probably associated with the cultivation of crops. It is comparable with the arable/pastoral index at Old Buckenham Mere (Godwin 1968).

The taxa included in each category are as follows.

PASTORAL	ARABLE
Plantago lanceolata	*Triticum*
Artemisia	Compositae
Rumex	Chenopodiaceae
Ranunculaceae	Cruciferae
	Vicia
	Polygonum
	Centaurea cyanus
	Knautia
	Trifolium
	Centaurium

Four distinct pollen assemblage zones have been recognised on the diagrams and these have been given local designations, SSa, SSb, SSc, and SSd. They clearly correspond with late VIIb and VIII of the standard British pollen zonation. They indicate that there have been two periods, SSd and SSb, when the area was reasonably well wooded and two periods, SSc and SSa, when most of the land was either pasture or heath, possibly with some under cultivation.

Peat from two levels has been radiocarbon dated by the Gashukuin Radiocarbon laboratory with the following results.

1 Peat from 41 cm containing high values of Compositae Lig., *Calluna*, Gramineae, and *Plantago lanceolata* pollen types. GaK-3/033. 2,060 B.P. ± 120, i.e. 110 B.C. ± 120.

2 Peat from 33 cm containing low values of the pollen types mentioned above and a higher value for tree pollen. GaK-3/032. 840 B.P. ± 100, i.e. A.D. 1110 ± 100.

These dates indicate that the first period when the area was not well wooded, SSc, began just before the time of Christ and that the second period with trees, SSb, began about 1100 A.D.

The mean rate of peat growth during SSc can be calculated on the basis of these dates. Eight cm of peat formed during 1,210 years, giving a mean rate of 1 cm in 150 years. Similarly the mean rate in periods SSb and SSa was 1 cm in 25 years. These rates are only, of course, approximate but they do indicate a much slower rate of peat growth and drier conditions on the bog surface in the first millenium A.D. than in the second.

Bollihope

Bollihope is situated to the south of the main Wear valley (Fig. 1) and the bog, like that at Steward Shield Meadow, is extremely small, 40 × 25 m. It has developed in the course of one of the minor tributaries of the Bollihope Burn (Fig. 6) and lies on the sloping hillside some 625 m to the north of the burn and 250 m west of the B6278 Stanhope to Middleton road.

The surface of the bog is wet consisting of a relatively even carpet of *Sphagnum papillosum*, *S. rubellum*, and *S. cuspidatum* with *Aulacomnium palustre* and a fair amount of *Calluna vulgaris*, *Empetrum nigrum*, *Erica tetralix*, *Vaccinium oxycoccus*, *Narthecium ossifragum*, and *Eriophorum angustifolium* growing in it. There is also

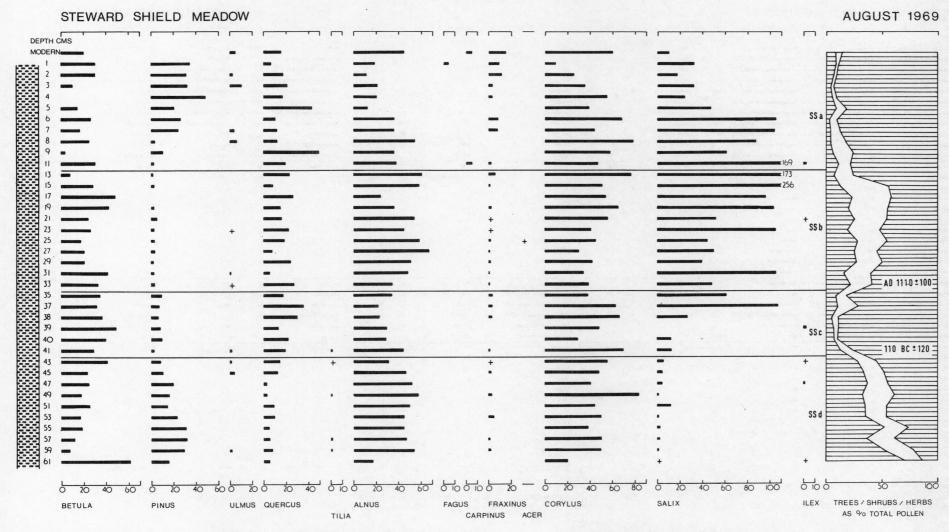

Figure 3. Tree pollen diagram from Steward Shield Bog. Pollen frequencies are plotted as a percentage of the total tree pollen. The ratio of tree/shrub/herb pollen is shown on the right of the diagram. Radiocarbon dates are given in calendar years.

some *Carex echinata*, *C. panicea*, *Nardus stricta*, and *Juncus squarrosus*. To the edge of the bog, near the stream, there is a patch of *Juncus effusus*.

The hillslopes around the bog are largely *Pteridium* dominated *Agrostis-Festuca* grassland with *Sphagnum palustre* growing in the damper places. The only tree in the vicinity is a single rowan lower down the tributary although in the main valley of the burn there is some juniper scrub.

The stratigraphy of the bog was investigated by means of two lines of borings at right angles to each other (Fig. 6). The results of these borings are given in the same figure and show that the bog developed on top of a complex series of silts, sands and clays deposited by the stream in a comparatively wide and flat part of its course. The peat varies in humification but tends to be more highly humified in the lower part where there is a large amount of wood associated with it. It varies in

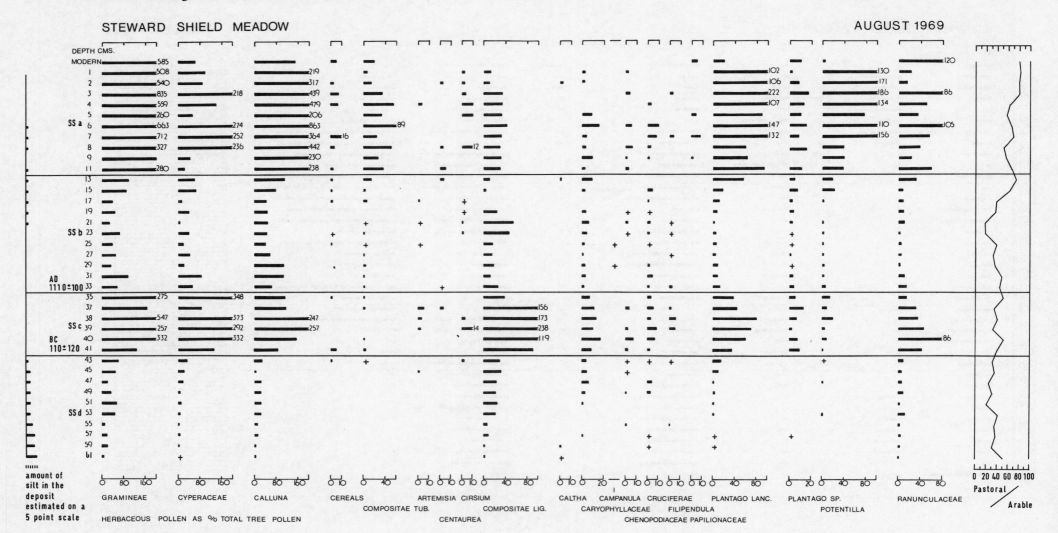

Figures 4 and 5. Herbaceous pollen diagram from Steward Shield Bog. Pollen frequencies are plotted as a percentage of the total tree pollen. Radiocarbon dates are given in calendar years.

colour from black to light khaki-brown and in the upper half of most borings remains of *Sphagnum* were easily recognisable.

Samples for pollen analysis and later for radiocarbon dating were collected from the deepest part of the bog in the same way as at Steward Shield Meadow. The same methods were also used for extracting the pollen and presenting the results except that at this site only a few samples from near the base required treatment with HF.

Where possible over 500 grains were counted but at some levels, namely 136, 151, 177, 184, and 192 cm, totals were less than 200, but over 100.

Six, rather than four, distinct pollen assemblage zones can be recognised on the diagrams (Figs. 7, 8 and 9) and again have been given local designations, Ba, Bb, Bc, Bd, Be, and Bf. Of these three, Bf, Bd, and Bb, represent periods when the area was well wooded and three, Be, Bc, and Ba when it was not. All six assemblage zones, as the four at Steward Shield, are part of pollen zones late VIIb and VIII.

Peat from three levels has been radiocarbon dated with the following results.

1 Peat from 148 cm containing low values of tree pollen, the beginning of Be.

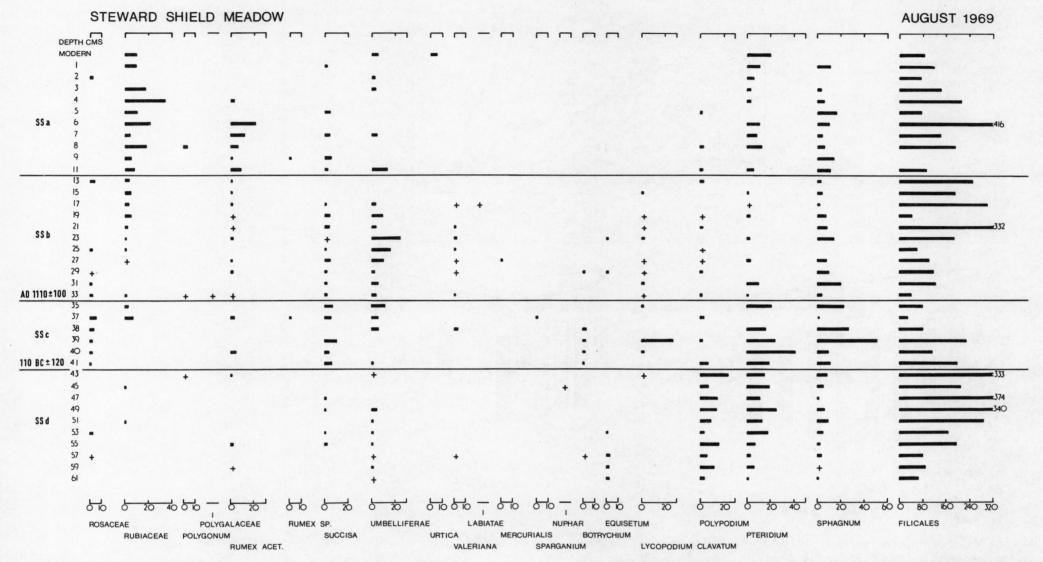

GaK-3/031. 1,730 ± 100 B.P., i.e. A.D. 220 ± 100, uncorrected calendar years.

2 Peat from 108 cm containing high values of tree pollen, the beginning of Bb. GaK-3/030. 80 ± 80 B.P., probably equivalent to A.D. 1700, corrected calendar years (Kigoshi, personal communication).

3 Peat from 80 cm containing high values of *Calluna* pollen and low values for tree pollen, the beginning of Ba. GaK-3/029. 170 ± 80 B.P., probably equivalent to A.D. 1780, corrected calendar years (Kigoshi, personal communication).

These dates show that the first period with high tree pollen frequencies, Bc, began just after the time of Christ, possibly a little but not substantially later than at Steward Shield Meadow and that the period Bb, again with high tree pollen frequencies, was relatively short, probably not more than 200 years, and that it occurred ± 100 years in the eighteenth century.

As at Steward Shield Meadow approximate mean rates of peat accumulation can be calculated from the radiocarbon dates for some of the periods, i.e. 1 cm in 37 years

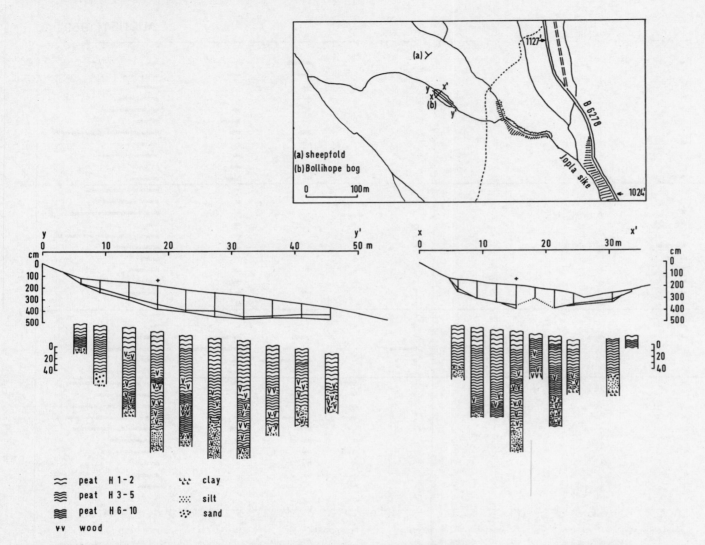

Figure 6. Map of Bollihope Bog showing its relation to the surrounding topography and the position of the transects of borings. Also sections showing the detailed stratigraphy of the bog.

for Be, Bd, and Bc, and 1 cm in 3 years for Bb and 1 cm in 2 years for Ba. This indicates comparatively slow growth in the first 1700 years A.D., when the more humified peat with wood remains was forming and a much more rapid mean rate during Bb and Ba when the less humified *Sphagnum* peat was forming.

Discussion

PRINCIPLES OF INTERPRETATION

In interpreting the pollen diagrams several assumptions have been made. The first is concerned with the area from which the pollen has been derived. For this there are two particularly relevant factors, the smallness of the bogs and the nature of the upland environment.

Both bogs are small, very small compared with the deposits palynologists usually

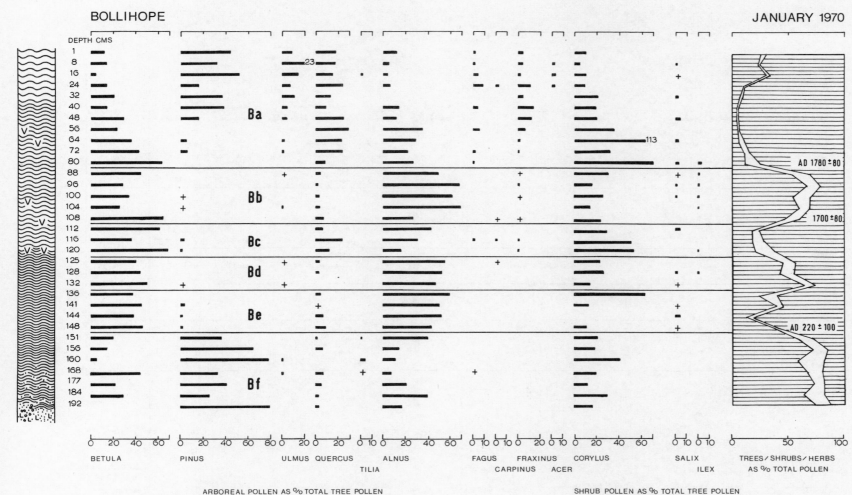

Figure 7. Tree pollen diagram from Bollihope Bog. Pollen frequencies are plotted as a percentage of the total tree pollen. The ratio of tree/shrub/herb pollen is shown on the right of the diagram. Radiocarbon dates are given in calendar years and the two upper A.D. 1780 and 1700 have been corrected in order to take account of the variations in the concentration of $C^{14}O_2$ in the atmosphere revealed by tree ring dating.

work with, and one is tempted to argue on the basis of work by Tauber (1965), Andersen (1970), and Janssen (1966) that with such small bogs nearly all the pollen is likely to have been derived from within a very short distance of the deposits. But at Steward Shield we think this is unlikely because the site is exposed to the west, south and east, and once substantial clearance was under way the topography is such that the local pollen would have had far less chance of drifting to the ground a short distance from its source under the influence of gravity than it does in lowland forest areas or even within more sheltered situations in the uplands. We are not arguing that there is *no* element of local pollen but that a considerably larger proportion is 'extra-local' than would be the case at either a lowland site or a more sheltered upland site of comparable size. By 'extra-local' we mean, likely to be coming from within a few kilometres; in other words we think the diagram will, by and large, record the vegetational history of the tributary valley within which the site occurs.

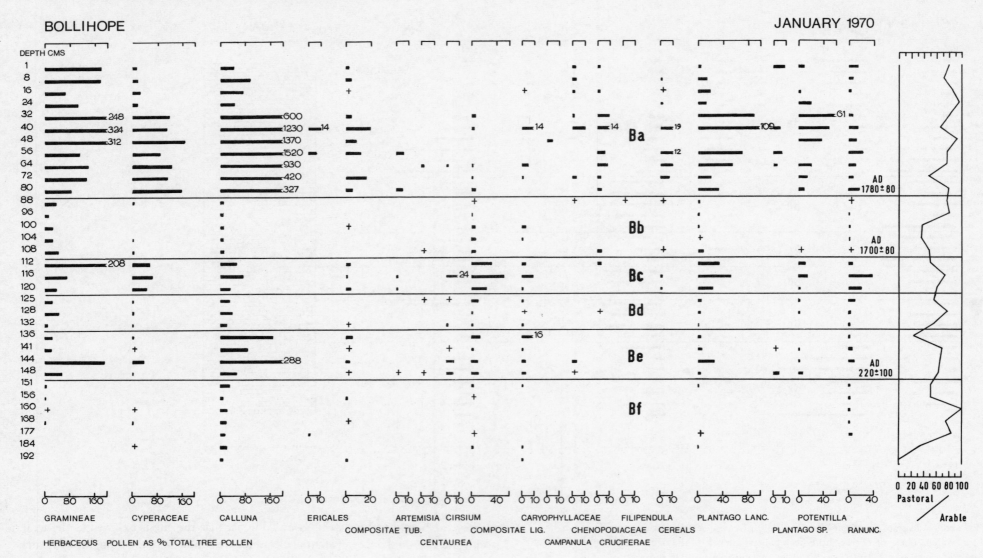

Figures 8 and 9. Herbaceous pollen diagram from Bollihope Bog. Pollen frequencies are plotted as a percentage of total tree pollen. Radiocarbon dates are given in calendar years and the two upper, A.D. 1780 and 1700 have been corrected in order to take account of the variations in the concentration of $C^{14}O_2$ in the atmosphere revealed by tree ring dating (Stuiver & Suess 1966).

In many ways the Bollihope bog is similar but it is slightly more sheltered, situated as it is in a narrow gully, and this makes a big difference to the strength of the wind locally. Those of us who have worked in the area are well aware of the fact that it is far less windy on the bog surface than on the valley slopes beyond the gully. We therefore think that, particularly during the wooded periods, much more of the Bollihope pollen will be really local in origin than is the case at Steward Shield, having been derived from within 20 to 30 m of the bog, that is from the gently sloping sides of the gully.

The second assumption is that some parts of the diagrams are more valuable than others. This is because the rate of peat accumulation has varied so much, from

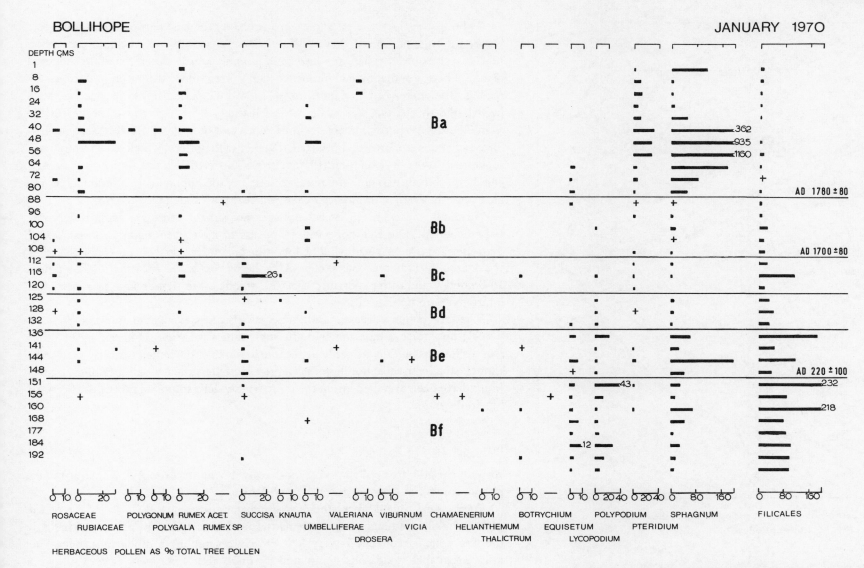

1 cm in 2 years at Bollihope during Ba to 1 cm in 150 years at Steward Shield during SSc. This means that some of the 1 cm thick samples that have been analysed contain the pollen produced during as little as two flowering seasons whereas others contain the pollen that has been produced over periods of a century or more. It follows that a series of the former samples could reasonably be expected to show how the vegetation changed, or did not change, over as little as a decade whereas a series of the latter samples could only show really long-term changes, those occurring over half a millenium or more. Of course these are the two extremes. Portions of both diagrams are from peat which accumulated at intermediate rates. Clearly for our purposes series of samples showing changes over decades are better than those showing them over half millenia. The former samples might have one drawback, however, and that is that they are likely to record fairly accurately random short-term fluctuations in the pollen rain such as those caused by particularly good or bad flowering seasons for individual species.

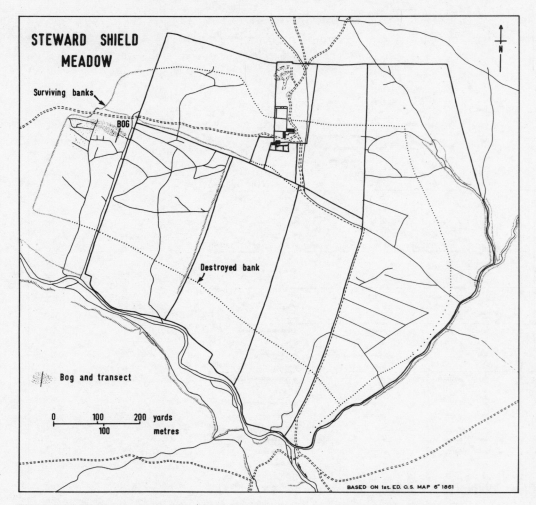

Figure 10. Map showing the relation between the medieval enclosure banks, the seventeenth century enclosure walls and the bog at Steward Shield Meadow.

POLLEN ZONES SSd AND Bf

There are no radiocarbon dates for the start of the local pollen zones SSd and Bf but it seems reasonable to suppose that peat began forming at these sites at the same time as it did over large areas of the lower slopes of the northern Pennines, that is, with the climatic deterioration some time after 1200 B.C. The end of SSd and Bf appear on the evidence of the radiocarbon dates to be broadly synchronous, especially when one takes into account the fact that the peat dated would itself have formed over as much as 150 years.

At both sites therefore for some time, probably the best part of a thousand years before the time of Christ, there was a reasonable amount of woodland containing pine as well as birch, alder, and hazel and possibly a little oak in the more sheltered places. These woodlands had almost certainly been affected to a limited extent by man as artifacts ranging in date from the mesolithic period to the Bronze Age have been found throughout Weardale, while Heathery Burn Cave, in the valley of the Stanhope Burn, approximately one mile south east of Steward Shield Meadow produced a famous late Bronze Age hoard, together with pottery, which, if contemporary, suggests its use as a habitation (Britton 1971). Nevertheless this impact was limited and it is interesting to note the presence of pine, the dominance of birch, alder, and hazel and the relative unimportance of oak at this altitude (350-380 m).

Although the SSd and Bf assemblages are similar there are some differences between them. The Bollihope valley appears to have been slightly more wooded than Steward Shield Meadow. The total tree pollen frequency is higher and there is considerably less herbaceous pollen. Also there is no *Salix*, *Fraxinus*, or *Ilex* pollen at Bollihope indicating possibly that the woods were denser and had been less interfered with by that time.

The tree pollen curves at Bollihope are also less even than those at Steward Shield; note for example the sporadic high values for *Pinus*. There may be other reasons for this but it is certainly explicable in terms of the pollen being derived from a much smaller area at Bollihope as suggested earlier. Such small local fluctuations would have been averaged out in the more widely derived pollen rain of the Steward Shield diagram.

POLLEN ZONES SSc AND Be

At both sites the tree pollen decreases sharply during the Iron Age/Romano-British period, sometime between 200 B.C. and A.D. 300.

At Steward Shield where the decrease occurs early, 110 B.C. ± 100 years, the pine and alder frequencies decrease as the total tree pollen drops and during SSc the tree pollen frequency is very low, closely resembling a modern sample. This must mean that woodlands dominated by *Pinus* and *Alnus* were cleared to give a landscape *as open as that of today*, a considerable change. A wide variety of herbaceous pollen taxa have very high frequencies during SSc; Gramineae, Cyperaceae, *Calluna*, Compositae Lig. all have frequencies of well over 100% indicating a large increase in grassland, heath, and some cultivation. Cereal pollen occurs in the sample at 41 cm, 5%, although not during the rest of SSc, and it seems likely that cereals were being grown, at least for a short while in the neighbourhood.

How long did these open conditions last? According to the radiocarbon dates over a thousand years, but some caution is needed. The mean rate of peat accumulation during SSc is 1 cm in 150 years. With only eight cm forming in over 1,000 years it is highly questionable whether or not one can meaningfully regard peat growth as continuous. There is admittedly no stratigraphic evidence for a hiatus in growth or for erosion, but we regard it as possible that at Steward Shield some of the sequence during SSc may be missing, while in Be the same may occur.

The documentary evidence may be outlined as follows: the period before 1200 is largely undocumented, but Boldon Book of 1183 and the evidence of place-names shows the presence of village-based farming communities on the better lands of the dale floor, notably at Wolsingham, Frosterley, and Stanhope (Fig. 1). That these settlements exploited the upland pasture reserves during the summer months is indicated by the occurrence of the Middle English place-name element *schele*, or *shield*, 'a shieling' or more rarely the Old Norse *erg* 'a summer pasture', traditionally occupied between April and August. In the border dales of Northumberland and Cumberland this practice continued into the seventeenth century, but there are indications from Weardale, for example Steward Shield itself, that these temporary occupations had become more permanent before the late fourteenth century. East of the hamlet of Eastgate our principal late fourteenth century source, the Hatfield Survey of 1381, demonstrates extensive piecemeal reclamation, resulting in a continuous block of improved land along the dale floor and sides, extending up to between 240 and 270 m, and this can be seen as a product of expansion during the late twelfth and thirteenth century, possibly terminating, if general European trends are a reliable guide, in the early decades of the fourteenth century. West of Eastgate, however, there is evidence to show that the tributary valleys and summits rising up to 650 m were reserved for the chase, the scene of great hunts involving the Bishops' tenants from all over west Durham, when the deer were 'driven with much din into the deep valleys' to be slaughtered, and carted down-dale to Bishop Auckland and Durham as provender for the episcopal household. It was, significantly, not until the late fourteenth or early fifteenth century that an enclosed park was created in the upper dale, perhaps as a 'conservation' measure to maintain the deer (See Appendix, note 1).

The two bogs, however, both lie within the common wastes grazed by the beasts of Stanhope and Frosterley, and in addition all of the long established dale farms possess grazing rights on the fells, and the importance of these is emphasised by the driftways or drove-roads leading upward and outwards to the grazings. During the thirteenth century population pressure led to the permanent occupation of sites such as Steward Shield Meadow and Bollihope Shield, which became islands of improvement within the sea of waste, at this stage bounded by bank and ditch enclosures. In addition on the moorlands to the north and south of Stanhope village there is evidence in the form of the physical remains of enclosures for the intaking of extensive areas up to 420 m. A single documentary reference in the Hatfield Survey to certain 'waste lands' at *Catryk*, surely the deserted fields near Catterick Moss above Bollihope Shield, suggests that these represent a late thirteenth century high watermark of development. This land was certainly ploughed; the plough ridges remain, and plough scratches in the subsoil have been traced by excavation. At neither bog is this high watermark of cultivation and occupation identifiable with any certainty within the assemblage zones SSc and Be, but this is hardly surprising because these were periods of slow bog growth, presumably caused by the same climatic amelioration that permitted arable cultivation at such altitudes (See Appendix, note 2).

The evidence from the Bollihope diagram is similar to that from Steward Shield; at the beginning of the period, here radiocarbon dated to A.D. 220 ± 100, marginally later than at Steward Shield, the pine frequency drops abruptly, virtually disappearing from one sample to the next as the total tree pollen frequency decreases and this presumably indicates the clearance of local pine woods. This is followed by an open phase of uncertain duration; uncertain because as at Steward Shield peat growth was slow or discontinuous during this period and we consider it dangerous to attach too much significance to the minor fluctuations visible from sample to sample in the upper part of Be. The most interesting evidence from Bollihope concerning phase Be, however, is the discovery of the remains of a number of huts, both circular and rectangular, on the south-facing slopes of the main valley of the Bollihope Burn (see Fig. 1) which by comparison with sites in Northumberland are almost certainly Romano-British in date. There are associated sub-rectangular enclosures, farmyards rather than fields, and there is little doubt that the valley was occupied, even if only seasonally, at that time. With this site in mind it is interesting to note that the low tree pollen frequency during Be is not as low as that at Steward Shield Meadow during SSc; neither are the increased herbaceous frequencies anything like as high. Only the odd sample has high *Calluna* or Gramineae frequencies although a higher proportion of pasture to cultivated land is indicated. It seems clear from this that for the areas represented in each diagram, the one at Bollihope was on the whole less cleared of its trees and less intensively used than at Steward Shield Meadow, and it may be possible for the former site to argue in terms of 'galleries' of woodland along the steeper valley sides, with clearance extending along the benches which lie some 50-100 feet (17-33 m) above the main river. In this connection the discovery about half a mile down-valley from the Bollihope bog of a third century altar dedicated to Silvanus, the god of the woods, erected by a local cavalry commander on the occasion of the bagging of a boar of 'outstanding size', is perhaps significant; C. Tetius Veturius Micianus and his companions indulged in 'pig-sticking' in the glades and groves of the main valley (See Appendix, note 3).

POLLEN ZONES SSb AND Bd

At Steward Shield Meadow, after the long period of SSc during which the area appears to have been virtually free from trees, alder and willow, and to a lesser extent the other trees, began to grow again and continued to do so for a few centuries. This resulted in a reduced area of grass and heathland. It is this recession which appears to be securely documented in the Hatfield Survey of 1381, which records under the heading 'waste lands' in Stanhope that '80 acres of land at Stewardhall' lies waste and outside the tenures, a situation confirmed by a second entry in another context which more specifically refers to '80 acres of land in Stanhop in Werdale, called there, Stewardshell', a spelling which explains how the scribal error occurred in the first case (Greenwell 1857, p. 73 and 188). It was the bank and ditch enclosures of this first farm, overlain discordantly by the stone walls of the present farm, which first attracted our attention, and as Fig. 10 shows the bog lies *within* the earlier enclosures. No firm date can be put on the beginning of this phase of desertion but it is not likely to fall before 1310 or after 1350, and this, and other reversions to waste in Durham county, must be attributed to the great agricultural depression beginning in the second decade of the fourteenth century. It must be noted that a radiocarbon date of A.D. 1100 for the commencement of phase SSc is not inconsistent with this conclusion; there is a standard deviation of 100 years and in addition the slow rate of peat accumulation at this level in the bog means that the dated peat may itself have covered 150 years.

We would correlate this new, more wooded, period SSb with Bd at Bollihope, although it is worth noting that the Bollihope tree pollen frequency is again higher than at Steward Shield. Another difference between the two assemblages is in the *Salix* frequency; *Salix* becomes important at Steward Shield Meadow and maintains this importance throughout SSb and well into SSa. In practical terms this record implies the development of scrub-woodland in the neighbourhood of both sites, and this is attributable to several interlocking causal factors; first the great depression of the fourteenth century commenced in the second decade, and throughout the whole of western Europe a widespread recession occurred, reinforced by the famine years of 1316-17 and the plague years of 1347-9; second, and in particular, Weardale was ravaged by Scottish armies in 1316 and 1327, and in the latter year Froissart recorded that no less than 500 cattle-carcasses were left behind in the Weardale camp of the retreating Scottish army. These depredations probably encouraged the desertion of outlying farmsteads out of hailing distance with neighbours, while the destruction of livestock led to sharp decreases in grazing pressure. At a time of expansion these would have been merely temporary setbacks, but in the economic context of the first half of the fourteenth century their impact was decisive, and shrub growth was initiated in former habited and grazed localities, giving rise eventually to stands of scrubby woodland. A third factor may have been the onset of wetter conditions during the earlier decades of the fourteenth century, and the rising of the water table would account for both the acceleration of bog growth and the prevalence of willow and alder, both of which may have been present already, preserved as stocks for coppicing, having a use for hurdles and basketwork (See Appendix, note 4).

At both sites this regeneration phase was ended by further clearance and at this point the history of the two diverge. However, in assessing the duration of the recession, and the dates and character of the re-occupations it is necessary to return to the general picture provided by the documentary sources. The main medieval headdyke of Weardale shows evidence of local fluctuations but no major, overall retreat; in short, in the main dale below Westgate the medieval advances were, by and large, maintained. In the upper dale above Eastgate a Master Forester's account roll of 1438 provides our first clear picture and two points are of particular importance; first, farms were scattered along the dale floor and sides, and the prevalence of *shield* names indicates that former temporary occupations had, by 1438, become permanent in this part of the dale; second, attached to each farm were grazing rights for a specified number of cattle and sheep, i.e. stinting arrangements were in force. Upland commons were rarely restricted at such an early date and these arrangements are attributable to the power of the Bishops of Durham and possibly their wish to conserve pasture for deer. In 1511 a schedule of leases shows that the *same* farms were occupied, although a few new ones had appeared, largely within the confines of the Great Park. By the last decade of the sixteenth century farms were increasing in area and number, and it is evident that the deer numbers were falling rapidly as grazing intensity by cattle, sheep and horses increased. During the seventeenth and eighteenth century farms multiplied, and by 1800 most of the farms now visible, either occupied or in ruins, were in being. This quickening of economic activity in the sixteenth century forms a background against which the individual site histories must be seen. Two further general comments relating to the dale are of relevance, in about 1540 Leland noted that 'Weredale ... is wel-woodid' but a survey of 1652 notes that there was then 'no timber remaining in the High Forest (i.e. above Eastgate) and no coppice or underwood saving scrubby land and brushwood near Wearhead' in the upper dale some eight miles west of Steward Shield. Lead accounts, roughly contemporary with Leland's description, record that ore from the Blackdene area of the dale, two miles west of Westgate, was being carted down-dale to Wolsingham some twelve miles, for smelting, using the protected timber resources of Wolsingham Park. There are grounds for believing that the assault on the woodland reserves of the dale had been continuous throughout the fifteenth century; in 1401 a lessee of the Bishop's lead mines in Weardale was granted 'Timber in sufficient quantities' and it may be no accident that in 1430 Robert Kirkhouse granted the right to win iron ore on 'the north syde of Stanhope Parke' was burning his charcoal in woods some twenty-four miles further north in the Tyne valley. The Bishop's own iron master, in 1408-9 worked a forge at Bedburn, on the south side of the Wear valley below Wolsingham, and in this case his charcoal was drawn from the protected timber resources of Bedburn Park. Clearly timber was a precious commodity and Kirkhouse's agreement shows that his charcoal burners were forbidden to touch 'ooke, esshe, holyn, apiltre and crabtre', and also 'all the wode that be felyes or beemes the whiche allewey shall be fellyd'. There are strong indications that by the mid-fifteenth century the woodland resources of upper Weardale were already wholly depleted, a fact not inconsistent with the presence of scrub at certain localities (See Appendix, note 5).

POLLEN ZONES SSa AND Bc, Bb, AND Ba

At both Steward Shield Meadow and Bollihope the transition to the next vegetational phase is characterised by the clearance of most of the woodland and only at Bollihope are radiocarbon dates available; the second regeneration phase, Bb, occurs between about 1700 and 1780, placing zone Bc before 1700, but at Steward Shield the shallowness of the bog made the dating of the clearance between SSb and SSa impracticable. Unfortunately our documentary sources fail to provide clear evidence: there is a brief but ambiguous reference to land at Steward Shield Meadow in a Collector's Account roll of 1504, but in 1532 the Halmote Court of Stanhope, puzzled by the precise tenurial status of the farm, ordered an enquiry to be made. This reinforces the argument for a phase of complete desertion, but Halmote courts were very human institutions and in spite of the surviving run of court books being complete, the inquiry was never made, and our next firm ground is 1665, the present owner's earliest deed, by which date the farm, still described in medieval terms as a holding of 80 acres, shows no sign of being newly re-occupied. The site is not mentioned in a survey of the dale in 1596, and it seems that re-occupation took place in the decades between 1600 and about 1650, a period of general economic expansion throughout the Durham dales, with farmers swinging to more specialised stock farming based on the grazing resources of the area. This would appear to be the date of sample 13 at Steward Shield Meadow and indicate a possible date for the transition period at Bollihope between Bd and Bc. The centuries after 1600 in Weardale have been characterised by accelerating economic activity and rising population occasioned by the extraction of lead, iron, and limestone and this was paralleled by the multiplication of farms and the development of the small 'onsteads' of miner-farmers on the moorland edges. In the last decade of the eighteenth century this activity culminated in an act for the enclosure of extensive tracts of the lower portions of the moorlands, the 'pastures', flanking the old enclosures in the dale although the actual award was delayed until 1815. Certainly, during the plough-up campaign of the Napoleonic War marginal land up to the 330 m contour was brought into cultivation, and these 'improvements' were paralleled by the planting of woodlands, involving primarily pine and spruce, although in the absence of detailed documentation precision is not possible. These plantations suffered during the two world wars, although their place has now been taken by both Forestry Commission woodlands and private ventures (See Appendix, note 6).

These general trends are visible in both pollen diagrams; at Steward Shield Meadow zone SSa cannot easily be subdivided and throughout woodland proportions are such as to indicate the presence of a landscape similar to that of today. A useful chronological pointer is present in sample 7, where the high cereal pollen count (16%) can be equated with the first and second decade of the nineteenth century; in 1801 conditions seem to have been particularly favourable for encouraging the expansion of arable for the rector of Wolsingham wrote 'the corn is better filled and the crops of everything more abundant than ever were before remembered'; in Stanhope 'the crops of wheat, barley, and oats are very abundant and of the best quality that can be expected from the providentially kind season, of nature, of the soil'. A more balanced view of the cropping of marginal land was taken by the curate of St. John's Chapel (Fig. 1) who noted that 'such small quantities of barley and oats as are grown in the higher parts of this district (which is very high) are not good', adding 'it is almost the first trial that has been made so high up the vale for raising these two sorts of grain, and the result is not very promising. The barley is a thin crop, and the oats, tho' more abundant will be late in ripening'. Optimism was such, however, that extensive enclosure took place, and field evidence in the form of ridge and furrow within abandoned fields attests extensive arable. The curate's caution was justified, for changing economic circumstances after the peace of 1815 led to a recession. In 1842 the Tithe Map records 1,800 acres of arable, 21,000 acres of meadow and pasture, 660 acres of woodland and 31,400 acres of uncultivated moor within Stanhope parish. The sharp rise in pine pollen in SSa must relate to the increased plantations and if the date for sample 7 suggested above is accurate i.e. 1800-1816 then this development occurred primarily after 1816 and may well be linked with landlords' attempts to obtain a return from marginal land unwisely intaken in 1815 (See Appendix, note 7).

The story at Bollihope is more complex for we have the additional question of what appears to be a major recession between 1700 and 1780. The tree pollen increases substantially. Given the sheltered location, however, we feel that this change may have resulted from the growth of trees, particularly alder, within the tributary valley where the bog is sited. Alder timber is particularly valuable for use in wet locations, for the manufacture of pumps, water-pipes, and sluices, and although there is no direct proof, it may have been allowed to grow for such purposes. Certainly in 1696 the Blackett-Beaumonts acquired the Weardale lead-mining leases, and during the eighteenth century the industry was placed on a more rational footing (Raistrick & Jennings 1965, pp. 148-153). High counts of cereal pollen occur in samples 64 and 72, and the *Pinus* rise begins in sample 48; we suggest that a date of 1815+ is probable for this latter level. At Bollihope the tree pollen proportion increases notably in the last three samples, possibly because the easterly winds of spring bring pollen from the extensive pine plantations further east in Wolsingham parish. The elm is more difficult to account for; it is unlikely to be a local phenomenon and there is no parkland in the vicinity of the site.

A WIDER CONTEXT

These two pollen diagrams from sites near the margin of occupation within Weardale pose many questions concerning their detailed interpretation, but at both sites there is clear evidence for two distinct periods of occupation, one Iron Age/Romano-British, the other medieval/modern. While there are no grounds for arguing that in each case exactly the same site was re-occupied we would argue that the same piece of territory, within no more than half a mile radius from each bog, was involved in each period. The medieval/modern occupation of these upland farms has essentially been one of dependence, either direct or indirect. The term *shield*, 'shieling', implies a specialist upland settlement dependent upon a lowland grain-producing cluster; in both of the cases discussed this focus was probably Stanhope, and although these ancient bonds have now been largely destroyed the hill farms of Weardale while being integrated within a wider economic system involving specialist regional production are still dependent upon their lowland markets. How ancient is the dependency of these marginal locations on more favoured regions? There are grounds for believing that in the post-Roman period Weardale was part of a wider grouping *Aucklandshire*. This territory, embracing Weardale and part of the more fertile Wear lowlands focusing on Bishop Auckland, may well have very deep roots, and within the last few months a number of very substantial Iron Age or Romano-British farm sites have been found in the foothills of the lower dale—possibly the more prosperous grain-producing components of the multiple estate to which were attached the marginal and possibly seasonally occupied pastoral settlements of the uplands, represented by the earlier clearing phases at both Steward Shield Meadow and Bollihope (See Appendix, note 8).

Acknowledgements

We would like to acknowledge our debt to the Department of Palaeography and Diplomatic in Durham, in particular to Mrs Linda Drury for her help in searching the record collection. We would also like to thank Mr A. Collingwood, the present owner of Steward Shield, for letting us see his private deeds and Capt. Pease for access to the Bollihope site. The work was carried out with the financial support of a N.E.R.C. research grant which we gratefully acknowledge. Finally we wish to record our appreciation of Mrs Rita Ball's meticulous care in typing the manuscript.

References

ANDERSEN S.TH. (1970). The relative pollen productivity and pollen representation of north European trees and correction factors for tree pollen spectra. *Danm. geol. Unders.* Ser. 11, **96**, 99 p.

BRITTON D. (1971). The Heathery Burn Cave revisited. *Brit. Mus. Quarterly* XXXV (1-4), 20-8.

DEWDNEY J.C. ed. (1970). *Durham City and County with Teesside*. British Association for the Advancement of Science. Durham.

EGGLESTONE W.M. (1882-3). *Stanhope and its Neighbourhood*. Stanhope, via Darlington.

GODWIN H. (1968). Studies of the post-glacial history of British vegetation XV. Organic deposits of Old Buckenham Mere, Norfolk. *New Phytol.* **67**, 95-107.

GREENWELL W. (1857). 'Bishop Hatfield's Survey'. *Surtees Soc.* **32**. Durham.

JANSSEN C.R. (1966). Recent pollen spectra from the deciduous and coniferous-deciduous forests of northeastern Minnesota: a study in pollen dispersal. *Ecology* **47**, 804-25.

LAPSLEY G.T. (1905). *Victoria County History*, Durham, I. 259-422. London.

RAISTRICK A. & JENNINGS B. (1965). *A History of Lead Mining in the Pennines*. London.

RAMM H.G., McDOWALL R.W. & MERCER E. (1970). *Shielings and Bastles*. H.M.S.O. London.

SLICHER VAN BATH B.H. (1963). *The Agrarian History of Western Europe, AD 500-1850*. London.

STUIVER M. & SUESS H.E. (1966). On the relationship between radiocarbon dates and true sample ages. *Radiocarbon* **8**, 534-40.

TAUBER, H. (1965). Differential pollen dispersion and the interpretation of pollen diagrams. *Danm. geol. Unders.* Ser. 11, **89**, 69 p.

Appendix

1 Greenwell (1857), Lapsley (1905), V.E. Watts, 'Place-Names' in Dewdney (1970, pp. 251-65), Ramm *et al.* (1970, pp. 1-8). A summary of the economic background is to be found in Slicher van Bath (1963, pp. 132-51). The quotation is from B. Stone, *Sir Gawain and the Green Knight* (Penguin, 1964, p. 70). pp. 69-70, and 76-8 contain vivid descriptions of a deer hunt. As Egglestone (1882-3, pp. 15-17), makes clear 'we are not positively informed' when Stanhope Great Park was enclosed. Where possible the content of this volume has been checked against the original documentation. It is generally accurate, and frequently cites evidence now no longer available.

2   Greenwell (1857, p. 59). The plough ridges were excavated by B.K.R. by way of an experiment.
3   *Vide* G. Joby, 'A Field Survey in Northumberland' in A.L.F. Rivet, ed., *The Iron Age in Northern Britain* (1966, pp. 89-109); E.J.W. Hildyard, *Archaeology of Weardale, Summary of Research* (1950-52, item B5).
4   M.A. Richardson, *Local Historian's Table Book* (1841), Historical Division, I, pp. 98-9 and 107-10; Sir J. Froissart, *Chronicles of England, France and Spain*, ed. T. Johnes, I (1806, pp. 45-68), where it is stated that in addition to 500 large cattle the Scots left behind 1,000 spits of meat and 300 cauldrons made of the skins of slaughtered beasts, although it must be noted that such estimates are apt to be notoriously unreliable. Ramm (1970, p. 64) emphasises the importance of having neighbours within hailing distance. H.H. Lamb in *The Changing Climate* (1966, pp. 208-210) demonstrates the onset of wet summers during the early decades of the fourteenth century.
5   The primary sources used in this study are lodged in the Dept. of Palaeography and Diplomatic, Durham; of particular importance is a collection known as the *Weardale Chest*, documents relating to the Forest of Weardale and Stanhope Park, which includes such individual items as the survey of the Bishop of Durham's lands and rights in the parishes of Stanhope and Wolsingham in 1595-6 (items 42 and 44), further survey material dated 1652 (items 38 and 39) and a schedule of leases in 1511 (item 1); the Master Forester's account rolls are amongst the Ecclesiastical Commissioners, *Ministers' Accounts*, and the roll for 1438-9 is no. 190,030; for the lead accounts see Ecclesiastical Commissioners, *Ministers' Accounts* nos. 190,012-16; L. Toulmin-Smith, *Leland's Itinery in England, c.* 1535-1543 (1907, pp. 70-1); G.T. Lapsley, 'The Account Roll of a Fifteenth Century Iron Master', *Eng. Hist. Rev.* XIV (1899, 509-29); Egglestone (1882-3, pp. 26-36).
6   Ecclesiastical Commissioners, Ministers' Accounts, Darlington Ward, no. 188785; Halmote Court Books, I, no. 6, 52; Halmote Court Enclosure Docs., Copy Awards, no. M7; J. Bailey, *General View of the Agriculture of the County of Durham* (1810, pp. 86-99, 181-98); F.J. Lewis, 'Geographical Distribution of Vegetation of the Basins of the Rivers Eden, Tees, Wear and Tyne, Part II', *Geog. J.* (1904, pp. 267-85).
7   The acreage returns for Durham Diocese are in the Public Record Office, H.O. 67/8; the Stanhope Tithe Apportionment and Plan is lodged in the Dept. of Palaeography and Diplomatic, Durham.
8   G.R.J. Jones, 'The Multiple Estate as a Model Framework for Tracing Early Stages in the Evolution of Rural Settlement', in *L'Habitat et les Paysages Ruraux d'Europe*, Les Congrès et Colloques de l'Université de Liège, 58 (1971, pp. 251-67).

The anthropogenic factor in East Anglian vegetational history: an approach using A.P.F. techniques

RICHARD E. SIMS *The Botany School, University of Cambridge**

Introduction

The aim of this paper is to present new data on certain events in the pollen record of East Anglia. The discussion will mainly centre on the 'elm decline' of north-west Europe and its associated 'landnam' phase, with a brief excursion into a phase of decreased forest cover during the Mesolithic. The information presented includes absolute pollen frequency determinations which throw a slightly different light on the interpretation of the changes in the pollen record around the elm decline.

The decline of elm pollen was used as a pollen zone boundary for many years prior to the publication of 'Landnam i Danmarks Stenalder' by Iversen in 1941. This publication provided a new interpretation for the elm decline, previously thought to have a climatic cause. With the use of high pollen counts and closely spaced samples Iversen discovered the elm decline to be a more complex event than a simple decrease in *Ulmus* pollen frequencies. There were, in Denmark, two declines of *Ulmus* pollen frequencies. The first was accompanied by a fall in *Hedera* pollen frequencies. This marked the start of Iversen's (1941) pollen zone VIII and was thought to be caused by a change of climate. The second was accompanied by a fall in the pollen frequencies of all the 'mixed oak forest' species, and associated with an increase in the pollen frequencies of *Betula*, *Corylus*, and *Alnus*. This was followed by a transitory increase in the pollen frequencies of Gramineae and herbs, after which the pollen frequencies of the 'mixed oak forest' taxa, except *Ulmus*, regained their former levels. The interpretation placed upon this second fall in *Ulmus* pollen frequency was forest clearance by man for farming. Clearance came first and was followed by a period of farming after which the forest regenerated as the people moved away. This temporary clearance phase was termed 'landnam' or 'land occupation phase'.

Since then similar events in the pollen record have been discovered throughout north-west Europe. In most cases the *Ulmus* pollen frequencies do not regain their former level but in some regions, such as Ireland (Mitchell 1956), there seems to be almost complete regeneration. The radiocarbon dates available for the elm decline in north-west Europe lie between 5,800 C^{14} years B.P. and 4,500 C^{14} years B.P., the majority occurring at 5,100 C^{14} years B.P. Those sites that have been investigated in a manner similar to Iversen (1941) generally show a similar three-fold pattern interpretable as reflecting clearance, cultivation, and regeneration.

The publication of Iversen's work has prompted much discussion on the possible causes of the elm decline (see Smith 1961, 1970). The factors proposed include climate and man, through burning (Iversen 1941, 1960), man through fodder cropping (Troels-Smith 1960), and disease (Watts 1961). There is still much discussion as to which of these or other unknown factors are the causes of the elm decline. The present study is an attempt to clarify the picture by the use of absolute pollen frequency determinations to determine whether measurements of pollen influx can contribute to the resolution of this problem of the elm decline, for example whether it is *Ulmus* alone which is affected or whether other tree taxa are involved.

Prior to the elm decline it is possible that activities of Mesolithic man resulted in vegetational changes (Smith 1970). Evidence for such effects have now been found in East Anglia and will be discussed here.

In this paper reference will be made to other events in the pollen record from Hockham Mere and Seamere. These details are fully discussed in an unpublished thesis by the author.

Sites and methods

The sites chosen for investigation are two East Anglia meres, Hockham Mere (TL 935937) and Seamere (TG 038012). These sites, together with Old Buckenham Mere (TM 049921) (Godwin 1968), allow comparisons to be made between three sites of close proximity. The distance between them is about ten km (Fig. 1). This in turn gives a chance for elucidating local differences between the sites, such as might result from the sites being in areas of differing geology. Hockham Mere is in the sandy Breckland, whilst the others are situated in chalky boulder clay. All the lakes have diameters greater than 200 m and thus the pollen spectra may largely represent the regional vegetation, as opposed to the local mire vegetation.

The pollen preparation method used was one which would lend itself to absolute

*Present address: North East London Polytechnic, Romford Rd., London E15 4LZ

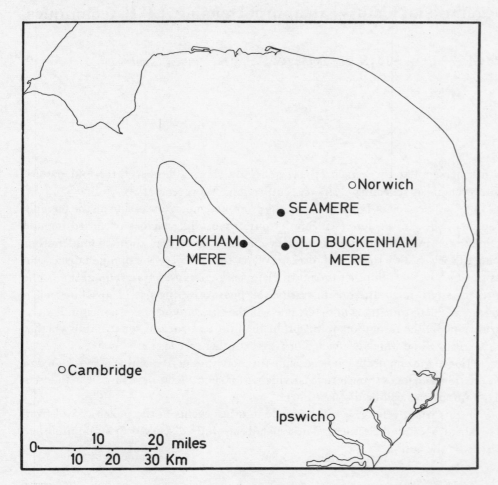

Figure 1. Map of E. Anglia showing sites and outlining the extent of the Breckland.

frequency (A.P.F.) determinations, as well as the determination of relative pollen frequencies. The advantage of A.P.F. data is that the pollen counts are independent of the interrelations of the taxa within the pollen sum, with the pollen frequency curves giving true fluctuations in the pollen concentration of each pollen taxa. The A.P.F. pollen data presented here are, as yet, uncorrected for sediment accumulation rates and are thus presented as pollen concentration figures. The radiocarbon dates were not available in time to allow the corrections to be made.

The analysis, using the Troels-Smith (1955) method, of the Hockham sediment from 250–350 cm shows it to be a homogeneous algal lake mud ($Ld^1 4$). A plot of sediment depth with age (radiocarbon years) gives a straight line relationship with a regression coefficient of 1 cm/12.3 C^{14} years and a correlation coefficient of 0.986 ($P<0.01$). The pollen influx figures can be obtained by dividing the pollen concentration by 14.02.

At Seamere there is a change of sediment lithology at 638 cm from a shell marl to a marl just after the elm-decline. This has an uncertain effect on the sediment accumulation rate.

The pollen preparation method is, briefly, as follows. A known volume of sediment is taken and a known concentration of exotic pollen (*Nyssa sylvatica*) is added. The whole is then prepared in the usual manner (Faegri & Iversen 1964). After acetolysis the sediment is suspended in a known volume of trimethyl methanol from which measured aliquots are then taken and placed on a drop of silicone oil on a slide. The pollen count was, in most cases, greater than 1,000 land pollen and spores. The raw count was used as the basis for the relative pollen diagram (expressed as a percentage of total pollen spores, excluding obligate aquatics). The raw count was corrected for loss of exotic pollen to provide data for the A.P.F. diagram (Davis 1969).

Counting was carried out using a Leitz S.M. microscope at a counting power of ×600; critical determinations were made with an oil immersion objective at a magnification of ×1187. Cereal pollen identifications were made with the aid of a Leitz 'Phaco' phase contrast objective and condenser at a magnification of ×1,500.

Hockham Mere: 420–460 cm

RESULTS

This section deals with a phase of decreased forest cover during Mesolithic times at Hockham Mere. The pollen diagram (Fig. 2) can be related to zone VI of Godwin and Tallantire (1951), before the increase in *Alnus* pollen frequencies which marks the start of zone VIIa. The main features to note are the temporary decrease in the arboreal pollen frequencies between 438 cm and 430 cm, resulting mainly from decreased frequencies of *Ulmus*, *Quercus*, and *Betula* pollen. Associated with this is a temporary increase in the pollen frequency of *Corylus*. The frequency of *Hedera* pollen shows a slight rise from 440 cm to 437 cm above which there is a sharp fall. There are two peaks in the pollen frequencies of those herbs usually associated with open treeless areas (Group A herbs—*Plantago lanceolata*, *P. major/media*, Compositae (Lig.), *Rumex acetosella*, Chenopodiaceae, and *Urtica* pollen and *Pteridium* spores). These peaks fall within the phase of reduced arboreal pollen frequencies, being separated by a temporary increase in Gramineae pollen frequencies. Also of note is the occurrence of *Fraxinus* pollen for the first time and that, while clustering in two parts of the pollen diagram, the herb pollen taxa are scattered throughout.

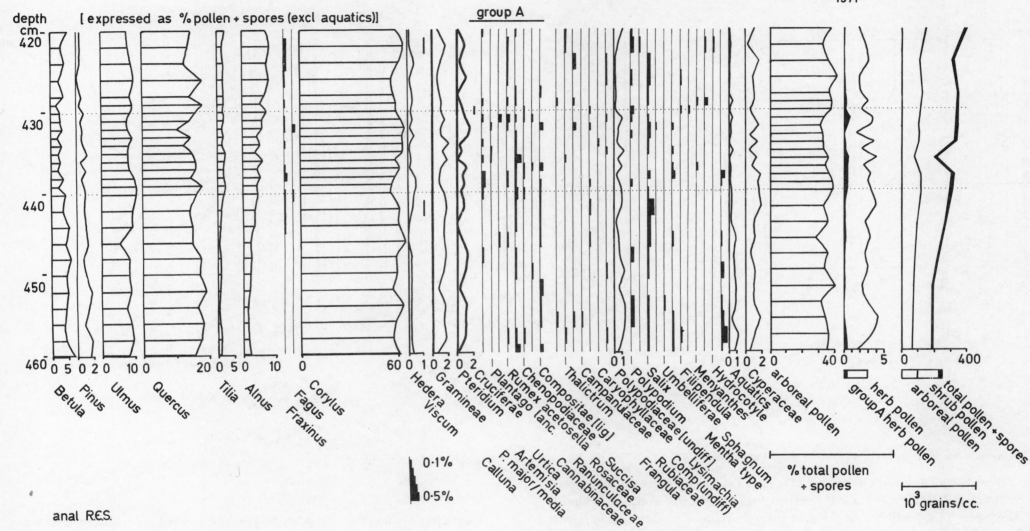

Figure 2. Pollen diagram showing Mesolithic clearance phase from Hockham Mere.

The pollen concentration figures are from widely spaced samples and show little change, except for a drop in pollen concentration of all taxa at 435 cm. This could be related to a change in sediment accumulation rate rather than a response of pollen concentration to a vegetational change.

INFERRED VEGETATIONAL CHANGES AND THEIR CAUSE

The inferred vegetational changes that took place during the time represented by the pollen diagram are as follows. The mixed oak forest was still maturing in floristic composition and structure just prior to the expansion of *Alnus*, c 7,000 C^{14} years B.P. The forest would probably be one of a fairly open canopy with *Quercus* and *Ulmus* as the dominants with *Corylus* either as an understorey or forming part of the canopy proper. The remaining tree taxa would be of scattered or local distribution only.

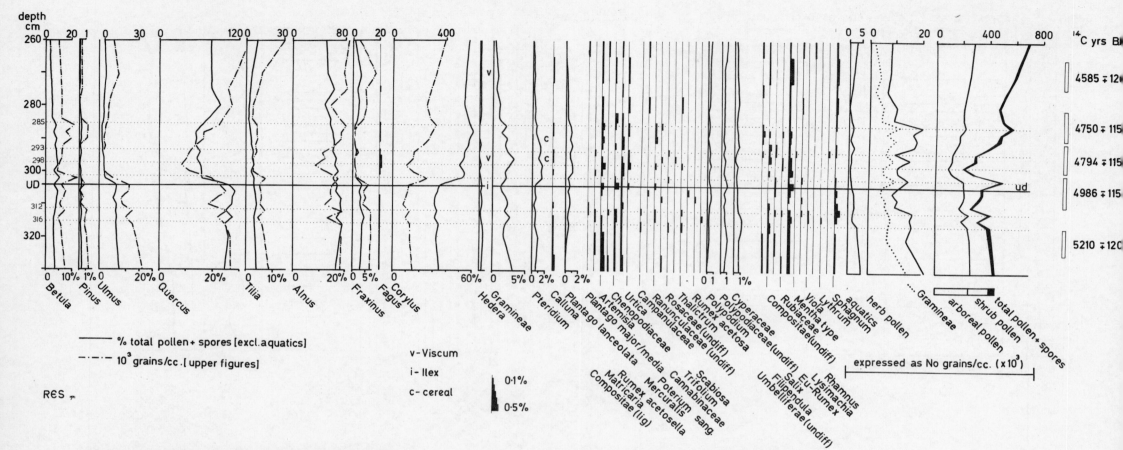

Figure 3. Pollen diagram showing elm decline from Hockham Mere.

The changes seen in the pollen record suggest a temporary reduction in the forest cover. That *Betula* is also reduced in extent suggests that this would be in already fairly open areas, possibly near the mere margin. There is archaeological evidence for such openings (Mosby 1935). A number of worked flints have been found from the area on the northeastern fringes of the mere and Clarke (1960) suggests that Hockham Mere was a temporary camp site of Magelemosian hunters. He suggests a camp size of about 200 × 50 m. It is possible that the felling of trees on the drier upper margins by the lake could have produced the changes observed in the pollen diagrams. The evidence from the pollen diagram, whilst indicating that these were temporary encampments, might suggest that they were periodically revisited. Future radiocarbon dating, it is hoped, will assist in determining the duration of the events envisaged here.

Hockham Mere: elm decline

RESULTS

The pollen diagram (Fig. 3) has not been formally zoned as certain difficulties are posed by the rapidity of the changes in the pollen frequencies and concentrations and also by their complexity. Instead a number of horizontal guide lines have been drawn at 285, 293, 298, 302, 304, 312, and 316 cm. These are only intended to aid in discussion of the pollen diagram.

Relative pollen diagram

The elm decline at Hockham is seen at 304 cm (Fig. 3) but before dealing in detail with the changes, the situations prior to the elm decline must be described. The analysis starts at 330 cm. Here the pollen record is dominated by arboreal pollen (60% total pollen and spores) of which the pollen of *Alnus* and *Quercus* predominate. The pollen of *Betula*, *Tilia*, *Ulmus*, and *Fraxinus* form the remainder of the arboreal pollen in frequencies at 5-10%. The majority of the remainder of the pollen is that of *Corylus*, the herb pollen values being very low.

316-304 cm. Changes occur in the pollen record prior to the elm decline but these, in relative terms, are of uncertain significance. One point of note is that at 315 cm there is a cluster of herb pollen taxa, such as *Plantago major/P. media*, *Scabiosa*, and Campanulaceae. The arboreal pollen frequencies at this level hardly change. The possible significance of this event will be shown in relation to the A.P.F. data.

304-285 cm. The elm decline at 304 cm sees a drop in the frequency of *Ulmus* pollen from 7% to less than 2% within 2 cm. This fall is accompanied by similar falls in the pollen frequencies of *Quercus*, *Fraxinus*, and *Hedera*. There are increases in the pollen frequencies of *Plantago lanceolata*, Gramineae, and *Corylus*, and of *Pteridium* spores. *Corylus* values rise from 35 to 60% and remain at this level for the remainder of the diagram. At 285 cm the values for all the pollen taxa, except *Corylus* and *Ulmus*, revert to their former levels.

The above portrays, in outline, the events in the pollen record around the elm decline. In more detail at 293 cm there is a change in the slope of the curves for the pollen frequencies of *Tilia*, *Fraxinus*, and *Quercus*, and there is a change in the curves for *Plantago lanceolata* pollen and *Pteridium* spores.

Turning to the non-arboreal pollen taxa, it can be seen that they show a scattered presence before the elm decline. They increase in frequency immediately after the elm decline and mirror the changes seen in the curves for *Plantago lanceolata* pollen. Gramineae pollen of the *Triticum* type is present in two samples. The curve for *Hedera* pollen undergoes a number of changes. The maximum value at any one level is 0.5%. There is a decline at 315 cm to 0.1% which is followed by an increase at 310 cm before a decline at 304 cm above which it is represented by scattered occurrences until 285 cm where it again forms a continuous curve.

Absolute Pollen Frequency Diagram

The absolute pollen frequency diagram is expressed in terms of pollen concentration as number of grains/cm^3 of sediment. As previously mentioned division by 14.02 will convert the pollen concentration figures to pollen influx rates (grains/cm^2/C^{14} year). The herb pollen taxa are expressed only in relative terms as there is little difference between the relative and the absolute changes in the frequencies with which these taxa occur.

316-312 cm. The base of the diagram shows the same general features as seen in the relative pollen diagram. The first difference to emerge is the temporary decline in tree and shrub pollen concentrations associated with a peak of Gramineae pollen concentration, at 315 cm. At the same level herb pollen concentration increases, including the pollen of Campanulaceae, *Scabiosa*, and *Plantago major/P. media*.

312-304 cm. Immediately above at 310 cm, there is a second temporary decline of tree and shrub pollen concentration. This corresponds to peaks in the pollen concentrations of *Hedera*, Gramineae, and *Plantago lanceolata*.

304-285 cm. The major decline in *Ulmus* pollen concentration is accompanied by declines in the pollen concentrations of *Quercus* and *Fraxinus*. At this level the pollen concentrations of *Betula*, *Corylus*, *Tilia*, and *Alnus* rise as do those of the herb pollen and *Pteridium* spores.

At 302 cm the pollen concentrations of *Betula*, *Corylus*, *Alnus*, and *Tilia* decrease. There is a decline in the herb pollen concentration but a rise in that of *Plantago lanceolata*. At 298 cm the tree and shrub pollen concentrations start to increase, mainly as a result of the increased pollen concentrations of *Quercus*, *Alnus*, and *Corylus*. This rise continues until 285 cm. However, within these 12 cm other changes occur. There is a change of slope of the tree and shrub pollen curves at 293 cm, and associated with this is the rise of herb pollen concentration to its main peak from 294-285 cm, along with temporary increases in the pollen concentrations of *Betula* and *Pinus*.

285-260 cm. At 285 cm another series of changes can be seen. The peaks in pollen concentrations of herbs, *Pinus*, and *Betula* die away. The pollen concentrations of *Alnus*, *Corylus*, and *Tilia* show a slight fall, whilst those of *Ulmus* and *Fraxinus* start to increase for the first time since the elm decline, and *Hedera* resumes its continuous curve. From 280 cm the character of the pollen diagram reaches a fairly stable level. Most of the tree pollen taxa have reached their former pollen concentrations. The main difference is that the *Ulmus* pollen concentration is half its former value whilst that of *Corylus* is twice its former value.

INFERRED VEGETATIONAL HISTORY

The vegetation of the area represented at the base of the pollen diagram may best be considered as a mosaic of vegetation types from dense mixed oak forest to areas of open canopy, as evidenced by the occurrence of pollen of both light demanding herbs and the importance of *Fraxinus* in the pollen record.

316-312 cm. The event at 315 cm would appear to represent a temporary recession in the canopy of the mixed oak forest. There are present pollen of plants of calcareous habitats (*Scabiosa*, *Campanula*) as well as those of grassland habitats (*Plantago major/P. media*, *P. lanceolata*). It is suggested that this reduction of woodland canopy was in an area with a fairly calcareous substrate.

312–304 cm. The second event, at 310 cm, again seems to represent a temporary recession in the woodland canopy. The main increase occurs in Gramineae, *Hedera*, and *Plantago lanceolata* and may possibly represent a reduction of woodland canopy on a less calcareous substrate. Both these events are of a temporary and localised nature.

304–298 cm. The elm decline occurs next. The pollen record suggests the following pattern of events. A reduction occurred in the extent of woodland containing *Quercus*, *Ulmus*, and *Fraxinus*. If the reductions in pollen concentration are representative of the reduction in areal cover of the species then each taxon was reduced to 30–40% of its original cover. At the same time there is a temporary increase in the extent of *Alnus*, *Tilia*, *Corylus*, and *Betula*. The first two named increase by about 20%, while the last two increase by about 100%. This suggests that in response to decreased *Quercus*, *Ulmus*, and *Fraxinus* cover these other trees acted as colonisers in the now open areas, *Corylus* and *Betula* being the dominant colonisers. This colonising scrub is then itself reduced in cover.

298–260 cm. After the reduction in cover of the colonising scrub gradual regeneration of the forest trees begins, including the regeneration of *Corylus* probably in place of *Ulmus* which did not regenerate until somewhat later. This feature is discussed below.

During this regenerative phase there is a temporary readvance in the extent of open canopy species. After this readvance of open ground *Fraxinus* and *Ulmus* begin to expand in area and this regeneration continues to form the mature secondary forest ecosystem.

In response to this fluctuation in woodland cover the herb flora also undergoes fluctuations of a similar but reversed nature. Consequent to the elm decline there is an expansion of *Pteridium* and grasses and of such herbs as *Artemisia*, Chenopodiaceae, and *Urtica*. There is a decline in the extent of *Hedera*, a point discussed below.

At the same time as the invading scrub is cleared there is a further increase in grassland cover which is accompanied by an increase in the extent of *Plantago lanceolate*, *Artemisia*, and *Rumex acetosella*. The *Artemisia* which has been present throughout the Flandrian at Hockham Mere might reflect communities with *A. campestris*. This species still occurs in the Breckland. However, this species does not appear distinct in its pollen morphology.

The next phase sees a reduction in *Pteridium*, from 298 cm, at the start of the regenerative phase of the woodland. It is at this time that the pollen of *Triticum* type is observed, indicative of cultivation occurring fairly near the site. Heim (1963) has shown that the frequency of cereal pollen only reaches 2–3% A.P. when the stands are closer than 2 km radius of the site. Consequent to the temporary halt in the regeneration of woodland there occurs a major peak of herb cover, including *Triticum*. From then the herb cover is reduced as the mature secondary forest nears its climax and it is joined by *Ulmus*, *Fraxinus*, and *Hedera*, these being released from the flowering constraint imposed at the elm decline.

The time scale for this must, at present, remain provisional as the dates came too late for full analysis. Tables 1 and 2 give the provisional time scale.

Table 1. Radiocarbon dates from Hockham Mere spanning the elm decline

Lab. Number	Position of Sample (cm)	C^{14} years B.P.
Q-1045	265–274	4,585 ∓ 120
Q-1046	281–289	4,750 ∓ 115
Q-1047	290–298	4,794 ∓ 115
Q-1048	300–308	4,986 ∓ 115
Q-1049	316–324	5,210 ∓ 120

Table 2. Age and duration of vegetational changes over the elm decline at Hockham Mere

Depth (cm)	Event	Radiocarbon date C^{14} years B.P.	Estimated date* C^{14} years B.P.	Duration* C^{14} years
304	Elm decline	4,986 ∓ 120	4,986	
302	Peak of scrub invasion		4,957	
302–298	Clearance of scrub		4,957–4,910	53
298	Start of woodland regeneration		4,910	
285	Regeneration of *Ulmus* and *Fraxinus*	4,750 ∓ 115	4,750	
272	Secondary forest at maturity		4,590	
304–298	Grassland phase		4,986–4,910	76
298–293	1st cultivation and grassland phase		4,910–4,850	60
293–285	Major cultivation and grassland phase	4,794 ∓ 115	4,850–4,750	100
304–285	Elm decline—end of herb pollen peak		4,986–4,750	236

*Calculated from regression equation (regression coefficient = 12.297 ∓ 1.192 correlation coefficient = 0.986)

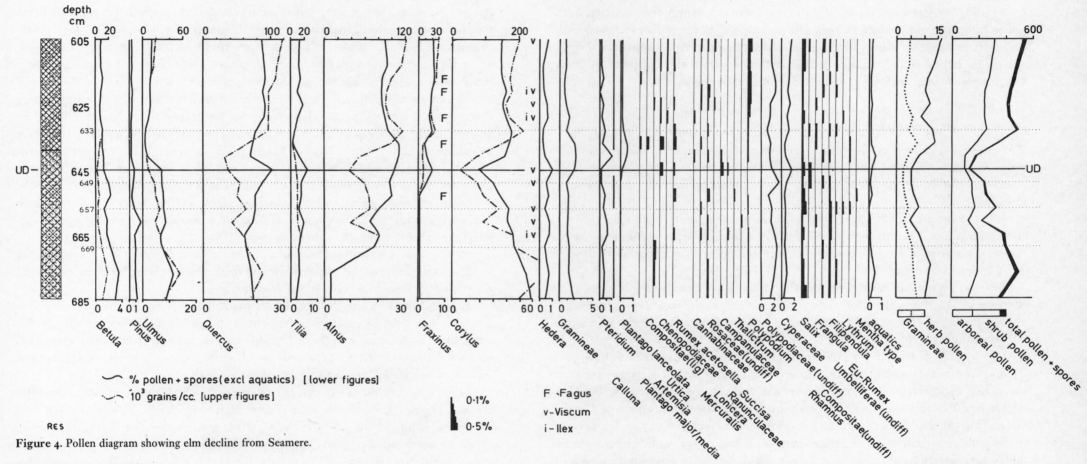

Figure 4. Pollen diagram showing elm decline from Seamere.

Seamere: elm decline

RESULTS

As with the pollen diagram across the elm decline at Hockham Mere the Seamere diagram has not been formally zoned. Guide lines are drawn at 633, 645, 649, 657, and 669 cm.

Relative pollen diagram

At the base of the diagram (Fig. 4) the dominant pollen taxa are *Corylus*, *Quercus*, and *Ulmus*. The *Alnus* pollen curve starts to increase at 677 cm, thereby marking the opening of pollen zone VIIa (Godwin & Tallantire 1951).

669–645 cm. The first feature of note is the temporary increase in the frequency of *Hedera* and *Viscum* pollen, and of *Pteridium* spores at 669 cm. At 649 cm *Fraxinus* pollen, previously present in low frequencies, rises to 5%. Associated with this there are temporary increases in the pollen frequencies of *Hedera*, *Viscum*, Gramineae, and Cyperaceae.

645–633 cm. The elm decline occurs at 645 cm, the reduction in pollen frequency being from 10% to 2% over 4 cm. Reductions in the pollen frequencies of *Quercus*, *Tilia*, *Fraxinus*, *Hedera*, and *Viscum* also occur. There is an increase in the pollen frequency of *Corylus* and in the spore frequency of *Pteridium*. At 637 cm the herb pollen frequencies decrease and there are increases in the pollen frequencies of *Quercus*, *Ulmus*, *Tilia*, and *Fraxinus*. Those of *Quercus* and *Fraxinus* are more rapid than either *Tilia* or *Ulmus*.

633–605 cm. Between 629 cm and 621 cm there is a temporary reduction in the pollen frequency of *Corylus* and *Tilia* with corresponding temporary increases in the pollen frequencies of *Hedera*, *Viscum*, *Ilex*, and Gramineae. At the top of the diagram the pollen record is very similar to that at the base with the exception of higher *Corylus* pollen frequencies and lower *Ulmus* pollen frequencies.

Absolute Pollen Frequency Diagram

The changes seen here in absolute concentration terms differ in many ways from those seen at Hockham Mere, and from the relative diagram from Seamere. The decline of elm pollen begins at 677 cm and continues to 640 cm, there being no sharp fall.

669–657 cm. From 665 cm to 657 cm there is a temporary decrease in the tree and shrub pollen concentration, particularly of *Ulmus*, *Quercus*, *Alnus*, and *Corylus* pollen. This is associated with an increase in the pollen concentration of *Hedera*, *Viscum*, and Gramineae.

657–645 cm. At 649 cm the slopes of the tree pollen taxa undergo a more rapid decrease in pollen concentration. There is an increase in the pollen concentration of *Fraxinus*, *Viscum*, *Hedera*, and Gramineae as well as a general increase in herb pollen concentration.

645–633 cm. At 645 cm the pollen concentrations of *Corylus*, *Alnus*, *Betula*, and *Plantago lanceolata* increase, whereas those of *Viscum* and *Hedera* decrease. At 641 cm the pollen concentration of *Quercus* increases and the concentration of *Pteridium* spores peaks at this level. The increases in the pollen concentrations of *Corylus*, *Alnus*, *Quercus*, and *Betula* continue to 633 cm and are joined at 637 cm by increases in the pollen concentrations of *Fraxinus* and *Tilia* and at 633 cm by increases in the pollen concentrations of *Hedera* and *Ulmus*. The herb pollen concentration reciprocates with these changes. There is an increase in herb pollen from 645 cm to 637 cm from whence it declines until 629 cm, the main peak being at 637 cm.

633–605 cm. Between 633 cm and 621 cm a temporary decline is registered in the pollen concentrations of *Corylus*, *Alnus*, *Betula*, and *Fraxinus* with corresponding increases in the concentrations of *Viscum*, *Ilex*, *Hedera*, and herb pollen. The pollen concentration at the top of the diagram is roughly similar to that at the bottom except in the values of *Corylus*, *Fraxinus*, and *Ulmus*.

INFERRED VEGETATIONAL HISTORY

Here the changes represented in the pollen diagrams are more difficult to interpret than at Hockham. One complicating factor is the change in sediment lithology. This has an uncertain effect on the sediment accumulation rate and this must be taken into account when the changes in the pollen diagrams are interpreted. The inferences drawn here are thus tentative and they are made with reference to the changes occurring in the pollen diagram from Hockham Mere (Fig. 3) for which there is much more confidence concerning the changes described.

The base of the pollen diagram shows the expansion of *Alnus*. At Hockham Mere this may have resulted from a decrease in the water level of the mere, followed by colonisation of *Alnus* over the newly exposed area. It may have expanded at Seamere in a similar manner. The *Alnus* rise has been shown to be a non-synchronous event in western Europe by Hibbert *et al.* (1971) and its seemingly late rise at Seamere, just prior to the elm decline, may be a reflection of local conditions. Just after the expansion of *Alnus* the vegetation appears to have differed from that at Hockham Mere in that with the low frequencies of *Fraxinus* and herbs there would appear to be, generally, a more completely closed woodland.

669–649 cm. The feature described at 661 cm would suggest an opening of the forest of sufficient magnitude to allow the increased flowering and/or dispersal of the pollen of the shrubs *Viscum* and *Hedera*.

649–633 cm. The elm decline, whilst marked at 645 cm in the relative pollen diagram, would best be placed at 649 cm in the absolute pollen diagram. At this point there is an increase in the reduction of shade tolerant trees and an expansion of light demanding plants (*Fraxinus* and herbs). The occurrence of *Hedera* and *Viscum* pollen would support this (cf. 669–649 cm).

There then follows, at 645 cm, regeneration of a scrub woodland with *Alnus* and *Corylus*, joined a little later by some regeneration of *Quercus*. This regeneration continues and is further joined by *Fraxinus*, *Tilia*, and *Ulmus*. The fluctuations in the extent of woodland is reflected in the fluctuations of herb cover. During the period immediately after the elm decline there is increased cover of grassland, along with a transitory *Pteridium* peak. There then follows, during the regeneration of the forest, the major peak of herb cover, as also occurs at Hockham. The evidence from the *Hedera* pollen curve supports this sequence of events.

633–605 cm. After this regeneration of the woodland there came a third temporary recession in woodland cover, mainly resulting from a loss of *Alnus*, *Corylus*, and *Fraxinus*. This is accompanied by an extension of herb cover and the greater representation of *Hedera* and *Viscum*. These last two behave in a similar manner to that seen at 669–649 cm. The mature secondary forest ecosystem is formed after this clearance. In the secondary forest *Corylus* and *Fraxinus* have, to some extent, replaced *Ulmus*.

The gradual decline in tree and shrub pollen concentrations from 655 cm could represent one of three possibilities. Firstly the gradual reduction in woodland cover from a site far away to one nearer the mere. This is not supported by the herb pollen concentration figures as these also decline. The second possibility is the reflection, in the pollen diagram, of an increased trunk space component (*sensu* Tauber 1965), or thirdly there is a change in sediment accumulation rate. It is the latter which is thought most likely to have occurred, although the other factors may have played some part.

Discussion

CAUSAL FACTORS OF THE ELM DECLINE

Before a discussion of the pollen diagrams across the elm decline is started, mention must be made of the limitations of the methods used here. The cores used for radiocarbon dating were 3 inch (8 cm) diameter cores from a Livingstone piston corer. In order to provide the requisite amount of carbon for dating purposes, samples of 8 cm depth were required. Ideally a much larger diameter core should have been used in order to provide samples which more nearly represent the sampling interval used for pollen analysis. However, to get samples containing sufficient carbon from a depth of 0.6 cm (thickness of pollen sample) a core of about 11 inches (*c* 28 cm) is required. The other inherent error is in the uncertainty values attached to the radiocarbon dates.

For the sake of simplicity the midpoint of the sediment samples has been used, as have the radiocarbon dates without the errors, to calculate a regression equation for sediment depth against radiocarbon years. This has provided the data in Table 2 and it is these which are used in the discussion. It is realised that this can be criticised but it is believed that the errors involved do not invalidate the conclusions reached.

The discussion will centre largely on the pollen diagram from Hockham Mere and only draw from that of Seamere when appropriate.

The possible causes of the elm decline have been widely discussed (Iversen 1941, 1960, Troels-Smith 1954, 1960, Smith 1961, 1964, 1970, Smith & Willis 1962, Oldfield 1963, Godwin 1956). The causes usually suggested are varied and include climate, man (lopping for fodder and/or clear felling), and disease. The last named is discussed by Heybroek (1963). He rules out disease as a factor and, in the absence of any evidence to the contrary, this is also ruled out for East Anglia. If this is accepted then the remaining causes are man and/or climate, or some, as yet, unknown factor.

The picture is complicated by the fact that Iversen (1941) considered climate to be the cause of the *Ulmus* fall at the zone VII/VIII boundary. The next fall of elm, which occurs a short way above this boundary was associated with the 'landnam' (*sensu stricto*) and its cause was ascribed, by Iversen, to human activity. However, Troels-Smith (1960) ascribed the changes at the zone VII/VIII boundary in Denmark to man whom, he envisaged, lopped *Ulmus*, *Hedera*, and *Viscum* for fodder. This is also mentioned in connection with the Ertebølle culture (Troels-Smith 1954). Thus it is really the initial elm decline, not associated with any increase in indication of human activity, over which the conflict of climatic versus anthropogenic interpretation occurs. The 'landnam' itself is fairly widely accepted to be the result of human activity.

The pollen diagrams over the elm decline in Britain, with certain exceptions (Oldfield 1963), generally show one elm decline and this is usually associated with a 'landnam' type feature (e.g. Fallahogy-Smith & Willis 1962). Oldfield's (1963) diagram from Thrang Moss shows a primary elm decline before a decline associated with a 'landnam'. The evidence from Hockham Mere and Seamere is, however, in keeping with that generally found in the British Isles (e.g. at Fallahogy).

Evidence for climatic change often stems from botanical evidence and one must be very careful to judge its worth. Pennington (1973) has suggested there was increased precipitation at this time, based on chemical analyses of sediments from lochs in north-west Scotland. Iversen (1960) envisaged a fall in winter temperature to account for the initial elm decline, and Frenzel (1966) has suggested a cold phase from 3400–3000 B.C. Oldfield (1963), however, does not associate the primary elm decline in the south-east Lake District with any such climatic shift.

It is believed that the evidence from both Hockham Mere and Seamere would support an anthropogenic interpretation as a cause of the elm decline, in East Anglia at least. The mere at Hockham has a very small catchment area (*c* 1,300 ha) (Mosby 1935). Mosby (1935) states the mere to be spring fed but since such springs would flow through chalk and hence be of a calcareous nature one would expect a similar sediment to be laid down as is found at the other sites (a calcareous marl). This is not the case and it is suggested that either the springs have no connection with the chalk or that the mere is fed only by atmospheric precipitation. If the latter is the case then the mere would be very sensitive to variation in the precipitation:evaporation ratio. This sensitivity can be seen in the main pollen profile. The absence of any change in sediment lithology over the elm decline would suggest that no great change in the precipitation:evaporation ratio. This reduces the effectiveness of involving a major climatic shift in East Anglia at that time. The effect of man would thus seem to be more likely. The presence of cereal pollen would also suggest that man was cultivating the ground around Hockham Mere. There is, in the main pollen profile from Hockham Mere, evidence for a change in the precipitation:evaporation ratio just prior to the elm decline. This has yet to be radiocarbon dated.

The evidence for a climatic cause of the elm decline is often based on the changes in the pollen record of *Hedera*, *Ulmus*, *Ilex*, and *Viscum* (Iversen 1944). Again this is referring to the first elm decline seen in Danish pollen diagrams. In the Hockham Mere and Seamere diagrams there is a transitory peak of *Hedera* at the elm decline, and, at Seamere, this is accompanied by a peak of *Viscum*. Shortly afterwards these peaks die away. It is felt that a climatic implication is hard to apply here and hence the agency of man is again invoked, in a manner proposed below.

The species of *Ulmus* concerned has important implications concerning a climatic cause. *U. carpinifolia* is a more warmth demanding tree than *U. glabra*. Stockmarr (1970) has shown that, in Denmark, the *Ulmus* population in zone VII was predominantly *U. glabra*. If this is true for East Anglia then a change of climate would have to be greater than if *U. carpinifolia* were involved.

One major point for invoking a climatic cause is the fact that it would seem that it is *Ulmus* alone of the forest trees that suffers at the elm decline. However, from the absolute pollen frequency diagrams at Hockham and Seamere it is seen that the decrease of *Ulmus* is accompanied by a reduction in the frequency of a number of other forest trees, notably *Quercus*. The decrease of these trees amounts to 60% of their former contribution to the pollen rain. The different behaviour of *Fraxinus* at both sites would also make a climatic interpretation more difficult to accept. At Hockham Mere it is an important member of the pre-elm-decline forest but at Seamere it does not form an important part of the forest until the elm decline, when, at Hockham, it decreases in importance.

From these facts it is felt that climate cannot be involved as the main cause of the elm decline and its associated 'landnam' in East Anglia. However, reservation must be placed such as not to exclude it from having a synergistic effect with man's activities. This theme will be developed below.

Assuming man to be the causal agency of the elm decline then the sequence of events, referring mainly to Hockham Mere, may be as follows: The elm decline is caused by the clearing of an area, or areas, of woodland within the pollen catchment area of the site. This clearing was created by the cutting of young trees and allowing these to dry. The older trees were killed by ringing the bark and not by felling. After a period of time had elapsed, to allow for the drying of the timber, the area was burnt. This cleared the area and at the same time allowed the nutrients bound up in the vegetation to be returned to the soil. This sequence of events can be discerned from the Hockham pollen profile. The initial felling and ringing of trees is marked by the elm decline and the burning stage by the reduction of the colonising woodlands growing up between the initial felling and firing of timber (Iversen 1956).

Following the burning of these areas, which may have been restricted to certain parts, the remainder being left for the feeding of livestock and for future use as arable land, cultivation was carried out until soil nutrient exhaustion set in. Other areas were then used, the previous ones being left for pasturage. This sequence continued and eventually the areas of reduced nutrients would reach a certain critical level. A predominately pastoral economy was then practised until the peoples moved away from the area, or changed their effects on the vegetation.

This sequence can be followed on the pollen diagram from Hockham Mere. The gradual regeneration of woodland occurs from the end of the burning stage. This might allow the suggestion of restrictions being imposed on the movement of livestock as, if unrestricted, they would greatly impair the chances of forest regeneration. Cultivation can be seen to occur during this period. The feeding of livestock is by the collection of leaves for fodder (Troels-Smith 1960), and these used during the winter months. This can be seen from the non-regeneration of *Ulmus*, *Fraxinus*, and *Hedera*, the species usually taken for fodder (Troels-Smith 1960, Heybroek 1963). This fodder collection only ceased when the people moved from the area. It is suggested that the two phases of decreased woodland cover at Hockham prior to the elm decline could have involved the use of these species for fodder, especially *Ulmus*, as well as the felling of trees to provide timber for their temporary habitations. These would be on a much smaller scale than occurred at the elm decline.

In summary, the suggested sequence of events recorded in the pollen profile over the elm decline would fit into a cutting or ringing of timber followed by clearing by fire; a period of cultivation and pasturage; a period of predominate pasturage; and, finally, the regeneration of secondary woodland, consequent to the emigration of the people.

The definitive proof of man, in the stratified presence of his artifacts, is missing but the same event at a number of other sites has been tied in with early Neolithic cultures (Shippea Hill – Clark & Godwin 1962, Newferry – Smith 1964). It is felt, therefore, that the above explanation is reasonable and provides the best fit with the available evidence, at least with our present knowledge.

TEMPORAL DIMENSIONS OF THE ELM DECLINE

The next point to discuss is the time scale applied to the elm decline. Table 2 sets out a tentative time scale for the events. The occupation phase is estimated to have lasted for about 230 C^{14} years, from the elm decline to the end of the major peak in herb pollen. However, within this the regeneration of woodland starts after 70–80 years and continues throughout. The initial clearance took less than 25 years to complete. The scrub clearance was probably undertaken throughout the period of occupation. The main, initial phase took about 70 years to complete. During this initial phase pasturage seemed to predominate. The main cultivation and pasturage phase lasted about 100 C^{14} years, with pasturage predominating towards the end.

The scale of events at Hockham is much larger than that naturally envisaged for the 'landnam' of about 50 years (Iversen 1956), and is more like that presented by Pilcher et al. (1971) from Irish sites, that lasted for some centuries with differing phases of importance in cultivation and pastoralism.

SPATIAL DIMENSIONS OF THE ELM DECLINE

The areal extent of the clearances associated with the elm decline is another feature under discussion. The following is an attempt to obtain some idea of the magnitude of the possible area involved, there being no other such quantitative estimates available. The traditional qualitative view is of a fairly restricted area of clearance.

The former mere at Hockham is about 1 km in diameter. The sampling point is 300 m from the nearest shore. From the work of Tauber (1965) and Berglund (1973) the pollen catchment is estimated to be within a 10 km radius from the site. This gives a pollen source area of about 300 km^2. A second assumption is that the cumulative percentage of pollen represented in the diagram increases exponentially with distance reaching 90% of total pollen at 10 km. Fitting an equation to this curve in the form of $P(x) = 100(1-e^{-\alpha x})$ [x = distance, P = % total pollen, α = constant] gives the data described below. It is realised that the figures are only an approximation to the real area involved but it is believed that they will probably be out by less than an order of magnitude.

From this equation, a 60% decrease in tree pollen concentration at Hockham would represent a decrease of tree pollen from an area of at least 36 km^2. This would be an area immediately surrounding the mere. If the area was further from the mere then it must be increased such that at 500 m from the mere the area will be about 80 km^2, reaching 300 km^2 at a 2 km radius from the mere.

If there were extensive clearings involving fire within the catchment area at Hockham then two changes could be expected to be observed in the lithology of the mere. The first would be the presence of charcoal fragments. The second would be some evidence of mineral inwash from the exposed soil such as is seen in the Lake District (Pennington 1970). In the Lake District, at Barfield Tarn, there is a change in lithology, from an organic lake mud to a clay-mud, at the same time as the pollen stratigraphy shows a marked elm decline. This is interpreted as resulting from inwashing of the exposed soil from the cleared area. There is no evidence for this or for charcoal at Hockham Mere; the former would certainly be expected if there were any great exposure of soil within the catchment area as the lake seems to be largely fed by run off from the slopes. If this is so, and support might be taken from the absence of charcoal fragments, then the cleared area would be much larger or could result from a series of smaller areas being cleared within the pollen catchment area. Such a suggestion has been hinted at independently by Turner (1964).

However, the apparent absence of mineral inwash and of charcoal is surprising if these people did settle by the meres in the region, and two reasons can be put forward as a possible explanation for the absence of any changes in the sediment lithology. The first is that the slopes of the catchment area are too shallow to allow much soil erosion due to rain and sheet wash and any that did occur would have been trapped by the fringing fen before it reached the centre of the mere. One would expect, however, to observe microscopic charcoal fragments. The second possibility is that there was a spatial separation of crops, livestock, and settlement. The last named would be near the mere felling only a sufficient area for habitation. The area for cultivation and/or pasturage would be further away. This would be suggested if the interpretation placed on the elm decline presented above is correct. As the soil nutrients became exhausted other areas would be cleared of scrub woodland and these would be successively further from the habitation site.

There is no estimate of population size required to create a clearance of this size. The following is an attempt to suggest possible sizes of population required, and is thus liable to wide confidence limits. Using the figures presented by Clark and Piggott (1970) a population of 150 persons would require about 1 km^2 of cereal cultivation per year. If the cereal crop returns failed after the second year (Iversen 1956) then with the figure of 160 C^{14} years settlement cultivation, a population of 150 persons would require the clearing of some 80 km^2 of land. This gives some idea of the area of land required. Thus with a minimum cleared area of 36 km^2 a minimum population size of about 70 persons is envisaged for the site. However, it is strange that there is no archaeological evidence for such a habitation in the Hockham area.

HYPOTHETICAL SEQUENCE OF EVENTS OVER THE ELM DECLINE

On accepting the role of man as the cause of the elm decline in East Anglia and possibly with the elm decline associated with the 'landnam' or similar phase in northwest Europe another problem emerges. This is the apparent synchroneity of radiocarbon data for the elm decline. A synthesis of all available dates shows that these range from 5,800–4,500 C^{14} years B.P. with the bulk between 5,400–4,800 C^{14} years B.P., with the mode at 5,100 C^{14} years B.P. Along with this synchroneity there appears to be no consistent trend in the radiocarbon dates from a central source area as one would expect if it were due to migrations of people. This certainly provides a stumbling block for the acceptance of man as the agent of the elm decline and 'landnam'. On application of the Suess radiocarbon calibration curve (Suess 1970) it is seen that within the span of radiocarbon dates there are two or three possible corrected dates for each radiocarbon date. This would conceal any trend from a central source, providing the Suess curve is valid in this form.

It is of interest to note at this point the results of the application of radiocarbon dating to archaeological sites. Clark (1965) has made a synthesis of these dates and has grouped them into three units. The first, with an age greater than 5,200 C^{14} years B.C., is found in Turkey and the Near East. The second, with an age between 4,000–5,200 C^{14} years B.C., occurs in central Europe, and the third with ages between 2,800 and 4,000 C^{14} years B.C. in north-west Europe and southern France. If these are taken as the earliest farming settlements, they thus reflect a similar spread of dates as is seen at the elm decline. The main spread of dates in the third group is from 3,400–2,800 C^{14} years B.C., a very close parallel to the spread of elm decline dates. These include settlements at Hembury, Shippea Hill, and Windmill Hill. These settlements suggest a nucleated population, a point which will be discussed below. This supports the anthropogenic interpretation which has been placed on the 'landnam' and that early Neolithic peoples were concerned. The argument developed below is in accordance with this view.

Even with the Suess curve the synchroneity of the elm decline and 'landnam' is still striking. It raises the question as to the cause of man clearing the forest at this time. It must be noticed here that the next event at which man has anything like the effect of the elm decline and 'landnam' at Hockham, occurs at about 4,400 C^{14} years B.P. This is also seen in other diagrams from Britain and is usually ascribed to late Neolithic people clearing the woodland. However, this clearance is of a less temporary nature in that the mature forest is never allowed to return in extent as it does after the 'landnam'. There is, therefore, a gap of 200–300 years where the activity of man is limited to very small localised clearings such as are seen at Hockham and Seamere prior to the elm decline. This raises yet another question as to why this should be so.

It has been shown that from Mesolithic times man was able to create, at Hockham, small localised clearings for his temporary encampments and any livestock that he might have had. Why, then, should he, at one particular moment in time, have a relatively great effect on the vegetation, thereafter subsiding to his original level? A number of hypotheses can be put forward to explain this. This is done below but only one of these is developed in detail. The first hypothesis assumes incipient agricultural knowledge to be present throughout the people of north-west Europe before the time of the elm decline. This might be possible as the Danubian culture of central Europe was established before this, reaching Holland between 4470 and 4060 B.C. The Danubian culture was a farming culture and it might be possible for its knowledge to have spread over the remainder of Europe in the time before the elm decline. If there was incipient agriculture in Britain, especially in East Anglia, at that time a threshold must be passed in order for it to become important in the lives of the peoples concerned. This threshold could be a reduction in food reserves of some kind, such as hunted animals. This, however, would seem unlikely since the mature forest ecosystem would have a fairly stable animal population even allowing for a certain amount of exploitation by man.

The population density of the peoples could have risen and forced a threshold to be crossed when hunting could no longer be the mainstay of the food requirements of the culture. This would force farming to become more important in order to feed the extra population. There is, however, the problem of the farming intensity returning to its pre-elm decline level. Soil nutrient depletion could provide an answer to this in that, after a certain time, the area could no longer support the increased population and groups would have to break off from the original population and move elsewhere. This would necessitate the presence of a stable population centre similar to a village or township which was occupied for about 150–200 years, in the case of the Hockham area.

When the crop yield was diminished the population returned to keeping livestock and hunting from this centre, and, finally, moved away to another site. In fact it might be possible to see the gradual spread and utilisation of the farming technique bringing a migrant low density population group to form a more stable high density population group as the farming technique would allow the initial population group to support larger numbers and natural extension of this would eventually provide for a non-migrant population of fairly high density. The failure of the crop, a thing not envisaged in a migrant population of low density moving on after each year, forced a splitting up of this group and a return to a nomadic life with farming still playing a part but based on this low population size of the group. This would have a far smaller effect on the vegetation and could be represented in the same way as is seen at Hockham and Seamere before the elm decline. One advantage of an explanation such as this is that there need be no visible migratory trend as the population groups would expand individually from the inception of agriculture. In other words there could have been migration of the farming habit from a central region, but it is not this which is registered by the elm decline rather the culmination of this habit.

There remains the problem of the apparent synchroneity of the elm decline and perhaps here a climatic cause can be invoked. This is not to say that the elm decline and climatic shift are synchronous. It is suggested that the latter occurred first. The change in direction would be in the direction suggested by Frenzel (1966) towards a colder phase at c 3400–3000 B.C. Pennington (1973) has evidence for higher precipitation in Scotland at about 3000 B.C. There is evidence of a change in climate at Hockham before the elm decline and this would be about 5,500 C^{14} years B.P. continuing, with decreasing emphasis, until 5,300 C^{14} years B.P. (These are estimated dates and are liable to correction when further radiocarbon dates become available). This evidence takes the form of lowered lake levels, suggesting a lowered precipitation/evaporation ratio. The elm decline occurs just above this horizon.

If there was a deterioration in climate at about 5,400 C^{14} years B.P. then this would force the population from its original centre if conditions there became adverse and intolerable. A change towards a cooler, more continental climate would be felt more in central Europe than in Britain or in Denmark. Frenzel (1966) also suggests a cold phase between 5500–4500 B.C., and this would be contemporaneous with the spread of the Danubian farming culture to the Central European plain.

This enforced migration of the farming culture from its original area would continue until other, more suitable, areas were found. This migration would be likely to consist of a number of low density population units migrating each year, possibly after harvesting any cereals cultivated.

As the farming habit can support a greater population density than a hunting habit, in a given area, when more suitable environments were met, the population size would then increase. As this happened there would be a change from a nomadic existence to one where a stable community was built up, the change occurring when a certain population size was reached. The population may then have cleared a large area and cultivated the land. However, they then encountered the problem of soil nutrient exhaustion occurring in their cultivated plots after a year or two. This is a problem not generally associated with a nomadic culture. When this occurs the population created more cleared areas for cultivation. However, there is a limit to the extent of land which can be cultivated away from a single central population unit. Hence the population must either move on in bulk or as groups. The latter is probably the more likely in that groups would split off as the cereal yield diminished and these groups returned to a nomadic way of life, but still using cereal cultivation. This mode of life would remain nomadic until the immigration of late Neolithic culture into the region.

This hypothesis does not require there to be absolute synchroneity of radiocarbon dates but with a climatic initiation one would expect it to be broadly synchronous. This hypothesis does not demand the elm decline to mark the immigration of the Neolithic culture into north-west Europe but rather the attainment of the actual maximum population density. This, therefore, overcomes the very rapid immigration rate required if the elm decline was synchronous with the immigration.

In summary, the sequence of events suggested here consists of the following: (1) the spread of an incipient farming culture, which was initiated by a change in climate, (2) the population units then increased in density until a threshold was reached and forced them to become non-migrant population units, and it was this that caused the elm decline and 'landnam', (3) the exhaustion of the soil caused the gradual break up of this high density population unit and a return to a nomadic farming existence, and (4) this is continued until the late Neolithic migration some centuries later.

This is an hypothesis and, as such, is liable to have a limited approximation to the true sequence of events. It is one which, it is believed, is a closer fit than any that has been put forward previously, as it is in the light of new evidence provided from these investigations. In evaluating its closeness to the absolute true sequence of events, notice must be taken of the earlier references to the limits of the systems used. However, stimulation of discussion on this subject, towards the closest fit to the true sequence, is one of the aims of this paper.

A.P.F. DETERMINATIONS AND THE ELM DECLINE

One of the principal aims was the use of A.P.F. determinations in aiding in the elucidation of this problem. The evidence is portrayed and must be judged by those who wish to use the technique. The use of A.P.F. determinations has proved useful in the British Late-Devensian (Pennington & Bonny 1970) and in North American (Davis 1969). Its major use has been in the elucidation of major changes in pollen influx, such as occur at the end of the Late-Devensian. It has yet to be fully applied to Flandrian pollen diagrams from the British Isles and hence this approach is exploratory and is an attempt to determine if it will be of use in contributing to our understanding of the vegetational changes that occurred in the Flandrian. In the limited stratigraphical context of the elm decline it is believed to have given useful information which has led to the compilation of the hypothesis presented above.

The main points which became evident from the A.P.F. determinations are as follows:
(i) the reduction of arboreal pollen influx at the elm decline,
(ii) the parallelism of *Corylus* with the arboreal pollen,
(iii) highlighting the colonisation of *Corylus* and *Betula* after the elm decline,
(iv) the reduction in pollen-influx of other tree taxa as well as *Ulmus*, e.g. *Quercus*, and *Fraxinus*, and
(v) it has allowed some indication of the extent of the clearings to be made.

Mention must be made, however, of the overall similarity of the relative pollen frequency diagrams to the A.P.F. diagrams.

Acknowledgements

I should like to express my appreciation to all those who have assisted in the formulation of this paper in word and deed. My thanks are especially due to the late Dr Johs. Iversen for most profitable discussions at the D.G.U. in 1971, to Dr R.G. West for his courageous supervision of this research, to Dr V.R. Switsur for the radiocarbon dates, and to Dr Judith Turner for introducing me to the subject.

References

Berglund B.E. (1973) Pollen dispersal and deposition in an area of southeastern Sweden—some preliminary results. (this volume).

Clark J.D.G. (1965) Radiocarbon Dating and the Spread of Farming Economy. *Antiquity* 39, 45-8.

Clark J.D.G. & Godwin H. (1962) The Neolithic in the Cambridgeshire Fens. *Antiquity* 36, 10-23.

Clark J.D.G. & Piggott S. (1970) *Prehistoric Societies*. Pelican. London.

Clarke R. Rainbird (1960) *Ancient Peoples and Places: East Anglia*. Thames and Hudson, London.

Davis M.B. (1969) Climatic Changes in Southern Connecticut Recorded by Pollen Deposition at Rogers Lake. *Ecology* 50, 409-22.

Faegri K. & Iversen J. (1964) *Textbook of Pollen Analysis*. Blackwell Scientific Publications, Oxford.

Frenzel B. (1966) Climatic change in the Atlantic/Sub-Boreal transition on the Northern Hemisphere: botanical evidence. In *World Climate from 8000 to 0 B.C.* (ed. by J.S. Sawyer), pp. 89-123. Royal Meteorological Society, London.

Godwin H. (1956) *The History of the British Flora*. Cambridge University Press, London.

Godwin H. (1968) Studies of the Post-Glacial history of British Vegetation. XV Organic deposits of Old Buckenham Mere, Norfolk. *New Phytol.* 67, 95-107.

Godwin H. & Tallantire P.A. (1951) Studies in the post-glacial history of British vegetation. XII Hockham Mere, Norfolk. *J. Ecol.* 39, 285-307.

Heim J. (1963) Recherches sur les rélations autre la végétation actuelle et le spectre pollinique récent dans les Ardennes Belge. *Bull. Soc. r. Bot. Belg.* 96, 5-52.

Heybroek H.M. (1963) Diseases and lopping for fodder as possible causes of a Prehistoric decline of *Ulmus*. *Acta. bot. neerl.* 12, 1-11.

Hibbert F.A., Switsur V.R. & West R.G. (1971) Radiocarbon dating of Flandrian pollen zones at Red Moss, Lancashire. *Proc. R. Soc. B.* 177, 161-76.

Iversen J. (1941) Landnam i Danmarks Stenalder. *Danm. geol. Unders.* Ser. II, 66, 68p.

Iversen J. (1944) *Viscum*, *Hedera* and *Ilex* as climatic indicators. *Geol. För. Stockh. Förh.* 66, 463-83.

Iversen J. (1956) Forest Clearance in the Stone Age. *Scient. Am.* 194, 36-41.

Iversen J. (1960) Problems of the early Post-glacial forest development in Denmark. *Danm. geol. Unders.* Ser. IV, 4(3), 32p.

Mitchell G.F. (1956) Post-boreal Pollen Diagrams from Irish raised bogs. *Proc. R. Ir. Acad. B* 57, 185-251.

Mosby J.E.G. (1935) Hockham Mere. *Trans. Norfolk Norwich Nat. Soc.* 14, 61-7.

Oldfield F. (1963) Pollen analysis and man's role in the ecological history of the south-east Lake District. *Geogr. Annlr.* 45, 23-40.

Pennington W. (1970) Vegetation history in the north-west of England: a regional synthesis. In *Studies in the vegetational history of the British Isles* (ed. by D. Walker & R.G. West), pp. 41-50. Cambridge University Press, London.

Pennington W. (1973) Absolute pollen frequencies in the sediments of lakes of different morphometry. (this volume).

Pennington W. & Bonny A.P. (1970) Absolute Pollen Diagram from the British Late-Glacial. *Nature, Lond.* 226, 871-73.

Pilcher J.R., Smith A.G., Pearson G.W. & Crowder A. (1971) Land Clearance in the Irish Neolithic: New Evidence and Interpretation. *Science, N.Y.* 172, 560-2.

Smith A.G. (1961) The Atlantic Sub-Boreal Transition. *Proc. Linn. Soc. Lond.* 172, 38-49.

Smith A.G. (1964) Problems in the study of the earliest agriculture in northern Ireland. *Rep. VI Intern. Congr. Quater. Warsaw* 1961, Sect. 2, pp. 461-71.

Smith A.G. (1970) The influence of Mesolithic and Neolithic man on British vegetation: a discussion. In *Studies in the vegetational history of the British Isles* (ed. by D. Walker & R.G. West), pp. 81-96. Cambridge University Press, London.

Smith A.G. & Willis E.H. (1962) Radiocarbon dating of the Fallahogy landnam phase. *Ulster J. Archaeol.* 24-25, 16-24.

Stockmarr J. (1970) Species identification of *Ulmus* pollen. *Danm. geol. Unders.* Ser. IV, 4(11), 19p.

Suess H.E. (1970) Bristle-cone pine calibration of the radiocarbon time-scale 5200 B.C. to the present. In *Radiocarbon Variations and Absolute Chronology* (ed. by I.U. Olsson), pp. 303-11. Almqvist & Wiksell, Stockholm.

Tauber H. (1965) Differential pollen dispersion and the interpretation of pollen diagrams. *Danm. geol. Unders.* Ser. II, 89, 69p.

Troels-Smith J. (1954) Ertebøllekultur-bondekultur. *Aarb. Nord. Oldkynd. Hist.1953*, 1-62.

Troels-Smith J. (1955) Karakterisering af løse jordater. *Danm. geol. Unders.* Ser. IV, 3(10), 73p.

Troels-Smith J. (1960) Ivy, Mistletoe and Elm: Climatic indicator-fodder plants. *Danm. geol. Unders.* Ser. IV, 4(4), 32p.

Turner J. (1964) Surface sample analysis from Ayrshire, Scotland. *Pollen Spores* 6, 583-92.

Watts W.A. (1961) Post-Atlantic forests in Ireland. *Proc. Linn. Soc. Lond.* 172, 33-8.

A comparison of the present and past mire communities of Central Europe

KAMIL RYBNÍČEK *Botanical Institute of the Czechoslovak Academy of Sciences, Brno*

Introduction

The plant communities of aquatic and mire habitats have a distinctive position amongst other components of the vegetational cover. The majority of the natural dry-land communities in Central Europe are forest communities, and are chiefly affected by climatic factors. In contrast the water and mire stands are influenced and conditioned primarily by edaphic and hydrological factors. The rate of supply and trophic level of the water are decisive factors in influencing the composition of these communities, whereas climate generally plays a secondary role only. The influence of climate is indirect—it is mediated in most cases through the hydrological environment. Another important difference between these two groups of ecosystems is the character of the energy flow: dry-land communities are characterised by a high rate of solar energy conversion, i.e. by a high rate of turnover of organic matter produced in which individual elements return to the energy cycling, whereas aquatic and mire communities show essentially a retardation of energy flow, i.e. most of the matter produced is accumulated and only a negligible part is returned to the system by turnover.

This specific property of peat-forming plant communities offers a unique opportunity to study, with relatively direct methods, the processes of community development, successional trends and, to a considerable degree, the species composition of former peat-forming communities. The study of species composition is not only of practical value, for example in the establishment of peat properties of different layers, but also (especially for phytocenologists) of considerable theoretical interest. Unfortunately, we are still at the beginning of our researches in this field and our present knowledge is very fragmentary. Nevertheless, this study attempts to summarise the present knowledge on species composition of the principal peat-forming phytocenoses in the past and to discuss their distribution in time and space within Central Europe, and to compare them with the peat-forming plant communities of the present day.

Methodological problems

Although this paper is called 'A comparison of the present and past mire communities', it should be borne in mind that the comparison can be only partial because of the different and unequal character of the material compared. The study of present vegetation can be based on complete species analyses of chosen plots within more or less homogeneous parts of stands, and the area of the plots usually corresponds to the minimal area (in our case, an area of several sq m). On the other hand, the study of past mire communities can only be based on analyses of one or, under the most favourable circumstances, of several points in space obtained at random from the total area of the past community and representing a few sq cm at the most. This areal restriction is only slightly overcome by the fact that the sample analysed is of some vertical thickness, revealing something like a temporal projection of the floristic relations found in that small segment of vegetation. This apparent advantage, however, turns out, in some cases at least, to be disadvantageous when the sample has been taken from a cenotic complex, i.e. from a plant community of mosaic character.

A further problem in the reconstruction of past mire communities is the well-known fact that the macroscopic remains of some plants are not preserved or distinguishable in limnic or peat sediments. In this case, pollen analyses of the same sediments can sometimes be of assistance, as the finds of pollen grains can supplement, to a certain degree, the picture of the plant assemblages of the past communities. This is, for example, the case with *Utricularia*, *Drosera*, *Parnassia*, etc. (cf. the list of the main pollen types of local origin in Rybníčková & Rybníček 1971). In the present study this approach is used.

In the reconstruction of past mire communities another factor should also be taken into consideration, namely the unreliability of the quantitative relations between various members of the past plant assemblage. The abundance of seeds of a particular species does not necessarily mean that the species in question dominated in the respective plant community. Sediment often contains unripe, sterile seeds, produced by plants of low vitality. On the other hand, the seeds may be absent because they were very well developed and most of them germinated under suitable habitat conditions. Another problem in the comparison, especially of aquatic communities, is the possibility of seed transportation and their allochthonous origin in the sediment. All these problems are discussed in detail by Birks (1973).

We return to the expression of the degree of comparability between present and past mire communities. From the above discussion it follows that a full species composition with meaningful quantitative relations between individual members of the present community can be compared only with partial species composition with limited knowledge of quantitative interrelations in the case of subfossil communities.

In this situation a reasonably accurate comparison can be based on the phytocenological methodology widely used in Central Europe. This approach has the following important features:

1 It emphasises the presence of the so-called 'syntaxonomic indicators' or their groups in the communities as well as the full floristic composition of the stands.
2 The quantitative estimation of the occurrence of the species in the community need not always be decisive for the determination of the community.
3 When, owing to the established (although not always satisfactory) syntaxonomy, and, owing to the partial vegetational synthesis, we know comparatively well the cenotic affinities of the plants and their grouping into plant communities (Vergesellschaftung), it is possible to employ (to some degree at least) this knowledge in the reconstruction of past mire communities.

Survey of higher syntaxonomic units of the present peat-forming communities

Although the use of the Central Europe methodological approach to the reconstruction of past communities has considerable advantages, it has also one, fortunately only formal, drawback, viz. the chaotic nomenclature and classification of established units (cf. Koch 1926, Nordhagen 1936, 1943, Tüxen 1937, Hadač 1939, Schwickerath 1940, Duvigneaud 1949, Du Rietz 1954, Klika 1955, 1958, Oberdorfer 1957, Holub et al. 1967, Oberdorfer et al. 1967, Malmer 1968, Moore 1968). The necessary nomenclatural revision and a thorough synthesis of European mire communities cannot be attempted here. It is, however, possible to offer at least an approximate syntaxonomic outline of the units to be used in the following discussion and to give the necessary characterisation of at least the higher units in order to clarify the meaning and content of the terminology employed.

This characterisation and the system used here (see Table 1) is based on partial syntheses of the present mire communities of Central Europe. Wherever possible the moss constituents of the species assemblages have been used here in the floristic determination and delimitation of higher units, because mosses are much more specific indicators of habitat conditions than the vascular plants; that is, they are very reliable syntaxonomic indicators. The groups of syntaxonomic indicators by which the units are defined comprise not only the so-called 'character' and 'differential' species, but also constant companions. The selection of mire plants, listed in Table 1, is obviously not complete; the species which can be determined in a subfossil state are primarily considered, while from the remainder only the most significant taxa have been included. Table 1 also shows the usual cenotic affinities and cenotic amplitude of the species concerned.

Past communities and their comparison with the present ones

The material presented here concerning the past mire communities comes mostly from Czechoslovakia, but data available from other countries of Central Europe, especially from Poland, have also been considered. It does not represent all subfossil mire assemblages studied up to now; only those past assemblages are discussed here which were found at least in two localities. They may be thus regarded as representatives of more or less stabilised homogeneous communities distributed over a considerable area. This selection may suggest that heterogeneous transitional types have been excluded.

The floristic composition and the typification of past mire communities are presented in Table 2. The types established here are of unequal syntaxonomic character. Some of them are broad collective units, which will probably divide into several smaller units after further investigation; others, more or less, have the same rank as associations. All the data have been synthesised and, in Table 2, only the constancies of their occurrence are given for individual taxa. To make the comparison of past mire assemblages with the present cenotic conditions easier, the taxa are grouped, whenever possible, in a comparable way to the arrangement in Table 1.

The following notes should give more detailed explanation to Table 2 and provide the necessary phytocenological commentary on the individual types of past mire assemblages.

(i) The first group of subfossil plant communities is characterised by predominating floating-leaved plants, i.e. euhydatophytes and hydatoaerophytes *sensu* Hejný (1957), especially by the representatives of the genera *Potamogeton* and *Nymphaea*. These plant communities are easy to classify and can be identified with present associations of the alliance Potamion. The subfossil plant assemblages of this unit are divided into the following types.

Batrachium-Potamogeton subfoss. comm. In Czechoslovakia it is found at the base of mire profiles in non-calcareous regions. According to the present habitat requirements of the species (see Table 2, Column 1) we can suggest that it was restricted to shallow non-calcareous waters. The presence of some swamp taxa, such as *Menyanthes*, *Equisetum*, *Calliergon giganteum*, etc., can be explained by the transport of dead plant remains in water. The community (or communities) closely resemble some of the present subalpine and subarctic water plant communities. One plant assemblage from the Bohemian-Moravian Uplands (see Rybníček & Rybníčková 1968, Table 2, Columns 1, 2) included in this type with dominant *Potamogeton filiformis* can be identified with the present association Potametum filiformis Koch 1928. This assemblage could probably be established as a separate type of past mire community after further investigation.

Najas marina-Ceratophyllum demersum subfoss. comm. Unlike the preceding type, the existence of this plant community appears to have depended on water rich in calcium. The same species combination (see Table 2, Column 2) occurs in the present community recorded by Nedelcu (1970) from Roumania under the name Najadetum marinae.

As a rule, the subfossil communities of the alliance Nymphaeion can be hardly distinguished from the communities of the Potamion. Their differentiation can cause difficulties even today. Subfossil assemblages which are related to the present communities of this alliance are mainly characterised by the predominance of *Nymphaea* and *Nuphar* seeds. The communities with *Trapa natans*, whose nuts are found in European sediments, may also belong to this alliance. Unfortunately, more detailed species analyses are not available.

(ii) Another group of subfossil communities is characterised mainly by the dominance of *Phragmites communis* remains, sometimes with seeds of *Scirpus lacustris*, and of some other less significant indicators. The total number of species in this past assemblage is very low. All these features suggest that the subfossil communities can be included in the complex of the present communities of the alliance Phragmition. The following types can be recognised.

Phragmites communis-Scirpus lacustris subfoss. comm. This is very rare in Czechoslovakia, as the environmental conditions for its existence, i.e. suitable stable water-bodies, were lacking. (For the species composition see Table 2, Column 3.) However, it is abundant in sediments of Poland and northwest Germany. As suggested by its species composition, it is closely analogous to the present Scirpo-Phragmitetum Koch 1926.

Cladium mariscus subfoss. comm. This has not been analysed in Czechoslovakia. Its probable existence is known only from a find of *Cladium mariscus* in one layer of the profile (probably no longer existing) of a mire near Bělohrad Spa, Czechoslovakia (Purkyně 1925). However, the past community is again adequately documented by finds in northwest Germany and Poland. It is characterised by the predominance of *Cladium*, while most of the other species are represented sporadically (see Table 2, Column 4). The past floristic composition suggests that subfossil *Cladium mariscus* stands can be fully identified with the present Cladietum marisci Zobrist 1935.

(iii) The following (third) group of subfossil communities is closely analogous to some present associations within the all. Magnocaricion. These communities are usually characterised by the dominance of one species, in this case of one of the large sedges. The occurrence of past communities of this group in Czechoslovakia has not been supported by much evidence; the reasons are the same as in Phragmition, namely a lack of suitable habitats. In Poland, however, their former presence has been fully demonstrated. Two types have so far been established—one with *Carex acutiformis* and the other with *Carex rostrata*. This does not mean that other types do not exist. On the basis of the analogy with the present peat-forming communities, Tołpa *et al.* (1967) established Magnocaricioni-peat, formed by *Carex elata*, *Carex paniculata*, or *Carex gracilis*. These authors, however, did not publish their full macrofossil analyses.

Carex acutiformis subfoss. comm. This is practically defined by one dominating species only—*Carex acutiformis*. (For other species, see Table 2, Column 5). Present communities with a similar species composition were described as Caricetum acutiformis Sauer 1937. The present vegetation of this plant community has been recorded several times, but usually from the north eastern part of Central Europe (cf. Kobendza 1930, Jeschke 1959, etc.).

Carex rostrata-Menyanthes trifoliata subfoss. comm. For species composition see Table 2, Column 6. All the species mentioned there (*Carex rostrata*, *Menyanthes trifoliata*, *Lycopus europaeus*, *Eleocharis palustris*, etc.) form the present association Caricetum rostratae Rübel 1912, or Caricetum rostrato-vesicariae W. Koch 1926. Only one analysis comes from Czechoslovakia. Even now *Carex rostrata* has rather different cenotic bounds in this country, as its optimum lies within the communities of Scheuchzerio-Caricetea fuscae.

(iv) A distinctive group of past communities is represented by alder-swamp forests. Their peat-forming communities formed distinct layers in the Flandrian profiles of Central Europe. The assemblages of macroscopic remains are extremely uniform and the past alder communities can be clearly classified into the present higher units—Alnion and Alnetalia glutinosae. According to the two leading and constant macroscopic remains, most of the known subfossil assemblages can be denoted as:

Table 1. Floristic characterisation of the higher phytosociological units of present European mire communities.

	Class	A		B		C	D	E						F						G
	Order	I		II		III	IV	V		VI		VII		VIII	IX	X				XI
	Alliance	1	2	3	4	5	6	7	8	9	10	11	12	13a	13b	14	15	16	17	18	19	20

C,O: Potametea, Potametalia
- *Characeae* — +, + (cols 1,2); + (col 6); + (col 9)
- *Potamogeton natans* L. — +, +
- *Najas marina* L. — +, +
- *Najas minor* All. — +, +
- *Ceratophyllum demersum* L. — +, +
- *Ceratophyllum submersum* L. — +, +
- *Batrachium* spec. div. — +, +

A: Potamion
- *Potamogeton perfoliatus* L. — +
- *Potamogeton obtusifolius* Mert. & Koch — + −
- *Potamogeton pectinatus* L. — +
- *Potamogeton gramineus* L. — +
- *Potamogeton filiformis* Pers. — +
- *Potamogeton alpinus* Balb. — +
- *Potamogeton lucens* L. — +
- *Potamogeton praelongus* Wulf. — +

A: Nymphaeion
- *Zannichellia palustris* L. — +
- *Nymphaea candida* Presl — −, +
- *Nymphaea alba* L. — −, +   (− in later col)
- *Nuphar lutea* (L.) Sm. — +
- *Myriophyllum verticillatum* L. — −, +
- *Myriophyllum spicatum* L. — +
- *Trapa natans* L. — +
- *Utricularia vulgaris* L. — +   ; + (col 9)
- *Polygonum amphibium* L. — −, +

C,O: Phragmitetea, Phragmitetalia
- *Equisetum fluviatile* L. — +, +, +, −, +, −, −, +, −
- *Sparganium emersum* Rehm. — +, +, +
- *Alisma plantago-aquatica* L. — +, +
- *Iris pseudacorus* L. — +, +, +

240 K. Rybníček

Table 1. (continued)

Class	A		B		C	D	E							F						G
Order	I		II		III	IV	V		VI			VII		VIII	IX	X				XI
Alliance	1	2	3	4	5	6	7	8	9	10	11	12	13a	13b	14	15	16	17	18	19	20

C,O: Phragmitetea, Phragmitetalia (continued)
    Ranunculus lingua L.   + +
    Lycopus europaeus L.   + + + − + − − − − −
    Lythrum salicaria L.   − + −
    Rumex hydrolapathum Huds.   + + +

A: Phragmition
    Phragmites communis Trin.   − − + − + + − − − + + −   +
    Typha latifolia L.   + −
    Typha angustifolia L.   + −
    Glyceria maxima (Hartm.) Holmberg   + −
    Cladium mariscus (L.) Pohl.   + − −
    Scirpus lacustris (L.) Palla   +
    Hippuris vulgaris L.   − − + − − −

A: Magnocaricion
    Drepanocladus aduncus (Hedw.) Mönkem.   + −
    Carex acutiformis Ehrh.   + +
    Carex elata All.   + −
    Carex appropinquata Schum.   + − +
    Carex acuta L.   + +
    Carex vesicaria L.   + +
    Carex rostrata Stokes   + + + + + + + − +   +
    Carex pseudocyperus L.   + +
    Carex paniculata L.   +
    Carex aquatilis Wahlenb.   + −
    Eleocharis palustris (L.) Roem. & Schult.   + − −
    Naumburgia thyrsiflora (L.) Rchb.   + − + + + −
    Lysimachia vulgaris L.   + + + − − − − −
    Scutellaria galericulata L.   + + + − −
    Galium palustre L.   + + − + − − − −
    Bidens cernua L.   − +
    Bidens tripartita L.   − +

Table 1. (continued)

Class		A	B	C	D	E				F		G
Order		I	II	III	IV	V	VI	VII	VIII	IX	X	XI
Alliance		1 2	3 4	5	6 7	8 9	10 11 12	13a 13b	14	15	16 17 18 19	20

C,O,A: Alnetea glutinosae, Alnetalia glutinosae, Alnion glutinosae

Species	1	2	3	4	5	6	7	8	9	10	11	12	13a	13b	14	15	16	17	18	19	20
Mnium spec. div.			−		+	−									−						
Sphagnum fimbriatum Wils.					+																
Sphagnum squarrosum Pers.					+																−
Dryopteris austriaca agg.					+																−
Thelypteris palustris Schott					+																−
Calamagrostis canescens (Weber) Roth			−	−	+																
Carex elongata L.					+																
Scirpus silvaticus L.					+								−	+							
Calla palustris L.					+	−															
Filipendula ulmaria (L.) Maxim.			−		+			−							−	+					−
Urtica dioica L.					+																
Cirsium palustre (L.) Scop.			−		+	−				−					−						
Rubus idaeus L.					+																
Ribes nigrum L.					+																
Salix spec. div.			−		+			−							−	−					+
Frangula alnus Mill.					+			−							−	−					+
Alnus glutinosa (L.) Gaertn.			−	−	+										−	−					−

C,O: Montio-Cardaminetea, Montio-Cardaminetalia

Species	1	2	3	4	5	6	7	8	9	10	11	12	13a	13b	14	15	16	17	18	19	20
Philonotis fontana (Hedw.) Brid.				+	+								+								
Caltha palustris L.				+	+	+	+						−	+	+						−
Myosotis palustris agg.				−	+	+							−								
Glyceria fluitans (L.) R.Br.				−	−	+	−						−								

A: Cardamino-Montion

Species	1	2	3	4	5	6	7	8	9	10	11	12	13a	13b	14	15	16	17	18	19	20
Brachythecium spec. div.					+																
Cardamine amara L.					+																
Chrysosplenium alternifolium L.			+		+																
Stellaria alsine Grimm			−		+										−						
Montia fontana agg.					+										−						

A: Cratoneurion commutati

Species	1	2	3	4	5	6	7	8	9	10	11	12	13a	13b	14	15	16	17	18	19	20
Philonotis calcarea Schimp.				+									−								
Cratoneurum commutatum (Hedw.) Roth				+									+								
Pinguicula alpina L.				+																	

Table 1. (continued)

	Class	A		B		C	D		E								F						G
	Order	I		II		III	IV		V		VI			VII		VIII	IX	X					XI
	Alliance	1	2	3	4	5	6	7	8	9	10	11	12	13a	13b	14	15	16	17	18	19	20
C: Scheuchzerio-Caricetea fuscae																							
Bryum pseudotriquetrum (Hedw.) Schwaegr.						−	+	−	+	+	+	+	+	+	+							
Polytrichum strictum Sm.									+	+	+	−	+	+	−		−	−	−	−	−	
Equisetum palustre L.						−	−		+	−	+	−	+	+	+							−
Carex panicea L.							−		+	+	+	+	+	+	+							
Carex nigra (L.) Reich.				−	+	+			+	+	+	+	+	+	+	−	−					−
Carex echinata Murr.						−	+		+	+	+	+	+	+	+	−						+
Eriophorum angustifolium Honck.							−		+	+	+	+	+	+	+	+	+					−
Menyanthes trifoliata L.			−	−		+	+		+	−	+	−	+	+	+	+						
Comarum palustre L.							−		+	−	+	−	+	+	+							
Valeriana dioica L. vel *simplicifolia* Kabat									+	+	+	+	+	+	+							
Galium uliginosum L.							−		+	+	+	+	+	+	+							
O: Sphagno-Caricetalia lasiocarpae																							
Sphagnum subsecundum Nees									+	+	+	−	−	−	−							
Sphagnum inundatum Russ.							−		+	−	+	−										
Sphagnum platyphyllum Lindb.									+	+	+											
Sphagnum contortum Schultz									+	+	+	−	−	−								
A: Caricion lasiocarpae																							
Meesia triquetra (Hook. & Tayl.) Ångstr.									+		−			−								
Cinclidium stygium Sw.									+		−											
Bryum ovatum Jur.									+													
Calliergon giganteum (Schimp.) Kindb.						+			+		+											
Drepanocladus exannulatus (Br., Sch. & Gmb.) Warnst.									+		+											
Sphagnum obtusum Warnst.									+													
Carex lasiocarpa Ehrh.							−		+	+	+	−	−	+		−						−
Carex chordorrhiza Ehrh.									+		+		−	+								
Carex diandra Schrank							−		+		−		+	−								
Eriophorum gracile Roth.									+		−											
Ranunculus flammula L.							−		+					+	−							
A: Rhynchosporion albae																							
Drepanocladus vernicosus (Lindb.) Warnst.									+	+												
Lycopodium inundatum L.									+	−						−						

Table 1. (continued)

	Class	A	B	C	D	E						F			G
	Order	I	II	III	IV	V	VI		VII	VIII	IX	X			XI
	Alliance	1	2	3	4	5	6	7	8	9	10	11	12	13a	13b	14	15	16	17	18	19	20

A: Rhynchosporion albae (continued)

	1	2	3	4	5	6	7	8	9	10	11	12	13a	13b	14	15	16	17	18	19	20
Rhynchospora alba (L.) Vahl.								−	+	+		−	−			+	−	+	−		
Rhynchospora fusca (L.) Ait.f.									+	+						−					
Drosera intermedia Hayne								−	+	−						−					
Drosera anglica Huds.								−	+	−						−					

O: Tofieldietalia

	1	2	3	4	5	6	7	8	9	10	11	12	13a	13b	14	15	16	17	18	19	20
Scorpidium scorpioides (Hedw.) Limpr.			−						−	+	+	−									
Campylium stellatum (Hedw.) Lange & C. Jens.									+	+	+	−									
Drepanocladus revolvens (Turn.) Warnst.									−	+	+	+									
Fissidens adiantoides Hedw.									−	−	+	+	−								
Equisetum variegatum Web. & Mohr											+	+									
Carex flava L.									−	−	+	+	+	−	−						
Carex dioica L.										+	+	+									
Eriophorum latifolium Hoppe									−	+	+	+	−								
Eleocharis quinqueflora (F.X. Hartmann) Schwarz									−	+	+										
Triglochin palustris L.										+	+	−									
Utricularia minor L.									−	−	+	+									
Utricularia intermedia Hayne									−	−	+	+									
Parnassia palustris L.									−	−	+	+	+	−	−						
Linum catharticum L.									−	−	+	+	+	−							
Potentilla erecta (L.) Räusch.	−							−	+	+	+	+	−								−
Pedicularis palustris L.									−	−	+	+	−								

A: Caricion demissae

	1	2	3	4	5	6	7	8	9	10	11	12	13a	13b	14	15	16	17	18	19	20
Calliergon trifarium (Web. & Mohr) Kindb.									−	+	−										
Calliergon sarmentosum (Wahlenb.) Kindb.										+											
Carex pulicaris L.									−	+	−	+									
Carex demissa Hornem.									+	+	−	−	−								
Trichophorum alpinum (L.) Pers.									+	+		+									

A: Caricion davallianae

	1	2	3	4	5	6	7	8	9	10	11	12	13a	13b	14	15	16	17	18	19	20
Drepanocladus intermedius (Br., Sch. & Gmb.) Mönkem.f.							−			+											
Drepanocladus sendtneri (Schimp.) Warnst.							−			+											
Drepanocladus lycopodioides (Schwaegr.) Warnst.									−	+											

Table 1. (continued)

	Class	A		B		C	D		E						F					G
	Order	I		II		III	IV		V		VI		VII		VIII	IX	X			XI
	Alliance	1	2	3	4	5	6	7	8	9	10	11	12	13a	13b	14	15	16	17	18	19	20

A: Caricion davallianae (continued)
    Scorpidium turgescens (Th. Jens.) Mönkem.   − +
    Carex davalliana Sm.   + −
    Carex lepidocarpa Tausch   +
    Carex distans L.   +
    Carex hostiana DC.   +
    Schoenus ferrugineus L.   +
    Schoenus nigricans L.   +
    Blysmus compressus (L.) Link   − +
    Tofieldia calyculata (L.) Wahlenb.   − + −
    Pinguicula vulgaris L.   − − + −
    Primula farinosa L.   +

A: Sphagno-Tomenthypnion
    Helodium blandowii (Web. & Mohr) Warnst.   +
    Tomenthypnum nitens (Hedw.) Loeske   − − +
    Paludella squarrosa (Hedw.) Brid.   + +
    Sphagnum warnstorfianum Du Rietz   +
    Epipactis palustris (L.) Crantz   − +

O,A: Caricetalia fuscae, Caricion canescentis-fuscae
    Carex canescens L.   + + − −
    Juncus filiformis L.   + +
    Viola palustris L.   − − − + + − −
    Epilobium palustre L.   − + − − + + +
    Stellaria palustris Retz.   − − − − + +

SA: Sphagno-Caricion canescentis
    Calliergon stramineum (Brid.) Kindb.   − + − − − − − −
    Sphagnum amblyphyllum Russ.   + − + + ? ? ? ? ? ?
    Sphagnum apiculatum H. Lindb.   − − + − − − − + + + +
    Sphagnum teres (Schimp.) Ångstr.   + + +
    Peucedanum palustre (L.) Moench   + + − − + −

Table 1. (continued)

	Class	A		B		C	D		E						F						G
	Order	I		II		III	IV		V		VI			VII		VIII	IX	X				XI
	Alliance	1	2	3	4	5	6	7	8	9	10	11	12	13a	13b	14	15	16	17	18	19	20
SA: Ranunculo-Caricion fuscae																						
Calliergon cuspidatum Kindb.										−	−		−	−	+						
Climacium dendroides (Hedw.) Web. & Mohr															+						
Juncus effusus L.															+						
Juncus conglomeratus L.															+						
Juncus articulatus L.															+						
C: Oxycocco-Sphagnetea																						
Aulacomnium palustre (Hedw.) Schwaegr.								−	−	+	+					+	+	+	+	+	−
Sphagnum nemoreum Scop.										−	−					−	+	+	+	+	+
Drosera rotundifolia L.							−	+	+	−	−					+	+	+	+	+	−	−
Calluna vulgaris (L.) Hull																+	+	+	+	+	−
O,A: Scheuchzerietalia, Scheuchzerion																						
Drepanocladus fluitans (Hedw.) Warnst.										−			−		+						
Sphagnum cuspidatum Schimp.															+						
Sphagnum lindbergii Schimp.															+						
Sphagnum dusenii Russ. & Warnst.									−						+						
Sphagnum pulchrum (Lindb.) Warnst.															+						
Sphagnum tenellum Pers.															+		−	−			
Carex limosa L.								+	−				−		+						
Scheuchzeria palustris L.									−				−		+						
O: Ericetalia tetralicis																						
Sphagnum compactum DC.																+						
Juncus squarrosus L.															−	+	−				
A: Ericion tetralicis																						
Erica tetralix L.																+	+				
Narthecium ossifragum (L.) Huds.																+	+				
Trichophorum caespitosum (L.) Hartm.												+				+	+				

Table 1. (continued)

	Class	A		B		C	D	E							F						G
	Order	I		II		III	IV	V		VI			VII		VIII	IX	X				XI
	Alliance	1	2	3	4	5	6	7	8	9	10	11	12	13a	13b	14	15	16	17	18	19	20

O: Sphagnetalia
    Eriophorum vaginatum L.

																−	−	−	+	+	+	+	+

    Andromeda polifolia L.   −   −   +   +   +   +   −
    Oxycoccus quadripetalus Gilib   −   −   −   +   +   +   +   +

A: Sphagnion atlanticum
    Sphagnum papillosum Lindb.   −   −   +   −
    Sphagnum imbricatum Russ.   −   −   +

A: Sphagnion continentale
    Sphagnum fuscum (Schimp.) Klinggr.   −   +   +   +
    Sphagnum magellanicum Brid.   −   +   +   +
    Sphagnum rubellum Wils.   −   +   +   +
    Carex pauciflora Lightf.   +   −

A: Oxycocco-Empetrion hermaphroditi
    Rubus chamaemorus L.   +
    Empetrum hermaphroditum Hagerup   +
    Oxycoccus microcarpus Turcz.   −   +   +
    Betula nana L.   −   −   +   −   +

A: Pino-Ledion
    Sphagnum angustifolium C. Jens.   +
    Chamaedaphne calyculata (L.) Moench. (E. Eur.)   +
    Vaccinium uliginosum L.   +
    Ledum palustre L.   −   +
    Pinus rotundata Link   −   +
    Pinus silvestris L.   −   −   +   −

C: Vaccinio-Piceetea
    Vaccinium myrtillus L.   −   −   +   +   +   +
    Vaccinium vitis-idaea L.   +   +   +   +
    Picea abies (L.) Karst.   −   −   +

Table 1. (continued)

	Class	A	B		C	D	E							F						G
	Order	I	II		III	IV	V	VI			VII			VIII	IX	X				XI
	Alliance	1	2	3	4	5	6	7	8	9	10	11	12	13a	13b	14	15	16	17	18	19	20

O: Vaccinio-Piceetalia	
Sphagnum girgensohnii Russ.	+
Equisetum silvaticum L.	+
Calamagrostis villosa (Chaix) Gmel.	+
Trientalis europaea L.	+
A: Betulion pubescentis	
Polytrichum commune Hedw.	— — — +
Molinia coerulea (L.) Moench.	— — — +
Betula pubescens Ehrh.	— — — +

Explanation to Table 1: C, Class; O, Order; A, Alliance; SA, Suballiance.

A. Class: Potametea Tx. et Prsg. 1942
 I. Order: Potametalia W. Koch 1926
  1. Alliance: Potamion eurosibiricum (W. Koch) Oberd. 1957
  2. Alliance: Nymphaeion Oberd. 1957

B. Class: Phragmitetea Tx. et Prsg. 1942
 II. Order: Phragmitetalia W. Koch 1926
  3. Alliance: Phragmition W. Koch 1926
  4. Alliance: Magnocaricion W. Koch 1926

C. Class: Alnetea glutinosae Br.-Bl. et Tx. 1943
 III. Order: Alnetalia glutinosae Tx. 1937
  5. Alliance: Alnion glutinosae (Malc.) Meyer Drees 1936

D. Class: Montio-Cardaminetea Br.-Bl. et Tx. 1943
 IV. Order: Montio-Cardaminetalia Pawl. 1928
  6. Alliance: Cardamino-Montion Br.-Bl. 1925
  7. Alliance: Cratoneurion commutati W. Koch 1928

E. Class: Scheuchzerio-Caricetea fuscae Nordh. 1936 ex Tx. 1937
 V. Order: Sphagno (subsecundi) Caricetalia lasiocarpae, nom. nov.
  8. Alliance: Caricion lasiocarpae Vanden Berghen in Lebrun et al. 1949
  9. Alliance: Rhynchosporion albae W. Koch 1926

VI. Order: Tofieldietalia Prsg. apud Oberd. 1949
  10. Alliance: Caricion demissae Rybníček 1964
  11. Alliance: Caricion davallianae Klika 1934
  12. Alliance: Sphagno-Tomenthypnion Dahl 1957
VII. Order: Caricetalia fuscae W. Koch 1926
  13. Alliance: Caricion canescentis-fuscae Nordh. 1937
   13a. Suball: Sphagno-caricion canescentis Passarge 1964
   13b. Suball: Ranunculo-Caricion fuscae Passarge 1964

F. Class: Oxycocco-Sphagnetea Br.-Bl. et Tx. 1943
 VIII. Order: Scheuchzerietalia Nordh. 1936
  14. Alliance: Scheuchzerion Nordh. 1936
 IX. Order: Ericetalia tetralicis Moore 1964
  15. Alliance: Ericion tetralicis Schwick. 1933
 X. Order: Sphagnetalia Pawl. 1928
  16. Alliance: Sphagnion atlanticum Schwick. 1940
  17. Alliance: Sphagnion continentale Schwick. 1940
  18. Alliance: Oxycocco-Empetrion hermaphroditi Nordh. 1936
  19. Alliance: Pino-Ledion Tx. 1955

G. Class: Vaccinio-Piceetea Br.-Bl. et Tx. 1955
 XI. Order: Vaccinio-Piceetalia Br.-Bl. 1939
  20. Alliance: Betulion pubescentis Lohm. et Tx. in Tx. 1955

Table 2. Synthetic table of past Central European mire communities.

	Name of past community	Batrachium-Potamogeton	Najas marina-Ceratophyllum dem.	Phragmites-Scirpus lacustris	Cladium mariscus	Carex acutiformis	Carex rostrata-Menyanthes	Alnus glutinosa-Rubus	Carex lasiocarpa-Sphagna sect. subsecunda-Meesia triq.	Carex chordorrhiza-Sphagna sect. subsecunda	Carex diandra-Meesia triquetra	Carex rostrata-Calliergon giganteum	Carex lasiocarpa-Scorpidium scorpioides-Calliergon trifar.	Carex rostrata-Scorpidium scorp.	Carex diandra-Drepanocladus sendtneri	Paludella squarrosa-Helodium blandowii	Carex echinata-Sphagna sect. cuspidata	Carex echinata-Climacium dendroides	Sphagnum cf. cuspidatum-Drepanocladus fluitans	Rhynchospora alba-Sphagnum cf. cuspidatum	Eriophorum vaginatum-Sphagnum magellanicum
	Column number	1	2	3	4	5	6	7	8	9	10	11	12	13	14	15	16	17	18	19	20
	No. of analyses	3	5	8	4	3	7	7	8	2	5	4	5	5	3	6	3	5	8	6	6
	Alliance Number	1	1	3	3	4	4	5	8	8	8	8	10	11	11	12	13a	13b	16	16	17
P	Sphagnum								II	1		3				III	3				
T,P	Equisetum	3	I	II				V	IV	2	II	4	I	I	2	II		IV			
T,S	Carex	1		II	3	3	V	V	V	2	V	4	II	II	2	I	3	II	I	III	
P	Galium	2	II				I	IV	1		II	4	III		2	IV		IV		I	
P	Epilobium						III	I				2			2	I					
S	Potamogeton natans	1	III				I								II	I					
S	Najas marina		V																		
S	Najas minor		II																		
S	Ceratophyllum demersum		IV				I														
S	Batrachium sp.	3					I														
S	Potamogeton obtusifolius	3	II				I														
S	Potamogeton praelongus	2	II																		
S	Potamogeton gramineus	2														I					
S	Potamogeton lucens		II																		
S	Nymphaea		II	I		I										I					
S	Nuphar luteum		II																		
P	Myriophyllum spicatum	1	III	I											I	2					

Table 2. (continued)

	Column number	1	2	3	4	5	6	7	8	9	10	11	12	13	14	15	16	17	18	19	20
	No. of analyses	3	5	8	4	3	7	7	8	2	5	4	5	5	3	6	3	5	8	6	6
	Alliance Number	1	1	3	3	4	4	5	8	8	8	8	10	11	11	12	13a	13b	16	16	17
S,P	*Lycopus*			II		I	II	II			I					2					
P	*Sparganium-Typha*			II			I	I													
T	*Phragmites communis*		II	V	4	2	II	II	V		I	1	I		1	IV					I
S	*Scirpus lacustris*			III																	
T,S	*Cladium mariscus*						4														
T	*Drepanocladus aduncus* agg.						III							III	1						
S	*Carex acutiformis*			I	1	3	III							I							
S	*Carex elata*			I			II														
S	*Eleocharis palustris* agg.			II	1	1	III				II										
S,P	*Lysimachia vulgaris*			II			I	II	II		I										
S	*Carex pseudocyperus*			I	I			III					I								
S	*Carex rostrata*	1	III	II				V	II	1	I	4		V	3	V	1	IV			
P	*Polypodiaceae*			III	2				V												
S	*Scirpus silvaticus*								III		I								II		
S,P	*Filipendula ulmaria*			II					V	IV	2	1	II			II		IV			
S	*Urtica dioica*						I		III												
S	*Rubus* sp.								V		I							1	I		
S	*Cirsium* sp.								II				II					III			
P	*Frangula*			I					II												
T,S,(P)	*Salix*							III	III	II		2	I	I		III					
T,S	*Alnus glutinosa*			I				I	V	II						I					

Table 2. (continued)

		1	2	3	4	5	6	7	8	9	10	11	12	13	14	15	16	17	18	19	20
	Column number	1	2	3	4	5	6	7	8	9	10	11	12	13	14	15	16	17	18	19	20
	No. of analyses	3	5	8	4	3	7	7	8	2	5	4	5	5	3	6	3	5	8	6	6
	Alliance Number	1	1	3	3	4	4	5	8	8	8	8	10	11	11	12	13a	13b	16	16	17
T	*Philonotis* sp.										II							III			
S	*Caltha palustris*						I	I			II	I									
S	*Myosotis palustris* agg.						I											III			
S	*Stellaria alsine*						I				I							II			
S	*Carex nigra*	I		I			I			I	III	3		III		I	3	IV			
S	*Carex echinata*								III		IV		III			I	3	V			
S,P	*Menyanthes trifoliata*	2	I	II	I	I	V		IV	I	III	I	V	III	3	II		I	II	I	
S	*Comarum palustre*							II	III	2			IV	III				I			
S,P	*Valeriana*							III		I	II							III			
T	*Sphagna* sect. *subsecunda*								IV	I								II			
T	*Meesia triquetra*	I		I			II		IV	I	V	I		I	I						
T	*Calliergon giganteum*		3									4	III	II		II		I			
S	*Carex lasiocarpa*								V	I	I		V		I					I	
S	*Carex chordorrhiza*									2			II								
S	*Carex diandra*	I		I	I			III	II		V							II			
S	*Ranunculus flammula*	I									I				I			I	IV		
S	*Rhynchospora alba*													II						V	
T	*Scorpidium scorpioides*	3											V	V							
T	*Drepanocladus revolvens*	I	I										I		3		I				
S	*Carex* sect. *flava*												I	II				III			
S	*Potentilla erecta*							III					I			I		III			
T	*Calliergon trifarium*	I	I										V	I	I						

Table 2. (continued)

	Column number	1	2	3	4	5	6	7	8	9	10	11	12	13	14	15	16	17	18	19	20
	No. of analyses	3	5	8	4	3	7	7	8	2	5	4	5	5	3	6	3	5	8	6	6
	Alliance Number	1	1	3	3	4	4	5	8	8	8	8	10	11	11	12	13a	13b	16	16	17
T	*Drepanocladus sendtneri*		I		I				II					II	3						
T	*Helodium blandowii*															IV					
T	*Tomenthypnum nitens*		I						I							III					
T	*Paludella squarrosa*															V					
S	*Carex* cf. *canescens*					I	I		III	2							1	III			
S	*Viola palustris*						I		1	II							2	II			
S	*Stellaria palustris*					I	I		I	1	II					I		I			
T	*Calliergon stramineum*																		III		I
T	*Sphagnum* cf. *recurvum* agg.			I										2					I		IV
T	*Sphagnum teres*						II							II							
T	*Calliergon cuspidatum*						I		I	I						III					
T	*Climacium dendroides*									III						III					
S	*Ranunculus* cf. *acris*								III							II					
T	*Eriophorum vaginatum*								1					I					III	V	V
P	*Calluna vulgaris*													I							IV
S	*Andromeda polifolia*																			II	I
S,T,P,	*Oxycoccus quadripetalus*							I											I	I	V
T	*Drepanocladus fluitans*			I															V		
T	*Sphagnum* cf. *cuspidatum*																		IV	V	
T	*Sphagnum lindbergii*																		III		
S,T	*Carex limosa*	1							1			III		1	II				III	I	
S,T	*Scheuchzeria palustris*																		II	III	

Table 2. (continued)

Column number	1	2	3	4	5	6	7	8	9	10	11	12	13	14	15	16	17	18	19	20
No. of analyses	3	5	8	4	3	7	7	8	2	5	4	5	5	3	6	3	5	8	6	6
Alliance Number	1	1	3	3	4	4	5	8	8	8	8	10	11	11	12	13a	13b	16	16	17

		1	2	3	4	5	6	7	8	9	10	11	12	13	14	15	16	17	18	19	20
T	Sphagnum magellanicum																		III	II	V
T	Sphagnum sect. acutifolia																		I		III
T,S	Betula nana								2		V	I							I		
T	Pinus	1							II	1		I			I			I	I	V	II
T	Picea abies								III	I											I
S,T	Betula cf. pubescens	1							V												IV
	Betula			II		II	II					II	II		I				IV		
S	Lychnis flos-cuculi			I	1			II			III			III							
S	Carex riparia						II														
S	Ranunculus cf. repens							II													
S	Sambucus cf. nigra							III													
S	Prunella vulgaris										III										
T	Sphagnum cf. compactum																		II		
	Cenococcum geophilum								III							2					

In addition with low constancy (I) occurs:
In column 1: Potamogeton filiformis (S), Myriophyllum verticillatum (P), Nuphar pumila (S), Alisma plantago-aquatica (S,P), Hippuris vulgaris (S), Alopecurus aequalis vel geniculatus (S), Salix myrtilloides (T).
In col. 2: Potamogeton alpinus (S), Potamogeton pusillus (S), Rumex hydrolapathum (S). In col. 3: Alisma plantago-aquatica (S,P), Scutellaria galericulata (S), Bidens cernua (S), Bidens tripartita (S), Parnassia palustris (P), Drepanocladus intermedius (T), Utricularia (P), Scirpus tabernaemontani (S), Oenanthe aquatica (S), Cyperus flavescens (S), Carex cf. davalliana (S), Meesia hexasticha (T), Sphagnum palustre (T).
In col. 6: Bidens cernua (S), Bidens tripartita (S), Drepanocladus intermedius (T), Rumex cf. acetosa (S), Pedicularis palustris (S). In col. 7: Solanum dulcamara (S), Calla palustris (S), Glyceria cf. fluitans (S), Cardamine (S), Abies alba (T), Ajuga reptans (S), Galeopsis sp. (S), Impatiens noli-tangere (S), Ribes sp. (P), Cornus sanguinea (P), Eurhynchium sp. (T). In col. 9: Pedicularis palustris (S). In col. 10: Mnium cf. longirostre (T), Glyceria cf. fluitans (S), Juncus sp. (S), Linum catharticum (S,P), Pedicularis palustris (S), Drepanocladus cf. intermedius (T), Rumex cf. acetosa (S). In col. 11: Bryum pseudotriquetrum (T), Salix myrtilloides (T). In col. 12: Utricularia (P), Drepanocladus exannulatus (T). In col. 13: Bryum pseudotriquetrum (T), Drepanocladus exannulatus (T), Carex cf. vesicaria (S). In col. 14: Myriophyllum verticillatum (P). In col. 16: Carex cf. panicea (S). In col. 17: Mnium cf. longirostre (T), Glyceria cf. fluitans (S), Carex cf. panicea (S), Drosera (P), Cardamine (S), Aulacomnium palustre (T), Ajuga reptans (S). In col. 18: Molinia, Trichophorum caespitosum (T). In col. 19: Drosera (P), Vaccinium sp. In col. 20: Aulacomnium palustre (T).

Explanation to Table 2:
T: Tissue, including leaves, wood, bark, mosses, etc.
P: Pollen and spores.
S: Seeds, fruits, fruit-stones, nutlets, etc.
Constancy classes: I—1-20%, II—21-40%, III—41-60%, IV—61-80%, V—81-100%. Nos. 1, 2, 3, 4 express absolute number of occurrence.

Sources of data for Table 2 are given in the Appendix.

Alnus glutinosa-Rubus subfoss. *comm.* For species composition see Table 2, Column 7. A special and highly detailed work is devoted to the study of past alder communities by Marek (1965) and it can hardly be supplemented here by anything new. Marek introduces the following species as the leading macrofossils: *Alnus glutinosa*, *Rubus idaeus*, *Urtica dioica*, *Solanum dulcamara*, *Carex elongata*, *Carex pseudocyperus*, *Calla palustris*, *Lycopus europaeus*, and *Hottonia palustris*. On the basis of species comparison the author identifies the subfossil communities with the present Carici elongatae-Alnetum W. Koch 1926. The past alder stands usually succeed the communities of Phragmitetea.

The Czechoslovak analyses do not essentially differ from the situation ascertained in Poland. Higher constancy has been found in Czechoslovakia for *Picea excelsa*, for spores of Polypodiaceae, and for pollen grains of *Filipendula*, while *Carex elongata*, *Calla palustris*, and *Hottonia palustris* are less frequent or even absent. Notice should be drawn to a possible connection with the present Piceo-Alnetum Rubner 1954, at least at higher altitudes. Our past alder communities do not succeed the preceding peat-forming communities. The layers of alder peat analysed up to now are almost always deposited directly on the mineral substrate, and they are no older than the older Sub-Atlantic period. Their thickness is not very great, and progressive peat-forming cannot be expected to take place in them by natural development. When this occurred in the not-too distant past, it was probably due to the activities of man, to the felling of alder, and to his efforts to convert waterlogged areas into meadows.

(v) Unspecified fragments of spring vegetation and the adjacent spring zone constitute an independent group of past mire communities. In the subfossil state, the plants indicating alliance Cardamino-Montion (on Ca-poor habitats) are necessarily accompanied by many species of the surrounding vegetation. The sampling techniques largely exclude the possibility of obtaining an uncontaminated sample from a community occupying a very small area. One fragment (from the younger Dryas period) was reported by Rybníček and Rybníčková (1968, Table 2, Column 6b). The other fragment is subrecent, dating from the younger Sub-Atlantic period (Němčice PS-7-A, 28–35 cm). Its species composition is as follows: *Montia fontana* agg., *Stellaria alsine*, *Myosotis palustris* agg., *Ranunculus flammula*, *Carex* cf. *fusca*, *Carex echinata*, *Carex panicea*, *Carex* sect. *flavae*, *Carex rostrata*, *Viola palustris*, *Lychnis flos-cuculi*, *Equisetum*, *Mnium longirostre*, *Philonotis fontana*.

Subfossil communities of spring areas with Ca-rich water (alliance Cratoneurion commutati) have not yet been found. Owing to permanent erosion of the surface, the possibility of their finds in peat sections is very low. They may be found, however, in fossil tufa.

(vi) The group of subfossil plant communities, which are characterised in the moss layer by a combination of *Sphagna* sect. *subsecunda*, *Meesia triquetra*, etc., is closely analogous to the present communities of the alliance Caricion lasiocarpae. The vegetation is characterised by the presence of some species of the alliance Magnocaricion and Phragmition and succeed either the reedswamp communities or the 'brown-moss' communities. The subfossil assemblages with these features are very frequent in Central European mire profiles. The following four types can be differentiated among them.

Carex lasiocarpa-Sphagna sect. *subsecunda-Meesia triquetra* subfoss. *comm.* Communities of similar species composition (see Table 2, Column 8) exist even today. In most cases, however, *Meesia triquetra* is now absent. The phytocenoses with the basic dominants *Carex lasiocarpa* and some *Sphagna* sect. *subsecunda* have been described frequently under the name Caricetum lasiocarpae (cf. Schwickerath 1942, Poelt 1954, Larsson 1960, Görs 1961, Jeschke 1963, Spence 1964). In Czechoslovakia similar phytocenoses have been recently found in the Bohemian-Moravian Uplands, in the Doksy Basin, and in southern Bohemia.

Carex chordorrhiza-Sphagna sect. *subsecunda* subfoss. *comm.* The species composition (see Table 2, Column 9) is similar to that of the preceding type, but *Carex chordorrhiza* is substituted for *Carex lasiocarpa*, and *Betula nana* seems to be another significant species. At the present time, a number of phytocenoses are known which could be similar, but *Betula nana* is almost always absent. Relic communities of this type have been recorded from the Bohemian-Moravian Uplands, others are known from Germany (Görs 1961, Passarge 1964) and Poland (Mądalski 1930). Stands with *Betula nana* are reported from Norway by Nordhagen (1943, Tab. 90:40) but other stands with *Carex chordorrhiza* in Scandinavia and Scotland usually have a different character.

Carex diandra-Meesia triquetra subfoss. *comm.* This is characterised by high constancy of *Carex diandra*, *Carex echinata*, *Menyanthes trifoliata*, and *Meesia triquetra* (for further species see Table 2, Column 10). Analyses come from Poland, southern Moravia and southern Bohemia and the samples are all practically subrecent. Although their origin was affected by human activities (if not conditioned by them), *Carex diandra* in combination with *Meesia triquetra* does not occur frequently at the present day. The restriction of *Meesia triquetra* undoubtedly results from the extensive drainage of mires at the beginning of this century, and this may also have influenced the development of the present phytocenoses with *Carex diandra*.

Carex rostrata-Calliergon giganteum subfoss. *comm.* Analyses come from profiles in the Bohemian-Moravian Uplands and from the Třeboň Basin, where this community formed a distinct layer on the bases of peat deposits. For the species assemblage see Table 2, Column 11. Similar types of subfossil communities are known from Poland (Wąs 1965, Tables 7, 11). At present, the communities with the two dominants occur only rarely, and are considered as 'relics'. Stands with *Carex rostrata* and *Calliergon giganteum* can be found in fragments in the Bohemian-Moravian Uplands and in northeast Slovakia, whereas better developed types are reported by Poelt (1954, Table 5, Column 1) from Austria as Caricetum rostratae. This type of plant community is, in some respects, related to some of the waterlogged communities of the Tofieldietalia (see below).

(vii) A very important group among past mire communities is formed by types characterised by the presence of 'brown-mosses' such as *Scorpidium scorpioides, Calliergon trifarium, Drepanocladus revolvens* s.l., *D. lycopodioides, D. sendtneri*, etc. The present communities of this character belong to the order Tofieldietalia and they are further divided into three alliances (Caricion demissae, Caricion davallianae, and Sphagno-Tomenthypnion). They usually formed thick layers of peat in the past (cf. Früh & Schröter 1904, Rudolph 1917, 1928, Jasnowski 1957, 1959, etc.). Under present conditions, these communities, having lost their distinct peat-forming function, survive only as relics.

Carex lasiocarpa-Scorpidium scorpioides-Calliergon trifarium subfoss. comm. In Czechoslovakia, this type of past community was found in the Třeboň Basin, and one record comes from Austria. In the past it was definitely more frequent than now, although in a rather primitive moss form (see Früh & Schröter 1904, Jasnowski 1957, 1959, Karczmarz 1963, etc.), but a few suitable analyses are available. For the species composition see Table 2, Column 12. Similar communities are very rare now but they exist, although they lack *Betula nana*. In our country, such phytocenoses can be presently found in the Doksy Basin, in one locality in the Třeboň Basin and, only in fragments, in the Orava Basin in Slovakia. The communities of this type were described as Caricetum lasiocarpae by Koch (1926) and other relevés were published, for example by Poelt (1954). In Scandinavia, phytocenoses of similar composition are more frequent and resemble more the subfossil types (cf. e.g. Booberg 1930, Heikurainen 1953, Lounamaa 1961, Mornsjö 1969, etc.). The above-mentioned phytocenoses can be most probably classified within the alliance Caricion demissae.

The more calcitrophic habitats are indicated by the group of present communities within the alliance Caricion davallianae. From the subfossil types of the communities the only ones that can be reconstructed are those with very strong peat-accumulating ability and which simultaneously indicate more or less inundated sites. Such typical modern associations of the alliance (like Caricetum davallianae and Schoenetum nigricantis, with very low peat accumulation ability) have not, so far, been found in the older periods. In addition, Moravec and Rybníčková (1964) suggest that the development of Caricetum davallianae in its present composition was influenced by man's activities and it may not have existed earlier. The following types of past communities belong to the range of this alliance:

Carex rostrata-Scorpidium scorpioides subfoss. comm. For the species composition see Table 2, Column 13. Fruit stones of *Potamogeton* are probably of allochthonous origin. At present, the communities of this species composition are very rare. They are concentrated to Scandinavia (cf. Booberg 1930. Carex rostrata-Amblystegium scorpioides soc.) and Scotland (McVean & Ratcliffe 1962, Carex rostrata-brown moss *nodum*).

Carex diandra-Drepanocladus sendtneri subfoss. comm. Analyses of this type come from Poland and southern Moravia. Though they are relatively young (older and younger Sub-Atlantic period), the author does not know any community of this type at present. For the composition of the past assemblage see Table 2, Column 14.

It is interesting to note that subfossil finds of *Drepanocladus sendtneri* are relatively frequent (Hungary—Boros 1968, Poland—Wąs 1965, E. Germany—Lange in litt.); it often forms thick layers in Weichselian or Flandrian profiles. Today this moss is regarded as relatively rare.

This group of subfossil mire communities also includes types which were listed by Wąs (1965) in his collective tables. They are as follows: Caricetum rostrato-vesicariae with *Drepanocladus sendtneri*, Caricetum diandrae with *Calliergon trifarium*, and other past phytocenoses with dominating *Scorpidium scorpioides, Campylium stellatum*, and *Drepanocladus intermedius*.

Paludella squarrosa-Helodium blandowii subfoss. comm. The present phytocenoses which include such species as *Paludella squarrosa, Helodium blandowii, Tomenthypnum nitens*, and *Sphagnum warnstorfianum* in the moss layer belong to the all. Sphagno-Tomenthypnion. The communities of this alliance are now relatively frequent both in Central Europe and in Scandinavia, and they are also known from Scotland (Booberg 1930, Dahl 1957, Ruuhijärvi 1960, McVean & Ratcliffe 1962, etc.). The present stands can be divided into many associations or lower units, but they have not yet been satisfactorily evaluated from the phytocenological point of view. Our knowledge of past phytocenoses of this type is also limited and the results of the species analyses are summarised in Table 2, Column 15. Outside Czechoslovakia, the past assemblages of this type are also known from Poland (Wąs 1965). Certain doubts about the classification arise with regard to past assemblages published by Rudolph and Firbas (1927:92) from profiles from the Krknoše Mountains and mentioned also by Firbas (1927:172) from the Doksy Basin. *Carex limosa* or *C. rostrata* appear in them together with *Paludella*, but other syntaxonomic indicators of this group are absent. At present such communities indicate, according to Finnish mire typologists, the so-called *Paludella*-Moore (Cajander 1913, etc.). It seems, however, that they should be included in the group of mesotrophic types from the alliance Caricion lasiocarpae.

(viii) Considerable difficulties arise in the reconstruction of the oligotrophic types of mire communities. The present stands are often characterised by numerous species of various low sedges, by the absence of acidophilous bog plants, and by the dominance of *Sphagnum recurvum* agg. (*Sphagnum apiculatum*, *Sphagnum amblyphyllum*), *Sphagnum teres*, *Calliergon stramineum*, etc. in the case of peat-forming communities (suballiance Sphagno-Caricion canescentis). The communities with a low ability for peat accumulation have poorly developed moss layers with species such as *Climacium dendroides*, *Calliergonella cuspidata*, etc. (suballiance Ranunculo-Caricion fuscae). In Czechoslovakia, the present communities of the above-mentioned composition are mostly influenced by cultivation. Most of them are also very young, and they have originated after the removal of original alder forests during the successive waves of colonisation of upland regions in the 11th to 13th centuries. The peat-forming types are better preserved and more natural in the northeast part of Europe. According to the analyses which are at our disposal from Czechoslovakia, we can establish two types of past communities in this group.

Carex echinata-Sphagnum cf. *recurvum* agg. subfoss. comm. This probably comprises several units belonging to the suballiance Sphagno-Caricion canescentis. The past assemblage, besides the above-mentioned species, consists of *Viola palustris* and *Carex fusca*. For details see Table 2, Column 16. Parallels can be found in the present associations Carici canescentis-Agrostidetum R. Tx. 37, Sphagno-Caricetum rostratae Steffen 1930, and other similar units.

Carex echinata-Climacium dendroides subfoss. comm. This again is a collective type corresponding to a group of the present units of suballiance Ranunculo-Caricion fuscae. This type is obviously subrecent, and is very rich in species (see Table 2, Column 17). Its sediment is characterised by the prevalence of mineral particles. Similar stands are now very frequent in Czechoslovak wet meadows. From the cenotic point of view they have not, unfortunately, received sufficient attention and no concrete associations can be mentioned here; it should be noted, however, they are not very 'natural' communities.

(ix) A very important part of the present and past mire vegetation is represented by the communities of oligo- and dystrophic raised bogs, and of similar habitat types. Several types of past assemblages have been differentiated. Unfortunately, only three types of them can be described here. The others can be only mentioned briefly because of the lack of analyses.

There is no doubt that the first two types of past assemblages belong to the group of communities of raised-bog pools (Scheuchzerietalia).

Sphagnum cf. *cuspidatum-Drepanocladus fluitans* subfoss. comm. The species assemblage is characterised in Table 2, Column 18. Most of the analyses were performed some time ago by Rudolph and Firbas (1927) from the Krkonoše Mountains. They are supplemented by several analyses from northwest Germany (Jonas 1934). These past assemblages do not substantially differ from the plant communities which are now denoted as Caricetum limosae drepanocladetosum fluitantis Kästn. & Flöss. 33, Cuspidato-Scheuchzerietum (R. Tx. 37) Prsg. & R. Tx. 58, and others. Some Polish analyses may also be included in this group (see the subfoss. comm. Caricetum limosae ombro-sphagnetosum in Pacowski 1965).

Rhynchospora alba-Sphagnum cf. *cuspidatum* subfoss. comm. This assemblage can be established from sections in southern Bohemia. Besides the principal dominants it was formed by *Scheuchzeria palustris*, *Eriophorum vaginatum*, *Andromeda polifolia*, *Menyanthes trifoliata*, and nuts of some *Carex* spp. (see Table 2, Column 19). This species composition is closely analogous to the present stands denoted as Scheuchzerietum palustris R. Tx. 37, Rhynchosporetum sphagnetosum cuspidati Diem. & R. Tx. 37, or Sphagnum cuspidatum-Rhynchospora alba nodum in Rybníček (1970).

From other raised-bog communities (Sphagnion continentale) the following has been well preserved in Czechoslovakian peat layers:

Eriophorum vaginatum-Sphagnum magellanicum subfoss. comm. Besides the two dominants, *Calluna*, *Oxycoccus quadripetalus*, and *Sphagna* sect. *cuspidata* (*Sphagnum apiculatum*?) are constant, and the remains of *Pinus* and *Betula* cf. *pubescens* are found sporadically (Table 2, Column 20). Grosse-Brauckmann (1962) and Pacowski (1965) have published analyses from Germany and Poland, and have also suggested *Andromeda polifolia*, *Aulacomnium palustre* and *Sphagna* sect. *acutifolia* as constants. Past phytocenoses of this type correspond well to the present *Sphagnetum medii* Kästn. & Flöss. 1933, which can be considered as one of the most frequent raised-bog types in Central Europe today.

Other raised-bog subfossil communities cannot be discussed in detail. No suitable analyses are available from Czechoslovakia, and, as far as we know, they differ from the preceding type mainly by the different composition of the moss layer. However, data in the literature (Jonas 1934, Grosse-Brauckmann 1962, Pacowski 1965, Casparie 1969, etc.) suggest that the past and present types may be regarded as being similar. Thus, for example, the past assemblage of layers composed of *Sphagnum fuscum* corresponds to the species composition of the present Sphagnetum fusci Luq 1926. Layers composed mostly of *Sphagnum papillosum* were formed by a community analogous to the present association Sphagnetum papillosi Schwickerath 1940.

Special attention should be devoted to the consideration of the age and degree of autochthony of the present Central-European phytocenoses with *Pinus rotundata*, *P. silvestris*, and *Ledum palustre* (alliance Ledo-Pinion). Unfortunately, detailed macroscopic analyses are lacking, because the deep drainage of most of our localities in the latter half of the last century has resulted in the aeration of the upper layer of peat and in the decomposition of almost all the macrofossils. The only clue is provided by pollen analyses (Salaschek 1935, Jankovská 1967, Rybníček & Rybníčková 1968), all of which show a substantial and abrupt increase of the curves of *Pinus*, *Ledum* (if present), and also *Calluna* in the youngest (actually subrecent) periods.

At the present moment, difficulties also arise in the phytocenological evaluation of the birch wood layers of peat. The preliminary analyses so far available suggest that *Pinus silvestris* occurs regularly with *Betula* cf. *pubescens*, and *Phragmites*, *Equisetum*, *Filipendula*, and *Carex* sp. (*C. rostrata*?). *Sphagna* sect. *cuspidata* (*Sphagnum apiculatum*?) are constant among the mosses. In southern Bohemia, the corresponding layers date back to the Atlantic period, but their distribution in time and space is definitely not fully indicated by these data. At present, the communities of this character are limited to the laggs of raised bogs or 'pseudo-raised' bogs and they are included within the alliance Betulion pubescentis.

Conclusions

In conclusion, we can see that there exists a relatively high uniformity in the species composition, structure, and habitat requirements in the past and present mire communities of Central Europe. The past Late-Weichselian and Flandrian mire communities can be classified, perhaps in all cases, into higher phytocenological units (alliances, orders, and classes) based on analyses of, and established for, the present stands. In many cases it is possible to relate the past species assemblages even to present associations or other lower units. However, the floristic criteria, used in this, as in other phytocenological classifications, cannot be based only on the presence of one component of the assemblage, or on the presence of a few so-called 'character species'. They should be based ideally on the analyses of the whole species composition, and on the presence or absence of groups of syntaxonomic indicators, whenever the results of macrofossil analyses make this approach possible.

The causes of relative uniformity of the floristic similarity and structure of both the Late-Weichselian and Flandrian mire communities and the present ones can be explained by their specific synecology, already mentioned in the introduction. The following simplified generalisation may be formulated here: certain combinations of environmental factors give rise to communities of more or less similar species composition, depending naturally on phyto- or palaeophyto-geographical situations. Considering that the decisive factors which determine the existence and composition of mire communities in the past and in the present are very similar, namely the hydrologic factor (the chemical properties and amount of water), the formation and grouping (Vergesellschaftung) of the plants in past and present mire communities is also very similar. Past mire communities appear to be most similar to the present ones whenever the complex of environmental conditions which influence them is comparable. So it is natural that the modern analogues to Late-Weichselian or early Flandrian mire communities are obviously found in the north or at high altitudes, where the peat-forming process began much later, where waters of corresponding trophic levels can be found, and where perhaps comparable climatic or thermic conditions prevail today. The suggestion that the climatic conditions are not the main and direct factor determining the origin of mire communities is supported by the fact that even under the present climatic conditions these old communities exist, though rarely, as 'relics' in suitable habitats (e.g., the existence of relic stands of the *Carex lasiocarpa-Scorpidium scorpioides-Calliergon trifarium* community in Central Europe, cf. p. 255, etc.). The progressive restriction and shifting of the occurrence of some communities or groups of communities can be explained by the loss of suitable hydric conditions in some parts of a former area of distribution. In this way the changes in range of our Rhynchospora alba-Sphagnum cf. cuspidatum past assemblage may be explained. It seems to have been frequent in the mires of southern Bohemia in the Atlantic period. Today, after the natural accumulation of peat and associated decrease of water table it has disappeared from there, but it still exists on the raised and blanket bogs of the oceanic and sub-oceanic zones, where, due to higher humidity, suitable habitats have persisted. Another example is the restriction of the Najas marina-Ceratophyllum demersum past community, which was also apparently abundant in Central Europe during the Atlantic period, to the still open waters of the Balkans after the infilling of our lakes. These shifts cannot be regarded as only the direct results of climatic change, or as a result of climatic (in our case thermic) influence on the individual plants. Most of the species from the above-mentioned communities survive in Central Europe, but their cenotic affinities and combination are rather different.

The changes in cenotic affinities and the changing of habitat requirements of plant species over the course of time should be also mentioned. The habitat and cenotic variability of mire plants is of different extent. The individual properties of plant species certainly change in time. This problem has been discussed several times by various palaeoecologists (cf. Iversen 1964, Lang 1967, Janssen 1970, etc.). These changes seem to proceed in two main ways: as ecological divergence, i.e. as adaptability of plant taxa, and as ecological convergence, i.e. as specialisation of plant taxa. In the first case many examples among mire plants can be mentioned here. For example, *Carex lasiocarpa*, *Carex chordorrhiza*, and other plants were found only in eutrophic brown-moss communities in the Late-Weichselian or early Flandrian. These plants can be found (with the exception of extremely dystrophic habitats) in nearly all cenotic and trophic levels of mire vegetation today. For ecological divergence not so many examples exist. Nevertheless, an interesting instance was recorded by Katz and Katz (1964). *Scheuchzeria palustris* appears to have had a much wider habitat amplitude than it does today. Today it is a species growing distinctly in oligo-dystrophic mires. This is also probably the case of most of mire mosses, which now have very distinct cenotic and habitat requirements. In Central Europe this is also valid for *Betula nana*, but it is not true in Northern Europe.

Due to the different extents of cenotic and habitat requirements of the plants as well as of the plant communities, unusual, irregular and sometimes unexpected species combinations can often be found, especially on the borders between two or more vegetational types, or in places in which the natural equilibrium of habitat and cenotic conditions has been destroyed to a certain degree. These situations apparently also existed in the past and reconstruction of these types present many difficulties. However, it should be pointed out that in palaeoecology many 'unusual' past assemblages could simply result from the heterogeneity of the analysed peat material, from the possibility of transport of plant remains, and from the mixing of layers during sedimentation.

Finally, the distribution in time of the past mire communities should be discussed. The available data are presented in Fig. 1.

The oldest (Late-Weichselian and early Flandrian) mire communities are those of the order Tofieldietalia, which took part in the overgrowing of former swamps and fens abounding in water, for example Carex lasiocarpa-Scorpidium scorpioides-Calliergon trifarium, Carex rostrata-Scorpidium, etc., and the Paludella-Helodium blandowii past community of the alliance Sphagno-Tomenthypnion which usually succeed them. Carex rostrata-Calliergon giganteum past community develops only a little later; it opens the appearance of mesotrophic mire vegetation types of alliance Caricion lasiocarpae. The centre of their occurrence is in the Atlantic period. In Czechoslovakia, they include Carex lasiocarpa-Sphagna sect. subsecunda-Meesia triquetra community, and Carex chordorrhiza-Sphagna sect. subsecunda community, which as a rule form a transitional zone between the minerotrophic and ombrotrophic developmental phases in the sections. The main part of the present mire vegetation has developed since the older Atlantic period. The period was generally characterised by intensive filling processes of previously open waters, in

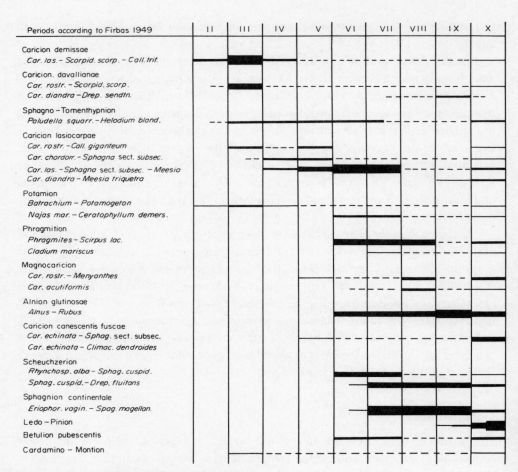

Figure 1. Time distribution of past mire communities.

which the sedimentation of various types of gyttja took place, and by intensive growth of mires, when better thermic conditions also enabled the spreading of new plants. It is the starting-point of most of the communities of Potamion and of the alliances Nymphaeion, Phragmition, Magnocaricion, and Alnion glutinosae. Only a little later, usually in regions with a lower nutrient supply and in which the accumulating peat reached the point when mineral water ceases to influence the vegetation (cf. Du Rietz 1949, 1954), the raised-bog communities of the class Oxycocco-Sphagnetea succeeded the meso- and oligotrophic communities of Caricion lasiocarpae. This radical and sudden change in the development of European mires seems to have taken place approximately at the same time, during the Atlantic period, throughout the whole of Europe (cf. Godwin 1956, Overbeck 1958, Ruuhijärvi 1963,

Casparie 1969), with the exception of the subarctic and subalpine regions, where among other factors the progressive oligotrophisation has not reached the critical point so far. No find of older raised-bog vegetation, not even in the interglacials, has been reported in Central Europe up to now (cf. also Katz & Katz 1964).

Most of the communities of Caricion canescentis-fuscae and of Ledo-Pinion are very young and some of them have resulted from (or at least have been influenced by) human activities.

Acknowledgements

The author would like to express his thanks to Dr H.J.B. Birks, University of Cambridge for several suggestions and for linguistic corrections of the manuscript.

References

BIRKS H.H. (1973) Modern macrofossil assemblages in lake sediments in Minnesota. (this volume).
BOOBERG G. (1930) Gisseläsmyren. *Morrland. Handbibl.* **12**, 329p.
BOROS Á. (1968) *Bryogeographie und Bryoflora Ungarns.* Akadémiai Kiadó, Budapest.
CAJANDER A.K. (1913) Studien über die Moore Finnlands. *Acta for. Fenn* **2**(3), 208p.
CASPARIE W.A. (1969) Bult- und Schlenkenbildung in Hochmoortorf. *Vegetatio* **19**, 146-80.
DAHL E. (1957) Rondane. Mountain vegetation in South Norway and its relation to the environment. *Skr. norske Vidensk-Akad. 1 Mat.-Naturv. Klasse.* 1956 (3), 374p.
DU RIETZ G.E. (1949) Huvudenheter och huvudgränser i svensk myrvegetation. *Svensk bot. Tidskr.* **43**, 274-309.
DU RIETZ G.E. (1954) Die Mineralbodenwasserzeigergrenze als Grundlage einer natürlichen Zweigliederung der Nord- und Mitteleuropäischen Moore. *Vegetatio* **5-6**, 571-85.
DUVIGNEAUD P. (1949) Classification phytosociologique des tourbiéres de l'Europe. *Bull. Soc. r. Bot. Belg.* **81**, 58-129.
FIRBAS F. (1927) Paläofloristische und stratigraphische Untersuchungen böhmischer Moore, IV. Die Geschichte der nordböhmischen Wälder und Moore seit der letzten Eiszeit (Untersuchungen im Polzengebiet). *Beih. bot. Zbl.,* B **43**, 145-206.
FIRBAS F. (1935) Die Vegetationsentwicklung der mitteleuropäischen Spätglazials. *Bibl. bot.* **112**, 1-68.
FIRBAS F., GRÜNIG G., WEISCHEDEL I. & WORZEL G. (1948) Beiträge zur spät- und nacheiszeitlichen Vegetationsgeschichte der Vogesen. *Bibl. bot.* **121**, 1-76.
FRÜH J. & SCHRÖTER C. (1904) Moore der Schweiz mit Berücksichtigung der gesamten Moorfrage. *Beitr. Geol. Schweiz.* Bern.
GODWIN H. (1956) *The history of the British flora.* Cambridge University Press, London.
GÖRS S. (1961) Das Pfrunger Ried. Die Pflanzengesellschaften eines oberschwäbischen Moorgebietes. *Veröff. d. Landesstelle f. Natursch. u. Landschaftspfl.* Baden-Württemberg **27-28**, 5-45.
GROSSE-BRAUCKMANN G. (1962) Moorstratigraphische Untersuchungen im Niederwesergebiet. *Veröff. geobot. Inst., Zürich* **37**, 100-19.
HADAČ E. (1939) Zur Nomenklatur und Systematik der Moorgesellschaften. *Stud. bot. čech.* **2**, 97-106.
HEIKURAINEN L. (1953) Die Kieferbewachsenen eutrophen Moore Nordfinnlands. *Ann. bot. Soc. 'Vanamo'* **26**(2), 189p.
HEJNÝ S. (1957) Ein Beitrag zur ökologischen Gliederung der Makrophyten der tschechoslowakischen Niederungsgewässer. *Preslia* **29**, 249-368.
HOLUB J., HEJNÝ S., MORAVEC J. & NEUHÄUSL R. (1967) Übersicht der höheren Vegetationseinheiten der Tschechoslowakei. *Rozpravy čs. Akad. Věd, Řada mat. přír. Věd* **77**(3), 3-75.
IVERSEN J. (1964) Plant indicators of climate, soil and other factors during the Quaternary. *Rep. VI Intern. Congr. Quater. Warsaw* 1961, Sect. **2**, 421-28.
JANKOVSKÁ V. (1967) Vývoj vegetace Třeboňské pánve na základě pylové a makroskopické analysy v pozdním glaciálu a holocénu. *Dissert. Brno,* MS. 191p.
JANKOVSKÁ V. (1970) Ergebnisse der Pollen- und Grossrestanalyse des Moors 'Velanská cesta' in Südböhmen. *Folia geobot. phytotax.* **5**, 43-60.
JANSSEN C.R. (1970) Problems in the recognition of plant communities in pollen diagrams. *Vegetatio* **20**, 187-98.
JASNOWSKI M. (1957) *Calliergon trifarium* Kindb. w ukladzie stratygraficznym i florze torfowisk holoceńskich Polski. *Acta Soc. Bot. Pol.* **26**, 701-18.
JASNOWSKI M. (1959) Czwartorzędowe torfy mszyste, klasyfikacja i geneza. *Acta Soc. Bot. Pol.* **28**, 319-64.
JESCHKE L. (1959) Pflanzengesellschaften einiger Seen bei Feldberg in Mecklenburg. *Feddes Repert., Beih.* **138**, 161-214.
JESCHKE L. (1963) Die Wasser- und Sumpfvegetation im Naturschutzgebiet 'Ostufer der Müritz'. *Limnologica* **1**, 475-545.
JONAS F. (1934) Die Entwicklung der Hochmoore am Nordhümmling. *Repert. nov. Spec. Regni veg., Beih.* **78**, 1-88.
KARCZMARZ K. (1963) Mchy Pojezierza Łęczyńsko-Włodawskiego, I. *Ann. Univ. Mariae Curie-Sklodowska,* C **18**, 367-410.
KATZ N. & KATZ S. (1964) Die Eigentümlichkeiten der Pleistozänen Pflanzengesellschaften. *Rep. VI Intern. Congr. Quatern., Warsaw* 1961, Sect. **2**, 439-45.
KLIKA J. (1955) *Nauka o rostlinných společenstvech (Fytocenologie).* Praha.
KLIKA J. (1958) K fytocenologii rašelinných a slatinných společenstev na Záhorské nížině. *Biol. Pr.* **4**(4), 1-34.
KOBENDZA R. (1930) Stosunki fitosocjologiczne Puszczy Kampinoskiej. *Planta pol.* **2**, 1-200.
KOCH W. (1926) Die Vegetationseinheiten der Linthebene unter Berücksichtigung der Verhältnisse in der Nordostschweiz. *Jb. st. gallen. naturw. Gesell.* **61**(2), 1-144.
KOTOUČKOVÁ V. (1963) Vývoj vegetace a stratigrafie rašeliniště Červené Blato. *Dissert. Praha,* MS. 122p.
LANG G. (1967) Über die Geschichte von Pflanzengesellschaften auf Grund quartärgeobotanischer Untersuchungen. *Ber. Intern. Sympos. Pflanzensoziologie und Palynologie, Stolzenau/Weser* 1962, 24-37.
LARSSON B. (1960) Bidrag till Västergötlands mossflora. *Svensk bot. Tidskr.* **54**, 423-38.
LOUNAMAA J. (1961) Untersuchungen über die eutrophe Moore des Tulemajärvi-Gebietes im südwestlichen Ostkarelien. *Ann. bot. Soc. 'Vanamo'* **32**(3), 1-63.
MCVEAN D.N. & RATCLIFFE D.A. (1962) *Plant communities of the Scottish Highlands.* H.M.S.O., London.

MĄDALSKI J. (1930) Krytyczne uwagi o występowaniu *Carex incurva* Ligleff i *Carex chordorrhiza* Ehrh. w poludniowo-wschodniej Europie. *Acta Soc. Bot. Pol.* 7, 205-14.

MALMER N. (1968) Über die Gliederung der Oxycocco-Sphagnetea und Scheuchzerio-Caricetea fuscae. *Ber. Intern. Sympos. Pflanzensoziologische Systematik, Stolzenau/Weser 1964*, 293-305.

MAREK S. (1965) Biologia i stratygrafia torfowisk olszynowych w Polsce. *Zesz. probl. Post. Nauk roln.* 57, 1-266.

MOORE J.J. (1968) A classification of the bogs and wet heaths of northwestern Europe. *Ber. Intern. Sympos. Pflanzensoziologische Systematik, Stolzenau/Weser 1964*, 306-20.

MORAVEC J. & RYBNÍČKOVÁ E. (1964) Die *Carex davalliana*-Bestände im Böhmenwaldvorgebirge, ihre Zusammensetzung, Ökologie und Historie. *Preslia* 36, 376-91.

MÖRNSJÖ T. (1969) Studies on vegetation and development of a peatland in Scania, South Sweden. *Op. bot. Soc. bot. Lund* 24, 187p.

NEDELCU G.A. (1970) Beitrag zum Studium der Vegetation des Mogoșoaia-Sees. *Arch. NatSchutz LandschForsch.* 10, 71-84.

NORDHAGEN R. (1936) Versuch einer neuen Einteilung der subalpinen Vegetation Norwegens. *Bergens Mus. Årb. (Naturv. Ser.)* 7, 88p.

NORDHAGEN R. (1943) Sikilsdalen og Norges fjellbeiter. En plantesociologisk monografi. *Bergens Mus. Skr.* 22, 607p.

OBERDORFER E. (1957) Süddeutsche Pflanzengesellschaften. *Pflanzensoziologie, Jena* 10, 564p.

OBERDORFER E. et al. (1967) Systematische Übersicht der deutschen Phanerogamen- und Gefässkryptogamen-Gesellschaften. *Schriftreihe f. Vegetationskunde* 2, 7-62.

OVERBECK F. (1958) Entwicklung eines Hochmoores in Niedersachsen. *Neues Arch. Nieders.* 9, 400-1.

PACOWSKI R. (1967) Biologia i stratygrafia torfowiska wysokiego Wieliszewo na Pomorzu zachodnim. *Zesz. problem. Post. Nauk roln.* 76, 101-96.

PASSARGE H. (1964) Pflanzengesellschaften des norddeutschen Flachlandes, I. *Pflanzensoziologie, Jena* 13, 324p.

POELT J. (1954) Moosgesellschaften im Alpenvorland, II. *S.-Ber. öst. Akad. Wiss. Wien* 163, 495-539.

PURKYNĚ C. (1925) Rašelinisko u lázní Bělohradu. *Čas. nár. Mus.* 99, 1-133.

RUDOLPH K. (1917) Untersuchungen über den Aufbau böhmischer Moore, I. Aufbau und Entwicklungsgeschichte südböhmischer Hochmoore. *Abh. k. zool. bot. Ges. Wien* 9, 1-116.

RUDOLPH K. (1928) Die bisherigen Ergebnisse der botanischen Mooruntersuchungen in Böhmen. *Beih. bot. Zbl.*, B 45, 1-180.

RUDOLPH K. & FIRBAS F. (1927) Paläofloristische und stratigraphische Untersuchungen böhmischer Moore, III. Die Moore des Riesengebirges. *Beih. bot. Zbl.*, B 43, 69-144.

RUUHIJÄRVI R. (1960) Über die regionale Einteilung der nordfinnischen Moore. *Ann. bot. Soc. 'Vanamo'* 31(1), 306p.

RUUHIJÄRVI R. (1963) Zur Entwicklungsgeschichte der nordfinnischen Hochmoore. *Ann. bot. Soc. 'Vanamo'* 34(2), 40p.

RYBNÍČEK K. & RYBNÍČKOVÁ E. (1968) The history of flora and vegetation on the Bláto-mire in south-eastern Bohemia, Czechoslovakia. *Folia geobot. phytotax.* 3, 117-42.

RYBNÍČEK K. (1970) *Rhynchospora alba* (L.) Vahl, its distribution, communities and habitat conditions in Czechoslovakia, II. *Folia geobot. phytotax.* 5, 221-63.

RYBNÍČKOVÁ E. & RYBNÍČEK K. (1971) The determination and elimination of local elements in pollen spectra from different sediments. *Rev. Palaeobotan. Palynol.* 11, 165-76.

RYBNÍČKOVÁ E. & RYBNÍČEK K. (1972) Erste Ergebnisse paläogeobotanischer Untersuchungen des Moores bei Vracov, Südmähren. *Folia geobot. phytotax.* 7, 258-308.

SALASCHEK H. (1935) Paläofloristische Untersuchungen mährisch-schlesischer Moore. *Beih. bot. Zbl.*, B 54, 1-58.

SCHWICKERATH M. (1940) Aufbau und Gliederung der europäischen Hochmoorgesellschaften. *Bot. Jb.* 71, 249-66.

SCHWICKERATH M. (1942) Bedeutung und Gliederung des Differentialartenbegriffes in der Pflanzengesellschaftslehre. *Beih. bot. Zbl.*, B 61, 351-83.

SPENCE D.H.N. (1964) The macrophytic vegetation of freshwater lochs, swamps and associated fens. In *The Vegetation of Scotland* (ed. by J.H. Burnett), pp. 306-425. Oliver and Boyd, Edinburgh.

TOŁPA S., JASNOWSKI M. & PAŁCZYŃSKI A. (1967) System der genetischen Klassifizierung der Torfe Mitteleuropas. *Zesz. problem. Post. Nauk roln.* 76, 9-99.

TÜXEN R. (1937) Die Pflanzengesellschaften Nordwestdeutschlands. *Mitt. flor.-soz. ArbGemein.* 3, 1-170.

WĄS S. (1965) Geneza, sukcesja i mechanizm rozwoju warstw mszystych torfu. *Zesz. problem Post. Nauk roln.* 57, 305-93.

Appendix

Col. 1: Rybníček & Rybníčková (1968)—1; Jankovská (1970)—2.

Col. 2: Marek (1965)—3; Rybníčková & Rybníček (1972)—2.

Col. 3: Grosse-Brauckmann (1962)—4; Marek (1965)—1; Rybníček & Rybníčková (1968)—1; Rybníčková & Rybníček (1972)—1; Jankovská, unpubl., Třeboň Basin, Švarcenberský rybník—1.

Col. 4: Grosse-Brauckmann (1962)—2; Marek (1965)—2.

Col. 5: Marek (1965)—3.

Col. 6: Marek (1965)—6; Rybníčková & Rybníček (1972)—1.

Col. 7: Firbas et al. (1948)—1; Jankovská, unpubl., Třeboň Basin, Švarcenberský rybník—1; Rybníček, unpubl., South Bohemia, Nahořany, Kraselov, Mladotice, Vacovice, Němčice—5.

Col. 8: Rybníček & Rybníčková (1968)—1; Jankovská (1970)—1; Rybníček (MA) & Rybníčková (PA), unpubl., Bohemian-Moravian Uplands, Řásná—2; Rybníček (MA) & Jankovská (PA), unpubl., Třeboň Basin, Borkovická blata—4.

Col. 9: Rybníček & Rybníčková (1968)—1; Rybníček (MA) & Jankovská (PA), unpubl., Třeboň Basin, Červené blato—4.

Col. 10: Marek (1965)—1; Rybníčková & Rybníček (1970)—1; Rybníček (MA) & Rybníčková (PA), unpubl., Bohemian-Moravian Uplands, Stálkov—1; Rybníček (MA) & Rybníčková (PA), unpubl., South Bohemia, Kraselov, Mladotice—3.

Col. 11: Rybníčková & Rybníček (1968)—3; Jankovská (1970)—1.

Col. 12: Rudolph (1918), Breites Moss—1; Firbas (1935), Kolbermoor—1; Rybníček (MA) & Jankovská (PA), unpubl., Třeboň Basin, Červené Blato—3.

Col. 13: Marek (1965)—4; Rybníček & Rybníčková (1968)—1.

Col. 14: Marek (1965)—1; Rybníčková & Rybníček (1972)—2.

Col. 15: Rudolph & Firbas (1927)—1; Salaschek (1935)—1; Rybníček & Rybníčková (1968)—1; Rybníček (MA) & Rybníčková (PA), unpubl., Bohemian-Moravian Uplands, Bláto—2; Rybníček (MA) & Jankovská (PA), unpubl., Třeboň Basin, Borkovické Blato—1.

Col. 16: Rybníček (MA) & Rybníčková (PA), unpubl., Bohemian-Moravian Uplands, Suchdol, Loučky—3.

Col. 17: Rybníček (MA) & Rybníčková (PA), unpubl., South Bohemia, Kraselov, Mladotice, Vacovice, Němčice—5.

Col. 18: Rudolph & Firbas (1927)—5; Jonas (1934)—2.

Col. 19: Kotoučková (1963)—4; Rybníček (MA) & Jankovská (PA), unpubl., Třeboň Basin, Červené Blato—2.

Col. 20: Rudolph (1918)—1; Firbas *et al.* (1948)—2; Rybníček & Rybníčková (1968)—1; Jankovská (1970)—1; Rybníček (MA) & Rybníčková (PA), unpubl., Bohemian-Moravian Uplands, Bláto—1.

Explanation: (MA)—Macrofossil analyses
            (PA)—Pollen analyses

Discussion on vegetational history and community development

Recorded by Dr Hilary H. Birks, Dr H.J.B. Birks,
Mr J.H.C. Davis, and Mr J. Dodd

Dr Markgraf informed Professor Watts that more recent work had shown that the dating of the Faulenseemoos sediments proposed by Welten was not reliable, and thus the laminations must be used with caution when inferring the rates of vegetational changes. Dr Bradbury commented that in the pollen diagram from Lake of the Clouds, there was a lower proportion in zone LC-6 of *Pinus strobus* than in LC-5, although the total amount of *Pinus* pollen remained constant. Mr Craig interpreted this change as perhaps representing increasingly moist conditions. There is little *Pinus strobus* around the lake at the present time. Dr Davis was interested in Professor Watts' comments, and said that studies from large and small lakes should be made to determine the population size of the forest components, and whether a species immigrated as a few individuals, or as an advancing front. Dr Davis also mentioned that the rise of *Pinus strobus* pollen at Rogers Lake, Connecticut was very sudden, and even more marked in terms of pollen influx, because the percentage values were depressed by high amounts of *Quercus* pollen. Dr Janssen said that the concept of the forest community should be clarified, in terms of dominance, structure, or floristic composition. If *Pinus strobus* invaded a *Quercus* forest, the overall structure of the community may remain the same. Dr H.J.B. Birks had obtained information on the early post-glacial rate of *Corylus* expansion on the Isle of Skye, Scotland. In sites only a few miles apart, the expansion rates were very different, and could be explained in terms of local factors of soil development and climate, rather than in terms of migration rate. Dr Cushing proposed that each species contained a certain amount of genetic variability, and this allowed them to be adaptable to each other's presence. Professor Watts did not wish to propose a mechanism, but observations had shown that two species in competition achieved a balance. In the case of the pines, the only way to investigate the mechanism would be to grow the species together in different proportions and densities, and see whether one gained dominance over the other after starting in the minority. Professor Hafsten stressed the importance of studying the sediment stratigraphy, as demonstrated by Troels-Smith. The different sequence of the arrival of trees in the Hoxnian interglacial may be due to sediment mixing rather than migration. Similarly, the phase of increased Gramineae pollen at Marks Tey may be due to sediment disturbance. Professor Watts said that the sediments at Marks Tey were laminated throughout this part of the diagram, and the laminations showed no sign of disturbance. Dr C. Turner added that the sequence of immigration and expansion of trees at Marks Tey was found in other Hoxnian sites in the region and was not a local phenomenon.

Dr Flenley would have expected contamination of samples for radiocarbon dating taken from shallow peats with *Juncus effusus* growing on the surface. Dr J. Turner replied that they had taken no sample shallower than 50 cm, and that there had been no visible penetration of living roots. In addition, the determinations gave consistent results. Dr Newey wondered if the type of land use could be detected by the occurrence of pollen of specific weeds. Dr Turner said that there was no detectable trend in the ratio of pastoral to arable weeds in the different phases, but there was an overall trend towards pastoral woods throughout. Mr Maltby asked if any soil differences could be detected between those in the enclosure and those outside. One could then perhaps put a timescale on the rate of processes of edaphic change, and see how long a soil retained signs of modification by husbandry. Dr Roberts regretted that no soil scientist had yet looked at the soils in detail. He had not detected any differential soil changes in the Pennines such as had been recorded for Scotland. Perhaps the rate of some geomorphological processes could be estimated, as one side of the thirteenth century bank round Stewarts Shield had been completely eroded away. Dr A.G. Smith noted that in the woodland regeneration phase, zone SSb, at Stewarts Shield, most of the tree pollen was *Salix*, which could be due to local growth of willows on the site. Dr Turner said that *Salix* was not included in the tree pollen sum, and that the statement was based on the general regeneration of all the trees.

Dr West asked Mr Sims if he envisaged different processes occurring in the Mesolithic or Neolithic clearances at Hockham Mere, or whether it was merely a difference in intensity. Mr Sims replied that the Mesolithic clearance was on the lake margin, and was very local, for the settlement only, and with no cultivation. The only tree felling would be for constructing the settlement. In contrast, the 'elm decline' and subsequent Neolithic settlements were more widespread, with more extensive felling for cultivation and pastoral activity away from the margin of the lake. Dr A.G. Smith warned that the conversion of pollen concentration to pollen influx could not be readily made unless radiocarbon dates were available, or unless a constant sedimentation rate could be proved. At Seamere, a change in sediment lithology was accompanied by a rise in the concentration curves of all the tree taxa, which may be an effect of changing sediment accumulation rate rather than a change in pollen influx. Mr Sims agreed that this was the case at Seamere, where, due to the calcareous nature of the sediments, no radiocarbon dating was possible. However, as he had previously mentioned, at Hockham Mere, the sediment had been carefully analysed, and there was no detectable change at this level. The lake has no inflow or springs, and the water level is determined by the precipitation:evaporation ratio. In other parts of the sequence, such as at the alder rise, stratigraphic evidence is present for changes in water level, but no such changes were observed around the elm decline.

Dr H.J.B. Birks asked Dr Rybniček about the nature of the interglacial fossil record for *Scheuchzeria palustris* reported by Katz and Katz. Dr Rybniček said that the authors did not state whether it was pollen, fruits, or vegetative parts. Dr Janssen asked about the criteria for establishing the vertical columns on Dr Rybniček's table of fossil communities; was it the floristic composition or the peat type? Dr Rybniček said that they were based on floristic composition, using the same principles used for constructing the table of present-day mire communities. This allowed a direct comparison as far as possible. After a question from Dr Cushing, Dr Rybniček explained that, on his table of fossil communities (Table 2), the roman numerals indicated standard constancy classes, while arabic numerals showed the number of occurrences in communities where the number of analyses available was too small to allow the calculation of constancy classes.

Section 6:
Limnological history

Preliminary studies of Lough Neagh sediments
I. Stratigraphy, chronology and pollen analysis

P.E. O'SULLIVAN, F. OLDFIELD AND R.W. BATTARBEE
School of Biological and Environmental Studies, New University of Ulster

Introduction

Lough Neagh, the largest lake in the British Isles has a surface area of some 367 km² and is 30 km along its longest axis. It lies at 50 feet (c 15 m) above the Irish Low Spring Tide Datum and has a catchment area covering some 5,700 km² that embraces over 40% of the total area of Northern Ireland in addition to neighbouring parts of the Irish Republic.

The present studies were stimulated by a growing awareness of the highly eutrophic state of the Lough (Gibson *et al.* 1971, Wood & Gibson 1972). These authors have documented the richness of the lake water in dissolved nutrients, its high primary productivity, and the overwhelming dominance of the lake's phytoplankton by blue-green algae. From 1968 onwards studies of the lake sediments have been carried out in order to determine when, by what stages, and in response to what factors the lake developed towards its present condition. This paper is a brief preliminary account of the lake's morphometry and of the stratigraphy, chronology, and pollen content of its fine sediments. It is intended to form a context for the second part by R.W. Battarbee on the fossil diatom record.

Fig. 1 shows the topography of the lake-bed as surveyed in 1835. The main elements are described below in terms of the 1835 depths*.

(i) From the former water level down to c 20 feet (7 m) there is a broad, usually sandy, gently sloping shelf round much of the lake margin. Rock also rises to this height near the centre.
(ii) From 20 feet to 38 feet (7–13 m) there is often a relatively steep slope especially parallel to parts of the north and north-west shorelines.
(iii) Between 38 feet and 48 feet (13–16 m) there is an extensive level area with muds forming the surface sediment. About 40% of the substrate of the lake lies between these limits.
(iv) Between 48 feet and 60 feet (16–20 m) shallow depressions can be traced converging on the north-west corner of the Lough from the south and east.
(v) Below 60 feet (20 m) lies the single 'Trench' feature in Toome Bay going down to about 100 feet (33 m).

Fig. 2 shows five seismic profiles across the Lough. Fig. 3 gives a series of histograms representing frequencies of sounding depths in the northern half of the Lough on the 1835 survey. It illustrates the sharp boundaries between features (ii) to (v) and especially the distinct break of slope at 38 feet (13 m) on the old survey which occurs at 34 feet (c 10.5 m) below present lake level. Fig. 4 shows the limits of the main central flat area (iii) in relation to the sampling sites referred to later.

The lake is very exposed and the open water is well mixed both vertically and horizontally with regard to its main biological, chemical, and physical characteristics. It remains broadly isothermal throughout the year except for rare, brief spells of calm, hot, summer weather. During these periods ephemeral oxygen gradients develop leading to severe oxygen depletion for short periods at the mud-water interface.

Early Flandrian sediments

The evidence obtained so far can best be considered in two parts with the dividing line in time falling close to 6,000 B.P. For the earlier period there is much less evidence. No pre-Flandrian sediments have been recovered in any samples taken so far from Lough Neagh.

POLLEN ZONES IV AND V

Pollen diagrams from below 2 m of water in Lough Beg (Fig. 5), a small lake representing little more than a widening of the Lower Bann between 2 and 7 km downstream of the Lough Neagh outfall, and from below c 10 m of water in Lough Neagh at AB11 (Fig. 6; located on Fig. 4) both show that early Flandrian gyttjas accumulated during pollen zones IV and V below a water level that cannot have been lower than that prevailing at present.

*Lake level at the time of this survey stood approximately 4 feet (1.3 m) higher than at present. Not only are present levels lower but they are artificially controlled; in the 19th century they fluctuated seasonally by up to 3 m.

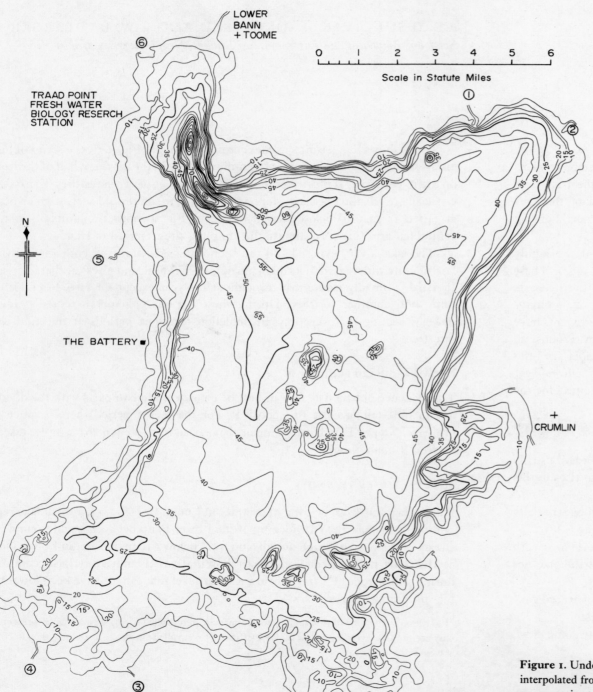

Figure 1. Underwater topography of Lough Neagh. Isobaths at 5 feet (1.6 m) intervals are interpolated from the survey of 1835 and are in terms of the depths recorded then. Present depths are approximately 4 feet (1.3 m) less than those shown here. The main inflowing streams are as follows:
1. Main 2. Six Mile Water 3. Upper Bann 4. Blackwater 5. Ballinderry 6. Moyola

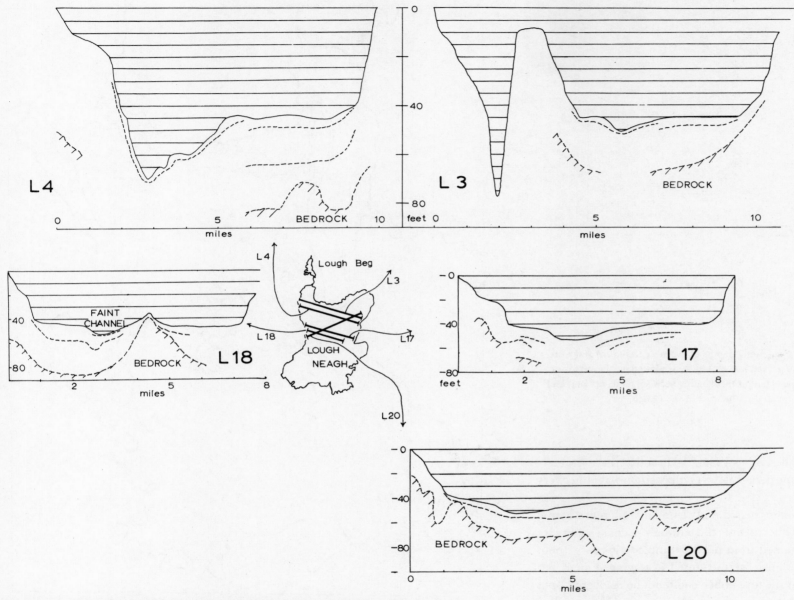

Figure 2. Selected seismic profiles across Lough Neagh. Profiles 3 and 4 have twice the vertical exaggeration of the others.

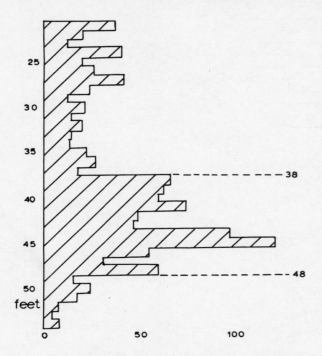

Figure 3. Histograms showing the frequency of sounding depths between 21 feet (7 m) and 54 feet (18 m) in the part of Lough Neagh north of the pecked line on Fig. 4, as plotted on the survey of 1835. This simple parameter underestimates the extent of the flat area between 38 feet (13 m) and 48 feet (16 m) since soundings are more widely spaced in that part of the Lough.

POLLEN ZONES VI AND VIIa

At Lough Beg, virtually the whole of pollen zones VI and VIIa and perhaps the early part of VIIb are missing. During the same period at AB11, deposition was either very contracted or interrupted or both.

In the open parts of Antrim Bay two cores penetrated a laminated silt representing all or part of this time interval. In AB9 (Fig. 10) only the bottom 1–2 cm of a 3 m core sampled this material. A 6 m core obtained from nearby sampled 1.75 m without bottoming it. Approximately 1,200 laminae were counted averaging c 7/cm but ranging from 4–5 at the base to 12–14 at the top. Pollen could not be recovered from this material except at the very top where despite very poor preservation a pollen spectrum was obtained which included fen and marsh taxa that are absent or poorly represented in the unlaminated muds immediately above the sharp transition. In the lower layers of the laminated sediment pebbles up to 4 mm size were found; diatoms were absent throughout. The laminae themselves show an alternation between narrow dark brown, apparently more organic bands and layers of clay, silt, and fine

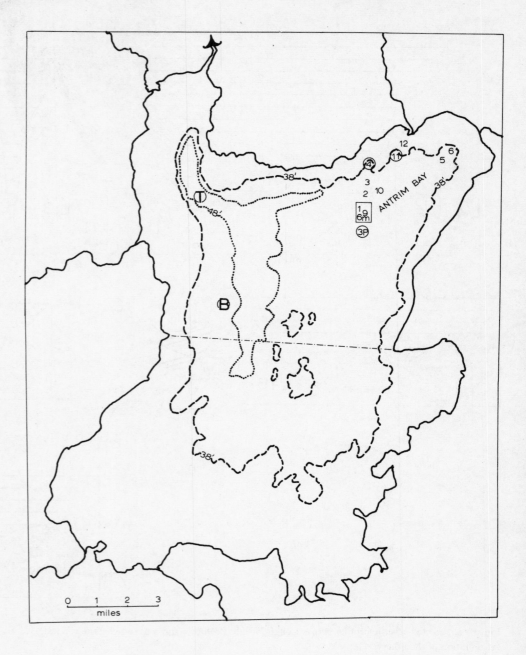

Figure 4. Lough Neagh. Sampling sites and areas in relation to the 38 feet (13 m) and 48 feet (16 m) isobaths on the 1835 survey. T = Trench; B = Battery. The numbers in Antrim Bay refer to the approximate location of 3 m and 6 m cores. 3P is the core from which radiocarbon dates (Fig. 7) and a pollen diagram (Fig. 12) were obtained.

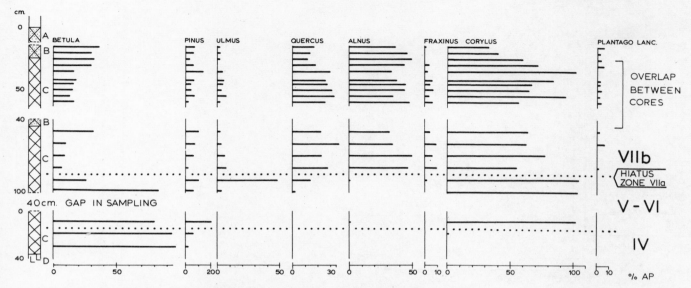

Figure 5. Pollen and stratigraphy from Lough Beg. Samples were obtained using a modified Livingstone and a Hiller borer. Selected pollen types are plotted as a percentage of the tree pollen sum excluding *Corylus* and *Salix*. The stratigraphy shows a transition from clay (D) to gyttja (C) at the base then from gyttja to sandy gyttja (B) becoming unconsolidated (A) at the top. The short pollen record above the main hiatus is discontinuous or truncated and no attempt has been made to subdivide or interpret it. Anal. R.W. Battarbee.

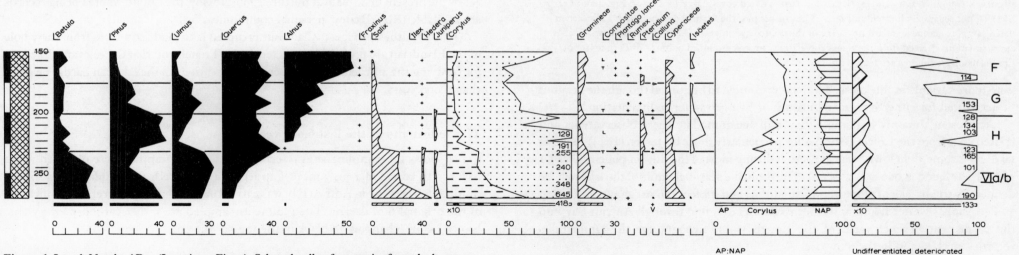

Figure 6. Lough Neagh; AB11 (Location—Fig. 4). Selected pollen frequencies from the bottom 120 cm of a 3 m core expressed as a percentage of the tree pollen sum excluding *Corylus* and *Salix*. Above pollen zone VI letters are ascribed to provisionally defined pollen assemblage zones. A gap in accumulation is thought to separate the zone VI sediments from those above. Anal. P.E. O'Sullivan.

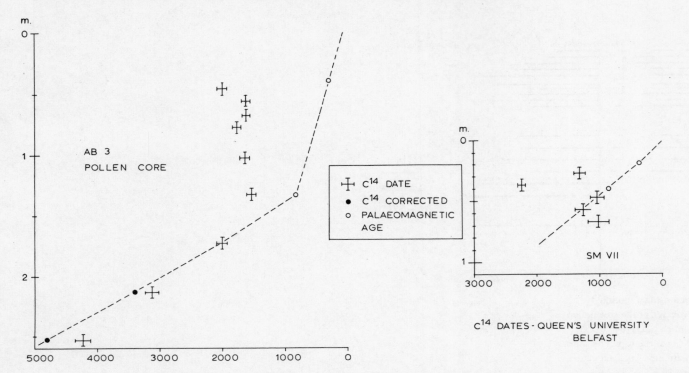

Figure 7. Lough Neagh, Antrim Bay. Radiocarbon dated cores AB3P located on Fig. 4 and nearby SMVII, not separately located on Fig. 4. The crosses plot the vertical depth of each radiocarbon dated sample and one standard deviation of the quoted age in years B.P. Corrected 'dates' are approximations derived from the Suess correlation curve reproduced on p. 31 of Berger (1970). For palaeomagnetic ages see text.

sand more variable in thickness. They are presumed to be annual though they cannot be accounted for either by glacial outwash or by deep water sedimentation.

At present the only hypothesis that will accommodate these observations and related ones from the Lower Bann Valley particularly those by Jessen (1949) requires lake levels some 12 m lower than at present during most of the time of pollen zones VI and VIIa (c 8,000–6,000 B.P. at least). The probable seasonal limits of the lake may be defined by the breaks of slope at 38 feet (13 m) and 48 feet (16 m) shown in Figs. 3 and 4. Annual winter flooding from the channel draining through Antrim Bay into the permanent 'Trench' area of the Lough would account for the laminations recorded in AB9 and in the 6 m core.

It seems likely that these deduced low levels were related to low early-Flandrian sea-levels at the mouth of the Lower Bann valley (Jessen 1949) which also contains a buried channel filled only with peat alluvium, gyttja, or diatomite to at least 12 m below the present floodplain at the three points below the Lough where boring records are available (R.A. Bazley, personal communication).

This hypothesis suggests that uninterrupted lake mud sequences from the whole of the Flandrian are probably restricted to the Trench and that the relevant timespan for studying the antecedents of present day conditions in the lake can hardly be more than 6,000 years.

The sediments of the last 6,000 years

Fig. 4 shows the sampling sites referred to below. All samples were obtained using Mackereth corers of 1 m, 3 m, or 6 m length (Mackereth 1958, 1969). Most of the routine stratigraphy in Antrim Bay was carried out using the 3m corer, a modification of the original 6 m design. The sediments sampled are rather carbon-poor gyttjas, brown below, changing to black in the top 50–100 cm.

CHRONOLOGY

Fig. 7 shows two sets of radiocarbon dates (Smith, Pearson & Pilcher 1973) each based on 10 cm thick slices from a pollen-analysed 3 m core (Fig. 12) and a minicore, SMVII, respectively, both from Antrim Bay. The dates from the upper levels of each core are substantially older than those from below. The samples were very poor in carbon and were prepared by a simple acid pretreatment only. It seems likely that the 'old' dates are the result of inwashed carbon from eroding soils and blanket bog within the drainage basin over the last thousand years. For the period between 5,000 and 2,000 B.P. it is possible to use the C^{14} dates obtained and corrected approximately for the 'de Vries effect' to determine sediment deposition rates. Other methods are needed, however, for the last two millenia.

Recent papers by Mackereth (1971) and by Creer et al. (1972) have shown how palaeomagnetic declination can be used as an indirect dating technique for the whole Flandrian when calibrated by means of radiocarbon. For the last millenium and a brief period during the Roman occupation of Britain, archaeomagnetic records of declination and inclination summarised in Aitken (1970) suggest a more direct chronological use of declination changes through time. Declination measurements were made by F.J.H. Mackereth on two minicores, one from the Battery and one from the Trench. These have made it possible to define the point in the pollen and diatom diagrams at which the maximum westerly swing around A.D. 1820 occurred. Declination measurements were also carried out by R. Thompson of the Department of Geophysics, University of Newcastle, on seven 3 m and one 6 m core from Antrim Bay. These have allowed rough comparisons of deposition rates between cores and together with the minicore data they permit a tentative timescale to be developed for the sediments deposited during the period between 2,000 B.P. and A.D. 1820 (cf. Molyneux et al. 1972). The later parts of the time/depth curves in Figs. 7 and 9 are based on this provisional palaeomagnetic chronology, which still requires much further confirmation in detail.

The latest date used in core SM111 (Fig. 9) is based on a correlation between a clearly marked peak in cereal pollen representation in the top few centimetres of each core pollen-analysed in sufficient detail (Fig. 13) and the well documented recent land-use history of the Province. Changes in cereal acreage during the period spanning the Second World War are noted by Symons (1963) as follows:
'the acreage under the plough expanded from 471,000 in 1939 (the lowest level then recorded) to 851,000 in 1943; it contracted from 701,000 in 1946 to 492,000 in 1953 and 344,000 in 1959, a new low level'.

The plough-up campaign altered the landscape of the vast and predominantly pastoral Lough Neagh drainage basin briefly but dramatically and appears to have had a distinctive effect on the pollen diagrams from the Lough.

From the above evidence a rather eclectic chronology can be applied to the sediments of the last 5,000 years. It is likely that further work will modify and refine this

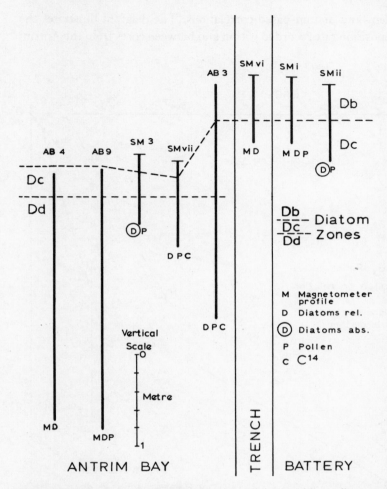

Figure 8. Lough Neagh, main analysed cores. Minicores with a well preserved mud/water interface are represented with a horizontal line across the top, all other cores are truncated below the present mud surface.

chronology, particularly for the last two millenia.

The main analysed cores from the three sampled areas in the Lough are summarised in Fig. 8 which also indicates the range of evidence available from each. Pollen-analytical and diatom analysis has allowed all of those shown to be correlated with each other and linked to the tentative timescale. Fig. 9 shows time/depth curves for three of the analysed cores from Antrim Bay. In each case the fixed points which allow rough accumulation rates to be plotted derive from the chronology already outlined. These points have been transferred between cores where necessary by means of both pollen- and diatom-based correlations. The diagram illustrates the sharp changes in deposition rate recorded within and between cores from the Antrim Bay sediments.

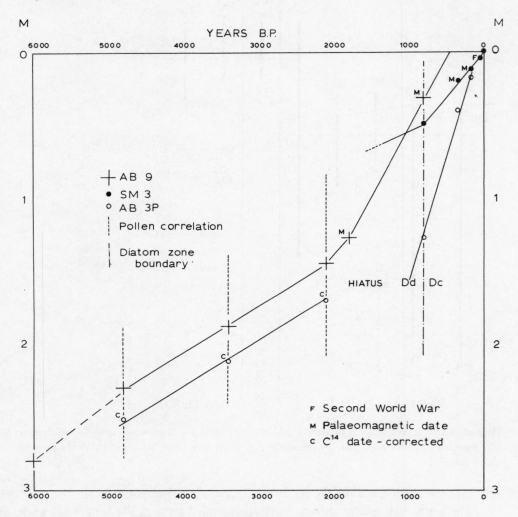

Figure 9. Lough Neagh, Antrim Bay. Selected deposition rates for 3 m cores AB9 (Fig. 10) and AB3P (Fig. 12) and for minicore SM3 (Fig. 13 and Part II). Radiocarbon dates where shown are derived from core AB3P (qv. Fig. 7); palaeomagnetic dates ascribed to core AB9 are based on direct measurement of declination in that core by R. Thompson, Department of Geophysics, Newcastle; palaeomagnetic dating ascribed to core SM3 are transferred by pollen and diatom correlation from two minicores SMI and SMII on which declination measurements were made by F.J.H. Mackereth, Freshwater Biological Association.

THE POLLEN DIAGRAMS

Figs. 6, 10, 11, 12, and 13 show part of the pollen-analytical information now available from Lough Neagh for the period since the end of the accumulation of laminated sediments in Antrim Bay. For this period, provisional pollen zones termed A to H are used simply as an aid to the accompanying text. Absolute ages are based on the evidence summarised above.

Bearing in mind the size of the lake and its catchment, it is reasonable to assume that most of the pollen in deep water is river rather than air borne. Inevitably it will present a rather generalised picture from a much larger area than is usually represented by pollen records from the British Isles. Whilst this imposes limitations on the value of the pollen-analytical evidence as a record of detailed changes in terrestrial ecology, it has some advantages in terms of the aims of the present study, for the effective pollen-source area (cf. Oldfield 1970) is likely to be no greater than the total drainage basin yet well representative of it, though perhaps over-emphasising riverside and wetland habitats. On this basis it would be tempting to regard the Lough Neagh sediments as potentially capable of providing reference pollen diagrams valid for a large area. However, three factors limit the value of the pollen-analytical data from Lough Neagh. Discontinuities in sedimentation coupled with abrupt changes in deposition rate within and between cores have already been noted. Secondly, there is the probable incorporation of derived pollen from eroding soils and blanket bog into the upper sediments especially those representing the last millenium and shown to be heavily contaminated by older carbon. Moreover pollen preservation is very poor; degraded grains (*sensu* Cushing, 1964) are abundant throughout, forming the bulk of the undifferentiated:deteriorated pollen plotted on Figs. 6, 10, and 11.

The pollen diagram from Antrim Bay Core AB9 (Fig. 10) covers the longest timespan of any analysed so far and begins immediately above the laminated silt/gyttja contact. The three lowest zones H, G, and F are also recorded in the AB11 diagram (Fig. 6) and the two lowest in the 6 m core diagram (Fig. 11) which spans the same clay/mud transition as that at the base of AB9. The high, then steeply falling pine frequencies of zone H are associated with unusually high proportions of deteriorated unrecognisable pollen and so may be, in part, a product of differential degradation. In so far as they are real they suggest a date late in pollen zone VIIa as defined by Jessen (1949; q.v. Smith 1970). The succeeding fluctuations in *Ulmus* frequency at the zone G/F boundary coupled with traces of *Plantago lanceolata* pollen in all three diagrams are suggestive of a date close to the conventional zone VIIa/VIIb or Atlantic/Sub-Boreal transition.

The first traces of *Plantago lanceolata* pollen occur in zone G in AB9 and AB6m immediately above the steep *Pinus* decline and there is some sign in the rest of the

Lough Neagh I. Stratigraphy, chronology and pollen analysis

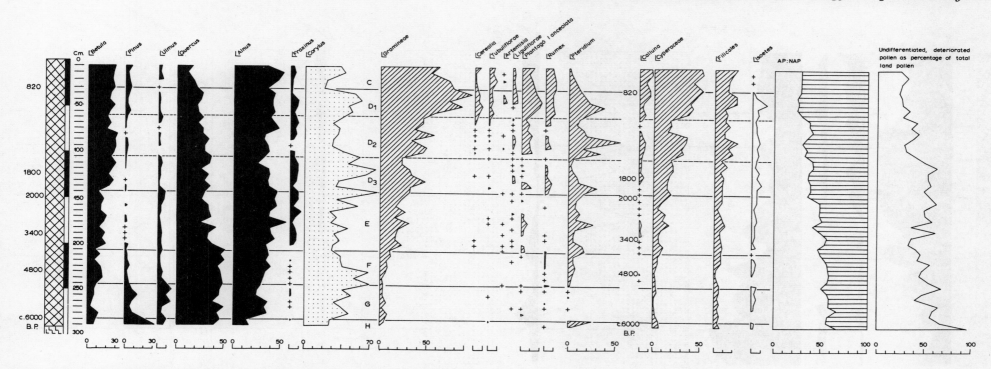

Figure 10. Lough Neagh, AB9 (Location Fig. 4; Chronology Fig. 9). Selected pollen frequencies from a 3 m core expressed as percentages of the tree pollen sum excluding *Corylus* and *Salix*. The stratigraphic boundary at the base is the laminated silt/gyttja contact discussed in the text. Anal. P.E. O'Sullivan.

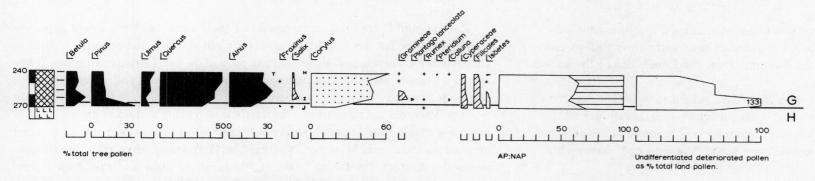

Figure 11. Lough Neagh, 6 m core (Location Fig. 4). Selected pollen frequencies from the base of the gyttja immediately above the highest laminated silts. Anal. P.E. O'Sullivan.

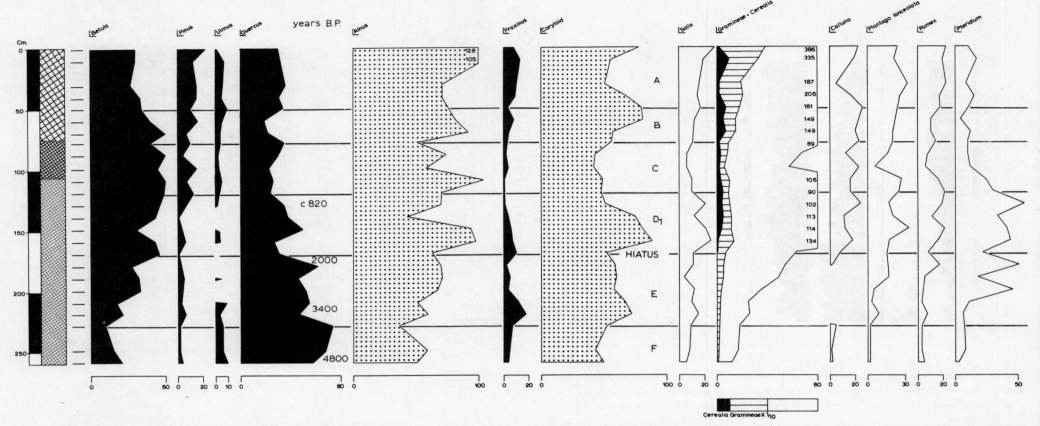

Figure 12. Lough Neagh, AB3P (Location Fig. 4; Chronology Figs. 7 & 9). Selected pollen frequencies from a 3 m core expressed as percentages of a tree pollen sum excluding *Corylus*, *Salix*, and *Alnus*. Anal. P.E. O'Sullivan.

pollen record, especially in the *Ulmus* and Gramineae curves, that this zone may record some disturbance in the forests of the drainage basin sufficiently widespread and synchronous to be reflected in the diagrams from the Lough before the succeeding landnam.

At the zone G/F boundary both AB9 and AB11 record most of the features associated with landnam-type clearance and subsequent forest regeneration. In AB9 this horizon also marks the beginning of steady increases in Gramineae and *Pteridium* values. General clearance around or just before 5,000 B.P. is indicated, presumably the result of early Neolithic activity.

From the above it would appear that little if any of the pollen record above the laminated sediments postdates the first stages of early prehistoric farming in the drainage basin and there would appear to be no possibility that a lake similar to the present one but unmodified by significant human activity in the drainage basin ever existed except perhaps briefly in the early Flandrian before the postulated fall in water level.

Zones F and E are both represented in Figs. 11 and 12. They would appear to span most of the last three millenia before the birth of Christ and cover the Middle to Late Neolithic, the whole Bronze Age, and most of the pre-Christian Iron Age. They record a gradual and almost uninterrupted increase in grass pollen representation alongside generally declining oak percentages. Cereal grains are few and the main weed pollen types are those more characteristic of pasture and waste ground than of extensive arable cultivation. A progressive extension of the grazed and de-forested area seems indicated. However, whereas the Gramineae curve rises almost continuously, discreet, coinciding peaks of *Plantago lanceolata* and *Pteridium* around 200 cm and 175 cm coupled with changes in *Quercus* and *Fraxinus* frequencies around the same levels suggest that the extension of the grazed and deforested area though

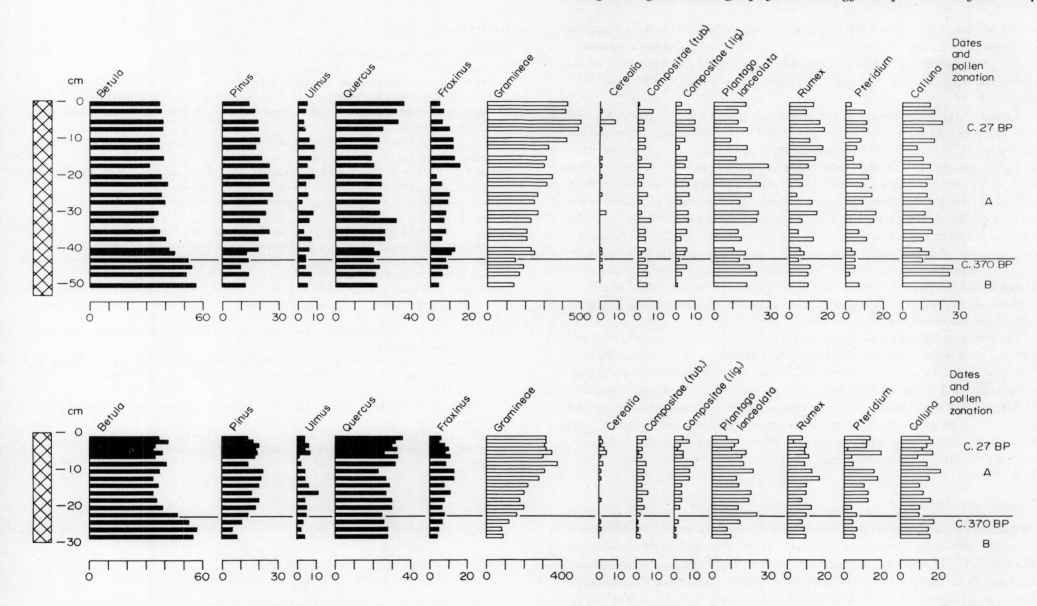

Figure 13. Lough Neagh, SMii and SMiii. Selected pollen frequencies from the upper levels of two minicores taken near the Battery (B on Fig. 4) and in Antrim Bay respectively. Percentages are based on a tree pollen sum excluding *Corylus*, *Salix*, and *Alnus*. (The Gramineae pollen curve is at one-tenth scale.) Anal. R.W. Battarbee.

cumulative, was discontinuous and took place during well defined periods of activity each sufficiently extensive to have affected pollen representation over a wide area. These two periods of inferred forest clearance and expanded farming appear on the present timescale to have begun around 4,000 B.P. and just before 3,000 B.P., respectively.

Zones D_3 and D_2 are present in Fig. 10 but are either missing or very contracted in Fig. 12. They record two further similar phases of forest clearance and expanded farming separated by what may have been an important regeneration phase just below the D_3/D_2 border. As with the forest clearance episodes in zones E and F, these phases lack evidence for extensive cereal cultivation. The earlier phase falls between c 2,000 and 1,800 B.P., the later one begins around 1,500 B.P. and is succeeded by the major extension of farming recorded by the pollen frequencies from zone D_1 (also well represented in Fig. 12) which for the first time includes high cereal pollen and some arable weed representation. The period covered by D_2 and D_1 falls within the mainly pre-Norman Christian period. The radiocarbon anomalies begin within zone D_1 which shows not only the increased pollen frequencies of arable indicators already noted but also increased *Calluna* and Cyperaceae pollen values which may reflect, in part at least, the influx of redbedded grains.

The pollen record from AB9 is truncated in zone C, a period of somewhat reduced weed and grass pollen representation and some 25-30 cm are probably missing from the record. Zones B and A, best represented in Fig. 13, show rapidly increasing grass pollen values and sustained high frequencies of most of the clearance indicators. Zone A, the final phase, postdates the seventeenth century Plantation of Ulster and represents for the most part the overwhelmingly cleared landscape of the last three centuries. It is during these later periods that the main changes in diatom frequency, which are the subject of the next paper, take place.

Acknowledgements

The authors would especially like to record their thanks to the late F.J.H. Mackereth for many kinds of help. They are also grateful to Mr R. Thompson of the Department of Geophysics, University of Newcastle and Dr A.G. Smith, Palaeoecology Laboratory, Queen's University, Belfast for palaeomagnetic declination records and radiocarbon dates respectively. A lot of the field and laboratory work has been aided by technicians in the School of Biological and Environmental Studies in the New University of Ulster.

References

AITKEN M.J. (1970) Dating by archaeomagnetic and thermoluminescent methods. *Phil. Trans. R. Soc. A* **269**, 77-88.

BERGER R. (1970) Ancient Egyptian radiocarbon chronology. *Phil. Trans. R. Soc. A* **269**, 23-36.

CREER K.M., THOMPSON R., MOLYNEUX L. & MACKERETH F.J.H., (1972) Geomagnetic secular variation recorded in the stable magnetic remanence of recent sediments. *Earth and Planet. Sci. Letters* **14**, 115-27.

CUSHING E.J. (1964) Redeposited pollen in Late-Wisconsin pollen spectra from East-Central Minnesota. *Am. J. Sci.* **262**, 1075-88.

GIBSON C.E., WOOD R.B., DICKSON E.L. & JEWSON D.H. (1971) The succession of phytoplankton in Lough Neagh, 1968-1970. *Mitt. Int. Verein. Limnol.* **19**, 146-60.

JESSEN K. (1949) Studies in Late Quaternary deposits and Flora-history of Ireland. *Proc. R. Ir. Acad. B* **52**, 85-290.

MACKERETH F.J.H. (1958) A portable core sampler for lake deposits. *Limnol. Oceanogr.* **3**, 181-91.

MACKERETH F.J.H. (1969) A short core sampler for subaqueous deposits. *Limnol. Oceanogr.* **14**, 145-51.

MACKERETH F.J.H. (1971) On the variation of the horizontal component of remanent magnetisation in lake sediments. *Earth and Planet. Sci. Letters* **12**, 332-8.

MOLYNEUX L., THOMPSON R., OLDFIELD F. & McCALLAN M.E. (1972) Rapid measurement of the remanent magnetisation of long cores of sediment and its application as a dating tool. *Nature, Lond.* **237**, 42-3.

OLDFIELD F. (1970) Some aspects of scale and complexity in pollen-analytically based palaeoecology. *Pollen Spores* **12**, 163-71.

SMITH A.G. (1970) Late- and Post-glacial vegetational and climatic history of Ireland: a Review. In *Irish Geographical Studies* (ed. by N. Stephens & R.E. Glasscock), pp. 65-88. Queen's University of Belfast, Belfast.

SMITH A.G., PEARSON G.W. & PILCHER J.R. (1973) Belfast Radiocarbon dates V. *Radiocarbon* **15** (in press).

SYMONS L. (1963) *Land use in Northern Ireland*. University of London Press, London.

WOOD R.B. & GIBSON C.E. (1972) Eutrophication and Lough Neagh. *Water Research* **6**, 9.

Preliminary studies of Lough Neagh sediments
II. Diatom analysis from the uppermost sediment

R.W. BATTARBEE *School of Biological and Environmental Studies, New University of Ulster and Institute of Quaternary Geology, Uppsala, Sweden.*

Introduction

The development of research into the problems of lake eutrophication has done much to focus interest on the importance of lake sediments (cf. Edmondson 1969, Züllig 1955). The sedimentary record contains indications of the lake's former status and evolution, and in a complex way reflects former conditions in the lake's catchment area. The results of sediment analyses, however, are often difficult to interpret and for this reason such results must be obtained and presented in a way which allows only a minimum of ambiguity. Investigations require detailed quantitative and qualitative information from the available sedimentary record with accurate estimations of the sediment chronology and the rate of sediment deposition. In this way analytical data can be placed in a historical or archaeological perspective, and fossil data can be expressed on an absolute basis.

This approach is especially necessary in the case of the most recent sediments where palaeolimnological results should be presented in a manner which enables them to be compared as closely as possible with limnological results. Since the limnologist defines and measures trophic status in terms of annual productivity (which can be quantified) rather than in terms of communities (which are difficult to quantify), it is logical to attempt to quantify sedimentary data in the same way, and present results in terms of annual deposition. Following from this, if organism counts from the sediment can be converted into terms which signify biomass (e.g. volume or dry weight) per annual deposition, direct comparisons can be made with the limnologist's figures for annual production.

The absolute method is most profitable if used in conjunction with the relative or percentage frequency approach normally used in diatom studies (e.g. Digerfeldt 1971, Haworth 1969), because the two sets of data are complementary. The former ideally reflects changes in the productivity of individual taxa and groups of taxa, and the latter expresses changes in community structures within taxonomical and/or ecological groups.

Both methods have been used in this consideration of the recent history of Lough Neagh using diatoms, as the best preserved algae, as an indicator group. The results of the percentage analyses are described first to serve as an introduction to the evidence for recent trophic change in Lough Neagh, and to serve as a basis for comparisons with the absolute and other diagrams. The main core analysed was taken from the Antrim Bay area of Lough Neagh (see O'Sullivan, Oldfield & Battarbee 1973, referred to subsequently as Part I, Fig. 4) with a Mackereth mini-corer (Mackereth 1969). The core had a clear mud-water interface, and contained 82 cm of sediment. It was designated SMiii.

The relative diagrams

THE PROPORTIONAL FREQUENCY DIAGRAMS

The percentage values for each taxon in the proportional frequency diagram (Fig. 1) were calculated from total sums of between 300 and 600 valves. The benthic and epiphytic forms were included in the total since their numbers in the samples were small and constant, thereby causing little or no distortion of the planktonic percentages. In the diagram (Fig. 1), however, only *Navicula cryptocephala* of the non-pelagic forms is represented. Four main zones have been distinguished.

1 The lowermost zone (Dd) from 82 cm to 49 cm shows a very stable diatom community dominated by *Melosira italica* subsp. *subarctica*. In this diagram it is only represented by 33 cm of sediment, but from other cores (Battarbee, in prep.) it can be seen to extend back to the laminated clay/mud boundary which occurs a little below the elm decline (see Part I). Since the upper boundary of the zone (Dd/Dc) has a date of about 800 B.P., it can be estimated that this stability existed for well over 4,000 years, through the late Neolithic period, and the whole of the Bronze and Iron Ages. Although this only specifically refers to the fossil diatom community it is probable that it reflects the state of the lake as whole during this period.

2 The boundary between zone Dd and zone Dc occurs at a depth of 49 cm. It is characterised by a marked decrease in *Melosira italica* subsp. *subarctica* frequencies, and increases in *Cyclotella* species and *Tabellaria flocculosa*, suggesting a possible

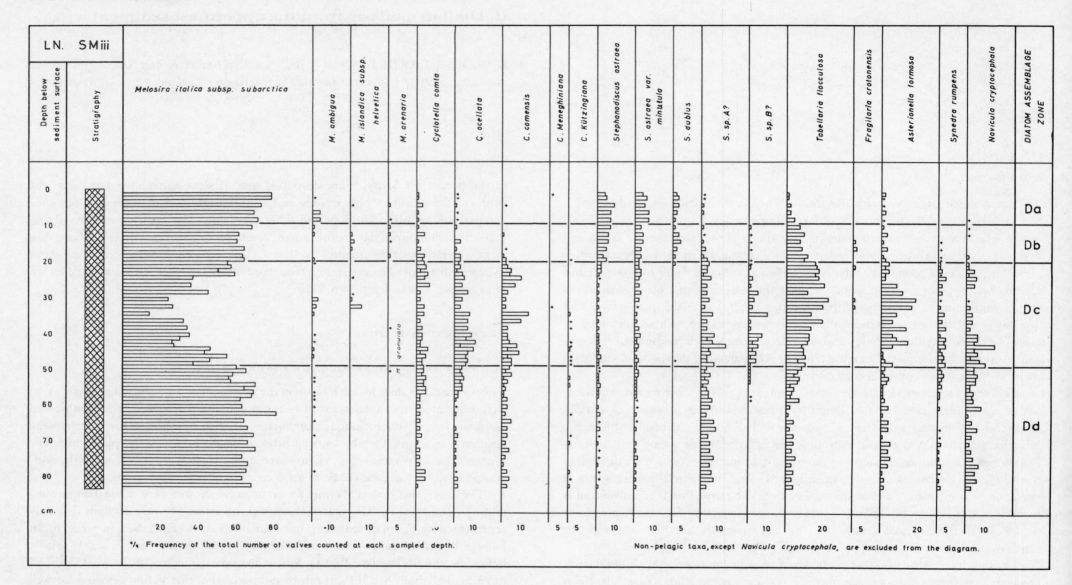

Figure 1. Proportional frequency diagram for diatom valves in Lough Neagh, Core SMiii. See text for further details.

decline in the nutrient status of the lake. It is not known whether this is consequent upon cultural change in the drainage basin or not, but an increase in the sediment density (Battarbee in prep.), an increase in the deposition rate (Fig. 6), and the beginning of erroneous radiocarbon dates slightly before the Dd/Dc boundary all possibly indicate an increase in drainage basin erosion. Since the diatoms suggest more acid conditions during this zone it is possible that the change was connected with peat erosion in the drainage basin, a phenomenon which would in addition help to explain the aberrant radiocarbon dates upwards from this level (see Part I, Fig. 9).

3   The third main zone (Db) begins at a depth of 20 cm, and shows the replacement of *Cyclotella* species by *Stephanodiscus* species, especially *Stephanodiscus astraea*, and

the beginning of a decline in *Tabellaria* values. These changes suggest that the lake was becoming enriched at this time, about the beginning of the 17th century. An increase in the quantity and proportion of *Chironomus* midge remains in the sediment at the same level is further evidence of enrichment (Binney, in prep.). The pollen diagram for SMiii from 22 cm upwards shows a rapid increase of grass pollen frequencies and a reduction in the A.P./N.A.P. ratio (Battarbee, in prep, and Part I, Fig. 13). This is probably related to forest clearance in Tudor times and, especially from the beginning of the 17th century onwards, to the wholesale destruction of forested areas in Ulster by Scottish and English colonists (McCracken 1944, 1947, 1971). It is suggested that the combined effects of this and subsequent agricultural developments were to a considerable extent responsible for the striking changes recorded in the sediment during this period.

4 A zone Da has been allocated to the top 10 cm of sediment to demarcate about the last 100 years of sedimentation. The zone is characterised by the almost complete absence of indifferent and acidophilous planktonic diatom forms. A further increase in lake enrichment is indicated. The rise and fall of *Melosira ambigua*, seen especially clearly in Fig. 2, during this period suggests the continuation of increases in enrichment through to the present time. Today the only diatoms found in quantity in the lake are *Melosira italica* subsp. *subarctica* and species of *Stephanodiscus*, and the dominant algae are blue-green algae, especially *Oscillatoria* spp. (Gibson *et al.* 1971).

The analysis of other cores from different areas of the lake reveal very similar results to those shown in Fig. 1. An example is Core SMii (Fig. 2). It was taken at a distance of 8 kilometres from SMiii at the Battery site (Part I, Fig. 4) in order to examine the spatial variation of the sediment record within the lake. Although it is apparent that the deposition rate is approximately twice that for SMiii the floristic changes, for those zones present, are identical, giving confidence for the assumed representivity of the sediment record from a single core in the lake.

THE PROPORTIONAL VOLUME DIAGRAM

Palynologists have sometimes been induced to correct pollen counts according to the relative pollen production of various trees and shrubs (Faegri & Iversen 1964, Andersen 1970). In this way Andersen (1973) has shown that the dominant tree pollen deposited may not be that of the dominant tree in the source community. In a similar manner, but with less error, a palaeolimnologist can weight diatom counts according to the mean volume of a taxon producing a fossil assemblage structure more closely resembling the structure of the living assemblage from which the fossils were derived. Table 1 shows the average cell volumes which were used to convert the data of Fig. 1 into the data shown in Fig. 3. A comparison of the volumes for Lough Neagh with those of Einsele and Grim (1938) for Swiss lakes shows the necessity of calculating mean volumes for individual communities.

Fig. 3 shows that the effect of the conversion is to stress the importance of *Stephanodiscus astraea* and *Tabellaria flocculosa*, and to show that the dominant diatom in terms of volume in the most recent zones is not *Melosira italica* ssp. *subarctica* but *Stephanodiscus astraea*. If the values for this species in Fig. 1 are compared with those in this diagram it can be seen that although only 10% by number it accounts for between 50 and 60% of the total volume.

Table 1. Diatom cell volumes of selected taxa with some comparisons from Einsele & Grim (1938).

Taxon	μm^3, Lough Neagh	μm^3, E & G
Melosira italica subsp.		
subarctica	494	860 (*M. italica*)
M. ambigua	280	
M. islandica subsp.		1,150
helvetica	3,825	2,500
Cyclotella comta	1,700	1,900
C. comensis	100	
C. ocellata	534	
Stephanodiscus astraea	11,536	25,000
S. astraea var.		
minitula	695	
S. dubius	511	
Tabellaria flocculosa	2,137	
Asterionella formosa	408	320–440

The absolute diagrams

NUMBERS PER UNIT VOLUME OF SEDIMENT

The total numbers of valves per unit wet volume of sediment were counted from slides prepared according to the evaporation tray method (Battarbee 1972). The method enables the production of random distributions of valves on coverslips, allowing the counts from sample traverses to be statistically reliable.

Fig. 4 shows the results for both cores SMii and SMiii in millions of valves/cm^3 of fresh sediment. In the core SMiii (Fig. 4b) counts from two instead of one mud sample were made every 10 cm. The mean values of these second samples are shown by a small plus (+) sign, and it can be seen that all have similar values to the main

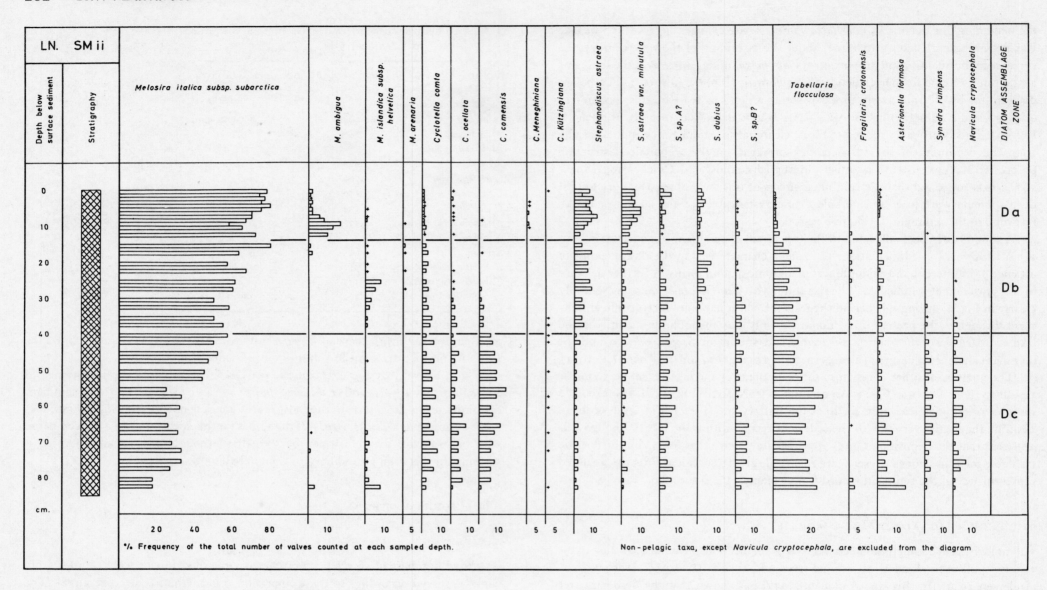

Figure 2. Proportional frequency diagram for diatom valves in LN. Core SMii. See text for further details.

samples, and often lie within the counting-error confidence-limits of the main sample.

A comparison of this curve with the one for SMii (Fig. 4a), after allowing for differences in the rate of deposition, shows that the close similarity of values within cores also exists between cores, although the cores are 8 km apart. The values change with time in the same way between cores, and the total number of valves deposited per unit time in each core can also be calculated to be similar. For example, at 50 cm in SMii (Fig. 4a) the number of valves/cm^3 is about 60×10^6. At the equivalent time in SMiii (25 cm) the number of valves/cm^3 is about 150×10^6. If the deposition rate of SMii is assumed to be twice that of SMiii this value would be about 75×10^6.

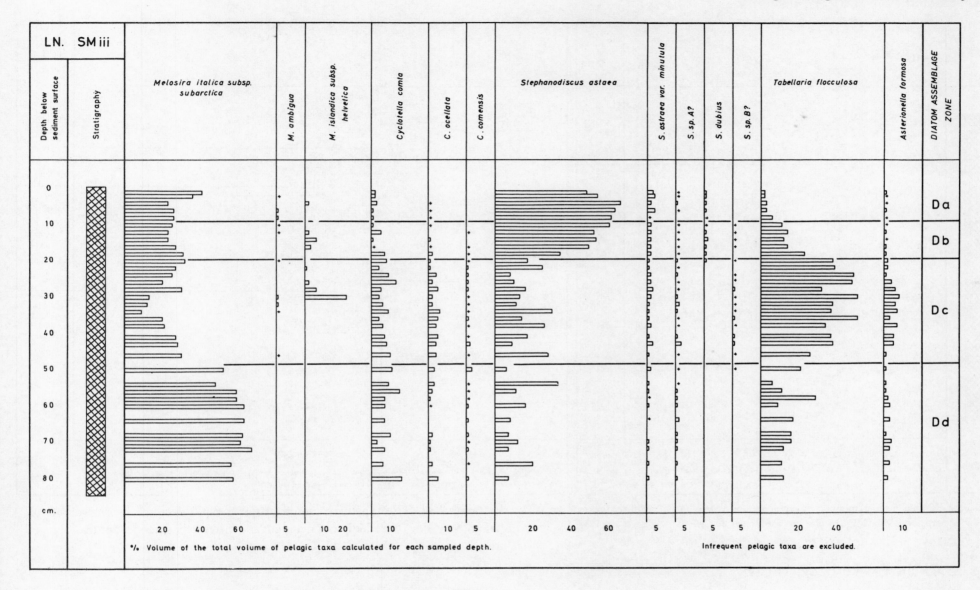

Figure 3. Proportional volume diagram for selected pelagic diatom valves in Core LN. SMiii. See text for further details.

Because of these similarities, and the similarity between the percentage frequencies (cf. Figs. 1 & 2), and because SMiii shows a longer history of sedimentation, it is necessary only to use the absolute counts of SMiii for further conversions.

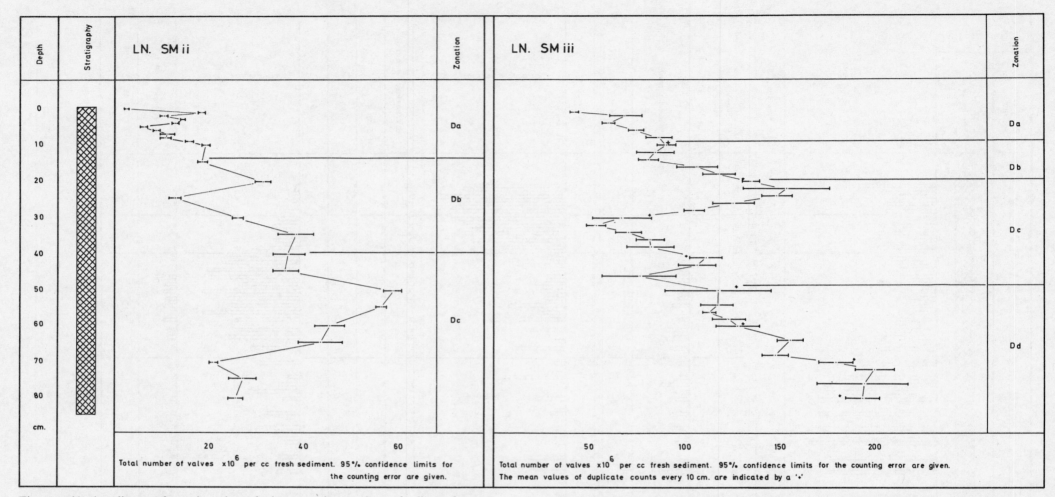

Figure 4. Absolute diagrams for total numbers of valves per unit wet volume of sediment for cores LN. SMii (Fig. 4a) and LN. SMiii (Fig. 4b). See text for further details.

the upper half. The top three samples show a rapid decrease in the volume of diatoms/cm³ due to the high water content and faster deposition rate of the topmost mud.

CELL VOLUME PER UNIT VOLUME OF SEDIMENT

The absolute numbers/cm³ fresh volume can be converted into absolute volumes by using the mean cell volume values in Table 1. Fig. 5a shows the mean number curve of Fig. 4b, without the confidence-limits, and Fig. 5b shows these values converted into volumetric terms. The effect of the conversion is to show that the upper part of the sediment, where *Stephanodiscus astraea* is dominant, contains a greater volume of diatoms/cm³ than the lower part, although there are less diatoms in number in

THE TIME-DEPTH CURVE

The volume curve (Fig. 5b) only represents an intermediate stage in the conversion of values. It is necessary to express the volumes in terms of amount deposited/cm²/annum by correcting for the number of years represented by one cm of sediment at each sampled depth. The accuracy of this depends on the reliability of the chronology.

A preliminary chronology for about the last 5,000 years has been worked out

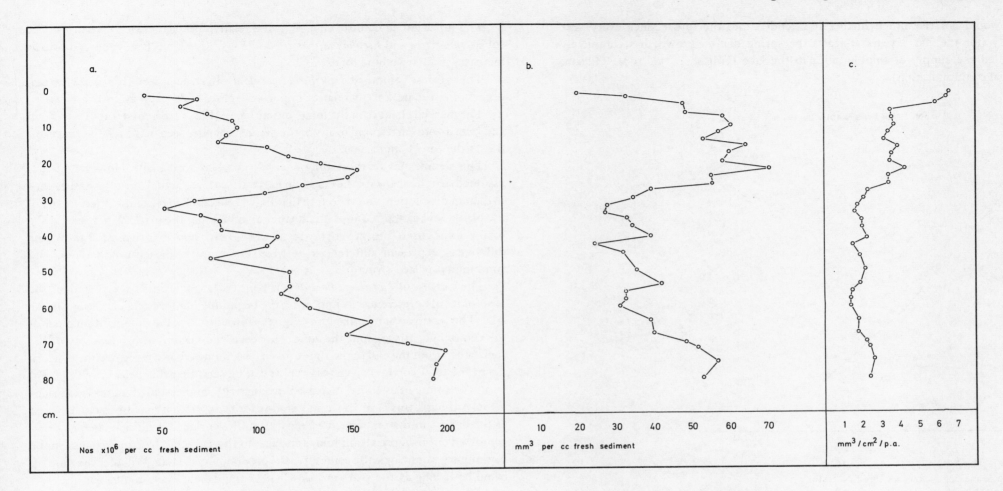

Figure 5. (a) Absolute numbers of diatom valves per unit wet volume of sediment for core LN. SMiii, (b) As above, but converted into volumetric terms, (c) Absolute volumes of diatoms deposited per annum. See text for further details.

using radiocarbon, palaeomagnetism, and pollen evidence (see Part I). All dates have been transferred to core AB3 which is used as a master core for date correlation. Dates are transferred to other cores using pollen and/or diatom features which the cores have in common.

In this way a chronology, or time-depth curve, for SMiii has been constructed (Fig. 6), but until the reliability of the master curve is established some of the values must remain tentative.

TOTAL VOLUME PER CM^2 PER ANNUM

The result of correcting the data of Fig. 5b by the deposition rate is shown in Fig. 5c. The volume of diatoms deposited per annum appears constant during most of the early part of the curve. In the upper part of the curve, however, from about 28 cm there is evidence for a significant increase in the volume of diatoms deposited per annum, and the top few centimetres show a second large increase (cf. Fig. 5b).

The trend of this curve is fully consistent with the increasing eutrophication of the lake during the last few centuries as suggested by the relative diagrams (Figs. 1, 2, & 3), and the large increase in volume deposited during the last 25 to 30 years is consistent with the much more rapid enrichment of recent times, associated especially with post-war enrichment by sewage and detergents. It is possible, however, that the curve would not rise much further if extended into the future since every year recently (for the last 4 years at least) the spring diatom growth has completely exhausted the supply of soluble silica in the lake (Gibson et. al. 1971, Gibson, personal communication).

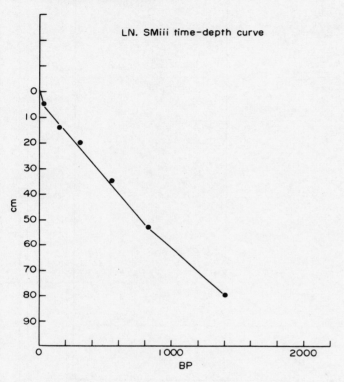

Figure 6. Time-depth curve for core LN. SMiii.

VOLUME OF INDIVIDUAL TAXA PER CM^2 PER ANNUM

The values for each taxa at each level can be calculated from the total curve (Fig. 5c) by using the percentage frequency values of Fig. 1. The result can be seen in Fig. 7. The total curve (Fig. 5c) is repeated in this diagram for purposes of comparison with the individual curves. Fig. 7 should also be compared with Figs. 1 and 3 for percentage frequency and percentage volume, respectively. In most instances a change in taxon totals is associated with a change in relative proportions, but there are a number of important differences.

1   The top 5 cm show little change in the relative proportions of *Melosira italica* subsp. *subarctica* and *Stephanodiscus astraea* (Figs. 1 & 3), but Fig. 7 shows dramatic increases in their volume totals.
2   The greatest volume of *Tabellaria flocculosa* valves deposited (Fig. 7, 28–20 cm) occurs *after* the peak in its relative importance (Figs. 1 & 3, 40–25 cm).
3   The marked changes in the total volume curve (Fig. 5c) discussed above at 26 cm and 5 cm do not correspond to any of the marked changes seen in the relative diagrams (i.e. at the zone boundaries).

The significance of these differences is not easy to explain. However, if it is assumed that the shapes of the curves in Fig. 8 are not artefactual due to past variations in diatom dissolution, and therefore that Fig. 7 reflects changes in the productivity of the planktonic diatoms, and if it is further assumed that the marked increases in the total volume curve (Fig. 5c) are the result of lake enrichment as suggested above, the differences represent differences between community 'palaeo-productivity' and community 'palaeo-structure'.

The increase of *Tabellaria flocculosa* in zone Dc (Figs. 1 & 3) is interpreted, on the basis of its pH preference, as indicating the beginning of more acid conditions in the lake. The relative volume diagram (Fig. 3) shows that it is the dominant member of the community throughout the zone. However, the maximum productivity of the species occurs at the end of the zone (30–20 cm), followed by a rapid decline (Fig. 7). From the total volume curve (Fig. 5c) it can be seen that this peak growth period occurs at a time of assumed increased productivity, and assumed increased enrichment. Although this is an apparent paradox for the species it may be that *Tabellaria* as the dominant diatom, benefited most from the enriched conditions before being competed out by more alkaliphilous species. In this case the structural change in the community would logically come after the productivity change. Whether the time lag could be as long as 100 years in a slowly changing lake is open to question.

The second marked increase in total volume deposited (5 cm) can be considered in the same way. Whilst both *Melosira italica* subsp. *subarctica* and *Stephanodiscus astraea* are both benefiting from increased enrichment there is the possibility that one of them would be replaced in the future, some years after the change in productivity. From the trend of the uppermost values (Fig. 7) the most likely species to be replaced would be *Stephanodiscus astraea*.

COMPARISON OF SURFACE SEDIMENT DATA WITH LIMNOLOGICAL DATA

The core SMiii was taken in just over 10 m depth of water. If the depth of water is taken as 10 m, figures for the total cell volume/l of lake water exactly correspond to figures of cell volume deposited per cm^2 of lake bottom. Gibson *et al.* (1971) have

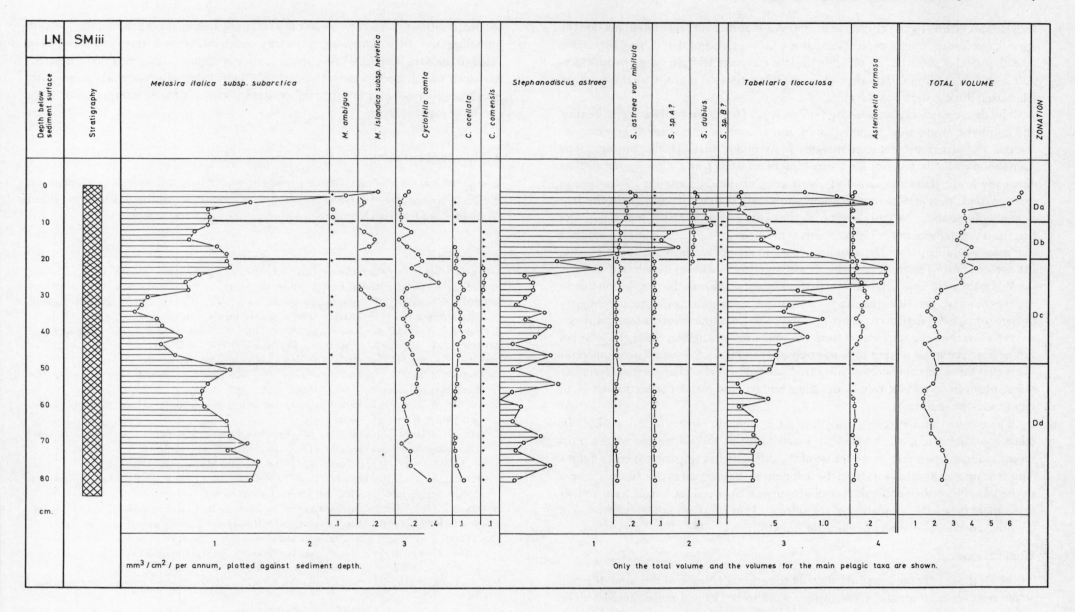

Figure 7. Absolute volumes of individual diatom taxa deposited per annum in core LN. SMiii. See text for further details.

shown that the measured diatom production in 1969 and 1970 was 6.25 mm³/l and 9.2 mm³/l respectively. The average yearly deposition for the top few centimetres of sediment has been calculated to just over 6 mm³/cm²/annum (Fig. 5c), suggesting a very close correlation between the two sets of figures. However, using the figures of

Einsele and Grim (1938), Gibson et al. showed that this volume of diatoms, for the species concerned, only represents about 2.5 SiO_2 mg/l, and that during the spring growth period of 1970 the silica level in the lake was reduced from 7 mg/l to undetectable levels. In addition, Battarbee (in prep) has estimated that the total amount of silica used during 1970 was 8.6 mg/l.

The discrepancy is discussed by Gibson et al. (1971), and they suggest benthic and epiphytic production, sedimentation, and grazing to be the most likely reasons for the low planktonic volumes measured. Assuming that benthic and epiphytic production can only account for a small fraction (10%?), and allowing for outflow down the lower Bann calculated at about 10% of the crop/annum (Battarbee, in prep., Wood, personal communication) the yearly diatom deposition on the lake floor should be about 80% of gross production (or perhaps 6-7 SiO_2 mg/l). The fact that the estimated amount in Fig. 5c is only at the most 30 to 40% of this figure suggests that a considerable part of the diatoms deposited are not retained in the sediment at the present time. The most probable causative factor is diatom dissolution.

When the sources of soluble silica in the lake are examined, the dissolution theory receives more weight. Battarbee (in prep.) has shown that during the summer months of 1970 over 4 SiO_2 mg/l arrived into the lake water from sources other than the major rivers, when the net contribution from the rivers was negligible. Since the total net inflow from the rivers during 1970 was only about 4 mg/l (cf. 8.6 mg/l used), it seems likely that this internal source of silica (redissolved?), if also characteristic for other years, plays an important role in the silica budget, and limits the accumulation of diatoms in the sediments.

The comparison between the water and the sediment figures cannot perhaps be taken too far because of the possible inaccuracies in the calculation of the most recent sedimentation rate, and because of the difficulty in comparing average figures representing a number of years from the sediment with a one year cycle of measurement in the lake. Nevertheless the data strongly suggest the need for longer term studies and a more detailed examination of the silica cycle in relation to the preservation of diatoms, and the availability of silica for growth.

Conclusions

Although the results presented are the first from Lough Neagh of this kind, the data so far are encouraging, and allow further work to be defined more precisely. The substantiation of the chronology is seen as the first most important requirement so that all fossil evidence (biological and biochemical) can be quantified according to annual deposition. Analyses of other forms of fossil evidence should help to locate past productivity changes more accurately, and assist in the further interpretation of the diatom and other results. A further requirement is an investigation into diagenetic and fossilisation processes in the lake as a means of making more precise links between limnology and palaeolimnology. This may not enable quantitative corrections to be made to the data from fossil assemblages, since it cannot be assumed that diagenesis has acted in the same way in the past, but knowledge of present day mud-water interface processes may produce information invaluable to the interpretation of the results of fossil analysis.

Acknowledgements

The investigation was carried out whilst the writer held a Research Studentship at the New University of Ulster, N. Ireland. I should like to thank Professor Frank Oldfield and Dr. Brian Wood for encouragement and discussion.

References

ANDERSEN S.TH. (1970) The relative pollen productivity and pollen representation of north European trees, and correction factors for tree pollen spectra. Danm. geol. Unders. Ser. 11, 96, 99 p.

ANDERSEN S.TH. (1973) The differential pollen productivity of trees and its significance for the interpretation of a pollen diagram from a forested region. (this volume).

BATTARBEE R.W. (1972) A new method for the estimation of absolute microfossil numbers. (in press).

BATTARBEE R.W. (in prep.) D.Phil. thesis. The New University of Ulster.

BINNEY C. (in prep.) D.Phil. thesis. The New University of Ulster.

DIGERFELDT G. (1971) The post-glacial development of Lake Trummen, Småland, Central South Sweden. University of Lund, unpublished manuscript.

EDMONDSON W.T. (1969) Cultural eutrophication with special reference to Lake Washington. Mitt. Int. Verein. Limnol. 17, 19-32.

EINSELE W. & GRIM J. (1938) Uber den Kieselsäuregehalt planktischer Diatomeen und dessen Bedeutung für einige Fragen ihrer Ökologie. Z. Bot. 32, 545-90.

FAEGRI K. & IVERSEN J. (1964) Textbook of Pollen Analysis, Munksgaard, Copenhagen.

GIBSON C.E., WOOD R.B., DICKSON E.L. & JEWSON D.H. (1971) The succession of phytoplankton in Lough Neagh, 1968-70. Mitt. Int. Verein. Limnol. 19, 146-60.

HAWORTH, E. (1969) The diatoms form a sediment core from Blea Tarn, Langdale. J. Ecol. 57, 429-41.

MACKERETH F.J.H. (1969) A short core sampler for subaqueous deposits. Limnol. Ocanogr. 14, 145-51.

MCCRACKEN E.M. (1944) The composition and distribution of woods in N. Ireland from the sixteenth century down to the establishment of the first Ordnance Survey. Unpublished M.Sc. thesis, Queens University, Belfast.

MCCRACKEN E.M. (1947) The woodlands of Ulster in the early seventeenth century. Ulster J. Archaeol. 10,

MCCRACKEN E.M. (1971) The Irish woods since Tudor times. Their distribution and exploitation. David & Charles, Newton Abbott.

O'SULLIVAN P.E., OLDFIELD F. & BATTARBEE R.W. (1973) Preliminary studies of Lough Neagh Sediments 1 Stratigraphy, Chronology and Pollen Analysis. (this volume).

ZULLIG H. (1955) Sedimente als Ausdruck des Zustandes eines Gewässers. Mem. Ist. Idrobiol. suppl. 8, 485-530.

The impact of European settlement on Shagawa Lake, Northeastern Minnesota, U.S.A.*

JOHN P. BRADBURY AND JEAN C.B. WADDINGTON
Limnological Research Center, University of Minnesota, U.S.A.

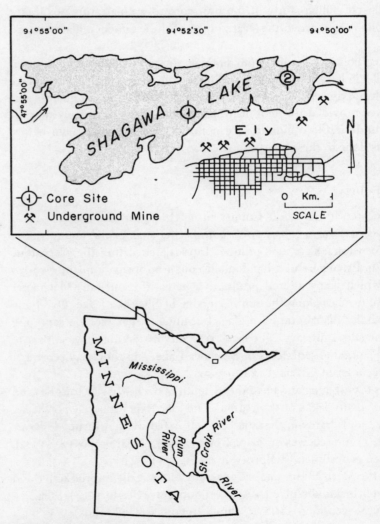

Figure 1. Location maps.

Introduction

The concepts and techniques of palaeolimnology can provide an important historical perspective about the kind and intensity of environmental changes caused by man when the sedimentary record in sensitive lakes can be related to human activities. Such a record exists in Shagawa Lake, northeastern Minnesota, and its detailed study has shown how diatoms, cladocerans, sediment composition, and the surrounding vegetation have responded to the arrival of the first European settlers and to the subsequent growth and development of Ely, a mining and logging town on the southern shore of the lake.

Shagawa Lake has been subjected to a wide variety of cultural disturbances, ranging from the destruction of its surrounding forests to the indiscriminate disposal of mine wastes and municipal sewage in its water. Today it is considered a highly polluted lake, and because of this it is the site of a Federal experimental sewage-treatment plant designed to measure the effect of variously treated municipal sewage on the algal life of the lake. Productivity and other limnological studies are now underway (Megard 1969).

These experimental studies, as well as the need to test stratigraphically the likelihood that human interference was responsible for the present condition of Shagawa Lake, inspired an initial study of the lake's trophic history in which diatoms, cladocerans, and sediment mineralogy were studied from a short core taken near the outlet of Ely's municipal sewage system (Bradbury & Megard 1972). A chronology was established by cursory examination of pollen to detect the rise in *Ambrosia* pollen, a weedy composite that invades cleared land, and by counting grains of hematite and limonite (iron ore) to indicate the beginning of mining next to the lake. The results of this study indicated a dramatic change in the diatom flora of the lake coincident with the beginning of mining and settlement around the shore. Cladoceran assemblages changed too, but not so dramatically. Because diatom preservation in the lower parts of the core was poor (possibly because of the marginal location of the core (Fig. 1, core site 1) it was suspected that some of the observed changes were artifacts of preservation, and the role of man in altering the limnology of the lake

*Contribution No 112 Limnological Research Center

could not be completely resolved until another core was studied—preferably from the deepest part of the lake, where sediment accumulation rates are high and where removal or destruction of microfossils would be less likely. Accordingly a second core was taken from the deep eastern basin of the lake (Fig. 1, core site 2), and a second detailed study was undertaken.

The first study was not designed to investigate changes in the forest vegetation around the lake; however, because man may have substantially influenced the lake indirectly through land clearance and lumbering, it was desirable to complete a detailed, comprehensive pollen study of the new core to evaluate this possibility and to relate the pollen sequence to the stratigraphy of the sediments, diatoms, and cladocerans.

This paper integrates the results from both studies to establish pre-cultural ecologic and limnologic base lines, and to document the history of Shagawa Lake since significant cultural development began between 1880 and 1890. The information and perspectives provided by such historical analyses may ultimately make it possible to detect undesirable trends in lake pollution and to arrest them before unnecessary harm is done to these ecosystems.

Geology and limnology

Shagawa Lake is located in St. Louis County in northeastern Minnesota, U.S.A. (Fig. 1). The bedrock in the area consists of Precambrian metamorphic and intrusive rocks, including greenstone, slate, and granite. Intercalated within the greenstone are discrete lenses and stringers of a banded siliceous iron formation (the Soudan Iron Formation), which when altered produces a hematitic iron ore (Machamer 1968). A rich lens of ore lies along the southern edge of Shagawa Lake. Overlying the bedrock in a patchy distribution are glacial sediments that include sand and gravel and lacustrine silts and clays. The lacustrine silty calcareous clays appear to continue beneath the organic sediments of Shagawa Lake, and they may represent the deposits of a precursor of Glacial Lake Agassiz.

Shagawa Lake is 6.5 km long and has a mean depth of 6.2 m. In size (964 ha) and depth it is typical of many lakes of the region. The long axis runs east-northeast, and the lake narrows and drains in that direction. It is fed by the Burntside River at the west end (Fig. 1) and receives drainage from about 92 square miles (238 km^2), about a fifth of which is occupied by Burntside Lake.

The lakes in northeastern Minnesota are generally biologically unproductive and characterised by water low in dissolved solids and nutrients (Bright 1968). Shagawa Lake is one of the few exceptions. It has a salinity (total dissolved solids) of 43 mg/l which is about twice that of neighbouring lakes. Most major and minor elements fit this pattern except for phosphorus which is 10 times higher (mean = 45 μg/l) in Shagawa Lake than in surrounding lakes.

Shagawa Lake is shallow and overturns easily in the spring, but in mid-summer it stratifies, producing an epilimnion about 3 m thick that may have very high chlorophyll *a* concentrations. The blue-green algae *Aphanizomenon flos-aquae* and *Anabaena* spp. are the dominant phytoplankters during the summer, but during the early spring and late fall diatoms are comparatively abundant. Dense populations of *Stephanodiscus minutus* Grun. (= *S. astraea* var. *minutula*) developed beneath clear ice during February and March, 1968 (Megard 1969), and *Fragilaria*, *Asterionella*, *Melosira*, and *Cyclotella* comprised more than half of the phytoplankton in November, 1965 (Minnesota Pollution Control Agency 1969). Productivity measured by oxygen light-dark-bottle experiments show that the daily gross photosynthesis during the summer is about 1.4 g carbon/m^2, while in other lakes in northeastern Minnesota the daily gross photosynthetic rate is only 0.4 to 0.8 g carbon/m^2 (Bradbury & Megard 1972).

Vegetation

Most of the land in the five townships surrounding Shagawa Lake has been disturbed since European settlement (post-1880) by fire and logging, and with the development of roads and houses ancillary to the town of Ely on the south side of the lake (Fig. 1).

The Federal Land Survey notes of 1880 for the townships around Shagawa Lake (then called Long Lake) were carefully scrutinised for vegetation indicators, fire history, and signs of European settlement. The burned areas noted around the lake in 1880 probably represent the 1875 fire that swept through this region, according to M.L. Heinselman (personal communication). Absence of roads, dwellings, and logged areas on the plot maps suggest that this area was not settled prior to the date of the survey (November 1880). Surveyor's records for these townships show that a total of 350 trees were blazed as witness trees, and of these 110 were pines, with 36 being white pine. White pine thus comprised 10% of the trees and 33% of the total pine population prior to logging or other disturbance by settlers.

The present vegetation was compared with sites in the nearby Boundary Waters Canoe Area (B.W.C.A.) that have not been logged or otherwise disturbed, and with the valuable survey of plant communities of the B.W.C.A. (Ohmann & Ream 1971) in order to understand how the vegetation mosaic that existed prior to 1880 around Ely has changed with cultural development.

The natural vegetation consists of a complex mosaic of many different plant species growing together in various assemblages at different locations depending on variations in drainage, soils, and nutrients. Curtis (1959) and Ohmann and Ream

(1971) indicate that stands within a community vary considerably but still resemble each other more than stands in other communities. Major components of this vegetation complex are listed in Table 1 (from Ohmann & Ream 1971). Vegetation patterns reconstructed from the pre-settlement pollen record suggest little change over the past 2,000 years with the exception of the changes caused by man since 1888. These changes are considered in the discussion of vegetation history.

Cultural history

Rich deposits of iron ore (hematite) were discovered in the Vermilion iron range near the southern shore of Shagawa Lake in 1886. Two years later mining began, and by 1900, five mines, all located near the lake, were in operation (Fig. 1). The tonnage shipped from these mines reached a maximum at this time (Fig. 2). Subsequent production fluctuated wildly, with a marked low occurring during the depression years, especially 1932. After World War II there was decreased production in many of the mines, and the last mine was shut down in 1967.

Local lumbering began at the same time as mining, partly in response to the demand for mine timbers. Most of the timber around Shagawa Lake and Ely was cut by 1895.

The population census of Ely shows an abrupt increase from 1890 to 1900, a peak population of about 6,000 in 1930, and a gradual decline thereafter. The early rise in population parallels the increasing ore shipments from the mines, but after 1900 there is little correspondence, and probably Ely's population changes have resulted from the interaction of several factors in addition to the mining economy. Much of today's economy is related to tourist use of the B.W.C.A. This summer, transient, tourist population is not measured by census statistics, but undoubtedly has a significant impact on Shagawa Lake and the surrounding environment. Between 1953 and 1963 visitation of the B.W.C.A. increased from 50,000 visitors to 230,000 visitors. A large majority of these are canoeists and campers and spend some time and money in Ely, a major outfitting centre for wilderness travel (U.S. Dept. of Agriculture 1964).

Ely has had a municipal sewage system since about 1901. Treatment was initiated in 1912 with the installation of two Imhoff tanks (in which particulate sewage settles and is anaerobically decomposed), and by 1954 most of the present sewage treatment facilities including a primary settling tank, a high-rate trickling filter, and a secondary settling tank were constructed (Brice & Powers 1969). The plant was remodeled in 1963 and since that time operates within its design capacity treating up to 1,000,000 gallons/day by reducing the biological oxygen demand (B.O.D.) from 100 mg/l to 20 mg/l (Minnesota Pollution Control Agency 1969).

Methods

To secure samples for stratigraphic analysis from Shagawa Lake a 1.64 m core was taken from the lake bottom with a piston sampler of 5 cm diameter (Cushing & Wright 1965). The coring site was located at the east end of the lake in water 12 m deep (Fig. 1). Duplicate, quantitative (0.5 cm^3) subsamples were taken every cm for pollen and diatom analysis from 0 to 50 cm, and every other cm from 50 to 160 cm. Quantitative samples in the loose, wet sediment (0–21 cm) were taken with disposable syringes; in the remaining stiffer section sediment was uniformly packed into a small piston core sampler, and a 0.5 cm^3 sample was extruded. After each quantitative subsampling the remaining sediment was put in a labelled plastic bag.

POLLEN PREPARATION AND COUNTING METHODS

Quantitative pollen samples plus 0.5 cm^3 of a standardised suspension of exotic pollen received the following chemical treatment to digest the sediment and concentrate the pollen: 50% acetic acid; hot hydrofluoric acid (40 min), followed by 95% ethanol prior to centrifugation to reduce the specific gravity; concentrated hot hydrochloric acid (30 min); acetolysis solution; and safranin stain. The residue was mounted in glycerine.

Microscope slides were prepared as thin as possible and scanned on a Wild M20 microscope equipped with ×6 oculars and a Fluotar ×50/1.0 oil-immersion objective. For identification of difficult grains a Fluotar ×100/1.30 oil-immersion objective was used. Especially thick slides were used for pine species determination. The ratio of pine species was established by counting 100 pine grains in which the furrow detail could be seen.

DIATOM PREPARATION AND COUNTING METHODS

0.5 cm^3 of sediment was oxidized in a 5:1 mixture of 30% hydrogen peroxide and concentrated nitric acid until the organic matter was gone. The residue was washed free of all traces of oxidants, and a known volume was mounted in Hyrax (refractive index 1.65). One thousand diatom valves were counted from each level with a Wild M20 microscope with ×10 oculars and a Fluotar ×100/1.30 oil-immersion objective.

SEDIMENT ANALYSIS

Permanent microscope slides were made of untreated sediment at each level investigated. Homogenised lake mud was dispersed on a cover slip with Clearcol mounting

Table 1. Occurrence of tree and shrub species in virgin plant communities of the Boundary Waters Canoe Area, Minnesota. (after Ohmann & Ream 1971)

			Presence		Frequency	
Common name	Scientific name	Family name	Number of stands	Percent of stands	Average	Maximum
TREES						
Balsam fir	*Abies balsamea* (L.) Mill.	Pinaceae	70	66.0	26.1	100
Balsam poplar	*Populus balsamifera* L.	Salicaceae	1	.9	.1	5
Bigtooth aspen	*Populus grandidentata* Michx.	Salicaceae	11	10.4	.8	10
Black ash	*Fraxinus nigra* Marsh.	Oleaceae	1	.9	.1	5
Black spruce	*Picea mariana* (Mill.) B.S.P.	Pinaceae	71	67.0	28.7	100
Jack pine	*Pinus banksiana* Lamb.	Pinaceae	60	56.6	31.6	100
Mountain-ash	*Pyrus americana* (Marsh.) DC.	Rosaceae	4	3.8	.5	30
Paper birch	*Betula papyrifera* Marsh.	Corylaceae	85	80.2	29.2	100
Quaking aspen	*Populus tremuloides* Michx.	Salicaceae	72	67.9	29.3	100
Red maple	*Acer rubrum* L.	Aceraceae	33	31.1	9.5	90
Red oak	*Quercus rubra* L.	Fagaceae	6	5.7	.8	35
Red pine	*Pinus resinosa* Ait.	Pinaceae	19	17.9	6.2	100
White-cedar	*Thuja occidentalis* L.	Pinaceae	22	20.8	10.8	100
White pine	*Pinus strobus* L.	Pinaceae	34	32.1	12.2	100
White spruce	*Picea glauca* (Moench) Voss.	Pinaceae	56	52.8	7.4	50
TALL SHRUBS						
Beaked hazel	*Corylus cornuta* Marsh.	Corylaceae	82	77.4	31.9	100
Bebb willow	*Salix bebbiana* Sarg.	Salicaceae	26	24.5	2.3	30
Bush honeysuckle	*Diervilla lonicera* Mill.	Caprifoliaceae	86	81.1	36.6	100
Chokecherry	*Prunus virginiana* L.	Rosaceae	8	7.5	.4	10
Downy arrow-wood	*Viburnum rafinesquienum* Schultes.	Caprifoliaceae	7	6.6	.8	20
Elderberry	*Sambucus* spp.	Caprifoliaceae	1	.9	.1	5
Fly honeysuckle	*Lonicera canadensis* Bartr.	Caprifoliaceae	63	59.4	8.3	45
Green alder	*Alnus crispa* (Ait.) Pursh.	Corylaceae	39	36.8	7.9	80
Ground-hemlock	*Taxus canadensis* Marsh.	Taxaceae	6	5.7	2.4	100
Hairy climbing honeysuckle	*Lonicera hirsuta* Eat.	Caprifoliaceae	10	9.4	1.5	42
Juneberry	*Amelanchier* spp. Medic.	Rosaceae	79	74.5	14.5	60
Low juniper	*Juniperus communis* L.	Pinaceae	4	3.8	.5	30
Mountain fly honeysuckle	*Lonicera villosa* (Michx.) R. & S.	Caprifoliaceae	1	.9	.1	15
Mountain maple	*Acer spicatum* Lam.	Aceraceae	62	58.5	27.3	95
Pincherry	*Prunus pensylvanica* L.f.	Rosaceae	14	13.2	.9	20

Table 1. (continued)

Common name	Scientific name	Family name	Presence		Frequency	
			Number of stands	Percent of stands	Average	Maximum
Red osier	*Cornus stolonifera* Michx.	Cornaceae	3	2.8	.3	20
Round-leaved dogwood	*Cornus rugosa* Lam.	Cornaceae	28	26.4	5.4	60
Speckled alder	*Alnus rugosa* (Du Roi) Spreng.	Corylaceae	2	1.9	.1	5
Spiraea	*Spiraea* spp. L.	Rosaceae	1	.9	.1	5
Sweet fern	*Comptonia peregrina* (L) Coult.	Myricaceae	6	5.7	1.3	70
Willow	*Salix* spp. L.	Salicaceae	8	7.5	1.4	45
LOW SHRUBS						
Bearberry	*Arctostaphylos uva-ursi* (L.) Spreng.	Ericaceae	5	4.7	0.5	15
Creeping snowberry	*Gaultheria hispidula* (L.) Bigel.	Ericaceae	11	10.4	.6	10
Currant	*Ribes* spp. L.	Saxifragaceae	5	4.7	.4	10
Dewberry	*Rubus pubescens* Raf.	Rosaceae	46	43.4	8.0	55
Labrador tea	*Ledum groenlandicum* Oeder.	Ericaceae	1	.9	.0	5
Late sweet blueberry	*Vaccinium angustifolium* Ait.	Ericaceae	65	61.3	17.0	100
Pipsissewa	*Chimaphila umbellata* (L.) Bart.	Pyrolaceae	19	17.9	1.2	15
Prickly rose	*Rosa acicularis* Lindl.	Rosaceae	27	25.5	1.9	20
Red raspberry	*Rubus strigosus* (Michx.)	Rosaceae	29	27.4	5.0	90
Thimbleberry	*Rubus parviflorus* Nutt.	Rosaceae	4	3.8	.8	35
Velvet-leaf blueberry	*Vaccinium myrtilloides* Michx.	Ericaceae	45	42.5	5.6	65
Wintergreen	*Gaultheria procumbens* L.	Ericaceae	28	26.4	9.0	90

Nomenclature follows Fernald (1950)

medium, or with water, dried, and mounted in Canada balsam or Hyrax. Such slides are an invaluable addition to multicomponent palaeoecological studies, because all microfossil and chemical profiles can be easily referred back to their original sedimentary matrix to check for variation in gross or minute sedimentary features. Because chemical treatments for specific analyses generally destroy unwanted sedimentary and palaeontologic components (e.g. diatoms are lost in pollen preparations by use of hydrofluoric acid), the slides are very important in helping the investigator relate the specialised studies to the complete record.

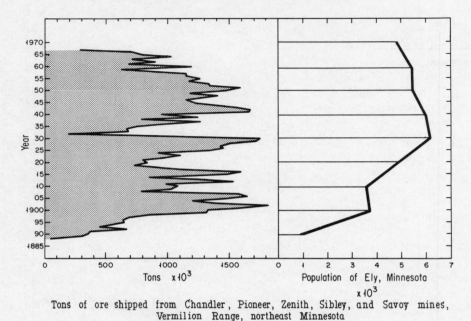

Figure 2. Population of Ely, and tons of ore shipped from Chandler, Pioneer, Zenith, Sibley, and Savoy mines, Vermilion Range, northeast Minnesota.

A method for semi-quantitative sediment description in the laboratory has been developed by the authors to be used routinely in palaeoecologic studies. Unlike the Troels-Smith (1955) method, which is field-oriented and classifies material according to its derivation, the method used here emphasizes microscopic analysis and the non-genetic description of sediment particles. The abundance of fossils and other sediment components (silt, clay, distinctive mineral grains, charcoal fragments, etc.) is estimated on a scale of 1-5 and recorded on data sheets (Table 2). The major components are presented diagrammatically as a stratigraphic column (Fig. 3). We support the use of Troel-Smith's graphic representation but have used his symbol for *limus detrituosus* in a less restricted sense to refer simply to the organic components of the sediment.

The water content was determined by weight loss at 110°C, and the organic component, here called biopel, by weight loss on ignition at 550°C. We introduce the term *biopel* (Gr.: *bios* = life, *pel* = mud) to describe the organic component of lake mud regardless of its origin. The terms available, such as *gyttja*, *mud*, *muck*, *ooze*, *slime*, etc. are all loosely used; they ordinarily include the inorganic as well as the organic component. *Sapropel* is restricted to the description of highly reduced, bituminous sediments (American Geological Institute 1960). Organic sediment must be consumed and excreted by arthropods before it becomes *copropel* (Swain & Prokopovich 1954), and this is not only difficult to prove but is seldom relevant to palaeoecological studies. *Biopel* is approximately equivalent to *gyttja*—a term that is not self-explanatory and yields no adjectival form.

Iron in the sediments was analyzed by digestion with concentrated hydrochloric acid, reduction with stannous chloride, and titration with potassium dichromate. In addition, grains of hematite, a common and distinctive mineral in the upper sediments, were counted in the diatom preparations.

Diatom silica (opal) was removed from the washed sediments by alkaline digestion with 4N sodium hydroxide, and the weight loss was determined after filtration on pre-weighed filter paper. This may be a more reliable estimate of diatom productivity than counts from volumetrically standardised sample preparations, because broken diatom fragments are commonly abundant in sediments and very difficult to identify or quantify by counting. Mild digestion (boiling for 10 min) is sufficient to remove the opaline silica and does not appreciably affect the clastic silicate minerals, although there was a slight weight loss after digestion of a powdered silt blank. Digestion with milder alkaline solutions would probably mitigate this effect.

Phosphorus was analyzed spectrophotometrically after the sediments were digested with potassium persulphate.

MISCELLANEOUS ANALYSES

Charcoal fragments of three size classes ($<50\ \mu m^2$, 50-$400\ \mu m^2$, and $>400\ \mu m^2$) were tallied from each pollen slide in five pre-set traverses. The projected area of each fragment was estimated by comparison with a graticule grid with squares 10 μm on a side (Waddington 1969). Exotic pollen grains were counted at the same time to compute the charcoal density, which is expressed as total fragment area per unit volume (Fig. 3).

Green algae were also counted in the pollen slides and their abundance indicated in Fig. 3. Blue green algae were essentially absent from the pollen and sediment slides.

Cladocera were counted in aliquots prepared from about 5 cm^3 of wet sediment disaggregated by boiling in 10% hydrochloric acid and then 10% potassium hydroxide. The fossils were concentrated by sieving through a no. 230 mesh screen (opening = 63 μm).

Table 2. Description of sediments by microscopic examination.

Sediment depth	Water %	Colour	Stratification	Elasticity	Tissues	Cladocera	Diatoms	Algae—other	Desmids	Pollen	Spicules	Phytoliths	Cysts	Rhizopods	Chitin	Lepid. scales	Organic ooze	Fungal elements	Charcoal	Other	Sand	Silt	Clay	NaCl	CaSO$_4$	CaCO$_3$	FeS	Fe$_2$O$_3$	Fe$_3$O$_4$	Chlorite	Other	Comments
											Biopel											Clasts				Minerals						
1	97		0	0	0	1	5	3	2	1	0	0	0	1	3	0	3	1	2		1	1					1	3		0		
6	95		0	0	0	3	5	3	1	1	1	0	3	1	3	0	3	0	0		1	1					1	3		0		
10	94		0	0	0	1	5	3	0	1	2	0	3	1	2	0	3	0	0		1	1					1	3		0		
16	92		0	0	0	1	5	3	1	1	2	0	3	1	2	0	4	0	0		1	1					1	3		0		
20	90		0	0	0	1	5	3	0	1	1	0	3	1	1	0	4	0	0		1	1					1	3		1		
24	89		0	0	0	1	5	3	0	1	1	0	2	1	1	0	4	1	2		1	1					1	3		2		
26	89	pink	0	0	0	1	5	3	1	2	1	0	2	1	1	0	4	1	2		2	1					1	3		3		
30	91	pink	0	0	0	1	5	3	1	2	1	0	2	1	1	0	5	1	2		3	1					1	4		3		
34	88		0	0	0	1	4	3	1	2	1	0	2	1	1	0	5	1	2		3	1					1	3		4		
36	89		0	0	0	1	4	3	0	2	2	0	2	1	1	0	5	1	2		2	1					1	2		4		
40	88		0	0	0	1	3	3	0	3	2	0	2	1	1	0	5	1	2		2	1					1	2		3		
50	88		0	0	0	1	3	3	1	3	2	0	2	1	1	0	5	0	2		2	1					1	2		2		
60	88		0	0	1	1	3	2	0	4	2	0	2	1	1	0	4	0	2		2	1					1	1		1		
70	88		0	0	1	1	3	2	0	4	1	0	2	1	1	0	4	0	2		2	1					1	1		1		
80	87		0	0	1	1	3	2	0	4	1	0	2	1	1	0	3	0	2		2	1					1	0		1		
90	87		0	0	1	1	3	2	0	4	1	0	2	1	1	0	3	0	2		2	1					1	0		2		
100	87	olive	0	0	1	1	4	2	0	4	2	0	2	1	1	0	4	1	2		2	1					1	0		3		
110	86		0	0	0	1	4	2	0	4	2	0	2	1	1	0	3	1	2		2	1					1	0		2		
120	87		0	0	0	1	4	2	0	4	2	0	2	1	1	0	3	1	2		2	1					1	0		1		
130	87		0	0	1	1	4	2	0	4	1	0	2	1	1	0	4	1	3		2	1					1	0		1		
140	86		0	0	0	1	4	2	0	4	1	0	2	1	1	0	4	1	3		2	1					1	0		2		
150	86		0	0	1	1	4	2	0	4	1	0	2	1	1	0	4	1	3		2	1					1	0		2		
160	86																															

0 = absent 3 = frequent
1 = rare 4 = abundant
2 = occasional 5 = very abundant

Palaeolimnology

CHRONOLOGY

In non-laminated lake sediments, an absolute chronology can only be established by carbon-14 dating and by correlation of sedimentary events with historical events of known date. We have used both techniques to establish a chronology for the 160 cm core from Shagawa Lake. Carbon-14 analysis (I-6329) on material from 158–160 cm yields a date of 1935 ± 125 years B.P. The stratigraphy of iron, hematite grains, and phosphorus (Fig. 4) can be correlated with the cultural history of Ely.

The graph of hematite grains (and to some extent of total iron) reflects the mining history of Ely. When water seeped into the underground mines there it had to be pumped out, and the wastewater contained many fine particles of iron ore (hematite), which eventually settled to the bottom of Shagawa Lake. Pumping was necessary in

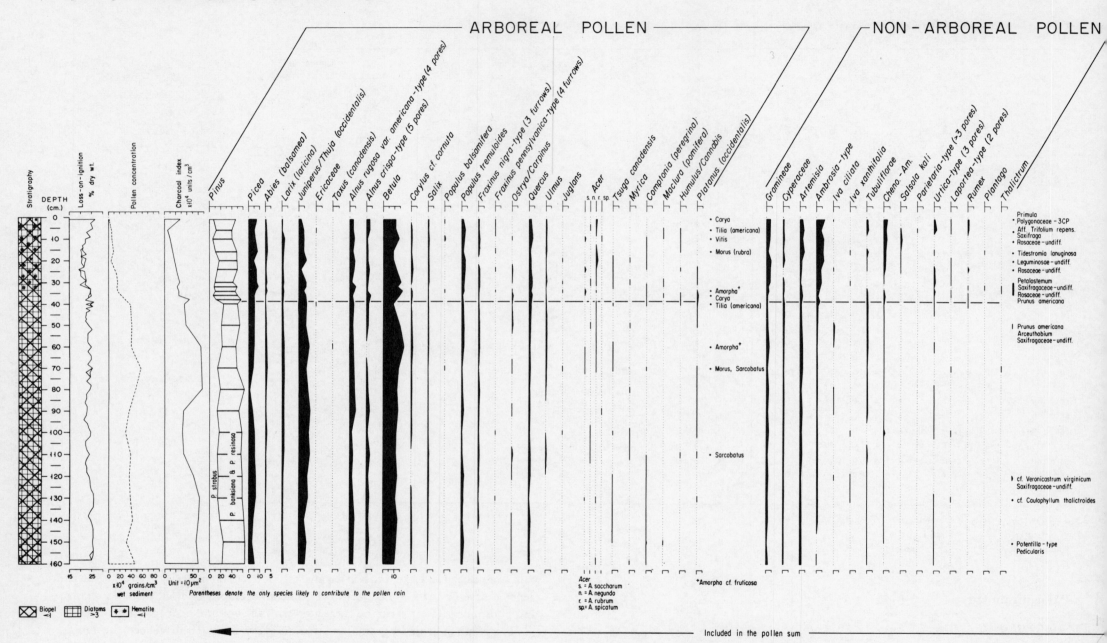

Figure 3. Lake Shagawa, St. Louis Co., Minnesota: Pollen Diagram.

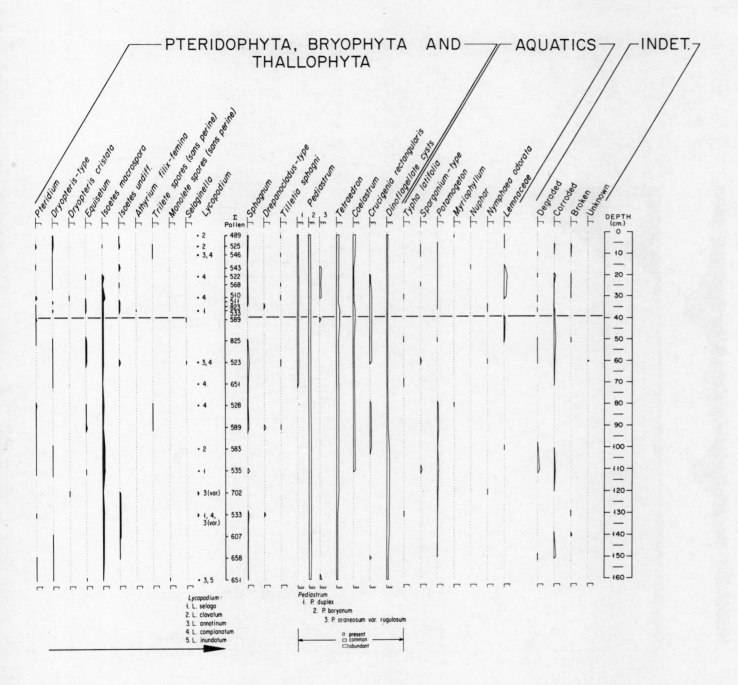

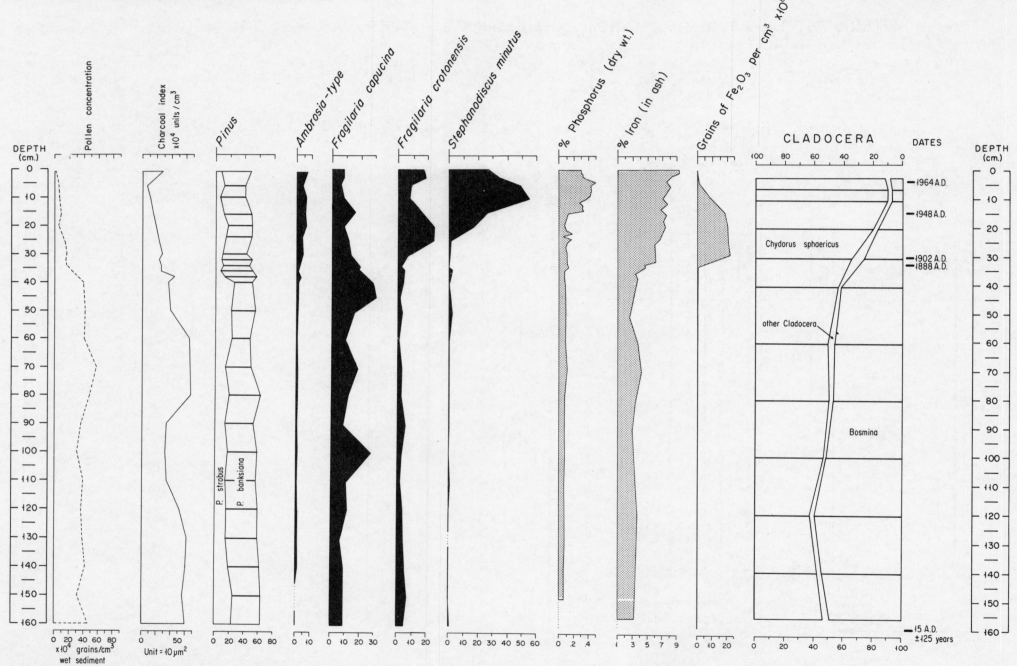

Figure 4. Selected stratigraphic profiles from Shagawa Lake.

both the Chandler and Pioneer mines in 1890 (Winchell & Winchell 1891) and possibly somewhat earlier, and it is reasonable to correlate the beginning of the sharp rise in the number of hematite grains (33 cm) with this date. In the same way the peak in ore production that occurred in 1902 (Fig. 2) may be correlated with the peak in hematite grains at 30 cm depth (Fig. 4). The wildly fluctuating production since 1902 is not recorded in the hematite stratigraphy, but the decline of mining activity, beginning in 1951 and ending in 1967, when the last mine closed, is probably represented by the diminishing number of hematite grains in the sediment above 15 cm depth. The curve for total iron continues to increase to the surface, and probably reflects increasing erosion and deposition of very fine-grained, iron-rich sediment (not optically recognisable) from the dumps near the mines and from the disturbed land around Ely.

The phosphorus profile also shows changes that can be tentatively related to cultural events. For most of the time represented by the core the phosphorus content of the sediment has remained relatively constant. There is a very slight increase—hardly greater than the background level, however—at a depth of 34 cm that may represent an increase in erosion due to logging or the earliest mining efforts. From that level to 15 cm it remains comparatively constant, aside from a double-peaked fluctuation of unknown cause between 25 and 20 cm depth. At 14 cm there is a sharp increase in the phosphorus content of the sediments. It increases again at 10 cm, reaches its maximum at 4 cm, and thereafter declines somewhat. Tentatively we ascribe the increase at 14 cm to the year 1948, when detergents containing phosphates became widely used in the United States. The steep increase between 10 and 4 cm may reflect the nearly five-fold increase in summer tourism in the area between 1953 and 1963. The decline in phosphorus after 4 cm may represent the remodelling and improvement of the sewage treatment facilities in 1963 (U.S. Dept. of Agri-

Table 3. Correlation of Shagawa Lake stratigraphy with historically documented events in the immediate area.

Stratigraphic evidence	Date	Years	Event	Depth in core (cm)	Stratigraphic interval (cm)	Deposition time (years/cm)
Water/sediment interface	1971		Surface	0		
		8			4	2.0
Phosphorus profile declines	1963		Remodelling sewage plant (decline in P)	4		
		9			6	1.5
Phosphorus profile shows steep increase	1954		Sewage plant completed; beginning of massive increase in tourism	10		
		3			2	1.5
Hematite profile declining	1951		Beginning of mining decline	12		
		3			2	1.5
Sharp rise in phosphorus profile	1948		Phosphates in detergents	14		
		46			16	2.9
Peak in hematite profile	1902		Peak in mine production	30		
		13			3	4.2
Beginning of hematite increase	1889		Beginning of extensive mining	33		
					127	15
Bottom of core C-14 date	15 A.D. ± 125		—	160		

culture 1964). These events are correlated with the stratigraphic evidence in Table 3. The calculated deposition time (years/cm) indicate that sediment was deposited nearly four times faster following the first cultural disturbance than before, and that with the introduction of phosphates in detergents the deposition rate again more than doubled.

The pre-disturbance deposition time (15 year/cm) is tentatively confirmed by a long core from Shagawa Lake. In this core there are about 8 m of biopelic sediments below the level where hematite grains indicate cultural disturbance. If it is assumed that Shagawa Lake began organic sedimentation about 10,500 years ago, as it did in Weber Lake about 40 miles to the south (Fries 1962), the deposition time becomes about 13 years per cm and reasonably approximates the rate established by the C-14 date of 1935 B.P. (15 A.D.) at 158-160 cm on the short core.

DIATOM STRATIGRAPHY

The diatom stratigraphy of Shagawa Lake (Fig. 5) can be divided into two zones by the behaviour of the profiles for *Fragilaria crotonensis* and *Stephanodiscus minutus*. A marked increase in *F. crotonensis* begins at 33 cm but is replaced by a dominance of *S. minutus* above 20 cm. These two diatoms are of only minor importance in levels below 33 cm. The lower levels are characterised by *Tabellaria fenestrata*, species of *Melosira* (notably *M. ambigua*), species of *Cyclotella*, and variable percentages of *F. capucina*. Benthic and epiphytic diatoms (principally species of *Navicula* and *Achnanthes*) exhibit a constant distribution in the levels below 33 cm and decrease in relative abundance above this level. This pattern is shown by the majority of diatoms and results from the proliferation of a few species above 33 cm. The number of species encountered in a count of 1,000 individuals (Fig. 5) is a simple measure of the great decrease in diversity in the upper levels. The increase in alkaline-soluble silica (largely diatom opal) from 33 cm to the surface suggests that on an absolute scale the diatoms characteristic of the lower 130 cm do not substantially decrease in number. This is emphasized by the fact that *Stephanodiscus minutus*, the principal dominant, is a very small (averaging 10 μm diameter) and weakly silicified diatom that makes a proportionally small contribution to the weight of diatom silica.

With the exception of *Fragilaria capucina*, *Melosira granulata*, and *M. granulata* var. *angustissima*, the dominant diatoms from 160 to 30 cm are characteristic of many of the undisturbed lakes of northeastern Minnesota (Bright 1968). However, many of these types, especially *Melosira ambigua*, *M. italica*, *Tabellaria fenestrata*, *Asterionella formosa*, and *Cyclotella comta* have a wide distribution that suggests they readily adapt to variable water chemistry and nutrient levels. Diatoms that are particularly characteristic of oligotrophic lakes, such as *Cyclotella stelligera*, play only a minor role in the plankton of Shagawa Lake and indicate that for the time represented by the core truly oligotrophic conditions seldom existed. The distribution of *Fragilaria capucina* throughout the core, and especially below 33 cm, argues strongly that Shagawa Lake was naturally mesotrophic or even eutrophic. This diatom is frequently a dominant in small eutrophic lakes in central and southern Minnesota (Bright 1968), and in Lake Michigan it responds positively to pollution (Stoermer & Yang 1970). The two peaks of *F. capucina* in the lower zone, one at 100 cm and another between 50 and 40 cm, must represent times when Shagawa Lake was even more enriched than usual. The presence of *Melosira granulata* and *M. granulata* var. *angustissima* below 33 cm also indicate comparatively eutrophic conditions.

The increasing abundance of hematite grains at 33 cm records the first cultural disturbance around Shagawa Lake, in particular the discovery and exploitation of iron ore in 1888-9. Changes in the diatom stratigraphy after 33 cm can be ascribed to cultural activities of various kinds. Natural causes for changes in the diatom stratigraphy in pre-cultural times (below 33 cm) are less obvious, however. Indeed, the diatom association for this time suggests a greater productivity than is typical for undisturbed lakes in this region, and it is not immediately apparent why this should be so. It is possible that drainage from the calcareous lake clays and silts of Pleistocene age around Shagawa Lake contribute natural nutrients that supported eutrophic diatoms in pre-cultural times. Another possibility is that nutrients reach Shagawa Lake from Burntside Lake during critical times, for example during the spring break-up. The residence time for water in Shagawa Lake is short (1 year) (Megard 1969), and a seasonal pulse of water from a large lake with an extensive drainage area might make an important nutrient contribution.

Such speculations are further complicated by the knowledge that some of the diatoms indicating eutrophy may come from the littoral area of lakes, where they live near the abundant source of nutrients in the sediments. This is particularly true of *Fragilaria capucina* (Stoermer & Yang 1970), and it may also be the case with the *Melosira* species (Stockner & Armstrong 1971).

Changes in the diatom stratigraphy after 33 cm are characterised first by a dominance of *Fragilaria crotonensis* and later by *Stephanodiscus minutus*. *Stephanodiscus hantzschii*, *S. hantzschii* var. *pusillus*, and *Cyclotella catenata* increase in the upper levels. Erosion following mining, timber cutting, and the development of Ely, with the disposal of municipal wastes from this town into Shagawa Lake, are almost certainly responsible for the initial dominance of *F. crotonensis* and the corresponding decline in *Melosira ambigua* as well as many other once characteristic diatoms. *Stephanodiscus minutus* begins its increase between 25 and 20 cm, coincident with the first major change in the phosphorus profile, and exhibits an explosive development at 14 cm at a time (1948) which we suggest corresponds to the introduction of phosphate

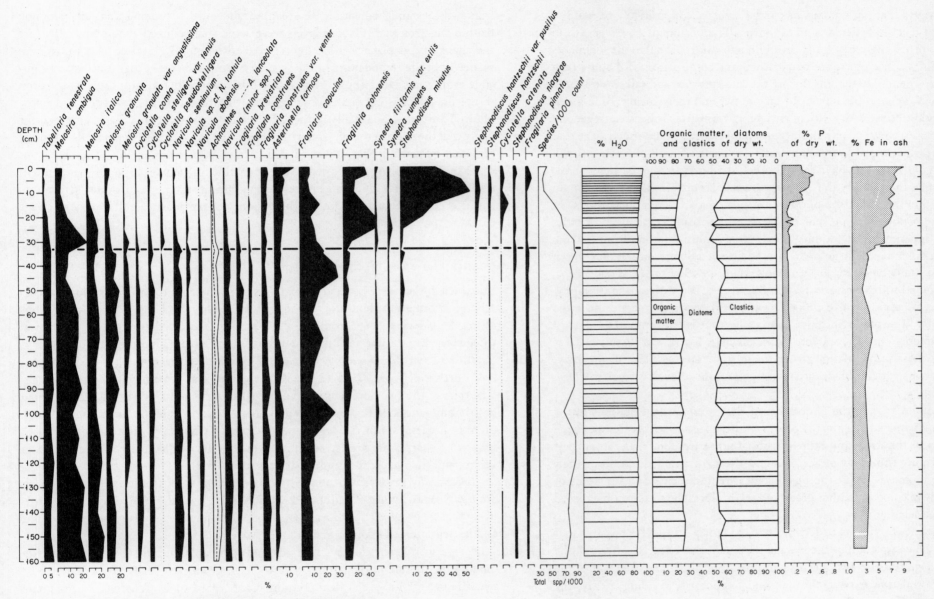

Figure 5. Diatom and Sediment Stratigraphy of Shagawa Lake, Minnesota.

detergents. It appears that both *S. minutus* and *F. crotonensis* positively respond to increased nutrient levels, particularly phosphorus. The decline of *F. crotonensis* in relation to *S. minutus* during the time of maximum phosphorus input to the sediments may reflect the different ecology of the two diatoms and competition with blue-green algae. *F. crotonensis* generally blooms during the late summer and fall (Stoermer & Yang 1970) while *S. minutus* blooms in the late winter and early spring. Blue-green algae become very abundant with nutrient enrichment, apparently because as general productivity increases, the concomitant increase in pH favours blue-green algae that can utilise CO_2 at the low concentrations present in water of high pH (King 1970). Once present, blue-green algae have several other attributes that assist their dominance. For example, their ability to regulate gas vacuoles enables them to maintain their position high in the photic zone, where they can maximise photosynthesis. In addition, when phosphorus is present blue-green algae can fix nitrogen. Lastly, because of their apparent impalatability they may be more immune from grazing zooplankton than other algae (Birge 1897). In the face of a dominance of blue-green algae, a diatom normally blooming at the same season, such as *Fragilaria crotonensis*, would be outcompeted and exhibit a corresponding decline in the diatom stratigraphy.

Stephanodiscus minutus, however, blooms early in the spring or even under the ice in the late winter. The pH in Shagawa Lake at this time is only around 8 (Megard 1969), and presumably this diatom can make effective use of the dissolved CO_2 present to achieve massive populations when the nutrient level is high. Accordingly, its contribution to the stratigraphic record is very great at those levels where *Fragilaria crotonensis* is low.

This hypothesis to explain the fluctuation of the two dominant diatoms in the horizon of cultural disturbance ignores obvious limiting factors such as temperature and light intensity or duration, and quite possibly these are among many important factors that determine the kinds of diatoms and the nature and duration of their blooms. As more is known about the physiological requirements and functions of diatoms, especially in contrast to blue-green and green algae, the stratigraphic record will become more meaningful.

Stephanodiscus minutus begins to decline at 10 cm (1954 A.D.). It is possible that this reflects the final completion of the sewage treatment plant at Ely. Although the phosphorus levels continue to rise, perhaps in response to the large increase in the summer tourist population passing through and temporarily residing in Ely, the nutrients delivered to Shagawa Lake during the summer would not be so important to the *S. minutus* population, whose preferred period for growth is in the winter or early spring. Perhaps during this season the treatment plant was capable of reducing the nutrient input, and in this way reduced the populations of *S. minutus*.

Fragilaria crotonensis begins to rise after 5 cm, and according to the interpretive framework established earlier, the reappearance of this summer and fall-blooming diatom suggests that nutrient levels were somewhat reduced at this time. Because the change takes place at a level correlated with 1963 A.D. it seems reasonable to suspect that the remodelling of the sewage treatment plant at that time was responsible for this change in the nutrient input. This supposition is supported by the drop in the phosphorus curve at the same level. Soluble nutrients (nitrogen and phosphorus) are not substantially reduced by secondary sewage treatment, and it seems likely that the drop in the phosphorus curve is due to a decrease in particulate sewage removed by settling that otherwise would have been sedimented in the bottom of the lake. If this is the case, a reduction in soluble nutrients (usable by algae) might result from a decrease in nutrient resolution during summer and winter stagnation, and this could occur by reducing the quantity of particulate, nutrient-rich sewage reaching the bottom.

CLADOCERA STRATIGRAPHY

Fossil Cladocera were studied from 12 levels in the core to provide additional correlative information about the trophic history of Shagawa Lake. The concentrations of fossil Cladocera range from $1.6 \times 10^4/cm^3$ at 160 cm depth to $4 \times 10^3/cm^3$ at the surface, but the decreasing trend upwards is irregular. The dominant cladocerans at all depths are *Bosmina longirostris* and *Chydorus sphaericus*. Species diversity is low, ranging between 8 and 14 species per count of approximately 500 individuals. The ratio of the two dominants and other species encountered (Fig. 4) shows about equal abundances of *B. longirostris* and *C. sphaericus* from 160 to 40 cm, although there is a maximum of *B. longirostris* at 120 cm. After 40 cm the proportion of *C. sphaericus* sharply increases, reaching a maximum of 90% of the total fauna at 10 cm, and then decreases somewhat thereafter.

Bosmina longirostris is probably the natural dominant in the undisturbed lakes around Shagawa Lake (Bradbury & Megard 1972). Under natural conditions it is likely that *B. longirostris* is dicyclic, with population maxima occurring in May–June and in September–October (Schindler & Noven 1971).

Chydorus sphaericus is not well adapted for planktonic existence, but it is frequently the dominant planktonic cladoceran in productive lakes, particularly those with blooms of blue-green algae (Bradbury & Megard 1972). It manages to maintain itself in the plankton by crawling along the floating algae and feeding on the detritus among the filaments. When blue-green or other large colonial algae are absent, *C. sphaericus* does not enter the plankton but remains in the littoral zone, and its fossils are sparingly deposited in profundal sediments.

The shared dominance of *Chydorus sphaericus* and *Bosmina longirostris* from 160 to 40 cm indicates that even during pre-settlement times Shagawa Lake was moderately productive, and probably *B. longirostris* only achieved its maximum population during the spring. *C. sphaericus* may have replaced it during the summer by taking advantage of blue-green algal blooms to move into the plankton. The high frequency of *C. sphaericus* above 40 cm reflects an increasing frequency, intensity, and duration of blue-green algal blooms. The peak in *C. sphaericus* coincides with the reduction of *Fragilaria crotonensis* between 20 and 4 cm (Fig. 4), which has also been attributed to a dominance of blue-green algae during the summer and early fall. Most likely the slight decrease in *C. sphaericus* at the surface of the core should be attributed to a decrease in algal-bloom intensity and duration that results from the recent drop in nutrient levels indicated by the phosphorus curve and the profile for *Stephanodiscus minutus*.

Correlation and comparison with a short sublittoral core

The first core studied from Shagawa Lake (Bradbury & Megard 1972) was a short core from the lower sublittoral zone adjacent to the outlet of Ely's sewage treatment plant (Fig. 1, core site 1). The hematite, diatom, and pollen stratigraphy (Fig. 6) of this core indicate that the top 25 cm were deposited during the post-settlement period, and that the peak in mine production (1902) can be correlated with the peak in hematite grains at 20 cm. As in the profundal core, *Stephanodiscus minutus* and *Fragilaria crotonensis* are the dominant diatoms in the culturally enriched zone, replacing *Melosira ambigua*, *M. italica*, and *Cyclotella comta* that are prominent below.

Differences between the diatom stratigraphies of the two cores, however, are as noteworthy as the similarities. In the first place the 100 cm sublittoral core has three zones (lower, middle, and upper) while the longer (160 cm) profundal core has only two. The lower zone of the sublittoral core, consisting mainly of benthic and epiphytic diatoms, is absent in the profundal core. Secondly, the stratigraphy of *Fragilaria capucina* is apparently reversed. It is one of the pre-cultural dominants in the profundal core, while in the sublittoral core it is a post-cultural dominant. Consequently *Melosira ambigua* and *Cyclotella comta* make up a greater percentage of the pre-cultural diatom assemblages in the sublittoral core than they do in the profundal core.

These discrepancies may have complex explanations that would require considerable study of the sedimentation patterns and of the neritic vs. limnetic diatom ecology in Shagawa Lake, and the following suggestions are necessarily speculative.

The greater number of diatom zones in the sublittoral core suggest that the rate of sediment accumulation in this core was slower than in the profundal core. The sediment production rate may have been the same or even greater, but it seems likely that currents, perhaps related to the comparatively short residence time of water entering and leaving Shagawa Lake, might winnow fine sediment from the sublittoral bottom and deposit it secondarily in the deeper profundal basins. The greater amount of silt in the diatom preparations from the sublittoral core also suggests sediment differentiation. The poor diatom preservation below 25 cm in the sublittoral core may be related to the lower sediment-accumulation rates. It can be reasoned that the longer a diatom frustule remains on the bottom without being buried the more likely it will be dissolved or broken by burrowing detritus-feeders. The feeding activity of benthic animals is proportional to the oxygen content of the bottom water, which was higher in pre-cultural times (especially in the sublittoral zone) than after cultural enrichment, and it follows that diatom destruction would have been greatest during pre-cultural times, and diatom preservation best during post-cultural times. Long fragile diatoms such as *Fragilaria capucina* would be selectively destroyed, while centric diatoms, (species of *Melosira* and *Cyclotella*), would be preferentially preserved, so *F. capucina* is of only minor importance in the sublittoral core below 25 cm, and the centric diatoms are more abundant.

In the post-cultural levels there is a notable difference in the stratigraphic pattern of *Stephanodiscus minutus* and *Fragilaria crotonensis*. In the sublittoral core they generally increase together with nutrient enrichment, while in the profundal core *S. minutus* replaces *F. crotonensis* with increasing enrichment. The explanation for this is not perfectly clear, but it may relate to a significant contribution to *F. crotonensis* from the littoral area. Studies of littoral diatoms in Minnesota lakes (Bright 1968) indicate that *F. crotonensis* is commonly an important member of the littoral association.

Our ability to match minor fluctuations in the diatom stratigraphy with particular events and trends in the trophic history of a lake, and with variations in the productivity and extent of different diatom habitats, is limited by a general lack of knowledge about the nutritional requirements and general ecology of many diatoms. As ecologic knowledge and biostratigraphic techniques improve, so will the potential of diatoms to produce detailed reconstructions of lake history.

Vegetation history

The most obvious feature of the pollen record (Fig. 3)* is the uniformity of pollen percentages through time with pronounced changes in some taxa limited to the upper 39 cm of the profile. A line denoting these changes is drawn at 39 cm, and

*For a table of pollen numbers, order document NAPS 01823, from ASIS, National Auxiliary Publications Service, c/o CCM Information Corporation, 866 Third Ave, New York, N.Y. 10022, U.S.A.; remitting $2.00 for each microfiche or $5.00 for each photocopy.

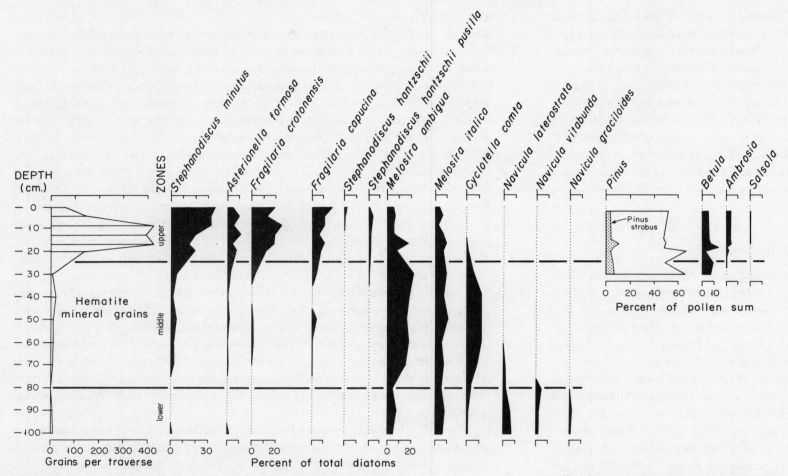

Figure 6. Shagawa Lake, Minnesota: Selected diatom and pollen curves from sublittoral sediments.

conveniently divides the diagram into pre- and post-settlement horizons. The boundary is based on a distinct decline in the percentage of *Pinus strobus* (white pine) pollen coincident with the increase in *Ambrosia* (ragweed) pollen. The increase in *Ambrosia* pollen, together with that of Urticaceae, Chenopodiaceae, *Rumex*, and *Plantago*, occurs in the upper part of most pollen profiles from Minnesota; it is interpreted as an increase in 'disturbance weeds' encouraged by logging and land clearance for agriculture. Pollen of introduced weeds, such as *Salsola kali* appears in the part of the diagram denoting post-settlement time. Invasion of upland grasses on disturbed sites probably accounts for the increased percentages of Gramineae pollen in post-settlement time, although in some cases this can be attributed to the spread of the aquatic wild rice (*Zizania aquatica*) (e.g. McAndrews 1969).

Changes in the pollen percentages are most pronounced in the case of the pines. At 40 cm white pine contributes approximately one-third of the total pine pollen. This value has also been found for other sites in the B.W.C.A. (Craig 1972) and reflects the ratio of white pine to total pine trees obtained from the surveyors' records of 1880. Although white pine pollen decreases, the total pine pollen percentages remain fairly stable, indicating that *P. resinosa* and *P. banksiana* are actually contributing more pollen to the pollen rain.

Datable changes in some of the other sedimentary components such as hematite and phosphorus (Fig. 4) suggest that the pine decline might reflect regional logging patterns rather than local events. Saw mills were operative in the area by 1893, and local logging would have started only shortly before that date when a railroad trunk line was established c 1890 to carry the lumber eastwards to ports on Lake Superior. The projected deposition rate for the core at the 39 cm level may lie somewhere between 4.2 and 15 years/cm and puts the pine decline at 1864 or earlier. We know from the survey records that the area was not logged by 1880. Fixing a date on such a regional event is difficult because historical records show that the first cutting of red (*P. resinosa*) and white pine occurred on the St. Croix River (130 miles south of Ely) about 1830, and that logging proceeded along the Rum River (150 miles southwest) and other tributaries of the Mississippi River about 1870, and it seems reasonable to suppose that the white pine decline is a reflection of this and similar events.

If this is so, the *Ambrosia* rise occurring at the same level is also regional, reflecting clearance and disturbance much farther south and before settlement around Shagawa Lake. The marked increase in *Ambrosia* pollen at 30 cm probably indicates the large population increase between 1890 and 1900 (Fig. 2).

Other changes in the tree pollen percentages are more subtle. *Larix laricina* pollen shows a decrease at the surface that could represent the effects of the larch sawfly (*Pristiphora erichsonii* (Hartig)). Aspen (*Populus tremuloides*) shows small fluctuations throughout the core that may be the result of severe fires. The increase in aspen pollen after settlement can be interpreted as reforestation following extensive logging and severe fires of 1875 and 1910.

The post-settlement increase in aquatic pollen is largely represented by the Lemnaceae, and may indicate eutrophication of Shagawa Lake. *Isoetes macrospora* tends to be less well-represented in post-settlement time. Possibly the littoral zone was reduced by shading from the increased duck weed Lemnaceae and/or masses of blue-green algae (buoyant because of their gas vacuoles) that are rafted shoreward where they accumulate and decay. In addition to shading, the decaying algae might convert the clear, silty bottom that provides a niche for *Isoetes* into an inhospitable, oxygen-poor mud.

The charcoal record is perplexing, because, although it shows discrete fluctuations, sharp peaks that might represent single fires are seldom represented, and there is no correlation between charcoal abundance and nutrient-loving diatoms (e.g. *Fragilaria capucina*) that would be expected if local fires increased the flow of nutrients to the lake. Examination of raw sediment every two cm did not reveal charcoal peaks at levels that should correlate with large fires in the area such as 1894 and 1910. All levels below 20 cm show a fairly high charcoal content and suggest that there is a high regional background of charcoal particles raining into the lake. The charcoal tally recorded particles one-fourth the size of the smallest pollen grain, and it seems reasonable that charcoal of the size of one pollen grain could be transported considerable distances, especially with the strong convective air currents associated with forest fires.

Rarity of discrete charcoal peaks may have resulted from the sampling method, which integrates as much as 6 years of sediment for each level studied, and charcoal-rich sediment may have been diluted by adjacent charcoal-poor sediment.

The lowest charcoal values occur at 4–10 cm, apparently representing a time when no fires swept the region. The relatively high value at the surface is interpreted as the record of the Little Sioux fire of 1971 in the B.W.C.A., less than 20 miles northwest of the coring site. M.L. Heinselman observed a cloud of black smoke hanging over Ely during the fire, and it is likely that charcoal was carried in this fashion to the lake.

The pollen concentration (Fig. 3) ranges from 2×10^4 to 57×10^4 grains/cm^3. The mean value below 39 cm is 40.3×10^4 grains/cm^3. The sharp decrease in pollen concentration during post-settlement time is attributed to the four- to ten-fold increase in sedimentation rate (Table 3).

Influx values

Influx values for pollen, diatoms, Cladocera, and fungal elements (Fig. 7) were calculated from their concentration and the deposition times shown in Table 3 (concentration/deposition time = number/cm^2/year = influx). The influx profiles for each of the sediment components indicate relatively constant background values that abruptly rise and reach maxima at or above 30 cm.

Influx maxima for diatoms and Cladocera clearly suggest increased productivity in Shagawa Lake, and their increased influx correlates with species changes that indicate a shift in dominance to species characteristic of eutrophic bodies of water (*Fragilaria crotonensis* and *Stephanodiscus minutus* among the diatoms, and *Chydorus sphaericus* of the Cladocera).

Pollen influx below 40 cm is fairly uniform, averaging 2.7×10^4 grains/cm^2/year. These values are in agreement with values obtained from the annually laminated sediments at Lake of the Clouds (Craig 1972) in a similar vegetation complex, and with values for the pine pollen zone in Minnesota (Waddington 1969) and Connecticut (Davis 1967). The explanation for high influx values for pollen from 34 to 23 cm requires a mechanism for selectively adding pollen grains to the sediment. Redeposition of sediment eroded from shallower parts of the basin (Waddington 1969, Davis *et al.* 1973) could easily increase the total sedimentation rate, but it would involve selective winnowing of pollen and other light particles to account for a differential increase in pollen sedimentation. Discrete changes in species composition of diatoms and Cladocera coincident with their increased influx argue against redeposition of previously deposited sediment.

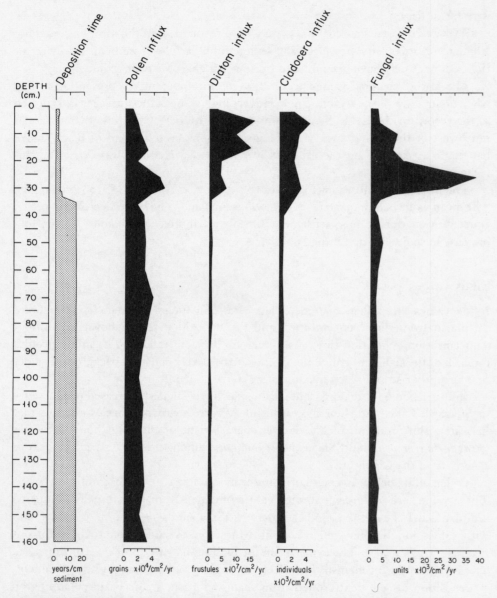

Figure 7. Shagawa Lake, Minnesota: Influx values for pollen, diatoms, Cladocera, and fungi.

The increase in clastic particles (Fig. 5) coincident with increased pollen influx suggests that increased soil erosion and runoff between 1889 and 1922 might be related to the selective deposition of pollen grains in the lake. Soil erosion and runoff have already been cited as probable causes of increased nutrient levels of Shagawa Lake, and this phenomenon is probably important in causing the increases in Cladocera and diatom influx. The fluctuating variety of fungal hyphae and spores in slides prepared for pollen analysis represent largely terrestrial organisms, although remains of aquatic fungi are probably also present. The terrestrial fungal elements are expected to increase with soil erosion and runoff, and their stratigraphic distribution (Fig. 7) indicates a strong increase paralleling the pollen influx maximum. These relations suggest that erosion is the most likely mechanism for increasing pollen sedimentation. The changes correlate well with the land clearance that accompanied the growth and development of Ely (Fig. 2).

The effective use of fungal remains in sediment to provide information about sedimentary processes will require considerable work in differentiating terrestrial and aquatic species and their ecological requirements, but the results presented here imply that such discrimination will be a valuable addition to palynological studies in the future.

Conclusions

The biostratigraphy of Shagawa Lake contains a clear illustration of increasing eutrophication and modification of the surrounding and regional vegetation that coincides with the advent of European settlement. A causal relationship is strongly suggested. Nevertheless, much remains to be learned about many facets of this biostratigraphy. The interrelationships of the phytoplankton and zooplankton presented here are still speculative, and both physiological and ecological studies will be required to refine palaeolimnologic interpretations based on diatoms and cladocerans. Charcoal stratigraphy appears to be a powerful tool to investigate natural changes in forest composition that might be reflected in a lake's nutrient balance, but little is known about mechanisms of charcoal dispersal and sedimentation, and consequently sampling and analytical techniques for charcoal may need refinement before its ultimate potential is realised.

The combination of sedimentologic studies with palynology and other biostratigraphic studies vastly increases the accuracy and extent of environmental reconstruction, and it is hoped that in the future even more techniques, such as isotope studies, pigment studies, palaeomagnetic analyses, complete chemical analyses as well as refined and more extensive palaeontologic investigations, can be united in stratigraphic reconstructions of past environments. As this is accomplished it will be increasingly possible to sharpen our historical perspective and thereby make the fullest use of the dimension of time in assessing the impact of man on the environment.

Acknowledgements

An interdisciplinary study of this nature depends on the willing assistance of many people, who help not only with analyses but also with the interpretation of results and the development of ideas. We are pleased to have the opportunity to express grateful acknowledgment to the following people who worked with us on this project. Helen Boyer provided the Cladocera analyses. Douglas Hall analysed phosphorus and iron. David Larson determined water content and weight loss on ignition of the sediments. Paul Smith and Robert Brice of the Environmental Protection Agency in Ely assisted sample collection by providing a boat and laboratory facilities. M.L. Heinselman kindly provided information on the fire and logging history of the region. Prof Neil Anderson, Plant Pathology Dept., University of Minnesota, gave helpful advice concerning fungal remains. We should also like to thank Richard Darling for his patient co-operation in drafting the figures. We have had many useful discussions with H.E. Wright, who also critically read the manuscript. Support for this study was provided by the Atomic Energy Commission (contract AT (11-1)-2046) and the Environmental Protection Agency (grant 1601 DXG).

References

AMERICAN GEOLOGICAL INSTITUTE (1960) Glossary of Geology and related sciences with supplement. *Am. Geol. Inst., Nat. Acad. Sci.*, Wash. D.C., 325p. (+ 72p. suppl.)

BIRGE E.A. (1897) Plankton studies on Lake Mendota, II. *Trans. Wis. Acad. Sci. Arts Lett.* 11, 274-448.

BRADBURY J.P. & MEGARD R.O. (1972) A stratigraphic record of pollution in Shagawa Lake, northeastern Minnesota. *Bull. geol. Soc. Am.* 83, 2639-48.

BRICE R.M. & POWERS C.F. (1969) The Shagawa Lake, Minnesota eutrophication research project. *Proc. of the Eutrophication-Biostimulation Assessment Workshop*, Hotel Claremont, Berkeley, California, June 19-21 1969, 258.

BRIGHT R.C. (1968) Surface-water chemistry of some Minnesota lakes, with preliminary notes on diatoms. *Limnological Research Center, Univ. of Minnesota, Interim report* 3, 59p.

CRAIG A.J. (1972) Pollen influx to laminated sediments: a pollen diagram from northeastern Minnesota. *Ecology* 53, 46-57.

CURTIS J.T. (1959) *The vegetation of Wisconsin*. University of Wisconsin Press, Madison.

CUSHING E.J. & WRIGHT H.E. (1965) Hand-operated piston corers for lake sediments. *Ecology* 46, 380-4.

DAVIS M.B. (1967) Pollen accumulation rates at Rogers Lake, Connecticut, during late- and postglacial time. *Rev. Palaeobotan. Palynol.* 2, 219-30.

DAVIS M.B., BRUBAKER L.B. & WEBB T. (1973) Calibration of absolute pollen influx. (this volume).

FERNALD M.L. (1950) *Gray's Manual of Botany* (8th edition). American Book Co., New York.

FRIES M. (1962) Pollen profiles of late Pleistocene and Recent Sediments from Weber Lake, Minnesota. *Ecology* 43, 295-308.

KING D.L. (1970) The role of carbon in eutrophication. *J. Wat. Pollut. Control. Fed.* 42, 2035.

MACHAMER J.F. (1968) Geology and origin of the iron ore deposits of the Zenith mine, Vermilion district, Minnesota. *Minn. geol. Surv., Spec. Publ.* 2, 56.

MCANDREWS J.H. (1969) Paleobotany of a wild rice lake in Minnesota. *Can. J. Bot.* 47, 1671-9.

MEGARD R.O. (1969) Algae and photosynthesis in Shagawa Lake, Minnesota. *Limnological Research Center, Univ. of Minnesota, Interim report* 5, 20p.

MINNESOTA POLLUTION CONTROL AGENCY (1969) *Report on investigation of pollution of the northern border waters from the mouth of the Pigeon River at Lake Superior westward through the Boundary Waters Canoe area and lower lakes to the outlet of Rainy Lake. Aug.-Nov. 1965*, 39p. and Appendices A-G.

OHMANN L.F. & REAM R.R. (1971) Wilderness ecology: virgin plant communities of the Boundary Waters Canoe Area. *North Central Forest Exp. Station, St. Paul, Minn.* 55p. (*U.S.D.A. Forest Service Res. Paper NC-63*).

SCHINDLER D.W. & NOVEN B. (1971) Vertical distribution and seasonal abundance of zooplankton in two shallow lakes of the Experimental Lakes Area, northwestern Ontario. *J. Fish. Res. Bd. Can.* 28, 245-56.

STOERMER E.F. & YANG J.J. (1970) Distribution and relative abundance of dominant plankton diatoms in Lake Michigan. *Great Lakes Research Division, Univ. of Mich., Ann Arbor, Publ.* 16, 64p.

STOCKNER J.G. & ARMSTRONG F.D.J. (1971) Periphyton of the Experimental Lakes Area, northwestern Ontario. *J. Fish. Res. Bd. Can.* 28, 215-29.

SWAIN F.M. & PROKOPOVICH N. (1954) Stratigraphic distribution of lipoid substances in Cedar Creek Bog, Minnesota. *Bull. geol. Soc. Am.* 65, 1183-98.

TROELS-SMITH J. (1955) Characterisation of unconsolidated sediments. *Danm. geol. Unders.* Ser. IV, 3(10), 38-73.

U.S. DEPT. OF AGRICULTURE (1964) Recreational use of the Quetico-Superior Area. *Lake States Exp. Station Publication. U.S.D.A. Forest Service Res. Paper LSJ, R.C. Lucas.*

WADDINGTON J.C.B. (1969) A stratigraphic record of the pollen influx to a lake in the Big Woods of Minnesota. *Geol. Soc. Am. Special Paper* 123, 263-82.

WINCHELL M.H. & WINCHELL H.V. (1891) Iron ores of Minnesota. *Geol. & Nat. Hist. Survey of Minnesota, Bull.* 6, 430p.

Discussion on limnological history

Recorded by Dr Hilary H. Birks,
Mr J.H.C. Davis, and Mr J. Dodd.

Dr C. Turner wondered why Professor Oldfield had associated the laminated sediments in the lower parts of the Lough Neagh sequence with a lowering of the water level. He normally associated such sediments with deep water, where there was little disturbance from animals. Professor Oldfield presented several reasons. The sediments are of early post-glacial age, and contain a higher proportion of pollen of marsh and fen plants than the sediments above or below. It is difficult to envisage pebbles up to 4 mm diameter and hazel nuts being transported across deep water a distance of 2 or 3 miles (3.2-4.8 km) from the shore. There is a complete absence of diatoms, which would be present in a deep water sediment. Where there is mineral inwash into deep water, it greatly dilutes the palaeomagnetism of the sediment, but in the laminated Lough Neagh sediments, the palaeomagnetism is an order of magnitude higher than in the sediment above. He suggested that the Lough Neagh basin was acting as a winter flood-plain, in which extensive fens were flooded annually, rather as in the annual overspill regions of the Danube today. Dr Berti asked what happened to the outlet, if the lake could be lowered by 40 feet (13 m) and subsequently raised again. Professor Oldfield explained that preliminary studies in the Lower Ban valley had revealed an alluvial infill of a buried channel, and that the mechanism was probably due to silting coinciding with the last stage in the eustatic rise in sea level. Dr G.H. Evans had not noticed the presence of *Asterionella formosa* in the diatom diagram from Lough Neagh, and thought this was surprising, as it is usually a good indicator of eutrophic conditions. Mr Battarbee said that it occurred in small amounts, and its behaviour coincided with that of *Tabellaria*, and it decreased in relative amounts at the rise of *Stephanodiscus*. He was not sure that it was a good indicator of eutrophication in Lough Neagh, as it was rather rare there today.

Dr Evans asked if Dr Bradbury had studied the diatoms in the effluent entering Shagawa Lake, and whether any pollution indicators were present. Dr Bradbury had some information for the lake itself: *Stephanodiscus minutus* bloomed under the ice in early spring, but in summer its numbers were reduced by blue-green algae. *Gomphonema parvula* is also present in the lake, but was only found in very low numbers in the sediments from the deep water, and was probably more important near the shore. Miss Haworth wondered if the diatom breakage observed was due to the redeposition of shore diatoms to sublittoral deposits. Dr Bradbury had supposed that it was, and that such a process would also account for the concentration of the heavy mineral, tourmaline, in the profundal area. Dr Meriläinen asked to what extent did benthic animals disturb the surface sediment, as he had observed animals in Wisconsin lakes moving daily through depths of 10 cm. Dr Bradbury agreed that such disturbance should always be kept in mind when making detailed stratigraphic studies of lake sediments. However, he doubted whether the disturbance in Shagawa Lake exceeded 1 cm, even though no laminations were observed, because the stratigraphy shown by the hematite was so very sharp; in one sample hematite was absent, but in the centimetre above, it was abundant.

Professor Frey then commented on the increasing interest in palaeolimnology. Most of the conference participants were interested in lakes as an environment of deposition for the recording of events in the catchment area—hence the frequent exclusion of aquatic pollen from the pollen sum. For pollen analysis, the sediments are chemically treated in such a way that traces of most limnic organisms are destroyed. However, untreated lake sediments retain fossils of diatoms, cladocera, etc., from which, in conjunction with chemical data, the palaeolimnologist reconstructs the lake ecosystem, and interprets changes in it in terms of the ecology of the organisms, and thence in terms of changes in the catchment area. Modern limnological studies aid in our interpretations; for example, the chemical effects of a hypolimnion may determine whether a fossil is preserved or dissolved. However, palaeolimnology is a young science compared with palynology, with problems of basic taxonomy and identification, and little information on the present-day ecology of the lake organisms. There is considerable potential for advances in the future.

Section 7:
Summation

Summing up: an ecologist's viewpoint

M.C.F. PROCTOR *Department of Biological Sciences, University of Exeter*

What has struck me most in this symposium is the growing emphasis on process in Quaternary studies. Questions of pollen dispersal and incorporation have seldom been far away, and one has constantly been conscious (as the contributors too have clearly been) of the reflection of vegetational processes by palynological and other evidence.

Dr Flenley's comments indicate clearly our awareness that the value of pollen analysis as a palaeoecological research tool rests largely on the habitual anemophily of the dominant species of temperate climates—for which there are good ecological reasons. The dispersal of pollen is, of course, an essential phase in the life-cycle of the flowering plant; the preservation of some of this pollen is an incidental bonus for the pollen analyst. In seeking an understanding of pollen dispersal there is a community of interest between the geneticist or genecologist interested in gene flow, and the student of Quaternary plant distribution concerned with questions of pollen representation in his samples, though the latter adds some problems of his own! As the contributions to the symposium have underlined, the subject is a many-sided and active field of study with a copious literature, but one in which we still have much to learn.

There are obvious parallels between the processes at work in the transport and deposition of pollen and macrofossils (with the flow of information they carry), and those involved in energy-flow into detritus-based ecosystems, or indeed in the interchange of material between ecosystems in general. One looks forward, with several of the contributors, to more studies in which palynological results are integrated into a broader ecological context.

Early pollen analytical investigations were concerned with providing zonation systems for peats and sediments, with broad regional vegetation development, and with floristics. We have come a long way in progressive refinement in the interpretation of the evidence, but this symposium shows that we still face insistent problems. Dr Rybniček's paper showed that, in some mire sites at least, subfossil plant communities can be characterised with considerable precision from macro remains. Many of us listening to his paper will have been able to picture modern representatives of many of the communities he described, in central or western Europe, Scandinavia or perhaps the American Arctic. In aquatic macrofossil assemblages the picture is less clear. As Dr Hilary Birks has shown, macrofossils from some aquatic plants are very localised, though different species have very different patterns of dispersal and representation is closely linked with ecological factors affecting seed production and with dispersal biology. The uncertainties in the interpretation of pollen evidence do not range over such wide extremes. Studies such as those of Dr Andersen on pollen production, of Dr Janssen and Professor Berglund on pollen dispersal and deposition, and diagrams based on absolute pollen influx of the sort presented by Dr Davis, put us on much firmer ground when we try to interpret pollen evidence in terms of vegetation. My own experience in other contexts with the kind of multivariate methods discussed by Dr Cushing makes me cautiously optimistic about their potential value in the interpretation of pollen diagrams, particularly in helping to elucidate the contribution to the diagram of pollen from different sources within the surrounding landscape. However, the characteristics that give pollen evidence much of its value for regional studies of vegetation mean that some uncertainty is inherent. As Dr John Birks' surface samples suggest, it is probably impossible to infer from the pollen spectra the full diversity of the landscape from which they are derived.

I think we must accept a middle course between too rigid interpretation of pollen diagrams in terms of present-day communities and the present-day behaviour of species, and the too-ready assumption that things were different in the past. There is no doubt that species behave differently in different areas. *Salix herbacea* is a good index of late snow-patch conditions in the Alps, but is ubiquitous in parts of central Iceland. *Fagus* reaches higher altitudes in the Alps than *Quercus*, but goes less far north in Scandinavia. The same genotype responds differently to different sets of physical conditions, and as the genecological literature abundantly testifies, the distribution of genotypes within species varies in both space and time. Variation is readily demonstrated in both morphological and physiological characters. However, there is no doubt that phytosociology can provide powerful and valid generalisations. It may be necessary to range over an unexpectedly wide area or in unexpected directions to find the right analogies, and it always has to be remembered that soil and aspect differences can produce apparently bizarre juxtapositions. Only the strongest evidence should compel us to reject the geologist's well-tried principle of actualism.

Interpretation of pollen diagrams and macrofossil assemblages in terms of vegetation brings us back to the importance of process. Absolute chronology and pollen influx data give us new information on immigration, establishment, and succession. The interplay of climatic change and ecosystemic adjustment provide a complete range of possibilities between the 'allogenic' and 'autogenic' extremes envisaged by Tansley. Perhaps we too easily allow ourselves to become the mental prisoners of the vantage point fixed in space and time by our human lifespan, our solidly immovable university buildings, and the apparent permanence of names on maps. We think readily enough in terms of years or decades, or in a more abstract way in terms of the millenia of Pleistocene or Flandrian chronology, but the forest lives, regenerates, and changes on a time-scale between these two. 'Climax' vegetation is not static, and there is no particular reason to think of the present or any other moment in the Pleistocene as 'normal' or otherwise privileged. Changes of the kind illustrated by Professor Watts embrace several generation times for the climax dominant trees. One thinks immediately of the experiments with combinations of field crops, and the generation-to-generation theoretical models for plant competition and coexistence developed by C.T. de Wit and his co-workers. It is of no consequence to the seed initiating a new generation whether it finds a congenial habitat in a gap in the forest ten metres or ten kilometres from its parent; migration falls naturally into place in this context of constant flux, and we then have to return to consider the statistics of seed dispersal!

Finally, palaeoecological evidence is giving us increasing insight into the influence of human activity, at scales ranging from the very local effects described by Dr Andersen, Dr Bradbury, and Mr Sims to the broad regional changes documented by Professor Oldfield from Lough Neagh. We see interactions between human activity and both climatic change and variation from place to place within past environments. The field ecologist is used to interactions between physical and human factors in the present-day landscape; the effects of climatic or edaphic boundaries are often sharpened and emphasised by the differences of land-use they have provoked. It is intriguing to see in Mr Sims and Dr Turner's papers the evidence of corresponding interactions in time.

List of participants

Dr Marjetta Aalto, Botany Department, University of Helsinki, Helsinki 17, Finland
Dr A. Agnew, Botany Department, University College of Wales, Aberystwyth
Miss U. Allitt, Botany School, Downing Street, Cambridge CB2 3EA
Dr S.Th. Andersen, Danish Geological Survey. Thoravej 31, 2400 Copenhagen NV, Denmark
Miss R. Andrew, Botany School, Downing Street, Cambridge CB2 3EA
Miss M.A. Atherden, Senior Common Room, The College, Ripon, Yorks.

Miss M.J. Bailey, Botany School, South Parks Road, Oxford
Mr K.E. Barber, Geography Department, University of Southampton, Southampton SO9 5NH
Mr B. Barnes, 'Swiss Lodge', Worden Park, Leyland, Lancs.
Mr R.W. Battarbee, School of Biological and Environmental Studies, New University of Ulster, Coleraine, N. Ireland
Mr P.W. Beales, Sidney Sussex College, Cambridge
Mr S.C. Beckett, Geography Department, University of Hull, Hull, Yorks.
Dr Frances G. Bell, Botany Department, University of Liverpool, P.O. Box 147, Liverpool L69 3BX
Professor B.E. Berglund, Laboratory of Quaternary Biology, Institute of Geology, University of Lund, Tornavägen 13, S 223 63, Lund, Sweden
Mrs B.E. Berglund, Valthornov 3, S 223 68, Lund, Sweden
Dr A.A. Berti, Botany School, Trinity College, Dublin 2, Eire
Dr Hilary H. Birks, Botany School, Downing Street, Cambridge CB2 3EA
Dr H.J.B. Birks, Botany School, Downing Street, Cambridge CB2 3EA
Miss A.P. Bonny, School of Biology, University of Leicester, Leicester LE1 7RH
Dr J.P. Bradbury, Limnological Research Center, Pillsbury Hall, University of Minnesota, Minneapolis, Minnesota 55455, U.S.A.
Mr A.P. Brown, Botany School, Downing Street, Cambridge CB2 3EA

Mr P. Callow, Gonville and Caius College, Cambridge
Mr P. Carter, Geography Department, University of Durham, Durham City
Mr E. Caulton, Geography Department, University of Edinburgh, High School Yards, Edinburgh EH1 1NR
Miss A.P. Conolly, School of Biology, University of Leicester, Leicester LE1 7RH
Mr A.J. Craig, National Parks and Monuments Branch, 10 Hume Street, Dublin 2, Eire
Mrs G. Crompton, 103 Commercial End, Swaffham Bulbeck, Cambridge

Dr Adèle Crowder, Biology Department, Queens University, Kingston, Ontario, Canada
Dr P.R. Cundill, Geography Department, University of Liverpool, P.O. Box 147, Liverpool L69 3BX
Dr E.J. Cushing, Botany Department, University of Minnesota, Minneapolis, Minnesota 55455, U.S.A.

Dr A. Danielsen, Botanical Museum, University of Oslo, Trondheimsvn 23B, Oslo 5, Norway
Mr T.C.D. Dargie, Geography Department, University of Sheffield, Sheffield S10 2TN
Dr Margaret B. Davis, Great Lakes Research Division, University of Michigan, Ann Arbor, Michigan 48104, U.S.A.
Mr J.H.C. Davis, Magdalene College, Cambridge
Mrs J.W. Deacon, Botany School, Downing Street, Cambridge CB2 3EA
Dr Winifred Dickinson, Botany Department, University of Leeds, Leeds 2
Mr J. Dodd, Peterhouse, Cambridge
Dr E.A.G. Duffey, The Nature Conservancy, Monks Wood Experimental Station, Abbots Ripton, Hunts.
Dr S.E. Durno, Macaulay Institute for Soil Research, Craigiebuckler, Aberdeen AB2 2QJ

Dr G.C. Evans, Botany School, Downing Street, Cambridge CB2 3EA
Dr G.H. Evans, Biology Department, Liverpool Polytechnic, Byrom Street, Liverpool L3 3AF

Professor K. Faegri, Botanical Museum, University of Bergen, Postbox 12, N 5014 Bergen, Norway
Dr J.R. Flenley, Geography Department, University of Hull, Hull, Yorks.
Dr Maj-Britt Florin, Institute of Quaternary Geology, University of Uppsala, Box 555, S 752 21 Uppsala, Sweden
Professor D.G. Frey, Zoology Department, University of Indiana, Bloomington, Indiana, U.S.A.
Mrs D.G. Frey, Zoology Department, University of Indiana, Bloomington, Indiana, U.S.A.

Mr S. Garrett-Jones, 64 Station Road, Llanishen, Cardiff
Mr P. Gibbard, Botany School, Downing Street, Cambridge CB2 3EA
Mrs A.N. Gibby, Prebend's Gate, Quarry Heads Lane, Durham City
Mr I. Goddard, Palaeoecology Laboratory, Queen's University, Belfast BT7 1NN

List of participants

Professor Sir Harry Godwin, Botany School, Downing Street, Cambridge CB2 3EA
Mr A.D. Gordon, Statistics Laboratory, University of Cambridge, Mill Lane, Cambridge
Miss P.M. Gough, Senior Common Room, Shenstone New College, Burcot Lane, Bromsgrove, Worcs.
Miss P.A. Greatrex, Trevelyan College, University of Durham, Elvet Hill Road, Durham City
Mrs K. Griffin, Institute of Geology, University of Oslo, Postbox 1047, Blindern, Oslo 3, Norway
Dr W. Groenman-van Waateringe, Institute for Prehistory, Nieuwe Prinsengracht 41, Amsterdam, Netherlands
Mr A.R. Gunson, Geography Department, University of Aberdeen.

Dr R. Harland, Institute of Geological Sciences, Ring Road Halton, Leeds LS15 8TQ
Professor U. Hafsten, Botanical Institute, University of Trondheim, NLHT, N 7000 Trondheim, Norway
Miss E.Y. Haworth, Freshwater Biological Association, Ferry House, Far Sawrey, Ambleside, Westmorland
Dr Kari Henningsmoen, Institute of Geology, University of Oslo, Postbox 1047, Blindern, Oslo 3, Norway
Dr Margaret Herbert-Smith, Geology Department, University of Birmingham, P.O. Box 363, Birmingham 15
Dr F.A. Hibbert, Biology Department, Liverpool Polytechnic, Byrom Street, Liverpool L3 3AF
Dr Sheila P. Hicks, Botany Department, University Oulu, Oulu 10, Finland
Mr M.T. Hine, Geography Department, University of Bristol, Bristol BS8 1SS
Dr J.M. Hirst, Plant Pathology Department, Rothamsted Experimental Station, Harpenden, Herts.

Miss J.L. Jackson, Botany Department, University of Manchester, Manchester M13 9PL
Mr R.M. Jacobi, Gonville and Caius College, Cambridge
Dr C.R. Janssen, Laboratory for Palaeobotany and Palynology, Heidelberglaan 2, Utrecht, Netherlands
Mr H.M. Jones, 102 Carisbrooke Way, Cardiff CF3 7HX
Dr R.L. Jones, Geography Department, University of Sheffield, Sheffield S10 2TN
Professor F.P. Jonker, Laboratory of Palaeobotany and Palynology, Heidelberglaan 2, Utrecht, Netherlands
Mrs F.P. Jonker, Utrecht, Netherlands

Mr P.E. Kaland, Botanical Museum, University of Bergen, P.O. Box 2637, Bergen N 5010, Norway
Mrs S.M. Kirk, Palaeoecology Laboratory, Queen's University, Belfast BT7 1NN

Professor H.H. Lamb, Climatic Research Unit, School of Environmental Sciences, University of East Anglia, Norwich NOR 88C
Mr K.D. Large, Botany Department, University of Aberdeen, Aberdeen
Mr M.J. Liddle, School of Plant Biology, University College of North Wales, Bangor, Caerns.
Dr P.S. Lloyd, Botany Department, University of Sheffield, Sheffield S10 2TN
Mrs P.S. Lloyd, Botany Department, University of Sheffield, Sheffield S10 2TN
Mr J.J. Lowe, Geography Department, University of Edinburgh, High School Yards, Edinburgh EH1 1NR

Mr D.J. Mabberley, Botany School, Downing Street, Cambridge CB2 3EA
Professor A. MacFadyen, School of Biological and Environmental Studies, New University of Ulster, Coleraine, N. Ireland
Mr E. Maltby, Geography Department, University of Bristol, Bristol BS8 2SS
Dr Kazimiera Mamakowa, Botanical Institute, Polish Academy of Sciences, Krakow, Lubicz 46, Poland
Dr Vera Markgraf, F-91 Bures-sur-Ivette, Inra, France
Mrs M.E. McCallan, 3 Portmore Road, Portstewart, Co. Londonderry, N. Ireland.
Mr P.R. Medhurst, c/o Palaeoecology Laboratory, Queen's University, Belfast BT7 1NN
Mrs E.M. Megaw, 11 Merchiston Gardens, Edinburgh EH10 5DD
Dr J. Meriläinen, Botany Department, University of Helsinki, Helsinki 17, Finland
Professor G.F. Mitchell, Trinity College, Dublin 2, Eire
Mr D. Moe, Botanical Museum, University of Bergen, P.O. Box 2637, Bergen N 5010, Norway
Dr P.D. Moore, Botany Department, King's College, 68 Half Moon Lane, London SE24 9JF
Mr R.J. Morley, Geography Department, University of Hull, Hull, Yorks.
Mr P.J. Mott, Biological Sciences Department, University of Dundee, Dundee DD1 4HN

Professor P.J. Newbould, School of Biological and Environmental Studies, New University of Ulster, Coleraine, N. Ireland
Dr W.W. Newey, Geography Department, University of Edinburgh, High School Yards, Edinburgh EH1 1NR

Professor F. Oldfield, School of Biological and Environmental Studies, New University of Ulster, Coleraine, N. Ireland
Dr H.A. Osmaston, Geography Department, University of Bristol, Bristol BS8 1SS
Dr P.E. O'Sullivan, School of Biological and Environmental Studies, New University of Ulster, Coleraine, N. Ireland

Dr N.R. Page, Hendon College of Technology, London NW4
Miss J.P. Paice, Department of Biological Sciences, University of Lancaster, Bailrigg, Lancaster
Dr N.V. Pears, Geography Department, University of Leicester, Leicester
Dr M.C. Pearson, Botany Department, University of Nottingham, University Park, Nottingham NG7 2RD
Miss R.M. Peck, Geography and Geology Department, University of Hong Kong, Hong Kong
Mrs S. Peglar, Botany School, Downing Street, Cambridge CB2 3EA
Mrs M.E. Pettit, Botany School, Downing Street, Cambridge CB2 3EA
Miss L. Phillips, Botany School, Downing Street, Cambridge CB2 3EA
Dr J.R. Pilcher, Palaeoecology Laboratory, Queen's University, Belfast BT7 1NN

List of participants

Dr M.C.F. Proctor, Department of Biological Sciences, Hatherly Laboratories, Prince of Wales Road, Exeter EX4 4PS

Dr Magdalene Ralska-Jasiewiczowa, Botanical Institute, Polish Academy of Sciences, Krakow, Lubicz 46, Poland
Dr R.E. Randall, School of Biological and Environmental Studies, New University of Ulster, Coleraine, N. Ireland
Dr B.K. Roberts, Geography Department, University of Durham, Durham City
Dr K. Rybniček, Botanical Institute, Czechoslovak Academy of Sciences, Stara 18, Brno, Czechoslovakia
Mr L. Rymer, Botany School, Downing Street, Cambridge CB2 3EA

Dr B. Seddon, Geography Department, University of Reading, Whiteknights, Reading RG6 2AF
Dr Kathleen Simpkins, Geology Department, University of N. Staffs, Keele, Staffs.
Mr R.E. Sims, Botany School, Downing Street, Cambridge CB2 3EA
Dr A.G. Smith, Botany Department, Queen's University, Belfast BT7 1NV
Mr D.G. Smith, Geology Department, University of Cambridge, Downing Street, Cambridge.
Mr J.P. Smith, Faculty of Humanities, The Polytechnic, Wolverhampton WV1 1LY
Dr M. Sonesson, Plant Ecology Department, University of Lund, Ö Vallgatan 14, S 223 61 Lund, Sweden
Dr R.H. Squires, Geography Department, University of Winnipeg, Winnipeg, Manitoba R3B 2E9, Canada
Mr R.N. Starling, 5 Winn Road, Southampton, Hampshire
Mr J. Stockmarr, Danish Geological Survey, Thoravej 31, 2700 Copenhagen N.V., Denmark
Dr V.R. Switsur, Botany School, Downing Street, Cambridge CB2 3EA

Mr P.A. Tallantire, Botanical Institute, University of Trondheim, NLHT, N 7000 Trondheim, Norway
Dr J.H. Tallis, Botany Department, University of Manchester, Manchester M13 9PL
Dr H. Tauber, National Museum, Department of Natural Sciences, Carbon-14 Dating Laboratory, Ny Vestergade 10, Copenhagen K, Denmark
Mrs H. Tauber, National Museum, Department of Natural Sciences, Carbon-14 Dating Laboratory, Ny Vestergade 10, Copenhagen K, Denmark
Dr J.A. Taylor, Geography Department, University College of Wales, Aberystwyth
Mrs H.M. Tinsley, Geography Department, University of Leeds, Leeds 2
Dr M.J. Tooley, Geography Department, University of Durham, Durham City
Dr C. Turner, c/o Botany School, Downing Street, Cambridge CB2 3EA
Dr Judith Turner, Botany Department, University of Durham, Durham City
Dr Winifred Tutin, School of Biology, University of Leicester, Leicester LE1 7RH

Professor Y. Vasari, Botany Department, University of Oulu, Oulu 10, Finland
Dr A. Voorrips, Institute for Prehistory, Nieuwe Prinsengracht 41, Amsterdam, Netherlands

Mr M.F. Walker, Geography Department, University College of Wales, Aberystwyth
Dr A.S. Watt, Botany School, Downing Street, Cambridge CB2 3EA
Professor W.A. Watts, Botany School, Trinity College, Dublin 2, Eire
Dr R.G. West, Botany School, Downing Street, Cambridge CB2 3EA
Dr R.B.G. Williams, Geographical Laboratory, University of Sussex, Falmer, Brighton
Mrs D.G. Wilson, Botany School, Downing Street, Cambridge CB2 3EA
Miss J. Witchell, Botany Department, University of Bristol, Bristol BS8 1UG

Author Index

Aario L. 32, 42, 144, 145, 146, 166
Aitken M.J. 273, 278
American Commission on Stratigraphic Nomenclature 195, 205
American Geological Institute 294, 307
Andersen S.Th. 9, 24, 37, 40, 42, 54, 59, 109, 110, 114, 117, 125, 129, 144, 166, 213, 220, 281, 288
Anderson J.A.R. 139, 140
Anderson M.C. 60
Andrews S.M. 60
Anthony R.S. 196, 205
Arber A. 178, 182, 188
Armstrong F.D.J. 300, 307
Ashton P.S. 141
Auer V. 144, 166

Bakker E.M. van Zinderen 136, 137, 140
Banky G.Y. 53, 59
Bartley D.D. 147, 166
Bassett I.J. 64, 74, 76
Battarbee R.W. 267, 271, 277, 279, 280, 281, 288
Behre K.-E. 91, 103
Beiswenger J.M. 17, 24
Benninghoff W.S. 63, 76
Berger R. 272, 278
Berglund B.E. 13, 24, 117, 129, 233, 236
Beschel R.E. 61, 74, 75, 76
Best C.A. 72, 73, 76
Binney C. 281, 288
Birge E.A. 302, 307
Birks H.H. 195, 205, 238, 259

Birks H.J.B. 80, 93, 103, 144, 152, 153, 154, 155, 156, 157, 158, 159, 164, 166, 167, 185, 198, 205
Blake V.B. 63, 76
Bonny A.P. 17, 24, 82, 83, 88, 90, 92, 93, 101, 103, 104, 196, 205, 235, 236
Booberg G. 255, 259
Boros A. 255, 259
Bourdo E.A. 12, 24
Bradbury J.P. 289, 290, 302, 303, 307
Brett D.W. 1, 3
Brice R.M. 291, 307
Brideaux W.W. 143, 166
Bright R.C. 173, 185, 186, 188, 189, 290, 300, 303, 307
British Standards Institution 59
Britton D. 216, 220
Brubaker L.B. 21, 24, 103, 307
Burges A. 198, 205
Burnett J.H. 260
Bye J.A.T. 46, 59

Cajander A.K. 256, 259
Cambray R.S. 104
Carrol G. 63, 76
Casparie W.A. 256, 259
Chafee F.H. 74, 76
Chamberlain A.C. 48, 54, 59, 110
Chanda S. 93, 103
Chandler D.C. 74, 76
Chaney R.W. 173, 188
Chapman L.J. 65, 77, 144
Chase C.E. 12, 24
Cheetham A.H. 159, 166
Chinery J.M. 60

Chorley R.J. 3, 54, 60
Clapham A.R. 166
Clapham W.B. 143, 166
Clark J.D.G. 232, 233, 234, 236
Clarke R. Rainbird 226, 236
Clough T.H.McK. 97, 103
Coe M.J. 140
Coetzee J.A. 136, 137, 140
Colby B.R. 53, 59
Colinvaux P.A. 147, 148, 166
Collins V.G. 80, 103
Collins-Williams C. 72, 73, 76
Comanor P.L. 144, 167
Copeland O.L. 53, 60
Cormack R.M. 159, 167
Corner E.J.H. 134, 140
Cottam G. 63, 76
Craig A.J. 16, 17, 24, 81, 103, 196, 197, 198, 200, 201, 205, 304, 305, 307
Creer K.M. 273, 278
Cross A.T. 43, 60
Crowder A.A. 236
Cuatrecasas J. 131, 141
Cushing E.J. 24, 25, 80, 81, 103, 185, 188, 189, 195, 198, 203, 205, 274, 278, 291, 307
Curtis J.T. 63, 76, 290, 307

Dahl E. 144, 164, 167, 255, 259
Daniel G. 98, 103
Dale, H.M. 75, 77
Davis M.B. 9, 10, 13, 14, 15, 17, 21, 22, 24, 47, 51, 60, 63, 75, 76, 79, 80, 81, 82, 96, 100, 103, 110, 114, 117, 125, 128, 129, 137, 138, 141, 143, 144, 147, 165, 167, 195, 196, 205, 224, 235, 236, 305, 307

Davis R.B. 10, 24
Deevey E.S. 9, 13, 23, 24, 79, 80, 82, 103, 163, 167
Degerbøl M. 167
Dewdney J.C. 220
Dickson E.L. 278, 288
Digerfeldt G. 279, 288
Dimbleby G.W. 53, 54, 60, 114
Doyle R.W. 10, 24
Durham O.C. 64, 74, 75, 76
Du Rietz G.E. 238, 258, 259
Duvigneaud P. 238, 259

Edmondson W.T. 279, 288
Egglestone W.M. 207, 220
Eglinton F. 3
Einsele W. 281, 288
Ekern P.C. 53, 60
Elgee F. 43, 60
Ellison W.D. 53, 60
Emmett W.W. 53, 60
Erdtman G. 40, 42
Escofier-Cordier B. 174, 188
Estes A.H. 143, 167
Evans G.B. 60
Evans G.H. 43, 60

Faegri K. 20, 23, 24, 31, 42, 65, 69, 76, 82, 88, 95, 103, 110, 113, 114, 131, 138, 141, 143, 148, 167, 199, 205, 224, 236, 281, 288
Fager E.W. 143, 167
Fagerstrom J.A. 2, 3
Dale, H.M. 75, 77
Federova R.V. 43, 58, 60, 69, 72, 76
Fernald M.L. 174, 188, 293, 307
Fisher W.R. 141

Flenley J.R. 132, 134, 141
Flint R.F. 131, 132, 141
Firbas F. 31, 42, 91, 103, 137, 141, 144, 145, 146, 157, 167, 255, 256, 259, 260
Fisher E.M. 104
Fowells H.A. 198, 199, 205
Franks J.W. 83, 103
Fredskild B. 147, 148, 149, 150, 151, 167
Frenzel B. 93, 103, 231, 234, 235, 236
Frey D.G. 76
Fries M. 300, 307
Früh J. 255, 259

Garwood A.E. 61, 68, 76
Germeraad J.H. 140, 141
Gibbs L.S. 134, 141
Gibson C.E. 267, 278, 281, 286, 288
Ginsburgh R.N. 43, 53, 60
Glasscock R.E. 278
Godwin H. 83, 84, 103, 208, 220, 223, 224, 229, 231, 232, 236, 258, 259
Gonzalez E. 131, 141
Goodlett, J.C. 22, 63, 75, 76, 110, 114
Gordon A.D. 159, 167, 198, 205
Görs S. 254, 259
Goulden C.E. 163, 167
Greenwell W. 218, 220
Gregory P.H. 48, 60
Grichuk M.P. 43, 60
Grichuk V.P. 147, 148, 167
Grim J. 281, 288
Groot J.J. 43, 53, 60
Grosenbaugh L.R. 12, 24

319

Author Index

Grosse-Brauckmann G. 256, 259
Grünig G. 259
Guppy J. 141

Hadač E. 238, 259
Hafsten U. 88, 91, 104, 132, 141
Hainault R. 76
Hamilton A.C. 134, 136, 137, 138, 141
Hammen Th. van der 93, 104, 131, 132, 135, 141, 157, 168
Hansen H.P. 63, 76
Harper J. 200, 201, 205
Harris W.F. 143, 167
Hawkes J.G. 206
Haworth E.Y. 87, 90, 91, 104, 279, 288
Hayes J.V. 32, 42
Hazel J.E. 159, 166
Hedberg O. 134, 136, 141
Heikurainen L. 255, 259
Heim J. 228, 236
Heinselman M.L. 32, 36, 42, 290
Hejný S. 238, 259
Heusser C.J. 144, 167
Heybroek H.M. 231, 232, 236
Hibbert F.A. 93, 104, 230, 236
Hodson F.R. 167
Holub J. 238, 259
Hopkins J.S. 53, 60, 125, 129
Horowitz A. 43, 60
Hueber F.M. 77
Hutchinson G.E. 75, 76
Hyde H.A. 137, 138, 141

Imeson A.C. 53, 57, 60
Ingram H.A.P. 43, 60
Iversen J. 20, 23, 24, 31, 42, 65, 69, 76, 79, 82, 93, 94, 95, 103, 104, 110, 111, 112, 113, 114, 129, 147, 148, 149, 151, 157, 166, 167, 223, 224, 231, 232, 233, 236, 257, 259, 281, 288

Jafri S. 61, 63, 76
James P.W. 144, 167
Jankovská V. 257, 259
Janssen C.R. 32, 37, 39, 40, 42, 54, 60, 79, 104, 109, 110, 114, 144, 167, 213, 220, 257, 259

Jardine N. 159, 160, 167
Jasnowski M. 255, 259, 260
Jennings B. 220
Jeschke L. 239, 254, 259
Jessen K. 79, 91, 104, 272, 274, 278
Jewson D.H. 278, 288
Johnson R.G. 143, 167
Jonas F. 256, 259
Jørgensen S. 118, 129

Kabailiené M. 109, 114
Kaesler R.L. 159, 161, 167
Kapp R.O. 63, 76
Karczmarz K. 255, 259
Katz N. 258, 259
Katz S. 258, 259
Kendall D.G. 167
Kendall R.L. 14, 24, 136, 137, 141
Kerfoot W.C. 15, 17, 24
King D.L. 302, 307
King J.E. 63, 76
Kirkby M.J. 2, 3, 54, 60
Kittredge J. 54, 60
Klika J. 238, 259
Kobendza R. 239, 259
Koch W. 238, 255, 259
Kohut J.J. 143, 167
Koreneva E.V. 43, 60
Kormondy E.J. 199, 201, 205
Kotoučková V. 259
Kruskal J.B. 162, 167
Küchler W.A. 12, 24, 196, 205
Kummel B. 60

Lance G.N. 159, 167
Lang G. 257, 259
Langmuir I. 46, 60
Lapsley G.T. 220
Larsson B. 254, 259
La Rush F. 74, 77
Leopold L.B. 53, 60
Leroi-Gourhan A. 113, 114
Lewis C.F.M. 69, 76
Leyton L. 54, 60
Lichti-Federovich S. 9, 16, 17, 24, 32, 42, 81, 104, 143, 144, 146, 147, 165, 167, 168
Lishman J.P. 83, 86, 87, 88, 90, 93, 104
Livingstone D.A. 110, 114, 136, 137, 141, 143, 144, 147, 148, 167

Lounamaa J. 255, 259
Löve A. 168
Löve D. 144, 167, 168
Lowdermilk W.C. 53, 60
Lull H.W. 60
Lund J.W.G. 80, 104
Lundqvist G. 125, 129
Lyell C. 131, 141

Maarleveld G.C. 104
Macdonald I.D. 76
Machamer J.F. 290, 307
Mackereth F.J.H. 79, 83, 84, 86, 93, 100, 101, 103, 104, 272, 273, 274, 278, 279, 288
Mądalski J. 254, 260
Maher L.J. 80, 104, 144, 167
Mallory B. 143, 168
Malmer N. 238, 260
Marek S. 254, 260
Martin P.S. 143, 167
Marschner F.J. 12, 24
Matthews J. 47, 60
Matthews J.V. 147, 148, 167
Maxwell J.A. 17, 24
McAndrews J.H. 32, 40, 42, 69, 70, 75, 77, 144, 165, 168, 173, 174, 185, 186, 188, 304, 307
McCallan M.E. 278
McClure H.E. 138, 141
McCracken E.M. 281, 288
McDowall R.W. 220
McNeely R.N. 69, 76
McQueen D.R. 173, 188
McVean D.N. 43, 60, 152, 153, 156, 157, 164, 168, 255, 259
Meacham J.H. 63, 68, 77
Megard R.O. 289, 290, 300, 302, 303, 307
Mercer E. 220
Merriam D.F. 167
Miller J.P. 53, 60
Milne P. 141
Minnesota Pollution Control Agency 290, 291, 307
Minshall W.H. 74, 77
Mitchell G.F. 223, 236
Molyneux L. 273, 278
Moore J.J. 238, 260
Moore P.D. 104
Moravec J. 255, 260
Mörnsjö T. 255, 260

Morrison M.E.S. 135, 141
Mosby J.E.G. 226, 231, 236
Mosimann J.E. 51, 60, 143, 167
Mott R. 63, 77
Moyle J.B. 176, 188
Muller J. 43, 53, 60, 138, 139, 140, 141
Murphy M.T.J. 3

Nedelcu G.A. 239, 260
Neuhäusl R. 259
Nordhagen R. 156, 168, 238, 254, 260
Norris G. 143, 167
Noven B. 302, 307

Oberdorfer E. 238, 260
Ogden E.C. 31, 32, 42
Ogden J.G. 166, 168
Ohmann L.F. 290, 291, 292, 307
Oldfield F. 81, 138, 165, 168, 231, 236, 274, 278, 279, 288
Olsson I.U. 24, 236
Orloci L. 175, 188
Osmaston H.A. 136, 141
Ostrom A.J. 11, 24
O'Sullivan P.E. 271, 275, 276, 279, 288
Overbeck F. 258, 260
Ovington J.D. 54, 60

Pacowski R. 256, 260
Pakczynski A. 260
Passarge H. 254, 260
Patten H.L. 137, 141, 152, 154, 168, 189
Pearsall W.H. 199, 205
Pearson G.W. 236, 272, 278
Peck R.M. 47, 60, 80, 92, 104
Pennington W. (Mrs T.G. Tutin) 14, 16, 17, 24, 79, 80, 82, 83, 85, 86, 87, 88, 90, 91, 92, 93, 94, 95, 96, 98, 100, 101, 102, 103, 104, 144, 166, 195, 196, 199, 205, 231, 233, 234, 235, 236
Percival M. 137, 141
Pfeifer R.E. 24
Pielou E.C. 164, 168
Pierce R.S. 53, 54, 60

Piggott S. 93, 103, 104, 233, 236
Pilcher J.R. 232, 236, 272, 278
Poelt J. 254, 255, 260
Pohl F. 109, 114, 157, 168
Polak, E. 139, 141
Polunin N. 144, 168
Post L. von 131, 132, 140, 141
Potzger J.E. 63, 77
Powell J.M. 132, 141
Power D.M. 69, 70, 75, 77
Powers C.F. 291, 307
Prokopovich N. 294, 307
Purkyně C. 239, 260
Putnam D.F. 65, 77

Ramm H.G. 220
Raistrick A. 220
Ratcliffe D.A. 43, 60, 152, 153, 156, 157, 164, 168, 255, 259
Raup D. 60
Raynor G.S. 31, 32, 42
Ream R.R. 290, 291, 292, 307
Rempe H. 109, 114
Reyment R.A. 164, 168
Reynolds E.R.C. 54, 60
Richard P. 63, 77
Richards C.M. 60
Ridley H.N. 180, 188
Ritchie J.C. 9, 16, 17, 24, 32, 36, 42, 81, 104, 143, 144, 146, 147, 165, 167, 168
Roberts B.K. 103, 104
Rolfe W.D.L. 1, 3
Rossignol M.L. 43, 60
Rudolph K. 137, 141, 255, 256, 260
Rune O. 144, 168
Rutter A.J. 54, 60
Ruttner F. 196, 205
Ruuhijärvi R. 255, 258, 260
Rybníček K. 173, 188, 237, 239, 254, 256, 257, 260
Rybníčková E. 173, 188, 237, 239, 254, 255, 257, 260
Rymer L. 148, 151, 152, 153, 168
Ryvarden L. 173, 188

Salaschek H. 257, 260
Salisbury E.J. 178, 188
Sangster A.G. 75, 77
Sawyer J.S. 103, 236
Schimper A.F.W. 131, 141

Author Index

Schindler D.W. 302, 307
Schröter C. 255, 259
Schwickerath M. 238, 254, 260
Settipane G.A. 74, 76
Shay C.T. 40, 42
Sibson R. 159, 160, 167
Sjörs H. 144, 168
Slicher van Bath B.H. 220
Smith A.G. 36, 42, 83, 93, 104, 198, 205, 223, 231, 232, 236, 272, 274, 278
Smith D.D. 53, 60
Sneath P.H.A. 158, 159, 160, 168
Sokal R.R. 158, 159, 160, 168
Sopper W.E. 60
Spence D.H.N. 254, 260
Spencer J.S. 24
Srodon A. 147, 148, 168
Stanley E.A. 43, 53, 60
Stapf O. 134, 141
Stephens N. 278
Stewart R.E. 188
Stockmarr J. 232, 236
Stockner J.G. 300, 307
Stoermer E.F. 300, 302, 307

Stuiver M. 14, 24, 196, 199, 205, 214, 220
Suess H.E. 214, 220, 233, 236
Sutton O.G. 31, 40, 42
Swain F.M. 294, 307
Switsur V.R. 93, 104, 236
Sylvio M. 77
Symons L. 273, 278

Tallantire P.A. 199, 205, 224, 229, 236
Tauber H. 13, 14, 24, 31, 32, 40, 42, 47, 48, 54, 60, 79, 81, 93, 96, 104, 117, 118, 122, 123, 129, 139, 141, 143, 168, 195, 205, 213, 220, 231, 233, 236
Tăutu P. 167
Terasmae J. 63, 68, 77, 147, 168
Thompson F.B. 54, 60
Thompson G.G. 43, 60
Thompson R. 273, 274, 278
Tippett R. 196, 205
Tołpa S. 239, 260
Trapnell, C.G. 148, 168

Trautman W. 36, 42
Traverse A. 43, 53, 60
Trimble G.R. 54, 60
Troels-Smith J. 49, 60, 93, 104, 174, 183, 188, 223, 224, 231, 232, 236, 294, 307
Turekian K.K. 104, 205, 206
Turner C. 201, 202, 205
Turner J. 3, 32, 42, 103, 104, 233, 236
Tutin W. (nee Pennington), see Pennington W.
Tüxen R. 238, 260

Ucko P.J. 114
U.S. Department of Agriculture 291, 299, 307
U.S. Geological Survey 56, 60

Valentine J.W. 143, 168
Vanderkloet S.P. 76
Vasari A. 157, 168
Vasari Y. 157, 168
Veatch J.O. 12, 24

Versey H.C. 60
Vogel J.C. 104

Waddington J.C.B. 15, 16, 17, 25, 81, 104, 294, 305, 307
Walker D. 3, 83, 86, 96, 104, 140, 141, 166, 168, 236
Warburg E.F. 144, 166, 168
Ward P.F. 103, 104
Wąs S. 255, 260
Wasylikowa K. 173, 189
Watts W.A. 104, 173, 185, 186, 188, 189, 199, 200, 202, 203, 204, 205, 206, 223, 236
Webb D.A. 199, 206
Webb T. 103, 307
Weischedel I. 259
Weitzmann S. 54, 60
Welten M. 156, 157, 168, 201, 202, 206
West R.G. 3, 79, 93, 104, 173, 189, 203, 204, 206, 236
Whitehead D.R. 17, 25
Wiedemann A.M. 138, 141

Williams C.B. 158, 163, 164, 168
Williams J. 141
Williams W.T. 159, 167, 168
Willis E.H. 231, 236
Winchell H.V. 299, 307
Winchell M.H. 299, 307
Winter T.C. 173, 185, 189, 204, 205
Wishmeier W.H. 53, 60
Wolman M.G. 53, 60
Wood R.B. 267, 278, 288
Worzel G. 259
Wright H.E. 9, 24, 25, 94, 96, 104, 137, 141, 144, 147, 152, 154, 168, 173, 185, 186, 188, 189, 198, 199, 200, 202, 203, 205, 206, 291, 307
Wright J.W. 42

Yang J.J. 300, 302, 307

Zagwijn W.H. 104
Zaiteff J.B. 43, 60
Zavitz C.H. 76
Züllig H. 279, 288

Subject Index

Abies, pollen representation of 110
Acaena, pollen representation of 131–2
Acer, pollen representation of 19–21, 68, 76, 125, 169
Africa 134–7, 170
Agrosto-Festucetum communities 154–6, 160, 165
Akureyri, Iceland 151
Allerød stade 82, 84, 91–3
Alnetalia glutinosae 239, 242, 248
Alnetea glutinosae 242, 248
Alnion glutinosae 239, 242, 248, 258
Alnus, pollen representation and dispersal of 37, 109, 110, 147–9
Alnus glutinosa-Rubus subfoss. community 249–54, 258
alpine summit vegetation, pollen rain of 153–6, 161–3
alpine zone 144, 153–6, 161
Ambrosia, pollen dispersal 31, 32, 34–6, 63–4, 74, 105
Ambrosia, pollen rise of 10, 303–5
aquatic plants 45, 173–189, 191, 238–9, 240–8, 305
arable/pastoral index 208, 263
arctic environments, pollen rain of 144–52
arctic zone 144–52

Bahama Bank 43, 53
Barfield Tarn 233
basal area 12, 20–1, 63, 76
Batrachium-Potamogeton subfoss. community 239, 249–53, 258
Betula, pollen representation of 18–21, 68, 76, 105, 109–10, 123, 125, 145, 149–52, 157–8

Betula pubescens-Cirsium heterophyllum communities 154–5, 157–8, 161, 165
Betula pubescens-Vaccinium myrtillus communities 154–5, 157–8, 160, 165
Betula pubescens woodlands, pollen rain of 154–5, 157–8, 161–2
Betulion pubescentis 248, 257–8
biopel 294
blanket bog development 94, 98, 103
Blea Tarn 80, 83–97, 101–2
pollen stratigraphy of 85–91, 95–8
Blekinge, Sweden 117–29
Blelham Bog 17, 80, 83–5, 89–93, 101–2
Blelham Tarn 81
Blind Lake, Michigan 10–11, 13–4
Bog D, Minnesota 40
bog vegetation 36–7, 68, 237–8, 240–53, 256–9
Bollihope, Co. Durham, see Weardale
Bølling stade 82, 93
Borneo 134
Bosmina longirostris, ecology of 302–3
Bövra River 85
Brasenia schreberi, seed representation of 179
Bronze Age 216, 276, 279
Buckle's Bog, Maryland 17

Calluna vulgaris, pollen representation of 51, 54, 56, 124–5, 128
Canada 61–78, 146–7
Cannons Lough 36
canonical analysis 27
Canyon Lake, Michigan 17

Cardamino-Montion 242, 248, 254, 258
Carex acutiformis subfoss. community 239, 249–53, 258
Carex chordorrhiza-Sphagnum sect. *subsecuada* subfoss. community 249–54, 258
Carex diandra-Drepanocladus sendtneri subfoss. community 249–53, 255, 258
Carex diandra-Meesia triquetra subfoss. community 249–54, 258
Carex echinata-Climacium dendroides subfoss. community 249–53, 256, 258
Carex echinata-Sphagnum cf. *recurvum* agg. subfoss. community 249–53, 256, 258
Carex lasiocarpa, macrofossil representation of 179–80
Carex lasiocarpa-Scorpidium scorpioides-Calliergon trifarium subfoss. community 249–53, 255, 257–8
Carex lasiocarpa-Sphagna sect. *subsecunda-Meesia triquetra* subfoss. community 249–54, 258
Carex rostrata-Amblystegium scorpioides association 255
Carex rostrata-Calliergon giganteum subfoss. community 249–53, 258
Carex rostrata-Menyanthes trifoliata subfoss. community 239, 249–53, 258
Carex rostrara-Scorpidium scorpioides subfoss. community 249–53, 255, 258

Carex rostrata-brown moss *nodum* 255
Caricetalia fuscae 245, 248, 258–9
Cariceto-Rhacomitretum lanuginosi association 153–5, 160, 165
Caricetum davallianae 255, 258
Caricetum diandrae 255
Caricetum lasiocarpae 254
Caricetum limosae 256
Caricetum rostrato-vesicariae 239, 255
Caricetum rostratae 239, 255
Carici canescentis-Agrostidetum 256
Carici elongatae-Alnetum 254
Caricion canescentis-fuscae 245, 248, 258–9
Caricion davallianae 244–5, 248, 255
Caricion demissae 244, 248, 255, 258
Caricion lasiocarpae 243, 248, 254, 256, 258
Carpinus, pollen representation of 109–10, 124–5
Carya, pollen representation of 18–21, 27, 68
cereal pollen identifications 113, 224
Chat Moss 93
Chydorus sphaericus, ecology of 302–3
Cladietum marisci 239
Cladium mariscus subfoss. community 239, 249–53, 258
cladocera 289, 294, 298, 302–3
cluster analysis 158–62, 175, 198
Colombian Andes 131–2
coniferous lake species 174, 176–8, 184–7
Coniston Water 85
coring methods 10, 48, 81–2, 208, 272, 279, 291

Corylus, pollen representation of 110, 113, 158, 201
Corylus, post-glacial history of 201, 263
Corylus avellana-Oxalis acetosella community 154–5, 158, 160, 165
Coulin Forest 98
Cratoneurion commutati 242, 248, 254
Cryptogrammeto-Athyrietum chionophilum 154–6, 165
Cs-137 dating 85
Cuspidato-Scheuchzerietum 256
Cyperaceae, pollen representation of 147, 149–50, 156

Danubian culture 234–5
deciduous lake species 174, 176–8, 187
Deschampsietum cespitosae alpinum 154–6, 160, 165
deposition time 9, 14–5, 83, 86–8, 100–1, 208, 215, 224, 264, 282, 284–6, 300, 306
diatom analysis 279–83, 291, 298, 300–6, 309
relation to limnology 279, 281, 286–8
dissimilarity coefficients 159–63
diversity of pollen rain 139, 158, 163–5
measurement of 163–4
Draved Forest 110, 129

East Anglia 223–35
Elan Valley 93
Eldrup Forest, Denmark 111–14

323

Subject Index

'elm decline' 79, 93–103, 223–35, 263–4, 274, 276
   causes of 223, 231–5
   spatial aspects 233
   temporal aspects 228, 232–3
Ely, Minnesota 289, 291
Empetrum, pollen representation of 149–52
Ennerdale Water 103
Ericaceae, pollen representation of 149
Ericetalia tetralicis 246, 248
Ericion tetralicis 246, 248
Eriophorum vaginatum-Sphagnum magellanicum subfoss. community 249–53, 256, 258
Ertebølle culture 231
Esthwaite Water 83
eutrophication 103, 267, 279–307
extralocal pollen component 37, 40
extraregional pollen component 32, 40

Fagus, pollen representation of 18–21, 27, 48, 109–10, 112, 114, 124–5
Fallahogy Bog 36, 231
Farskesjön, Sweden 118–21, 124–9
Faulenseemoos, Switzerland 201, 202, 263
fen vegatation 36–7, 237–59
Festuca ovina-Luzula spicata community 153–5, 160, 165
Finland 32, 144–5, 199
Floating Bog Bay, Minnesota 39
floods, effect on pollen transport 56–8, 80
fluviatile environments 2, 43–77
forest vegetation 10, 11, 43–5, 111–14, 117, 125, 129, 137–40, 154–5, 157–8, 196–205, 207–21, 225, 292–3
   measurement of 10–12, 17–21, 43–5, 61–4, 109, 292–3
   representation of 2, 16–17, 19, 20, 27, 28, 33–5, 109–14, 154–5, 157–8
fossil assemblages, derivation of 1–3, 143, 176–8
Fragilaria crotonensis, ecology of 302–3

Frains Lake, Michigan 11, 13–15, 27, 80–1
Fraxinus, pollen representation of 48, 54, 109–11
fungal remains in sediments 306

Gantekrogsö, Denmark 81
Godthaab Fjord, Greenland, vegetation of 148
   pollen rain of 148–9
Gortian interglacial 203
Gramineae, pollen representation of 134, 147, 149, 151, 156–7
Grand Marais, Minnesota 174
Great Langdale 97
Greenland, pollen rain of 148–51, 157

Hembury 234
Hockham Mere 223–35, 263–4
Hoxnian interglacial 202, 263
Huron Mountains 10, 11, 21

Iceland, pollen rain of 151–3
Igelsjön, Sweden 118, 120–1, 124
'indicator species' 37, 143–4, 147, 156, 238, 313
influx, see pollen influx
interglacial floras and vegetation 201–3
Ipswichian interglacial 203
Iron Age 216, 220, 276, 279
Itasca, Minnesota 39, 174

Jacobson Lake, Minnesota 174, 186, 199–203
Juniperetum nanae 160–1, 165
Juniperus, pollen representation of 88, 90–3, 124–5, 157
Juniperus-Thelypteris nodum 154–5, 157, 160, 165

Kirchner Marsh, Minnesota 174, 185–8

lake chemistry 173–5, 267, 286, 288, 290
Lake Inim, New Guinea 134

Lake Louise, Florida 81
Lake of the Clouds, Minnesota 16, 28, 81, 196–201, 205, 263, 305
lake size, importance of 13, 79, 80, 83, 100–3, 117, 122–3, 169–70, 195, 274
Lake Victoria 137
laminated organic sediments 14, 28, 92, 196, 201–2, 263
'landnam', see 'elm decline'
Langdale Combe 83, 96
late-glacial pollen stratigraphy 79, 80, 82–93, 144, 203, 235, 257–9
Late-Devensian, see late-glacial pollen stratigraphy
Late-Weichselian, see late-glacial pollen stratigraphy
Late-Wisconsin vegetation 21–4, 185
Ledo-Pinion, see Pino-Ledion
Leirvatn, Norway 80, 87–92, 102
limnic environments 2, 9–24, 27–8, 43–59, 69, 79–103, 117–29, 173–89, 267–307
local pollen component 2, 31–42, 92, 117, 121–3, 153, 170, 214
Loch a 'Chroisg 81, 86, 94–6, 98, 100–2
Loch Clair 81, 86, 94–6, 98–103
Loch Ness 80
Loch Sionascaig 102–3
Lommagölen, Sweden 118, 120–1, 124
Lough Neagh 267–88, 309
   chronology 272–4
   diatom stratigraphy 279–309
   morphometry and limnology 267–70
   pollen stratigraphy 267, 270–8
   productivity 285–8
   sediments 267, 269–70, 272, 309
Low Wray Bay, see Windermere
Lycopodium, spore representation of 149, 151–3

macrofossil assemblages 2, 3, 84, 173–89, 191, 200, 202–5, 237–59
   interpretation of 2, 3, 173, 176–8, 185–9, 191

Najadetum marinae 239
Najas flexilis, macrofossil representation 179
Najas marina-Ceratophyllum demersum subfoss. community 239, 249–53, 257–8

macrofossil deposition 182–5
macrofossil dispersal 173, 180–6, 191
macrofossil representation 173, 177–9, 185–6
Magelemosian culture 276
magnetic remanence of lake sediments 83, 272–4, 285, 306
Magnocaricion 239, 241, 248, 258
Malaysia 138–40
Manitoba 16, 17, 32, 81, 147
marine environments 2, 43, 53
Marks Tey 201–2
mathematical models in palaeoecology 27, 166
Meathop Moss 81
Melosira italica ssp. *subarctica*, ecology of 279, 281, 286
Mesolithic culture 96, 223–6, 234, 263
Michigan 9–24, 27
Minneapolis, Minnesota 174
Minnesota 15–17, 28, 33–9, 173–88, 196–204, 289–307
mire communities, ecology of 237, 257–9
   past 237–9, 249–59
   present-day 238, 240–8
Mirror Lake, New Hampshire 17
Missouri Pond, Minnesota 184–5
Montio-Cardaminetalia 242, 248
Montio-Cardaminetea 242, 245
Moorthwaite Moss 83
moss polsters for pollen trapping 63, 125, 131, 134–5, 149, 151, 153, 170
   effects of different species 65
Mount Kinabalu, Borneo 134, 137
multidimensional scaling 162–3
multivariate techniques 7, 27, 83, 86–7, 91, 94, 158–63, 166, 170, 174–8, 198, 313
Myrtle Lake, Minnesota 33–8, 105

Nardus stricta-Vaccinium myrtillus community 154–6, 160, 165
Neolithic culture 93–4, 96, 114, 169, 223–35, 263–4, 276
New Guinea 132–4, 137, 170
New Zealand 170, 173
Newferry 237
Nicolet Creek, Minnesota 40
North Dakota 174
Nymphaeion 239–40, 248, 258

Oakdale, Yorkshire 43–59
   floods at 56–59
   hydrology and limnology 45–6
   pollen budget 47–59
   pollen stratigraphy 48–50
   sediments 48–50
   stream discharge 48–59
   topography 43, 45–6
   vegetation 43–6
Old Buckenham Mere 223–4
ordination methods 162–3, 174–8, 191
Orinoco Delta 43
Oxycocco-Empetrion hermaphroditi 247–8
Oxycocco-Sphagnetea 246, 248, 258

palaeolimnology 267–307, 309
Paludella squarrosa-Helodium blandowii subfoss. community 249–53, 255, 258
Paramo de Palacio 131
park meadows 117, 125, 129
peat stratigraphy 207–10, 212, 237
Phragmites communis-Scirpus lacustris subfoss. community 239, 249–53, 258
Phragmitetalia 240–1, 248, 254
Phragmitetea 240–1, 248, 254
Phragmition 239, 241, 248, 258
phytosociology 37, 238–59, 264
Picea, pollen representation of 34–5, 37, 110, 124–5, 128, 145, 147–8
Picea abies, post-glacial migration 199
Piceo-Alnetum 254
Pickerel Lake, Dakota 174, 185–9
Pino-Ledion 247–8, 257–9

Pinus, pollen representation of 19–21, 32, 34–5, 37, 40, 42, 54, 68, 76, 110, 123, 125, 145, 148, 301
Pinus, specific pollen identification 200
Pinus banksiana, ecology and history 21–2, 198, 200–1
Pinus resinosa, ecology and history 198, 201
Pinus strobus, ecology and history 21–2, 198–201, 203–4, 263
plant communities 3, 11–12, 36–45, 61–3, 148–66, 203, 237–59, 264, 290, 313
  reconstruction of 3, 7, 9, 23, 40–2, 111–14, 131–2, 134, 143–4, 195, 203, 237–59, 313
pollen accumulation rates, see pollen influx
pollen deposition 2, 3, 12–14, 27–8, 31–60, 109, 117–29, 131–40, 143, 213–4
pollen dispersal and transportation 2, 3, 7, 9, 19–20, 27, 31–76, 80, 117–29, 137, 143, 213–14, 233
pollen filtration 37, 54, 105, 118, 122
pollen influx 9–24, 27–8, 47–50, 79–103, 105, 118, 121, 125–7, 135, 138, 146, 195–201, 204, 223–35, 291, 305–6
  at 'elm decline' 95–103, 224–35
  calibration of 9–24, 27–8
  in different vegetation types 15–17, 19, 27–8, 80, 96, 121–4, 135, 146–8
  interpretation of fossil data 21–3, 27–8, 49, 82–103, 105, 224–35
  late-glacial 17, 21, 23, 82–93, 105
  methods of determination 47, 82, 118, 146, 223–4, 291
  variation of 9–24, 27–8, 79–80, 82–6, 91–2, 95–6, 100–3, 105, 121, 125–7, 135, 138, 195–201, 223–35, 305–6
  within lake 13–15, 27–8, 100–3, 125–7
pollen preservation 75, 103, 138, 274

pollen production 2, 3, 9, 15–21, 27–8, 54, 109–14, 117–29, 131–40, 143–66
pollen redeposition and resuspension 2, 12, 14–15, 27–8, 49, 51, 80, 84–6, 122, 125–9, 305
pollen representation 2, 3, 15–21, 27–8, 31, 61–76, 109–14, 117–29, 131–40, 143–66, 169–70, 195–6
pollen sum 79, 121–2, 137, 153–4
pollen transportation by streams 2, 43–59, 61–76, 80, 92, 274
pollen trapping methods 47–8, 63–5, 118–19, 138
pollen zones, concept of 195, 198
population ecology 21–3, 195–205, 233–5, 314
post-glacial 80–1, 111–14, 196–201, 207–21, 223–35, 257–9, 267–307
Potametalia 240, 248
Potametea 240, 248
Potametum filiformis 239
Potamion 238–40, 248, 258
prairie lake species 174–8, 182–7
principal component analysis 83, 86, 91, 94, 162, 174–8
Pteridium, spore representation of 48–9, 51, 54, 56–8

Quebec 21
Quercus, pollen representation of 18–21, 27, 34–5, 48, 54, 68, 76, 109–10, 123, 125, 128, 169

R-values 110, 143–4, 148–9
radiocarbon dates 9, 14–5, 28, 83–5, 91–3, 95, 97, 101, 112, 196, 208, 210–11, 213, 223, 228 233–4, 263–4, 272–4, 278, 300
Ranunculo-Caricion fuscae 246, 248, 256
rational limit of species 23
recurrent groups of fossils 143
Red Tarn, Wrynose 83

regional pollen component 2, 31–42, 79, 92, 96, 105, 117, 121–3
regression analysis 20, 54–5
representation, see pollen representation, macrofossil representation
Rhynchospora alba–Sphagnum cf. *cuspidatum* subfoss. community 249–53, 256–8
Rhynchosporetum sphagnetosum cuspidati 256
Rhynchosporion albae 243–4, 248
Rogers Lake, Connecticut 9, 14, 17, 21–3, 80–1, 91, 196, 263
Romano/British period 216, 220, 273
Rumex acetosa, pollen representation of 157
Rutz Lake, Minnesota 15, 17, 81

St. Cloud, Minnesota 174
Salix, pollen representation of 149–50, 152, 157
Salix lapponum–Luzula sylvatica community 154–6, 160, 165
sapropel 294
Saskatchewan 147
Scheuchzeria palustris, ecology and history of 258, 264
Scheuchzerietalia 246, 248, 258
Scheuchzerio-Caricetea fuscae 239, 243, 248
Scheuchzerion 246, 248, 258
Schoenetum nigricantis 255
Scotland, modern pollen rain 152–64
  vegetational history 94–103
Seamere 223–4, 229–35, 264
sediment accumulation rates 15, 27–8, 49, 80, 83–8, 101–2, 208, 211–12, 215, 224, 274, 284
sediment characterisation 49, 83, 174, 208, 210, 224, 291–4
sediment-pollen relationships in streams 43, 53–4, 59, 73–4
sediment supply to streams 2, 53–4
sedimentary processes within lakes 2, 12–15, 20, 27–8, 49, 51, 85
Sedum rosea–Alchemilla glabra community 154–6, 160, 165

Senoren, Sweden 119, 121–4, 129
Sermermiut, Greenland 149–50
Shagawa Lake, Minnesota 289–307, 309
  charcoal stratigraphy 294, 298, 305
  chemical stratigraphy 294, 298–9
  chronology 289, 295, 298–300
  cladoceran stratigraphy 289, 294, 298, 302, 305–6
  cultural history 291, 294–5, 299
  diatom stratigraphy 289, 291, 298, 300–6
  fungal stratigraphy 306
  influx values 305–6
  morphometry and limnology 290–309
  pollen stratigraphy 291, 296–8, 303–6
  sediments 291–4, 301, 309
  vegetation 290–3
Shippea Hill 232, 234
single-link cluster analysis 159–61
Singletary Lake, N. Carolina 17
Small Water 83
snow-bed vegetation 88, 90, 154–6, 160, 165
soil changes in late-glacial 91, 101, 103
soil changes in post-glacial 93–4, 96, 98, 101–3, 233–5
Sörviken, Sweden 119, 121
South America 131–2, 137
Sphagnetalia 247–8
Sphagnetum modii 256
Sphagnion atlanticum 247–8
Sphagnion continentale 247–8, 256, 258
Sphagno-Caricetum rostratae 256
Sphagno-Caricetum lasiocarpae 243, 248
Sphagno-Caricion canescentis 245, 248, 256
Sphagno-Tomenthypnion 245, 248, 255, 258
Sphagnum cf. *cuspidatum–Drepanocladus fluitans* subfoss. community 249–53, 256, 258
Sphagnum cuspidatum–Rhynchospora alba nodum 256
stability in vegetation 202

Stephanodiscus astraea, ecology of 280–2, 286
Stephanodiscus minutus, ecology of 300, 302–3, 309
Stevens Pond, Minnesota 40–1, 105
Steward Shield Meadow, see Weardale
stream-borne pollen, see water transport of pollen
stream discharge measurements 45, 54–9, 69, 72
sub-alpine zone 144, 154–7, 161–3
sub-arctic zone 144
surface samples, classification of 7, 27, 158–63, 170, 176–8
  study of 9–24, 31–59, 88, 105, 109, 131–40, 143–66, 173–85
  use of in interpretation 137–8, 143–4, 164–6, 185–9, 195

Tabellaria flocculosa, ecology of 279–80, 286
taphonomy 1
terrestrial environments 2, 31–42, 68, 109–14, 131–40, 143–66, 193–205, 207–21, 223–35, 237–59
Tilia, pollen representation of 109, 111–12, 114, 125, 169
Tofieldietalia 244, 248, 255, 258
Tristan da Cunha 132
tropical forest pollen rain 138–40
tropical mountain pollen rain 131–8
tropical vegetation 131–40, 170
  pollen representation 131–40
Tsuga, pollen representation of 18–21
tundra vegetation, pollen representation 2, 16–17, 80, 89–90, 143–66
Two Creeks interval 82
Two Harbors, Minnesota 174
Typha, macrofossil representation of 179–80

Ulmus, pollen representation of 68, 95, 109, 227
  specific pollen identification 232
  use as fodder plant 223, 232

Vaccinio-Piceetalia 248
Vaccinio-Piceetea 247-8
Vågåvatn, Norway 85, 92
Valle de Lagunillas, S. America 131
varved clays 83-5, 90
Vieskär, Sweden 121, 123-4
Volga River 43, 69, 72

water transport of pollen 2, 43-76, 80, 274
Waubun Prairie Pond, Minnesota 182-3
Weardale, Co. Durham, cultivation 216-20, 263
  forest history 216-21, 263
  land-use history 216-21, 263
  lead mining 219-20
  pollen stratigraphy 207-21
  radiocarbon dates 208, 210-11
  vegetation 207-10
Weber Lake, Minnesota 300
Wilton Creek, Ontario 61-77, 105
  discharge 69-75
  pollen composition 63-76
  vegetation 61-3
Windermere 80, 83-6, 89-90, 93, 95, 100-3, 105
  pollen stratigraphy 83-6, 95
  sediments 83-6, 101, 103
Windmill Hill 234
Wisconsin 309

Younger *Dryas* period 85-6, 91-2
Yukon 185